THE
TRUE PRAIRIE
ECOSYSTEM

US/IBP SYNTHESIS SERIES

This volume is a contribution to the International Biological Program. The United States effort was sponsored by the National Academy of Sciences through the National Committee for the IBP. The lead federal agency in providing support for IBP has been the National Science Foundation.

Views expressed in this volume do not necessarily represent those of the National Academy of Sciences or the National Science Foundation.

US/IBP SYNTHESIS SERIES ▌16

574.5264
R496t
1981

THE
TRUE PRAIRIE
ECOSYSTEM

P. G. Risser
Illinois Natural History Survey

E. C. Birney
University of Minnesota

H. D. Blocker
Kansas State University

S. W. May
Indiana University Northwest

W. J. Parton
Colorado State University

J. A. Wiens
University of New Mexico

Hutchinson Ross Publishing Company

Stroudsburg, Pennsylvania

Copyright © 1981 by *The Institute of Ecology*
Library of Congress Catalog Card Number: 79–19857
ISBN: 0–87933–361–8

83   82   81   1   2   3   4   5
Manufactured in the United States of America.

Library of Congress Cataloging in Publication Data

Main entry under title:
The True Prairie ecosystem.
  Bibliography: p.
  Includes index.
  1. Prairie ecology—United States.   I.   Risser, Paul G.
QH104.T78   .   574.5′264   79–19857
ISBN 0–87933–361–8

Distributed world wide by Academic Press,
a subsidiary of Harcourt Brace Jovanovich,
Publishers.

CONTENTS

FOREWORD

This book is one of a series of volumes reporting results of research by U.S. scientists participating in the International Biological Program (IBP). As one of the fifty-eight nations taking part in the IBP during the period of July 1967 to June 1974, the United States organized a number of large, multidisciplinary studies pertinent to the central IBP theme of "the biological basis of productivity and human welfare."

These multidisciplinary studies (Integrated Research Programs) directed toward an understanding of the structure and function of major ecological or human systems have been a distinctive feature of the U.S. participation in the IBP. Many of the detailed investigations that represent individual contributions to the overall objectives of each Integrated Research Program have been published in the journal literature. The main purpose of this series of books is to accomplish a synthesis of the many contributions for each principal progam and thus answer the larger questions pertinent to the structure and function of the major systems that have been studied.

Publications Committee: US/IBP

PREFACE

In 1968 a group of scientists met to formulate a research plan for the Grassland Biome of the International Biological Program that had been formally organized earlier in the year. The first assignment for this group was an information synthesis; that assignment was accomplished, presented, and discussed at a series of retreats held in various areas of Colorado, Wyoming, and Nebraska. From those meetings the group decided to establish an intensive grasslands research site (Pawnee) and other "comprehensive sites where data were to be gathered on the structure, function, and utilization of grasslands." The True Prairie Site, also referred to in this book as the Osage Site, is located in northeastern Oklahoma a very short distance south of the Kansas state line. The clipping of the first quadrats late in 1969 signaled the beginning of data acquisition at Osage. With only two exceptions, the authors of this volume were entirely responsible for the data acquisition. The first year's data on soil microorganisms and small mammals were taken by non-authors. The author group, now located in Colorado, Illinois, Indiana, Kansas, Minnesota, and New Mexico, have met periodically since 1971 for the purpose of data synthesis and writing. In addition to this volume, data from the Osage Site have been used in forty-two publications.

Coverage of all subjects in this book is not purported to be even or complete. Differences in emphasis reflect differences in the state of our knowledge about components and processes in the True Prairie, as well as the different perspectives of the authors. We feel the attempt to integráte knowledge from the various biotic and abiotic components of the True Prairie is unique. We now present this information for your review. We hope it is pleasing and stimulating, but one of our major objectives will be met if this volume inspires additional research effort and if it presents the potential researcher with a good starting point.

The authors, their students, and their colleagues, who have labored in the True Prairie, have grown to respect it. We will be gratified if we can instill this same respect in our readers.

ACKNOWLEDGMENTS

Most of the authors have been associated with the Grassland Biome/U.S. International Biological Program since its organization in early 1968 under its director, George M. Van Dyne. We have, since that time, been associated with thousands of scientists from all over the world, most of whom have added a suggestion, a constructive criticism, or even a helping hand. We cannot possibly adequately acknowledge everyone who has contributed, and we ask the forgiveness of anyone who has been omitted. The latter was entirely unintentional and we are grateful for the former.

The Osage Site become available as a research site through the generosity of the late Mr. "Boots" Adams. We especially acknowledge the Adams family for this assistance. Mr. Dick Whetsell, manager of the Adams Ranch, spent many hours with us during the initial planning stages at the site. His knowledge of the True Prairie and his assistance throughout the study were invaluable. He and his staff provided accommodations for researchers on numerous occasions. Mr. Bob Bourlier of the Soil Conservation Service gave generously of his time and provided soil coring equipment, which was essential to those of us engaged in root analysis, soil analysis, and belowground invertebrate analysis.

Members of the Natural Resource Ecology Laboratory who made significant contributions include: George Van Dyne, M. I. Dyer, Normar French, David Swift, Vicki Keith, Marilyn Campion, Barbara Carlson, Leila Menges, Nancy Wilson, and Janie Downs.

Drs. W. R. S. Hoffman and J. Knox Jones, Jr., of the University of Kansas conducted the small mammal studies, and Dr. John O. Harris of Kansas State University conducted the microbiology studies during the initial year of research. We thank these scientists for their major contributions.

The comparative small mammal studies conducted in the northern True Prairie were supported by a grant from the Graduate School, University of Minnesota.

The reviewers at Natural Resource Ecology laboratory, D. C. Coleman, J. L. Dodd, M. I. Dyer, G. M. Van Dyne, W. H. Hunt, and R. G. Woodmansee, gave the first outside criticism of an early manuscript. Their suggestions resulted in a major improvement in the version submitted to external reviewers, Dr. Clair Kucera of the Univer-

sity of Missouri, Dr. Clenton Owensby of Kansas State University, and Dr. Dick Wiegert of the University of Georgia. These scientists added ideas, drew attention to inconsistencies, and made us much more comfortable about the volume. We carefully addressed each of their comments and each added significantly to whatever strength this volume possesses. The authors, however, assume complete responsiblity for the volume's content.

Our Graduate Students deserve special consideration. These young scientists labored diligently in the laboratory and field and generated much of the data included herein. Without them, the volume you are holding would not exist. These students include: Rodman Reed and R. L. Stepanich (Kansas State University); David Adams, Shelia Conant, Anthony Dvorak, Forrest Johnson, Robert Kennedy, Barb Klopatek, Janice Perino, and Joyce Sheedy (University of Oklahoma); John Rotenberry and John Ward (Oregon State University); David Byman and Merlin Tuttle (University of Minnesota).

Finally, we acknowledge the True Prairie for its frustrations and fascinations.

THE
TRUE PRAIRIE
ECOSYSTEM

1

Introduction

To those who know grasslands comes a unique perspective, which, though not explainable, is recognizable and indelible. Perhaps it is because the prairie, like people, can experience so many moods, each of which is all-encompassing. Nothing is more peaceful than a warm July morning in the surroundings where a cowboy sits comfortably in the saddle quietly watching a red-tailed hawk soar above a herd of Hereford cattle. Nothing is more expectant than a May morning in a prairie that is covered with showy pasqueflowers and shootingstars. Nothing is more ferocious than a driving winter blizzard on the prairie where the relentless cold wind drives sleet horizontally and no places of refuge are available.

A portion of the grassland intrigue arises from the difficulty of knowing it. Unlike the ocean and majestic forests, which openly display their characteristics, the prairie is more subtle. The prairie is frequently dismissed as uninteresting if not desolate, a barrier to be crossed rather than appreciated. Although grasslands are worldwide in distribution with certain consistent characteristics, each type has its own personality. This book is about one grassland type, the North American True Prairie. We hope that this treatment will not only provide a comprehensive presentation and interpretation of the relevant information, but that the reader will feel the acquisition of a new or better friend, the True Prairie.

Grasslands evoke an array of emotions in individuals as functions of the varied moods of prairies. This summons alone is perhaps sufficient justification for studying grasslands, to attempt an understanding of the phenomena that account for the "feel" of the prairies. In addition, several other reasons come to mind why scientific studies of prairie ecosystems are important. Grasslands are the foundation of grazing agriculture, which in North America forms an especially central position in human dietary demands and thus in agricultural economics. Grazing of prairie vegetation by large herbivores like cattle has profound effects upon the vegetation, but these influences may extend through a web of other components of the prairies. Without an understanding of these influences and interconnections, proper management of prairie

grazing systems for long-term yield can only be developed through costly trial and error procedures.

Grazing, however, is not the only use of grasslands. In the more mesic portions of the prairies, or where ground water permits the development of irrigation systems, much of the native prairie has been converted to cropland. The "breadbasket" regions of North America represent former prairies, and while people have substituted monocultures of cereal grains or other crops for the native vegetation and imposed some control over other features of the environment (such as the amount of available nutrients) the replacement crop systems are still subject to a wide range of environmental influences that affected the prairie system before them. Understanding these environmental influences in native systems provides the keys for enhancing our success in prairie restoration or converting prairies to agriculture, as well as indicating the consequences of doing so.

Some justifications for studying grasslands go beyond their economic importance to people. Grasslands are relatively simple systems. While the number of species in some groups, such as insects, is very large in the grasslands, the diversity of major dominant species is considerably smaller than that of many other terrestrial habitats. Further, the physical diversity of prairies is simpler than in many other systems; the multilayered structuring of forest vegetation, for example, stands in stark contrast to the relative uniformity of grassland expanses. This relative simplicity in species composition and physical structure of the prairies increases the likelihood that relationships among species or linkages between populations and other environmental factors will be apparent, or at least amenable to study.

Responses of populations or other components of the grasslands to features of the physical environment such as weather should be more obvious than in many other terrestrial vegetation types. For example, the very structure of the forest, with the sheltering canopy and the tangles of fringing vegetation, tends to reduce the amplitude of many environmental variations, so that organisms within the forest are usually exposed to a much more benign set of conditions than the conditions that occur outside the forest. Little buffering of environmental conditions occurs in the grasslands, since most organisms are exposed directly to a wide range of variation and extremes of temperature, humidity, wind, precipitation, and a host of other factors. Further, the range of daily, seasonal, or long-term variation in climatic factors on local or regional scales, is great in the prairies. Because the environment is so variable, and because the organisms are directly exposed, we can expect the relationships between biota and environment to be close and apparent.

GRASSLANDS OF THE WORLD

Grasslands are biological communities containing few trees or shrubs, characterized by mixed herbaceous vegetation, and dominated by grasses. They are found in some form on every continent and large island throughout the world. Types of grasslands range from

the dense bamboo of the tropics to northern steppes, from dry plains to arctic grasslands.

Roughly 46 million km^2 of the earth's surface are covered with grasslands (Shantz, 1954; Whyte, 1960) and the major types of grasslands may be compared according to size as shown here:

Grassland Type	Area (km^2 $\times$ 10^6)
High Grass Savanna	7.25
Tallgrass Savanna	10.10
Tallgrass Prairie	4.09
Shortgrass Prairie	3.11
Desert Grass Savanna	5.96
Mountain Grassland	2.05

These grasslands contain approximately 600 genera and 7,500 species of grasses, Poaceae, which grow from the poles to the tropics and comprise about 15 percent of the total flowering plants (Hartley, 1950, 1964). The Poaceae is one of the largest families of flowering plants and ranks third in number of genera behind the Compositae and Orchidaceae and fifth in number of species behind the Compositae, Orchidaceae, Leguminosae, and Rubiaceae. The Poaceae far surpasses all other groups in worldwide geographical occurrence and percentage of total world vegetation (Gould, 1968).

Grasslands usually develop in areas with 25 to 100 cm of annual precipitation and are found primarily on plains in the interiors of great land masses. They occur from sea level to elevations of over 4,900 m, as in the Andes. The wide diversity of habitats and climates suggests that a wide range of environmental conditions supports the array of grassland types. These biological communities are not controlled by any one simple environmental factor but by many interacting abiotic and biotic factors.

Climate

In general the climate of grasslands is one of extremes. Whether the vegetation on any portion of the landscape is dominated by grasses or woody plants may depend on the soil water content and its variations throughout a year and among many years. Soil water content may be in constant flux throughout the year because of precipitation patterns and blowing wind that greatly influences the loss of water by evapotranspiration. Temperatures fluctuate widely, both diurnally and seasonally. Extreme winter cold does not preclude grasslands since they occur in some of the coldest regions of the world, and the success of grasslands in Mediterranean climates shows that marked summer drought is not prohibitive either. Yet most midcontinent grasslands are characterized by summer precipitation patterns. Grasslands are not controlled simply by total annual precipitation, evaporation, or maximum and minimum temperatures, but rather by complex relationships like the precipitation-evaporation ratio and the seasonality of precipitation in relation to the temperature regime. These elements will be discussed further in Chapter 5.

Soils

Soils of grasslands generally are deep with well-developed but simple profiles. They are usually neutral to basic, are high in organic matter, contain a large amount of exchangeable bases, and are highly fertile. Almost all true grasslands are in the soil orders Mollisols and Aridisols, although a few soils characteristic of the drier grasslands are classified in the order Alfisols, and some coastal grasslands are in the Vertisols (U.S. Soil Conservation Staff, 1970).

In some grasslands the soil may be dry throughout the profile for a large portion of the year. Because of their dense fibrous root system in the upper layers of the soil, grasses are apparently better adapted than most trees to make use of frequent light showers during the growing season. When compared with forest soils, grassland soils are generally subjected to higher temperatures, greater evaporation, and more transpiration per unit of total plant biomass.

Vegetation

Rangelands found in temperate regions with reduced precipitation in the summer are generally dominated by warm-season grasses, though cool-season grasses are more important farther north. Woody plants are relatively few in number or unimportant, although these rangelands which are an intergradation of prairie and forest [savanna] or an intermixture of prairie and forest [parkland], do contain considerable woody components. Usually a long dormant period occurs each year; near the poles the dormancy may be caused by low temperatures, whereas nearer the equator it may be attributed to low rainfall and soil drought.

Prairies are found on all continents and compose almost 24 percent of the plant cover of the world (Harlan, 1956; Thornthwaite, 1933). The proportion of each continent that is composed of prairie differs widely, from about 44 percent in Europe to slightly less than 10 percent in Australia. Land area occupied by Tallgrass Prairie in South America is 33 percent; in Africa, 29 percent (but of this most is savanna and only about 5 percent is prairie); in Asia, 17 percent; and in North America, about 15 percent. The Tallgrass Prairie or high veld is found primarily in eastern South Africa (Harlan, 1956), but smaller areas of grassland are found well into east and central Africa. The prairie in Europe lies to the south of the taiga (northern coniferous forests) in the USSR and includes the productive farmlands of the Ukraine, Hungary, and Transylvania. Most of the Asian grasslands are steppes, although the wheat-producing regions of China and Manchuria were probably prairies at one time. The prairie in Asia generally exists, as in North America, between a steppe vegetation and a temperate deciduous forest. In South America, the most prominent prairies are the pampas in Argentina, but tallgrass areas also exist in Uruguay and in central Brazil. The topography of most prairies of the world is relatively level and rolling, which contributes to their success as agricultural lands.

UTILITY OF WORLD GRASSLANDS

Human evolution and present position of domination in the biological world have been affected by grasses. Most civilizations developed in grassland regions, and were it not for the abundance and widespread distribution of grasses, the human population of the world likely would not have attained its present level (Gould, 1968).

The cereal crops are grasses, and most of our domestic animals graze on grasses and live in grasslands. The most fertile and productive soils in the world developed under grassland cover. In these areas, cultivated grasses have replaced native grasses. Where vast seas of native grass once dominated the landscape, people have replaced the natural plants with hybrid grains, controlled competition, nutrient subsidization, and increased production of edible grass. Grasslands were an important segment in the world's productivity long before the advent of humans, and they may eventually control diet, population, and habits as they have for many other animals (Box et al., 1969).

Grasses and grasslands form the basis of the modern grazing industry. Direct human consumption of grain foods is a more energetically efficient means of resource utilization than is consumption of animal products. Meat and dairy foods, however, are basic constituents of the American diet and are necessary if high nutritional levels are to be attained and maintained.

The health and management of grasslands is of worldwide concern, especially since grasslands are being called upon to feed a hungry world. Yet, in most areas of the world, historical records indicate that the amount of grassland area is declining (Auclair, 1976).

THE INTERNATIONAL BIOLOGICAL PROGRAM
AND ITS RELATION TO THIS VOLUME

The initial impetus for the International Biological Program (IBP) can be traced back to October 1960, when an Italian presentation was made to the Executive Committee of the International Union of Biological Science (IUBS) and the Executive Board of the International Council of Scientific Unions (ICSU). The general theme was "The Biological Basis of Productivity and Human Welfare." Discussions about United States involvement occurred, primarily through the United States National Committee/IUBS, the National Academy of Science, and the National Science Foundation, but no positive steps were taken in this country until the winter of 1964, when a United States National Committee for the International Biological Program (USNC/IBP) was appointed. Evolution of the IBP occurred from 1965 to 1968 during which time the decision was made to develop integrated research programs relating to biological productivity, and the ecosystem approach was accepted as the focus. In 1966, the Analysis of Ecosystems Program (previously called Drainage Basins and Landscapes) was established by the National Science Foundation, and in 1967, the biome concept was finalized. The director of the Grassland Biome, Dr. George M. Van Dyne, and the director of the Biological and Medical Sciences Division of NSF,

Dr. Harve J. Carlson, finalized the organizational structure of the IBP Grassland Biome early in 1968 (National Academy of Science, 1975).

The Grassland Biome was designed as an integrated project to study the structure, function, and utilization of grasslands. Its research organization consisted of one Intensive Site and eleven Comprehensive Network Sites. Detailed studies were conducted at the Intensive Site and less extensive studies at the Comprehensive Network Sites. The latter represented different grassland types throughout the western United States (Table 1.1).

Field work was conducted at the Pawnee Site (Intensive Site) during the period from 1969 to 1974 and at many of the Comprehensive Network Sites from 1970 to 1972. Data were not collected at all Comprehensive Network Sites over the total duration of the three-year period. All sites were designed to include an ungrazed and a grazed treatment, and in addition, the Pawnee Site was used to evaluate the consequences of additional water and nutrients.

An attempt was made to use the same methods and collect the same types of data on each Comprehensive Network Site. During 1969, considerable effort was expended in determining what measurements should be made, what the most appropriate techniques were, and what methods should be used in the subsequent data analysis and

TABLE 1.1   Site Name, Location, and Grassland Type of the 12 Research Sites in the US/IBP Grassland Biome.

Site name	Location	Grassland type
ALE	Washington	Shrub steppe
Bison	Montana	Mountain
Bridger	Montana	Mountain
Cottonwood	South Dakota	Mixed-grass
Dickinson	North Dakota	Mixed-grass
Hays	Kansas	Mixed-grass
Hopland	California	Annual
Jornada	New Mexico	Desert
Osage	Oklahoma	True Prairie or Tallgrass
Pantex	Texas	Shortgrass
Pawnee	Colorado	Shortgrass
San Joaquin	California	Annual

interpretation.   Initially the focus was on measuring the dynamics of biomass and nutrients and then comparing these values with climatic and physical-chemical measurements.   Later, as the study evolved, a greater effort was directed toward the measurement of processes in an effort to explain these dynamics.

THE OSAGE SITE

This book centers around the True Prairie and particularly, a site in northeast Oklahoma called the Osage Site.   Information used in this book originates from the IBP project itself as well as previous studies conducted throughout the Tallgrass or True Prairie. This endeavor is unique because simultaneous data are available on abiotic variables, plants, invertebrates, birds, mammals, and decomposers.   These data were collected by the authors of this book on the Osage Site from 1970 to 1972.

The Osage Site is located on the K. S. Adams Ranch, a rangeland of 14,000 ha located 7.2 km north and 3 km east of the town of Shidler, Osage County, northeastern Oklahoma (32°N, 96.5°W).   The sample areas are on flat topography at an elevation of 375 m, and the ranch itself is rolling upland with occasional sharp limestone breaks around the rims of the valleys.   This location is in the southern end of the Flint Hills and represents the largest expanse of unbroken True Prairie in the United States (Barker, 1969).   The site is dominated by little bluestem* (Schizachyrium scoparius), big bluestem (Andropogon gerardi), switchgrass (Panicum virgatum), and Indiangrass (Sorghastrum nutans), with tall dropseed (Sporobolus asper) and, the adventive, Japanese brome (Bromus japonicus) becoming more important in moderately to heavily grazed conditions (Gardner, 1958).   Dwyer (1958) compiled a plant species list of the ranch and listed 315 species, of which 19 were trees.   He listed 66 families and 197 genera; the Poaceae, Compositae, and Leguminosae comprised 18 percent, 17 percent, and 9 percent, respectively, of the flora.

The ungrazed (control) treatment had been ungrazed by large herbivores for at least the last twenty years, though it may have been occasionally mowed for hay as late as 1968.   The grazed treatment was grazed primarily during the calving season in late fall and early spring.   Stocking rates on the ranch are relatively light (3.2 ha/cow and 1.6 ha/steer) and the range condition is mostly good to excellent (Anderson, 1940; Harlan, 1960a).   In both the steer and the cow-calf operation at this ranch, animal gains are expected to be about 300 kg per year.

Mean annual temperature is 15.2°C with a 27.3°C July mean and a 2.7°C January mean.   The 205-day growing season has a mean temperature of 19.1°C.   Mean monthly soil temperatures (at 5 cm) during the growing season range from 14.0° to 19.8°C.   Annual net radiation is 48,600 g cal cm^{-2}.   The annual precipitation is 100 cm with 60 percent falling during the growing season.   The soils are

---

*Common names for plants are taken from A Guide to Plant Names in Texas-Oklahoma-Louisiana-Arkansas (Soil Conservation Service, 1954).

Brunizems of the Labette-Summit-Sogan series. Specifically, these soils belong to the order Mollisol, suborder Ustoll, group Haplustolls, and family Fine Montmorillinitic (Gray and Roozitalab, 1976; Soil Survey Staff, 1975). These are dark-colored soils mostly with clayey subsoils developed on shales, sandstones, and limestones. The ungrazed and grazed sample plots are on a Labette soil with a dark, silty clay, 0-40 cm A horizon. The B_1 is dark brown, 40-60 cm; the B_2 is reddish brown, 60-80 cm; the B_3 is a brown, silty clay, 80-100 cm, and most of the bedrock is discontinuous limestone at 1-2 m.

For the ungrazed treatment, field capacity (0.1 bar) of the top 10 cm is 53.6 percent and the wilting point (15.0 bars) is 22.7 percent. In the top 5 cm the soil pH is 5.9, carbon is 7.6 percent, organic phosphorus is 14.0 mg g^{-1}, inorganic phosphorus is 3.7 μg g^{-1}, and phosphatase activity (μm nitrophenol g^{-1} soil h^{-1}) is 7.3. Ammonium nitrogen (0-50 cm) is 3.0 ppm, available phosphorus (0-60 cm) is 308 ppm, and available potassium (0-60 cm) is 0.4 ppm. The cation exchange capacity, total nitrogen, and total organic matter at various soil depths are shown in Table 1.2.

The Osage Site is thus located on a typical deep prairie soil and has developed under climatic conditions characteristic of the southern True Prairie region. The vegetation is also typical of the southern True Prairie, and, especially because of the relatively large control area, the operational ranch represented an ideal location for the Comprehensive Network True Prairie Site.

TABLE 1.2   Cation Exchange Capacity, Total Nitrogen, and Total Organic Matter at Various Soil Depths on the Osage Site, Ungrazed Treatment.

Soil depth (cm)	Cation exchange capacity (meq/100 g)	Total nitrogen (Percent)	Organic matter (Percent)
0-10	10.74	0.21	4.15
10-20	10.40	0.17	3.20
20-40	10.88	0.16	2.70
40-60	12.98	0.11	2.05
60-100	--	0.08	1.25

2

Grasslands Across North America

Most of the early explorers in North America came from the European continent, which had no broad expanses of grassland. As a result, these explorers were awed by the extent and uniformity of the grasslands they encountered as they pushed westward from the deciduous forest out into the Great Plains. Charlevoix (1761 as noted by Curtis [1959]) observed

> Nothing to be seen in this course but immense prairies interspersed with small copses of wood, which seem to have been planted by hand; the grass is so very high that a man is lost amongst it, but paths are everywhere to be found as well trodden as they could have been in the best peopled countries, though nothing passes that way except buffalos, and from time to time some herds of deer and a few roebuck.

Similarly, from Ruggles (1835),

> . . . in some instances, prairies are found stretching for miles around, without a tree or shrub, so level as scarcely to present a single undulation; and others, those called 'rolling prairies' appear in undulation upon undulation as far as the eye can reach presenting a view of peculiar sublimity, especially to the beholder for the first time. It seems in verdure, a real troubled ocean, wave upon wave, rolls before you, ever bearing, ever swelling; even the breezes play around to heighten the illusion; so that here at near 2,000 miles from the ocean, we have a facsimile of sublimity, which no miniature imagination can approach.

The word prairie is of French origin and it means "meadow." The French explorers really had no precise term for the large grasslands, but meadow implied that it was an open, grass-covered, treeless landscape. When the English settlers arrived, they

9

accepted the term prairie since they, too, had no exact term for the huge expanse of grassland (Curtis, 1959).

GRASSLAND TYPES AND ACCOMPANYING CLIMATES

The grassland formation is the largest vegetational unit in North America (Gould, 1968). Grasses are dominant throughout an area extending from southern Saskatchewan and Alberta to eastern Texas and from Indiana and the western edge of the deciduous forest westward to an elevation of approximately 2,000 m in the southern Rocky Mountains at the woodland zone (Figure 2.1). In addition to this area other grasslands can be found: the Shrub Steppe in portions of Washington, Idaho, and Utah; the western Mountain Grasslands at higher elevations, especially in Idaho and Montana; the Annual Grasslands in California; the Desert Grasslands in New Mexico and Arizona; and finally, the Successional or Temporary Grasslands in the eastern and southeastern part of the United States.

The eastern transition from grassland to forest has an annual precipitation of 75-100 cm from Texas to Indiana and 50-65 cm farther north. Roughly 75 percent of this precipitation falls during the growing season, but westward, as the total annual precipitation decreases to about 25 cm near the Rockies, the proportion falling during the growing season also decreases (Oosting, 1956).

Temperature patterns demonstrate variations of large magnitude. In the southern part of the Great Plains and in the Desert Grassland, maximum temperatures of 38°C are not uncommon. On the other hand, at the northern extremities of the Great Plains, temperatures of -25°C have been recorded (Gould, 1968). In the north, the growing season is cool and short and sub-zero temperatures occur for long periods. In the south, frosts are infrequent and summer temperatures exceed 35°C for days at a time (Oosting, 1956).

All North American grassland climates are characterized by a wet season followed by a period of drought or dry conditions. In the Central Great Basin, spring and early summer rains are followed by a dry summer, frequently with hot winds. A summer-winter precipitation pattern brings limited moisture to the desert grassland, while the Annual Grasslands and Shrub Steppe have a Mediterranean-type climate with winter precipitation followed by dry summers.

The decreasing moisture from east to west is accompanied by changes in the dominant species across the central plains (Weaver, 1954). As a result, three major regions are distinguishable by climate and/or vegetation (Oosting, 1956). In the eastern part of the central grassland, where available moisture is greatest, the grasses are taller and this region is generally called the "Tallgrass Prairie" or the "True Prairie." In the western portion, where precipitation is lower, the grasses are short and this is called the "Shortgrass Prairie." In the center of the Great Plains are grasses of intermediate height and these make up the "Mixed-grass Prairie" or "Mid-grass Prairie." Somewhat arbitrarily, the grasses varying from 15-60 cm in height at maturity are

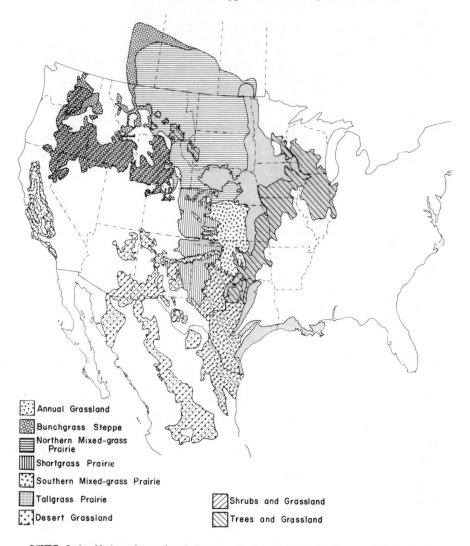

Annual Grassland
Bunchgrass Steppe
Northern Mixed-grass Prairie
Shortgrass Prairie
Southern Mixed-grass Prairie
Tallgrass Prairie
Desert Grassland

Shrubs and Grassland
Trees and Grassland

FIGURE 2.1 Major Grassland Types of the United States and Adjacent Canada and Mexico.

classified as shortgrasses, mid-grasses are intermediate, 60-120 cm, and grasses that exceed 120 cm are referred to as tallgrasses (Gould, 1968).

Many woody shrubs are also present in the prairie and under certain circumstances tree seedlings and a few trees may also be found. On the eastern border of the Tallgrass Prairie, the grasslands frequently grade smoothly through an oak savanna to a denser oak forest. On occasion a rather abrupt boundary may exist between grassland and forest, particularly at rivers or at places of rapid soil or topographic change (Bray, 1958; Buell and Facey, 1960;

Kilburn, 1959; Kilburn and Warren, 1963; Ovington et al., 1963; Weaver, 1960).

Various authors have subdivided the grasslands into compositionally homogeneous divisions. These divisions are made on the basis of species composition, presence or absence of dominant plant species, and prevailing climatic conditions. Although prairies are ecosystems composed of plants and animals, grassland classification and recognition have been historically based almost entirely on vegetation dominants.

The emphasis in this volume is on the Tallgrass or True Prairie, but some brief comments on the other grassland types will provide a broader perspective for these discussions.

True Prairie

Clements and Shelford (1939) described the True Prairie as being the most typical and maximally developed grassland of the Tallgrass Prairies. The designation "True Prairie" arose in 1920 and has been used extensively since (Bruner, 1931; Clements, 1920; Hensel, 1923; Sampson, 1921; Schaffner, 1926; Welch, 1929).

The True Prairie is a subset of the tallgrass prairie assemblage in the United States. According to Küchler (1964), there are three grassland and three deciduous forest ecotone types. The grasslands include the Agropyron-Andropogon-Stipa type of the Dakotas, the Andropogon-Calamovilfa-Stipa type of the Nebraska Sand Hills, and the most typical Andropogon-Panicum-Sorghastrum type that extends from eastern North Dakota and western Minnesota to eastern Oklahoma (True Prairie). The forest types include the Quercus-Andropogon type of the upper Midwest (Wisconsin, Minnesota, North Dakota), the Juniperus-Quercus-Sporobolus-Andropogon type of western Tennessee, Alabama, Missouri, and Arkansas, and the Quercus-Andropogon Cross Timbers area from Kansas to Texas. The three grassland types account for about 75 percent of the Tallgrass Prairie in Table 2.1. Although these forested types are not dominated by tallgrasses, the tallgrass community is the understory of mature systems and the major seral communities (Box et al., 1969; Bray, 1958; Kilburn, 1959).

The Tallgrass Prairie receives the most rainfall and has the greatest north-south diversity and the greatest number of major dominant species of any association within the grassland formation. Islands of Tallgrass Prairie are found throughout the northeastern deciduous forest as far east as Long Island, New York (Blizzard, 1931). In the north, the True Prairie merges with the aspen parkland and the boreal forest (Gould, 1968). According to Clements and Shelford (1939) the native True Prairie included northern and western Missouri and the eastern half of Oklahoma as far south as the coastal prairie of the Red River. Later studies defined the True Prairie as extending about halfway across Texas (Weaver, 1954). Because of the favorable climate and soil conditions, most of the northern part of the True Prairie is cultivated and relatively little of the original vegetation remains today (Auclair, 1976; Shelford and Winterringer, 1959).

The Tallgrass Prairie once extended eastward in the shape of a peninsula into Illinois and Indiana and in isolated patches much

TABLE 2.1   Grasslands of the Contiguous United States. Areas
Determined by Aggregation of Planimetered Area
Estimates of the Potential Natural Vegetation Types
of the United States (Küchler, 1964).

Grassland type	Area in contiguous United States (km^2)
Annual Grassland	52,688
Shortgrass Prairie	615,223
Mixed-grass Prairie	566,174
Tallgrass Prairie	573,511
Shrub Steppe	644,716
Desert Grasslands	207,565
Low Mountain Grasslands	267,731[a]

[a] An aggregation of the area of all ponderosa pine types in the
western United States is used as a best approximation of the
area of low mountain grasslands.

farther east, particularly in Ohio (Benninghoff, 1964; Bliss and
Cox, 1964; Davis, 1977; Morrissey, 1956; Thomson, 1940; Whitford,
1958; Wistendahl, 1975). Today these areas are almost entirely
cultivated, but their distribution has been well mapped (Gleason,
1923; Transeau, 1935). The predominating tallgrasses, as well as
other basic similarities, make it reasonable to consider the prairie
peninsula as a part of the True Prairie (Oosting, 1956). In fact,
the peninsula's presence may have been a barrier to the migration of
some mesic tree species, particularly beech, Fagus grandifolia, and
hemlock, Tsuga canadensis (Benninghoff, 1964).
    Under optimum conditions, some plants may reach a height of 2
to 3 m. The major dominants include tallgrasses such as big
bluestem (Andropogon gerardi), Indiangrass (Sorghastrum nutans), and
switch-grass (Panicum virgatum); mid-grasses such as little bluestem
(Schizachyrium scoparius), sideoats grama (Bouteloua curtipendula),
and procupine needlegrass (Stipa spartea); and shortgrasses such as
blue grama (Bouteloua gracilis) and hairy grama (B. hirsuta). These
species are not simply dispersed at random across the landscape, but
rather, occupy specific topographic positions within any local area
depending on soil and exposure.
    As might be expected from the large geographical range over
which the True Prairie is found, important differences exist along a
north-south gradient. Big bluestem, prairie dropseed (Sporobolus
heterolepis), needlegrass, and, the adventive, Kentucky bluegrass
(Poa pratensis) are commonly associated in the central portion of

the True Prairie. From Kansas northward, needlegrass assumes a
greater importance and may be associated with little bluestem, big
bluestem, prairie Junegrass (<u>Koeleria</u> <u>cristata</u>), and sideoats grama;
prairie cordgrass (<u>Spartina</u> <u>pectinata</u>) is a dominant in wet
lowlands. An intermediate community between the moist cordgrass and
drier bluestem communities is dominated by switchgrass and Canada
wildrye (<u>Elymus</u> <u>canadensis</u>). Switchgrass is more important from
Nebraska to the south and southeast, while Canada wildrye increases
in importance to the west and north (Gould, 1968).

Mixed-Grass Prairie

     The mixed-grasses occupy the area between the True Prairie and
the Shortgrass Prairie. Although the Mixed-grass Prairie dominants
are derived from both of the two adjacent communities, distinct
climatic and species compositional differences exist. Thus, the
Mixed-grass Prairie can be defined as a specific vegetational
association. The area forms a band from Saskatchewan through the
center of the Dakotas, Nebraska, Kansas, through western Oklahoma,
and into central and western Texas. On the western boundary of the
Mixed-grass Prairie tallgrasses disappear and shortgrasses occur as
dominants. Except in specific ameliorating topographic and edaphic
conditions the western edge of the mixed-grass prairie, roughly a
line from the western side of the Dakotas to central Texas, lacks
sufficient annual rainfall to provide adequate soil water to support
tallgrasses. The eastern limit is not as sharply defined but is
determined by the climatic conditions that support tallgrasses to
the degree that the shortgrasses are excluded from being dominant on
mesic sites. The eastern limit runs approximately from central
Texas and Oklahoma to just west of the North Dakota-Minnesota
border.
     Various species of needlegrass and grama grass are present
throughout the Mixed-grass Prairie, although the needlegrasses are
far more important in the north and the gramas become more important
in the south. The northern part of the Mixed-grass Prairie is
dominated by western wheatgrass (<u>Agropyron</u> <u>smithii</u>), thickspike
wheatgrass (<u>A</u>. <u>dasystachyum</u>), little bluestem, porcupine
needlegrass, needle-and-thread (<u>Stipa</u> <u>comata</u>), prairie Junegrass,
and blue grama. Farther south, the grasslands are dominated by
bluestems, particularly little bluestem, hairy grama, and sideoats
grama on the moist sites and blue grama on the drier sites.
     Because this association involves both tall- and shortgrasses,
the composition can be quite different depending upon the
topographic position or the prevailing climatic conditions. Wet
years and moist topographic positions favor the tallgrasses; drought
years and dry sites favor the shortgrasses. Thus, the grassland
type developed at any given location in this intermediate
"transitional" zone between the True Prairie and the Shortgrass
Prairie may fluctuate through time as climatic conditions vary.

Shortgrass Prairie

The Shortgrass Prairie is dominated by grasses adapted to xeric (dry) conditions. This prairie occurs westward from the Mixed-grass Prairie to the woodland zone of the lower Rockies and extends northward through Wyoming into the northeastern half of Montana. The southern portions occupy eastern Colorado, western Oklahoma and Texas, and parts of eastern and northern New Mexico and Arizona.

Rainfall ranges from under 50 cm per year to as low as about 25 cm per year. The northern portion may have a relatively short growing season of less than 150 days; in the south the growing season is over 200 days and is accompanied by much warmer temperatures, especially in the winter.

Blue grama is the most common species throughout the Shortgrass Prairie. In the north the prairie is associated with needle-and-thread and western wheatgrass while farther south blue grama and buffalograss (Buchloe dactyloides) are the codominant grasses. In the southern parts, especially in Texas, Oklahoma, and New Mexico, purple threeawn (Aristida purpurea), tobosa (Hilaria mutica), and hairy grama become important.

Desert Grassland

The Desert Grassland occupies a sizable area in the southwestern United States and north central Mexico. Specifically, this association occupies southwestern Texas as well as central and southern New Mexico and Arizona below about 2,000 m elevation. To the southwest the grassland grades into the Chihuahuan Desert and in the northwest it meets the woodland communities of the higher elevation in the Rockies. The eastern boundary represents a transition to the Shortgrass and Mixed-grass Prairies.

The Desert Grassland is the driest of all the grassland associations, with an average annual precipitation ranging from about 25-45 cm. Normally, this rainfall occurs in two seasons: summer (July or August) and winter (December or January). The summer precipitation frequently comes as localized heavy showers of short duration but high intensity. The winter precipitation is more general and comes over a longer period of time and with less intensity. Maximum summer temperatures frequently exceed 38°C for days at a time. These high temperatures and low rainfall amounts produce extremely dry conditions (Gould, 1968).

At lower elevations black grama (Bouteloua eriopoda) and tobosa dominate along with rothrock grama (Bouteloua rothrockii), poverty threeawn (Aristida divaricata), and purple threeawn. At higher elevations, perhaps 1,200-1,600 m, curly mesquite (Hilaria belangeri) and blue grama are the major grasses associated with other shortgrasses such as hairy grama, black grama, tobosa, and species of threeawn.

Annual Grassland

The Annual Grassland is largely confined to the Central Valley of California and is sometimes called the California Grassland. The

Mediterranean-type climate is typified by winter precipitation and hot, dry summers. Annual precipitation averages from 25-50 cm and the frost-free season is long, approximately 300 days per year. Since the summers are very dry, growth occurs primarily in the early spring and most plants are dormant during the summer.

The original vegetation consisted of perennial bunchgrasses dominated by purple needlegrass (Stipa pulchra) and nodding needlegrass (Stipa cernua) with associated species such as blue wildrye (Elymus glauca), pine bluegrass (Poa scabrella), and deergrass (Muhlenbergia rigens). Since settlement, these vegetation communities have been almost entirely replaced by vigorous annual species that are good competitors and are well adapted to the Mediterranean-type climate. The present grasslands include: wild oats (Avena fatua), slender oats (Avena barbata), soft cheat (Bromus mollis), ripgut brome (Bromus diandrus), foxtail brome (Bromus rubens), mouse barley (Hordeum murinium), little barley (Hordeum pusillum), false foxtail fescue (Festuca myuros), and foxtail fescue (Festuca megalura). Management systems of these grasslands are now based on these annual grasses rather than the original perennial grasses (Gould, 1968).

Mountain Grassland

The Mountain Grassland (excluding the ponderosa pine [Pinus ponderosa] mountain vegetation of the Rocky Mountain Plateau) extends northward along the eastern foothills of the Rocky Mountains from about central Montana to central Alberta, Canada. These grasslands are found from the lower foothills up to about 2,500 m in Montana. Rainfall is extremely variable, but ranges from 40-80 cm at higher elevations. Thirty percent or more of the precipitation may come as rain during the summer and the remainder falls as snow. Growing seasons are of short duration, frequently one hundred days or less.

This vegetation type might be termed the Fescue Prairie, because it is largely dominated by grasses belonging to the genus Festuca (Coupland, 1961). Rough fescue (Festuca scabrella) is the dominant species in most of the sites and its associates include bearded wheatgrass (Agropyron subsecundum), timber oatgrass (Danthonia intermedia), spike oat (Helictotrichon hookeri), and porcupine needlegrass. With disturbance Idaho fescue (Festuca idahoensis) increases at the expense of rough fescue.

Shrub Steppe

The Shrub Steppe is located in eastern Washington and Oregon, southern Idaho, northern Utah, and Nevada. This relatively arid grassland receives the major portion of its precipitation between November and February, much of it as snow. The precipitation ranges from 15 cm in the west to 60 cm in the eastern portion. Since much of the precipitation falls as snow, infiltration is frequently high, and excellent grass production is obtained relative to the total amount of precipitation. However, very little rainfall occurs during the summer, so most of the growth occurs either in the spring

or fall. The plants remain dormant for lack of soil water in the summer and because of low temperatures in the winter (Clements and Shelford, 1939; Gould, 1968; Stoddard, 1941).

In the original vegetation, bluebunch wheatgrass (Agropyron spicatum) was important with associated species like Sandberg bluegrass (Poa secunda), Cusick's bluegrass (Poa cusickii), and prairie Junegrass (Daubenmire, 1970). Idaho fescue is dominant especially at higher elevations. Much of the prairie has been grazed to the extent that big sagebrush (Artemisia tridentata) has become important and several annual species have invaded such as downy brome (Bromus tectorum), Japanese brome (Bromus japonicus), tumblemustard (Sisymbrium altissimum), tansymustard (Descurainea spp.), and filaree (Erodium cicutarium).

Other Grasslands

Other authors have delimited various grassland associations either not included in the previous seven types or as a result of different designations within these types (Küchler 1969). For example, Gould (1968) described seven grassland types, but combined the Mixed- and Shortgrass Prairies, and separated a Fescue Grassland, which corresponds roughly to the northwestern portion of the Mixed-grass Prairie and the Mountain Grassland. In addition, he defined a Coastal Prairie that is separated from the True Prairie by the abundance of Texas needlegrass (Stipa leucotricha). Other species of the Coastal Prairie included silver bluestem (Andropogon saccharoides), little bluestem, big bluestem, tanglehead (Heteropogon contortus), Indiangrass, and various species of Trachypogon, Bouteloua, Elymus, Manisurus, Paspalum, and Panicum.

In the southeastern portion of the United States, relatively large areas are covered by a successional grassland commonly dominated by broomsedge (Andropogon virginicus). These grasslands normally proceed successionally from the broomsedge stage through pines and eventually to southern hardwood forests (Oosting, 1942).

Alpine meadows are found in both the eastern and western higher mountains. The meadows are usually dominated by a large number of sedges (Carex) and marshlands characterized by Typha, Scirpus, and Spartina.

Since most grasslands are identified by their major grass species, the descriptions of the grassland types have been based primarily on the grass species currently present, although these communities include numerous forbs as well as woody species. Some grassland types are further subdivided on the basis of the presence or absence of nongrass species (e.g., the Mesquite [Prosopis] Grassland in the Mixed-grass or Shortgrass Plains and the Oak [Quercus] Savanna in the Tallgrass Prairie).

Grasslands may also be divided on the basis of differences in soil types. Two well-known examples are the Nebraska Sand Hills in the Mixed-grass Prairie and the Blackland Prairie in the Tallgrass Prairie.

As noted previously, grassland types are primarily identified by the major grass species that are recognized as the dominant species. This identification is a matter of convenience--that is, dominant plants are by definition plentiful, they do not move around

and are easy to find, and for the most part, are easy to identify. Clearly, however, not only do other plant species exist in the grasslands, but each grassland type has associated invertebrates, birds, mammals, fungi, and other groups of organisms. Like some plant species, animal species may be found in more than one grassland type, though they may be relatively more abundant in one type as opposed to another. Each grassland type can be described in terms of the plants and animals present as well as the associated environmental conditions, such as the soils and climate. These biological organisms and their environment are related in an intricate organization in which the components interact with and affect each other. The remainder of this book is an exploration of these interrelationships within the Tallgrass Prairie, and, more specifically, the True Prairie.

3

Origin and Biota of the True Prairie Ecosystem

GEOLOGIC ORIGIN

The geological attributes of an ecosystem are described by the age and mineralogical composition of rocks and their geographical and vertical distributions. These attributes are not caused by ecosystems, but rather the resulting ecosystems are determined in part by these characteristics. Although these geological materials are differentiated into soil under the influence of climatic and biotic factors, the availability of geological materials for subsequent development may be altered by such major events as mountain building, alluvial and aeolian deposition, erosion, and glaciation.

Landforms

The Rocky Mountains arose near the end of the Mesozoic era (135 million years ago). Erosion of these mountains eventually resulted in the deposition of a huge alluvial east-sloping plain that stretched from the Rockies across the central United States to the Mississippi River. By Oligocene times (30 million years ago) the mountains had been eroded into a peneplain, but the erosion was followed by a second uplift in the Miocene. The mountain construction initiated another erosional cycle that is still going on and is continuing to contribute material to the Great Plains and the parts of the Central Lowlands that lie west of the Mississippi River. Thus, one of the major events that operated as an independent controlling factor and influenced the geological materials eventually available for the True Prairie was the initiation of the Rocky Mountains. The mountains not only contributed deposited geological material but had a marked effect on the prevailing climate (Dix, 1964).

The rocks in the northern portion of the Central Lowland region, from the Canadian borders south to northern Arkansas (Figure 3.1), dip gently from three centers of uplift: (1) the Cincinnati Anticline that exposed Ordovician beds; (2) the Ozark Dome that

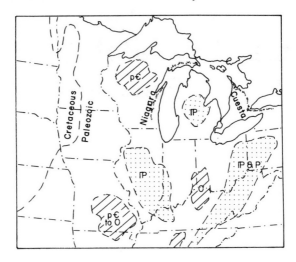

FIGURE 3.1 Three centers of uplift ⧄ indicated by the old
rocks exposed. These are the Superior Upland, the Ozark Dome, and
the Cincinnati Anticline. Also, three centers of depression ⬚
indicated by the presence of younger rocks, the Allegheny Plateau,
the Michigan coal basin, and the Illinois coal basin. Cretaceous
rocks at the west occupy the syncline of the Great Plains. Within
the limits of this map it may be assumed that all beds dip slightly
from the older outcrops to the younger. pε = Precambrian, O =
Ordovician, IP = Pennsylvanian, P = Permean. Redrawn from Fenneman,
1938.

brought the Precambrian to the surface; and (3) the Superior Upland
of Precambrian rocks, more specifically, the dome-like outlier in
northern Minnesota, Wisconsin, and Michigan (Fenneman, 1938). These
areas represent centers of uplift and the oldest rocks are rocks
noted in the list. Younger rocks outcrop in successive belts around
these centers with the older rocks closer to the center of the
uplifts.

Within this region, all dips in the rocks are toward three
synclines (see Figure 3.1): (1) the great Appalachian coal basin,
in which lie buried all the Paleozoic systems from the Permian down;
(2) the Michigan basin, deep enough that coal layers have been
preserved from erosion; and (3) the long, wide syncline of the Great
Plains, under which the Paleozoic rocks of the Central Lowland pass
beneath the Cretaceous (Fenneman, 1938). An eastward extension of
this trough reaches to Indiana and western Kentucky and can be
recognized by the Carboniferous rocks that have been perserved
within it.

Glaciation

The repeated advances of the glaciers were the second major
developmental influence on the region of the True Prairie. These
large sheets of ice not only rearranged the topography, but they

also generated masses of rocks that were deposited as the glaciers advanced and retreated. In some places this glacial debris or till was laid down on relatively level plains, but in other areas the material was pushed ahead as moraines or left behind as hills and ridges (drumlins and eskers). In all cases the material was heterogeneous and largely obliterated any effects of the previously exposed geological material. After the retreat of the glaciers, the topography was uneven, but no sharp breaks or escarpments were evident. As the glaciers melted, the resulting streams eroded or dissected the relatively level plains. Some of the previous drainage patterns had been altered by the glaciers and ice blocks that later resulted in the impoundment of shallow lakes and potholes. The flood waters from the melting glaciers carried heavy loads of debris that were deposited in wide streambeds along the major rivers. Subsequently, dry cycles occurred and the debris deposits were subjected to wind erosion that resulted in deep deposits of windblown material, loess, throughout the Great Plains and Central Lowlands.

In the northern part of the Central Lowland, the importance of the individual rock strata has been minimized by the glacial history, since parts of the area were at one time or another covered by several advances of glaciers (Figure 3.2). The glaciation advances have been classified into four glacial epochs: Nebraskan, Kansan, Illinoian, and Wisconsin, each of which was followed by a period of warmer climate and consequent deglaciation (Figure 3.3). The present landscape is largely a function of glaciation and subsequent erosion rather than the preglacial geology.

Two areas within this Central Lowland prairie were not glaciated. The Osage section, which stretches from Missouri and Kansas to Texas, is south of the limit of glaciation (see Figure 3.2). The Wisconsin Driftless section lies nearly in the center of the glaciated region (see Figure 3.3), but most of the area was apparently untouched by the ice sheets and was only affected by the resulting outwash. The northeast section, from central Ohio to the Mississippi, was glaciated relatively recently during the Illinoian and Wisconsin advances and is now characterized by plains with relatively small surface relief. The western area, from the Mississippi River through southern Iowa, eastern Nebraska, northeastern Kansas, and northern Missouri (Figure 3.3), was glaciated by the earlier Kansan advance, but later erosion has developed a submaturely dissected topography.

The Illinoian drift sheet, where not subsequently covered by the Wisconsin, is covered with loess, which is windblown material derived from the flood plains of large, south-flowing streams during the dry interglacial period. Since the prevailing dry winds are from the west, the deposit is deepest on the east side of these major rivers and the thickness of this mantle decreases eastward. Near the eastward edge of these windblown deposits, the loess (here sometimes called <u>white clay</u>) is of very fine texture, greater density, and has a higher clay content than usual. In the western portion of Illinois and parts of Iowa and Missouri the loess is the parent material of a very dark, rich soil, but the productivity of the soil decreases toward the south and east. The white clay of southern Illinois, Indiana, and Ohio, especially on flat, poorly drained remnants of the original plain, is an inferior soil and the

FIGURE 3.2 Map of North America showing the area covered by
Pleistocene ice sheets at their maximum extension; also the main
centers of accumulation (Redrawn from Fenneman, 1938).

greatest productivity is generally on the younger drift rather than
on the loess (Fenneman, 1938).

Much of the original Kansas till plain has been destroyed by
erosion, which is more advanced at the western edges of the
glaciated region. Near the Missouri and Kansas Rivers northwest of
Kansas City, the glacial drift remains only in patches or is covered
by Peorian loess from the postglacial period. Elsewhere the sheet
in Kansas may be more than 13 m thick and this area now resembles
the driftless Osage section and the dissected till plains of
southern Iowa.

The unglaciated Osage section stretches from Kansas to Texas
between the eastern Interior Highland of Missouri and the Great
Plains. The plain is of low relief, interrupted at intervals by
east-facing escarpments or cuestas that indicate the presence of
stronger strata in a great mass of relatively weak rocks dipping
gently west or northwest toward the syncline of the Great Plains.
The eastern boundary against these Ozark Plateau is essentially at

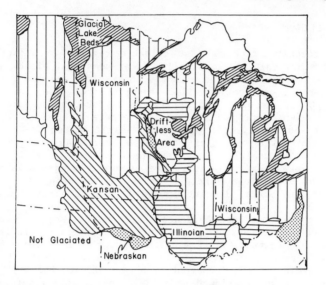

FIGURE 3.3 Map of north-central United States showing the several drift sheets, also areas of lacustrine plain. The southern areas indicated as Nebraskan are designated by Leverett as containing only scattered boulders (Redrawn from Fenneman, 1938).

the edge of the Pennsylvanian rocks where the rocks overlap the low Ozark dome.

The Flint Hills cuesta is located in southeastern Kansas. A large proportion of limestone is found in the first 130 m of the Permian beds and chert is common in several horizons. From central Kansas to northern Oklahoma, the Flint Hills upland has the distinct form of a cuesta, but northward from central Kansas, the height of the cuesta diminishes and its form becomes less distinct. Southward into Oklahoma, the limestone formation diminishes and sandstones become much more important (Fenneman, 1938).

The area that subtends the True Prairie is derived from geological materials of different ages, ranging from Precambrian rocks (more than 600 million years old) to material deposited subsequent to the latest glaciation, perhaps 10,000 years ago. The characteristics of the native geological material at any one place determine, to some extent, the soil that will be formed. However, at various locations, this parent material may have been covered by glaciers or materials deposited during wind and water erosion. The process is a dynamic one since the newly deposited material is also subjected to erosion, which, if sufficiently extensive, will expose the original parent material. The northern half of the True Prairie occupies an area that has been mostly glaciated; the southern portion was unglaciated. Most of the True Prairie region has received deposits of loess that vary from depths of 40 m along the Mississippi River to essentially zero farther east.

HISTORIC CLIMATE

The reconstruction of past climatic regimes is a fascinating game in mystery solving. Basically, it assumes that plants have the same climatic requirements now as had their close relatives from the past. The reconstruction also assumes that the layering or stratigraphy of geological materials can be dated by position of the layer or other dating techniques. Because they are relatively stationary and fossilized, plants and invertebrates are frequently utilized for dating purposes, but some larger animal groups also have been found useful. The fossilized remains of alligators and crocodiles, manatees, and tapirs are generally regarded as evidence of tropical conditions, whereas bones or skeletons of reindeer, walrus, or the boreal lemming indicate cold, subarctic conditions (Dorf, 1960).

From the accumulated evidence (Dorf, 1960), we can reconstruct the generalized temperature regime of the True Prairie region since the Cambrian (Figure 3.4). Temperature prior to the Pleistocene was generally warmer than present temperature. In North America a general warming trend apparently began before the end of the Paleocene Epoch and continued into the Eocene. As a result, most of the True Prairie region was subtropical about 30 million years ago (Figure 3.5a). By the beginning of the Miocene, about 25 million years ago, the temperature had cooled and the northern half of the present True Prairie region could be considered as warm temperate and the southern half as subtropical (Figure 3.5b).

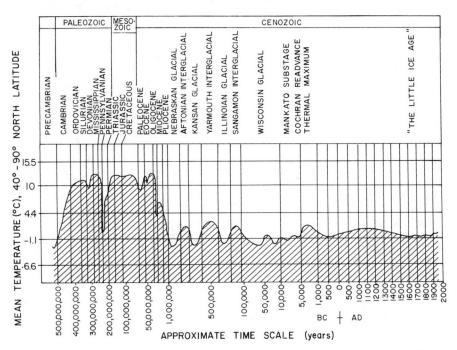

FIGURE 3.4 Generalized temperature variations during the geologic past (Redrawn from Dorf, 1960; copyright 1960 by American Scientist).

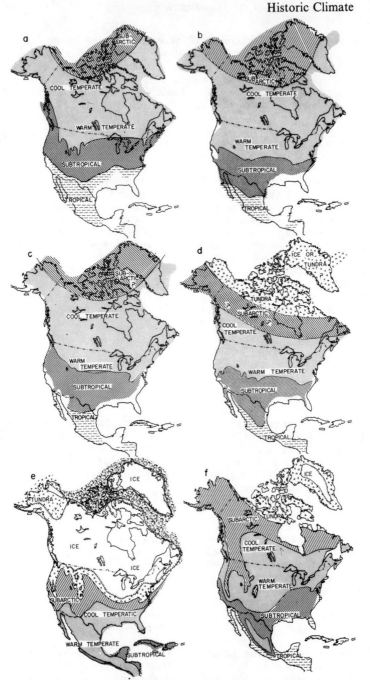

FIGURE 3.5   Generalized climatic zones of (a) late Eocene-early
Oligocene, (b) late Oligocene-early Miocene, (c) middle Miocene-late
Miocene, (d) late Pliocene, and (e) a composite of glacial and (f)
interglacial stages of Pleistocene (Redrawn from Dorf, 1960;
copyright 1960 by American Scientist).

Near the middle Miocene (Figure 3.5c) the forest migrations reversed and hardy, drought-resistant shrubs and chaparral moved north into the Great Plains. By late Miocene another cooling trend began and the Late Miocene fossil record indicates a gradual development and spread of grasslands and savannas. The cooling trend progressed, and by the late Pliocene the northern half of the True Prairie region was warm temperate and the southern half subtropical (Figure 3.5d). At this point in time, the grasslands were apparently very similar to their present conditions (Dorf, 1960).

The cooling trend of the Pleistocene began in the Oligocene and led to the formation of the glaciers that moved down from the north (Figure 3.5e) and then back during interglacial stages (Figures 3.5f and 3.6). Considerable discussion has taken place about whether the vegetation zones were shifted slightly south (Braun, 1950) with most of the vegetation response near the glacier edges or whether the complete vegetation zones moved south over large distances (Martin, 1958). Most evidence supports the contention that entire vegetation zones were essentially translated south (Dorf, 1960; Wright, 1974). For example, evidences of spruce and fir exist in north central Florida, south central Texas, northern Oklahoma, and southern Kansas. Records of musk-ox go as far south as Mississippi, Texas, Oklahoma, and Southern California.

PRE-PLEISTOCENE AND PLEISTOCENE BIOGEOGRAPHIC DEVELOPMENT

Plants

Fossil records indicate that during successive geologic eras vegetation of the world has been dominated first by one major plant group and then by another. Ferns, pteridosperms, and lycopods were the dominant plant forms during the Carboniferous Period of the Paleozoic. About 110 million years ago, flowering plants radiated to become the most successful form of plant life in the world today (Raven and Axelrod, 1974). As is true of the angiosperms in general, the fossil record of the grasses is too incomplete to be of much help in determining the phylogenetic relationship. Grasses probably came into being late in the Mesozoic after flowering plants were well diversified, but grasses do not appear frequently in the fossil record until the Lower Eocene (Raven and Axelrod, 1974). Well-preserved fossils of vegetative grass structures (rhizomes, culms, roots, buds, and basal parts of leaves) have been obtained from late Tertiary rocks in Europe, and carbonized grass roots, which have been found in the lower Miocene deposits of the Florissant Beds in Colorado, have been assigned to Stipa (Gould, 1968).

The origin of the North American prairie probably dates back about 25 million years to the Oligocene Epoch of the Tertiary Period. Earlier in the warm, moist Eocene a temperate forest occupied the Great Plains, but as the Rocky Mountains arose, they intercepted moisture from the prevailing western winds and low summer precipitation was accompanied by dry winters (Weaver and Albertson, 1956). This climatic change caused a more rapid

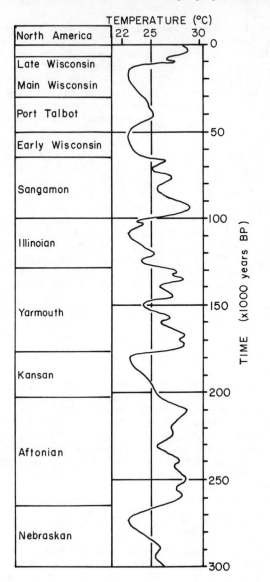

FIGURE 3.6 Temperature conditions during the Pleistocene glaciation. Time indicated as ×1000 years B.P. (before present). Redrawn from Watts, 1971.

evolution of the grassland in the upper Oligocene until the forerunner of the modern prairie existed in the Miocene (Clements, 1936).

Seeds of over thirty grass species have been found in the sands of the middle and late Miocene and Pliocene deposits, including needlegrass (Stipa), panicum (Panicum), and bristlegrass (Setaria).

Since none of these grass seeds have been found in the lower (older) beds of the same region, the fossil record seems to confirm the inference that originated from the adaptation of mammalian teeth--specifically, that grasses first spread over the plains in the Miocene (Schuchert and Dunbar, 1941; Weaver and Albertson, 1956). In the Pleistocene, the grasslands migrated south during glacial advances and northward during warmer periods. Actually the fossil record does not locate the prairies earlier than about 11,000 years ago.

The species that compose the modern prairie have different origins (Clements, 1936). For example, Junegrass (Koeleria), wheatgrasses (Agropyron), needlegrass, wildrye (Elymus), bluegrasses (Poa), and some sedges (Carex) are of northern origin and are all cool-season grasses that produce flowers early in the spring. Blue grama (Bouteloua gracilis), sideoats grama (B. curtipendula), buffalograss (Buchloe dactyloides), and some threeawns (Aristida spp.) migrated from the mountain plateaus of Mexico and Central America. Big bluestem (Andropogon gerardi), little bluestem (Schizachyrium scoparius), Indiangrass (Sorghastrum nutans), and switchgrass (Panicum virgatum) are semitropical in origin and are believed to have come from the east and southeast.

Invertebrates

The pre-Pleistocene and Pleistocene distribution and evolution of invertebrate fauna are not well known. Although the evolution of grasshoppers and leafhoppers could have occurred in the prairie and spread to other areas, the preponderance of evidence suggests most of the invertebrates originated in the early successional stages of the deciduous or boreal forests (Ross, 1970). Leafhoppers, common invertebrate herbivores, formed the basis of Ross's discussion. Leafhopper species diversity is much greater in certain seral stages of the deciduous forest than in comparable grass communities of the True Prairie. In addition, some of the leafhopper genera that feed on forest understory bluestem species are abundant in the forest but absent in the prairies.

The genus Diplocolenus is a holarctic group containing six species. The three American species occur in subclimax communities of the boreal coniferous forest, but one species is also abundant in the northern portions of the prairie. The simplest explanation is that the evolution of all species occurred in the grass subclimaxes of northern coniferous forests or in northern tundra areas and that the prairie species, subsequent to its evolution as a distinctive species, has evolved to inhabit the Great Plains grasslands. The genus Rosenus, however, appears to be a prairie species now occupying the tundra or seral grass stages of the boreal or subboreal forest (Ross, 1970).

Of the thirty known species of the leafhopper genus Flexamia, which is wholly North American, six occur only west and south of the Great Plains, four occur only east, and five appear to be Great Plains endemics. The remaining fifteen species occur commonly either in all or part of the Great Plains and also to the west, south, and east of it. The more primitive species of the Great Plains probably arose from western and southern ancestors, but the

more specialized members arose from bluestem-feeding species in eastern North America (Ross, 1970).

In summary, some groups, exemplified by <u>Diplocolenus</u> and <u>Rosenus</u>, represent a large group of more northern species resulting from an exchange between North America and Europe. <u>Flexamia</u>, however, represents another group that is restricted to North America and with a pattern of evolution and speciation that has occurred south of the boreal forest. Events of the Pleistocene influenced the speciation pattern of the more northern species, but the Pleistocene's influence on the southern endemic leafhoppers is unclear.

Although extensive leafhopper collecting has taken place in the southwest United States and Mexico, no evidence of relict populations exists, which would be expected as a consequence of major southward extension of the grasslands during the Pleistocene. On the basis of this evidence, Ross (1970) suggests that perhaps no large tracts of grassland existed, and the survival of grassland leafhoppers may have occurred in small patches of prairie occupying south or southwest-facing slopes or the open dunes in the Nebraska Sand Hills.

Birds

Like the invertebrates, relatively little can be said of the bird faunas characterizing the North American grasslands during past geological periods. Unlike mammals, few fossil remains can be found from which past faunal assemblages may be reconstructed, and birds simply do not enter the fossil record in a major way. The available fossil evidence indicates that most orders of contemporary birds were elaborated by the onset of the Eocene, and most modern families were in existence by the Miocene. By the Pleistocene many modern species had appeared, and often in excess of 75 percent of the species represented in middle and upper Pleistocene deposits are modern (i.e., living) forms (Brodkorb, 1971; Moreau, 1954). The apparent bias toward contemporary species in Pleistocene deposits may reflect the fact that most known deposits are in southern areas in which environmental changes during the Pleistocene were relatively moderate (Selander, 1966). The meager evidence indicates that Pleistocene avifaunas generally resembled avifaunas existing in comparable habitats today. One major change is the disappearance of many Falconiformes during the Pleistocene, apparently in conjunction with the extinction of many large grazing mammals. This extinction especially affected vultures. In addition to forms that represent predecessors of the three modern North American cathartids, four other species found in Pleistocene deposits failed to survive. Additionally, three of the eleven extinct species of the hawk group (Accipitridae) from North American Pleistocene deposits were members of the Old World vulture group, which persists today in Africa and Eurasia, but is absent from North America (Selander, 1966). Thus, at least this component of the terrestrial avifaunas was notably more diverse during the Pleistocene than today, perhaps especially in grassland environments.

Mammals

By the late Cretaceous period, when grasses apparently evolved, mammals had already been differentiated from their reptilian ancestors for roughly 100 million years but were still relatively small and did not include any specialized grazing herbivores, with the possible exception of some multituberculates. By the Paleocene, however, several families of the Condylarthra, Primates, and perhaps Insectivora, in addition to the Multituberculata, were largely herbivorous (Van Valen and Sloan, 1966). Competition for these diverse herbivore niches resulted in the remarkable success of the Condylarthra (and their descendants) throughout the Cenozoic and the extinction of the Multituberculata early in the Oligocene. Extant descendants of early condylarths include the whales, elephants, sirenians, hyraxes, and all of both the odd- and even-toed ungulates (McKeena, 1969). These early descendants are in addition to at least ten orders of extinct herbivores, several of which consisted exclusively of grassland species. Early Cenozoic radiation of grasses made possible much of the Cenozoic radiation of mammals.

During the Eocene, rodents became ubiquitous in North America. Although numerous in terms of individuals, the rodent fauna at that time still consisted of only a moderate number of species. Primitive artiodactyls appeared but were less abundant than were several related kinds of primitive perissodactyls. Several lineages of primitive carnivores emerged at that time also, probably as predators of the numerous rapidly evolving lineages of herbivorous mammals. Lemuroid primates and some primate-like insectivores first appeared and became relatively abundant. Multituberculates, marsupials, primitive insectivores, and several larger condylarths persisted into the Eocene, but the multituberculates apparently were extinct soon after the close of the epoch (Van Valen and Sloan, 1966).

Tertiary shifts in the mammalian fauna of North America continued and representatives of several groups (i.e., marsupials, primates, and condylarths sensu stricto) became extinct there; considerable faunal interchange occurred at times between North America and Eurasia (Simpson, 1947), and numerous herbivorous lineages shifted from a browsing to a grazing existence as the grasslands expanded in response to increasing aridity during the Miocene.

Grasses wear down grinding teeth, largely because of the opal ($SiO_2 \cdot nH_2O$) located in their cells. Opal is roughly as hard as quartz (7 on the Mohs scale) and much harder than the apatite of mammalian teeth. Thus, the rate of tooth wear was much greater on early grazing mammals than on their browsing ancestors (Sloan, 1972). In response, most grazing mammals have evolved high-crowned teeth that grow to replace much of the loss resulting from wear. Early Cenozoic mammals were characterized by low-crowned teeth, but, during the Miocene, horses, camels, and some other plains-dwelling mammals began to show rapid changes toward high-crowned grazing teeth (Schuchert and Dunbar, 1941).

Increase in body size and in cursorial specialization occurred at about the same time in numerous groups such as the ungulates, carnivores, and lagomorphs. Cursorial specializations (Vaughan, 1972), together with specializations associated with the evolution

of high-crowned teeth, probably represent the most spectacular mammalian adaptations associated with exploitation of the early grasslands. Specializations that served either to lengthen the stride or to increase the rate of the stride resulted in additional speed for both predators and prey and for greater cruising range between scarce water sources and between patches of available herbage. Lengthy seasonal migrations also became a way of life for some grassland species.

Whereas the Miocene was an epoch of worldwide upheaval resulting in rapid change in climate as well as in flora and fauna (through both extinction and adaptation), the Pliocene was by comparison a period of serenity. Kurten (1972) picturesquely described it as follows: "It is as if we were emerging onto a wide grassy plain under a strong sun, with a rising wind from distant snow-capped mountains. Such a landscape, populated by great herds of three-toed horses, mastodonts and antelopes, might indeed be taken as a symbol of the Pliocene. . . ." Grasslands and savannas of North America were widespread and relatively much richer in mammalian herbivores than at present. For example, seven genera of apparently sympatric horses were found in 12 million-year-old deposits from near what is now Valentine, Nebraska (Sloan, 1972). Of these genera, three were three-toed savanna species, one was a three-toed grassland horse, and two were one-toed grassland horses. Simpson (1947) listed some of the many taxa of carnivores, rodents, camels, peccaries, elephants, tapirs, antelopes, and rhinoceroses that were common and important elements of the recently evolved North American grasslands of the Pliocene. One endemic North American grassland group, the pronghorns, was at its peak during this epoch. Today only a single monotypic genus still exists, but some twelve other genera of New World antelopes, which flourished and then became extinct, can be found, all in North America (Koopman, 1968).

Quarternary mammalian faunas in the area occupied by the True Prairie were reviewed by Schultz and Martin (1970) and by C. H. Hibbard (1970). Detailed review of Hibbard's analysis of an upper Pliocene local fauna (the Rexroad) provides a picture of the Tallgrass Prairie at that time and provides a feel for the origin of the present grassland mammalian fauna. The Rexroad fauna consists of 2 insectivores, 1 bat, 16 carnivores, 20 rodents, 1 sloth, 1 mastodon, 4 rabbits, 5 artiodactyls, and 2 horses. All but one species, the modern badger, are now extinct. Twenty-seven of the fifty-two species are known only from the Pliocene. The sloth, raccoon, and cotton rat were genera of South American origin, demonstrating the influence of that biota on the grasslands of North America. C. H. Hibbard (1970) concluded with the following interpretation: "I picture the region at the time the fauna lived to have had well-developed gallery forests along the banks of braided streams. The valleys contained some marsh areas, and the remaining valley areas were savanna with tall grasses and scattered groves of trees and shrubs. Shortgrass would have occurred only on the driest of upland areas."

The Pleistocene was a period of pronounced faunal interchange between North America and Eurasia, the most extensive and intensive interchange that has occurred since the early Eocene (Simpson, 1947). Mammoths became abundant and modernized types of deer and

especially bovids had a profound influence that persists today. Nevertheless, even among important groups of migrants such as bovids and microtine rodents, the actual migrants were a very small minority of the total species living at the time (Simpson, 1947).

Pleistocene mammals are known from the grassland during the times of each of the four major glacial advances and the three interglacial periods. During this time the grassland mammalian fauna slowly became the fauna that we know today. Table 3.1 provides a list of Recent mammals known from south-central Kansas, just west of the presently recognized boundaries of the True Prairie, and compares modern species with their nearest relatives from the general area during the upper Pliocene and the Pleistocene.

The Nebraskan interval of time is poorly known, but available fossils indicate a fauna living under less severe climatic conditions than conditions experienced by later glacial faunas (see Figure 3.6). During the subsequent interglacial (Aftonian), large land tortoises in northern Nebraska indicate a mild year-round climate, but during the Kansan advance, these animals are known from no farther north than Knox and Baylor Counties, Texas, just south of the Texas Panhandle. Land tortoises occurred as far north as Kansas during the next interglacial period (Yarmouth, which is poorly known). The Illinoian fauna of the Great Plains takes on a distinctly Recent appearance for the first time and is indicative of a climate cooler in the summers but warmer in the winters than at present. Rainfall probably was similar or only slightly greater than today, but cooler summers caused less evaporation and that difference probably would have resulted in more tallgrasses and other mesic vegetation. The subsequent interglacial (the Sangamon) was warmer but gave way to the Wisconsin advance. Small vertebrates from that time are largely the same species living at present. For example, of the 134 Wisconsin mammal taxa listed by Hibbard, 78 are referred to as living species.

Mammalian evidence of coniferous forest in the Great Plains during the Wisconsin glaciation was provided by Guilday and Parmalee (1972) in the form of late Pleistocene fossils of Phenacomys, a genus of voles, from Missouri and Nebraska. Recent specimens are from no closer than the boreal coniferous forests in northeastern Minnesota and the Rocky Mountains of Colorado and New Mexico (Hall and Kelson, 1959).

Massive and relatively sudden changes took place in the North American mammalian fauna near the close of the Pleistocene. At the very time, about 11,000 years ago, when grasses were replacing the conifers of the prairie regions, horses, mammoths, and giant ground sloths disappeared, and the vast majority of the large mammals in North America became extinct over a relatively short time span (Martin, 1975). The glacial advance and associated encroachment of coniferous forest into prairie habitats may or may not be related to these extinctions, but we know that every biome in the hemisphere was affected. None was affected as greatly as the Grassland Biome.

Martin (1973, 1975) concluded that people acting as big game hunters were solely responsible for this multitude of extinctions. However, Grayson (1977) argues that both the magnitude and pattern of avian extinction are incompatible with the hypothesis that humans played a major role in causing the demise of numerous North American mammalian genera at this time. Whatever the cause, the grasslands

TABLE 3.1 Mammals that presently occur in south-central (Meade County) Kansas, or are believed to have occurred there within historic time, compared to nearest relatives (S = same species; F = same genus; F = same family) recovered there from the late Pliocene and glacial and interglacial periods of the Pleistocene. Recent species are given biogeographic affinity codes as follows: GR = grassland species of the Great Plains; D = steppe or desert species that have invaded from the southwest Sonoran Region or Great Basin; CF = coniferous forest species; DF = deciduous forest species; N = neotropical species; and W = widespread or Pan-American species. Data were compiled from Hall and Kelson (1959), C. H. Hibbard (1970), and Hoffmann and Jones (1970).

Species	Biogeographic code	Late Pliocene	Nebraskan	Aftonian	Kansan	Yarmouth	Illinoian	Sangamon	Wisconsin
Virginia opossum (Didelphis virginiana)	N	None	None	None	None	None	None	None	None
Least shrew (Cryptotis parva)	DF	F	F	F	None	F	S	F	F
Eastern mole (Scalopus aquaticus)	DF	F	None	None	None	None	S	None	None
Little brown bat (Myotis lucifugus)	W	F	None	None	None	None	F	F	None
Cave bat (Myotis velifer)	D	F	None	None	None	None	F	F	None
Small-footed myotis (Myotis leibii)	W	F	None	None	None	None	F	F	None
Silver-haired bat (Lasionycteris noctivagans)	W	F	None	None	None	None	F	F	None
Evening bat (Nycticeius humeralis)	DF	F	None	None	None	None	F	F	None
Big brown bat (Eptesicus fuscus)	W	F	None	None	None	None	F	F	None
Red bat (Lasiurus borealis)	DF	G	None	None	None	None	G	G	None
Hoary bat (Lasiurus cinereus)	W	G	None	None	None	None	S	S	None
Pallid bat (Antrozous pallidus)	D	F	None	None	None	None	F	F	None

33

Table 3.1 Continued.

Species	Biogeographic code	Status							
		Late Pliocene	Nebraskan	Aftonian	Kansan	Yarmouth	Illinoian	Sangamon	Wisconsin
Mexican free-tail bat (Tadarida brasiliensis)	N	None	None	None	None	None	None	None	None
Nine-banded armadillo (Dasypus novemcinctus)	N	None	None	None	None	None	None	None	None
Eastern cottontail (Sylvilagus floridanus)	DF	F	None	F	None	F	G	G	G
Desert cottontail (Sylvilagus audubonii)	D	F	None	F	None	F	G	G	G
Blacktail jackrabbit (Lepus californicus)	D	F	None	F	None	S	G	S	F
Thirteen-lined ground squirrel (Spermophilus tridecemlineatus)	GR	G	G	G	None	G	S	S	S
Spotted ground squirrel (Spermophilus spilosoma)	D	G	G	G	None	G	G	S	G
Black-tailed prairie dog (Cynomys ludovicianus)	GR	F	F	G	None	F	G	S	S
Eastern fox squirrel (Sciurus niger)	DF	F	F	F	None	F	F	F	F
Plains pocket gopher (Geomys bursarius)	GR	G	G	G	None	G	S	S	G
Plains pocket mouse (Perognathus flavescens)	GR	G	None	G	None	G	G	G	G
Silky pocket mouse (Perognathus flavus)	D	G	None	G	None	G	G	G	G
Hispid pocket mouse (Perognathus hispidus)	GR	G	None	G	None	G	S	S	G
Ord kangaroo rat (Dipodomys ordii)	D	F	None	F	None	F	S	S	F
Beaver (Castor canadensis)	W	F	F	F	None	None	S	S	None

34

Table 3.1 Continued.

Species	Biogeographic code	Status							
		Late Pliocene	Nebraskan	Aftonian	Kansan	Yarmouth	Illinoian	Sangamon	Wisconsin
Plains harvest mouse (Reithrodontomys montanus)	GR	F	F	F	None	G	S	S	G
Western harvest mouse (Reithrodontomys megalotis)	D	F	F	F	None	G	S	S	G
Deer mouse (Peromyscus maniculatus)	W	G	G	F	None	F	G	G	G
White-footed mouse (Peromyscus leucopus)	DF	G	G	F	None	F	G	G	G
Northern grasshopper mouse (Onychomys leucogaster)	D	G	F	F	None	G	S	S	S
Hispid cotton rat (Sigmodon hispidus)	N	G	F	G	None	G	F	F	F
Southern plains woodrat (Neotoma micropus)	D	G	F	F	None	G	G	S	F
Prairie vole (Microtus ochrogaster)	GR	F	F	F	None	F	G	S	S
Muskrat (Ondatra zibethicus)	W	F	F	F	None	G	G	S	F
Southern bog lemming (Synaptomys cooperi)	C	F	G	F	None	G	G	G	S
Norway rat (Rattus norvegicus)	I	None	None	None	None	None	None	None	None
House mouse (Mus musculus)	I	None	None	None	None	None	None	None	None
Porcupine (Erethizon dorsatum)	W	None	None	None	None	None	None	None	None
Coyote (Canis latrans)	W	G	None	None	None	S	S	S	F

Table 3.1 Continued.

Species	Biogeographic code	Status							
		Late Pliocene	Nebraskan	Aftonian	Kansan	Yarmouth	Illinoian	Sangamon	Wisconsin
Gray wolf (Canis lupus)	W	G	None	None	None	G	G	G	F
Swift fox (Vulpes velox)	D	F	None	None	None	F	S	S	S
Red fox (Vulpes vulpes)	W	F	None	None	None	F	G	G	G
Gray fox (Urocyon cinereoargenteus)	D	G	None	None	None	G	F	F	F
Raccoon (Procyon lotor)	W	G	None	None	None	None	None	None	None
Longtail weasel (Mustela frenata)	W	F	F	F	None	S	G	S	F
Black-footed ferret (Mustela nigripes)	GR	F	F	F	None	G	G	G	F
Mink (Mustela vison)	W	F	F	F	None	G	S	G	F
North American badger (Taxidea taxus)	W	S	F	S	None	F	F	F	S
Spotted skunk (Spilogale putorius)	W	G	F	F	None	S	F	S	F
Striped skunk (Mephitis mephitis)	W	F	F	F	None	F	F	G	S
River otter (Lutra canadensis)	W	F	F	F	None	F	F	F	F
Mountain lion (Felis concolor)	W	G	None	None	None	G	F	S	None
Bobcat (Lynx rufus)	W	F	None	None	None	F	F	F	None

Table 3.1 Continued.

Species	Biogeographic code	Late Pliocene	Nebraskan	Aftonian	Kansan	Yarmouth	Illinoian	Sangamon	Wisconsin
						Status			
Elk (Cervus elaphus)	W	F	None	None	None	None	F	F	F
White-tailed deer (Odocoileus virginianus)	W	G	None	None	None	None	G	S	G
Mule deer (Odocoileus hemionus)	W	G	None	None	None	None	G	G	G
Pronghorn (Antilocapra americana)	W	F	None	None	None	F	F	F	None
Bison (Bison bison)	W	None	None	None	None	None	G	G	G

37

of North America no longer contained a high diversity of large, grazing herbivores, a characteristic of most of their evolutionary development.

Although some herbivores such as bison (Bison bison) and the pronghorn (Antilocapra americana) survived, many more species became extinct. The full effect of these extinctions will never be known. The important point is that present North American grasslands are ecologically very different from the earlier grasslands that supported the diverse fauna. Without the toxodonts, glyptodonts, mastodons, mylodonts, and their extinct relatives, the earlier pathways of energy flow in the True Prairie or any other North American prairie are impossible to distinguish. An interesting comparison would have been the relative proportions of energy that flowed naturally from grass to native consumer and whether or not a greater proportion of energy now flows directly to decomposers. Thus, although useful in today's world, all of the research that has contributed to this book, as well as other publications on the subject, has only increased our understanding of the structure and function of a system that has undergone a tortuous history.

PRESENT BIOTA OF THE TRUE PRAIRIE

Plants

Late Tertiary grasses were probably adapted to semiarid situations, perhaps steppes or savannas. For example, bulliform cells permitted the leaves to become enrolled during periods of drought. A basal intercalary meristem enabled the leaves to recover from grazing. The intercalary meristems and the silica content evolved during the differentiation of the family, increasing resistance to grazers--particularly the relatively inefficient grazing mammals or reptiles--as well as many insects. The modification of petals into lodicules enabled the florets to open during periods of favorable moisture and close during periods of excessive moisture or drought. Finally, wind pollination was generally an adaptation to drought regions where pollinating insects are relatively scarce and winds are strong.

Adaptations originated primarily for asexual reproduction and greater seed dispersal especially by wind, which is aided by bristle-like structures that surround the dispersal unit. Usually the lemma, palea, and caryopsis structures have been modified into bristles. Mechanisms for seed dispersal by animals are even more diverse. The most familiar mechanisms are long rough awns or beards that penetrate the hair and feathers of mammals or birds or even the skin, especially near the mouth parts of grazing animals. In other instances, particularly among species in which the caryopsis surrounding lemmas are small and light, trichomes on the surface or the base of the lemma help the caryopsis adhere to animal fur. Some mechanisms for dispersal by animals render the caryopsis and its surrounding envelopes attractive for ingestion by birds and at the same time at least partially resistant to their digestive juices (Stebbins, 1972).

The prairies have developed with species originating from geographically diverse sources. For over 20 million years, the species have been mixed and stirred by the recurring periods of hot and cold, and wet and dry climatic conditions. The major species are characterized by a broad ecological amplitude or tolerance and have relatively large geographical ranges. Ninety-five percent of the species are perennial, many with long life spans (twenty years or more). As a result, the prairie is a highly integrated and developed community that shows variability as a result of climatic and edaphic conditions, but its physiognomy is relatively homogeneous throughout (Weaver, 1958).

To insure that the reader is familiar with the dominant plant species of the True Prairie, eleven of the most important grasses are described in the subsequent sections and much of the descriptive text follows Weaver and Albertson (1956).

Big Bluestem

Big bluestem (<u>Andropogon</u> <u>gerardi</u>) is an erect, warm season perennial that may reach a total height of 2-3 m under the best growing conditions (Figure 3.7). Growth begins in midspring with rapid development of the shoot, which results in abundant foliage, and is 1 m in height by late summer. The leaves are often 0.5-0.7 m long and spread so the area of the top of the clump may be twice the area occupied by the base. Flower stalk initiation occurs in midsummer with anthesis (full bloom) by early fall, but only a few tillers produce flower stalks in any year. The tall, sometimes purple or blue stalks terminate in three fingerlike racemes (therefore, the species is sometimes called <u>turkeyfoot</u>) that produce abundant seed. Big bluestem reproduces primarily asexually by short, stout rhizomes and has a determinant growth pattern in which the number of buds in the rhizome determines the tiller number of the current year. The roots of mature plants are quite coarse, 2-3 mm in diameter, and frequently reach a depth of 2-3 m.

Seedlings develop rapidly under favorable conditions and tillering begins early, seven or eight weeks after germination. Leaves remain green and function under light intensities of only 5 percent to 10 percent of full sunlight. The individual stems, usually about 1 cm apart, grow into mats of sod under moist conditions, but under dry conditions the clumps are much smaller and isolated. The growth characteristics contributing to its ability to dominate the prairie are its rapid growth, dense sod-forming habit, great stature, and tolerance of the plant and its seedlings to shade.

Together with little bluestem, big bluestem frequently constitutes 75 percent of the plant cover on upland sites in the eastern True Prairie. Big bluestem is palatable, nutritious, and productive forage. If not cut for prairie hay, the stalks become hard during the latter part of the summer and are not grazed.

FIGURE 3.7 Big bluestem (<u>Andropogon gerardi</u>). Plant, X ½; pair
of spikelets, X 5. (Am. Gr. Nat. Herb. 255, D.C.) (from Hitchcock
and Chase, 1950).

Little Bluestem

Little bluestem (<u>Schizachyrium scoparius</u>) is an erect, warm
season perennial, reaching a maximum height of about 1 m (Figure
3.8). The leaves are slender, about 0.4 m in length, and reach a
canopy height of 0.5 m. Flower stalks are produced in late summer
and bear many seeds.

In moist habitats little bluestem may form a loose sod, but
more often in drier conditions it forms clumps in which the bases

FIGURE 3.8 Little bluestem (<u>Schizachyrium</u> <u>scoparius</u>). Plant,
X ½; pair of spikelets, X 5. (Am. Gr. Nat. Herb. 268, D.C.) (from
Hitchcock and Chase, 1950).

are 0.1-0.2 m in diameter and the aerial foliage is more than twice
that size. The leaf bases are very flattened, and in the fall the
leaves turn a characteristic purple or reddish-brown. Propagation
is by seed, tillers, and sometimes by short inconspicuous rhizomes.
     Little bluestem becomes proportionately more important under
drier conditions. It is a palatable, nutritious forage that
provides excellent and durable range when grazed judiciously.

Indiangrass

Indiangrass (<u>Sorghastrum</u> <u>nutans</u>) is a tall, coarse, warm season grass with habitat requirements very much like big bluestem (Figure 3.9). The leaves are relatively broad and have a characteristic light green color. The leaf blades are oriented at a 45° angle from the stem and there is a prominent ligule with two large points. The large yellowish-green panicles, 0.1-0.3 m long, are at heights 1-2 m and are dazzling when viewed in the late autumn sun.

FIGURE 3.9 Indiangrass (<u>Sorghastrum</u> <u>nutans</u>). Plant, X ½; spikelet with pedicel and rachis joint, X 5. (Deam, Ind.) (from Hitchcock and Chase, 1950).

The seeds are usually viable and germinate if not buried deeply in the soil. The species may form patches of sod or occur in small bunches. Tillers are formed from rhizomes in late summer and these tillers overwinter and emerge to provide most of the early growth the following spring.

Although usually not as abundant as big and little bluestem, Indiangrass may form fairly solid patches, frequently increases after fire, and is a good colonizer of disturbed habitats. Like the previous two species, Indiangrass provides excellent forage, especially in the late spring and summer. Though found throughout most of the True Prairie, it is more abundant in the southern portions.

Switchgrass

Switchgrass (Panicum virgatum), a tall, coarse, warm-season species with a preference for relatively moist habitats in the True Prairie (Figure 3.10), has long rhizomes and usually forms colonies in the central to southern True Prairie, but it may form a sod under moist conditions or exist as a single stalk in drier habitats. The roots are coarse and very deep, frequently exceeding 2 m. Early growth, beginning in midspring, is rapid, and at maturity in late summer the plants may be over 2 m in height. Switchgrass derives its common name from the large, open, spreading panicles, 0.5 m across that develop in the latter part of summer. The seeds are shed in the winter or late fall and generally require stratification before germination.

This species is not particularly shade-tolerant but is a good forage species and prospers under cropping for prairie hay.

Sideoats Grama

Sideoats grama (Bouteloua curtipendula), named for the inflorescence that consists of thirty-five to fifty pendulous spikelets on a slender flower stalk, is a warm-season perennial with short, scaly rhizomes (Figure 3.11). It is intermediate in height, usually not more than 1 m including the flowering stalks. Growth begins in the spring and flowers are produced from midsummer until the end of the growing season. Growing as an open sod or in clumps, the abundant tillers and new shoots from rhizomes produce much foliage. This species is tolerant to shade and grazing and is more tolerant to drier conditions than the preceding four species.

Prairie Dropseed

Prairie dropseed (Sporobolus heterolepis) is a warm season perennial which forms distinctive clumps up to 0.5 m in diameter (Figure 3.12). Prairie dropseed is found in the northern portion of the True Prairie where growth begins in early spring. The narrow yellow-green leaves, which may become nearly 1 m in length, are numerous--as many as two hundred in one medium-sized clump. Abundant fibrous roots reach a depth of 1.6 m and the foliage is

FIGURE 3.10 Switchgrass (Panicum virgatum). Plant, X ½; two
views of spikelet, and floret, X 10. (V. H. Chase, Ill.) (from
Hitchcock and Chase, 1950).

usually about 0.5 m in height. The broad, spreading panicles may be
on stalks up to 1 m. Each spikelet bears a single large seed that
is released at maturity (therefore the name). The grass provides
good forage, especially in the first half of the summer.

Tall Dropseed

     Tall dropseed (Sporobolus asper), a warm-season perennial
bunchgrass, may reach a height of 1.3 m (Figure 3.13) and is much

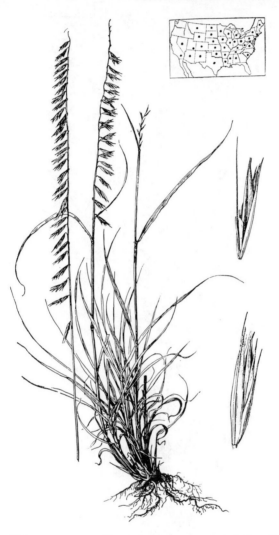

FIGURE 3.11 Sideoats grama (<u>Bouteloua</u> <u>curtipendula</u>). Plant, X ½; spikelet and florets, X 5. (Chase 5408, Colo.) (from Hitchcock and Chase, 1950).

more common in the southern part of the True Prairie. The stems are mostly unbranched and the compacted panicle, which is enclosed in the upper leaf sheath, may be up to 0.2 m in length. The long fibrous leaves characteristically curl as they mature and may remain on the bleached-white stem for a year.

Scribner Panicum

Scribner panicum (<u>Panicum</u> <u>scribnerianum</u>) is a low-growing perennial with small rosettes of broad smooth green leaves that

FIGURE 3.12 Prairie dropseed (Sporobolus heterolepis). Plant, X
1; spikelet and floret with caryopsis and split palea, X 10.
(McDonald, Ill.) (from Hitchcock and Chase, 1950).

frequently overwinters. In the summer the erect stems may reach a
height of 0.3 m.   The inflorescence is a conspicuous and
characteristic panicle with large spikelets (Figure 3.14).   Scribner
panicum is typically an understory species that is never dominant,
but it has a very broad ecological amplitude and is found frequently

FIGURE 3.13 Tall dropseed (Sporobolus asper). Plant, X 1; glumes and floret, X 10.   (Deam 42707, Ind.) (from Hitchcock and Chase, 1950).

throughout the prairie, though more abundantly in the southern portions.

Porcupine Needlegrass

Porcupine needlegrass (Stipa spartea) is a cool-season species of northern origin. It occurs primarily in the northern portion of

FIGURE 3.14 Scribner panicum (Panicum scribnerianum). Plant, X 1;
two views of spikelet, and floret, X 10. (Vernal phase, McDonald
32, Ill.; autumnal phase, Umbach 2365, Ill.) (from Hitchcock and
Chase, 1950).

the True Prairie (Figure 3.15) where its rapid growth begins early
in the spring; a height of 0.6 m may be attained by early summer.
The flower stalks develop by late spring and produce seeds with very
long awns (10 cm). These awns are twisted, and presumably drive the
seed in the ground by hygroscopic gyrations. Seeds germinate well
when buried at a depth of 4-5 cm. The grass produces high-quality
forage but is susceptible to overgrazing, especially in the early
spring.

Prairie Junegrass

    Prairie Junegrass (Koeleria cristata) is a cool season
perennial bunchgrass of boreal origin. It is more abundant in the
Flint Hills and northward in the True Prairie where the foliage
reaches a height of 0.3 m and the flower stalks may be 0.8 m tall
(Figure 3.16). Growth is in small bunches and the short leaves are
soft and conspicuously dark green.   The abundant flower stalks,
which appear in mid- to late spring, are densely flowered,
contracted, spike-like panicles, 5-10 cm long. The white color of
the mature inflorescences is quite attractive and abundant seed is
produced.   Like numerous other cool season species, a period of
summer dormancy is followed by vegetative regrowth in the fall.
This species is an excellent forage species and is selectively
grazed in the spring.

Kentucky Bluegrass

    Kentucky bluegrass (Poa pratensis), despite its name, was
introduced from Europe at the time of the white settlers. This

FIGURE 3.15 Porcupine needlegrass (<u>Stipa spartea</u>). Plant, X ½;
glumes and floret, X 2. (McDonald 16, Ill.) (from Hitchcock and
Chase, 1950).

small rhizomatous, cool season grass produces foliage about 10 cm in
height and flowering stalks about 20 cm (Figure 3.17). This species
grows very early in the spring and produces abundant seed by late
spring, so most growth occurs prior to the development of any
associated warm season species. Following a summer dormant period,
considerable vegetative growth may occur in the fall.

Kentucky bluegrass produces high-quality forage, but production
is lower than species of larger stature. In the southern portion of

FIGURE 3.16   Prairie Junegrass (<u>Koeleria</u> <u>cristata</u>).   Plant, X ½;
glumes and floret, X 10.   (Bebb 2862, Ill.) (from Hitchcock and
Chase, 1950).

the True Prairie, it is most abundant in moderately to heavily
grazed areas.   Farther north it is also common in lightly stocked
pastures.   Kentucky bluegrass produces early season forage but is
susceptible to spring burning.

Other significant grass species are in evidence, but these
descriptions include the most important ones.   Forbs are also
important, but they will not be described individually.   Figure 3.18
shows four typical forb species.   Weaver (1954, 1968) provides an
excellent discussion of the major True Prairie grasses and forbs.

FIGURE 3.17 Kentucky bluegrass (Poa pratensis). Plant, X ½;
spikelet, X 5; floret, X 10. (Williams, S. Dak.) (from Hitchcock
and Chase, 1950).

Phenology

Not all grassland plant species develop at the same time during
the growing season--that is, they do not all have the same phenology
(Dickinson and Dodd, 1976).
The mesic prairies, especially in the north, warm up more
slowly than dry prairies in the early spring and have a smaller

FIGURE 3.18   (a) Pale echinacea (<u>Echinacea</u> <u>pallida</u>) and (b) prairie
with patches of prairie fleabane (<u>Erigeron</u> <u>strigosus</u>). Below (c) is
a portion of plains wildindigo (<u>Baptisia</u> <u>leucophaea</u>) in blossom, and
(d) a top view of rosettes of field pussytoes (<u>Antennaria</u> <u>neglecta</u>).
(From Weaver, 1954.)

representation of the prevernal group including the pasqueflower (Anemone patens) and buttercup (Ranunculus fascicularis). Characteristic spring flowers of the Mesic Tallgrass Prairies include the shootingstar (Dodecatheon meadia), bedstraw (Galium boreale), hoary gromwell (Lithospermum canescens), downy phlox (Phlox pilosa), spiderwort (Tradescantia ohiensis), violet woodsorrel (Oxalis violacea), and prairie violet (Viola pedatifida) (Curtis, 1959).

The midsummer aspect of the Mesic Prairie is one of a continuously changing carpet of color--largely various shades of yellows and purples. Among the most showy members of this assemblage are the composites with tickseed (Coreopsis palmata), tall gayfeather (Liatris aspera), gray-headed prairie-coneflower (Ratibida pinnata), blackeyed Susan (Rudbeckia hirta), the rosinweed or compassplant (Silphium spp.), and pale echinacea (Echinacea pallida). Other species blooming in the summer are the buttonsnakeroot (Eryngium yuccifolium), floweringspurge (Euphorbia corollata), the purple-flowered leadplant (Amorpha canescens), Illinois tickclover (Desmodium illinoense), and purple prairieclover (Petalostemum purpureum) (Curtis, 1959).

In the autumn a large number of asters (Aster spp.), goldenrods (Solidago spp.), and sunflowers (Helianthus spp.) add to the flowering parade, but while these flowers are individually colorful, or even brilliant, their effect is diluted by the presence of the tallgrasses, especially in wet years. Such dominants as big bluestem, Indiangrass, and switchgrass begin to elongate flowering stems about mid-August and obtain full growth in late September or early October. Full growth means an average height of 1.8 m for big bluestem, with the tallest stands obtaining heights of 2.5-3.0 m in exceptionally favorable years, while the other two species range from 1.2-1.8 m (Curtis, 1959).

A correlation is evident between flowering date and average height of the species in the northern True Prairie (Butler, 1954). In dry mesic Wisconsin prairies, species that bloomed in May, June, July, August, and September averaged 13, 23, 32, 34, and 46 cm in height, respectively. The same progression was seen in all prairie species although the heights attained were greater with increasing soil water and reached a maximum of 91 cm in September plants of wet prairies. The average level of the leaf canopy of the entire community paralleled these flower heights but was 3-6 cm lower at a given season (Curtis, 1959). Additional phenology and height measurements on True Prairie species are given by Ahshapanek (1962) and Weaver and Fitzpatrick (1934). McKendrick et al. (1975) provide phenology data on big bluestem and Indiangrass.

In a detailed study of five grass species--big bluestem, little bluestem, Indiangrass, switchgrass, and sideoats grama--Rice (1950) examined the phenology during the 1948 growing season. In all species except sideoats grama considerable evidence showed that culms, which did not have the specified number of leaves by a certain date, did not initiate an inflorescence. Flowering occurred chiefly during the long days of early summer in sideoats grama, but in the other species flowering did not begin until the day-length had decreased a substantial amount below that at the time of floral initiation. The time between inflorescence initiation and exertion varied with each species: 43 days in little bluestem, 50 days in

big bluestem, 75 days in switchgrass, 64 days in Indiangrass, and only 15 days in sideoats grama. The total time between inflorescence initiation and flowering was just slightly less than 2 months in little bluestem and big bluestem, slightly over 2 months in switchgrass and Indiangrass, and about 1 month in sideoats grama.

Jones (1962) evaluated the phenology of grasslands near the Osage Site. As can be seen from Table 3.2, the sedges (Carex spp.) and Japanese brome (Bromus japonicus) accounted for most of the green material in the late fall, winter, and early spring. By August big bluestem and little bluestem contributed approximately 75 percent of the green biomass.

Floristic Similarities

Over the years numerous studies have been made on various portions of the True Prairie (Asby and Kelting, 1963; Barker, 1969; Buck and Kelting, 1962; Curtis, 1956; Evers, 1955; Kilburn and Ford, 1963; Lathrop, 1958; Murray, 1974; Ray, 1959; Ward, 1956; Weaver, 1958; and Weaver and Hemmil, 1931). These studies primarily report species composition of small areas, but collectively they comprise the data source for broad generalizations about species composition across the True Prairie.

In an early monograph Weaver and Fitzpatrick (1934) listed 225 important species of grasses and forbs in the True Prairie of the Missouri River Valley region; Steiger (1930) found 237 species in about 2,600 ha of prairie near Lincoln, Nebraska; Shimek (1931) indicated that about 265 species made up the bulk of the Iowa prairie flora (Weaver and Albertson, 1956); and Dwyer (1958) found 296 species on the ranch where the Osage Site is located.

Transeau (1935) compared the species of the Ohio Prairie Peninsula with species listed by Weaver and Fitzpatrick (1934) in the Missouri River Valley region and found:

Species category	Proportion on Ohio Prairie (percent)
Of the 11 major grasses	100
Of the 22 minor lowland grasses	73
Of the 67 lowland forbs	84
Of the 25 minor upland grasses	58
Of the 75 upland forbs	40

The prairies of the peninsula possess the same communities as the communities described for the Missouri Valley, though many species that are lowland in the west are found in dry prairie habitats in southern Ohio (Thompson, 1939; Weaver and Albertson, 1956).

Although many dominant species are found throughout the Tallgrass Prairie, one of the recognizable floristic features is the increasing dissimilarity with increasing distance. Curtis (1959) compared the Wisconsin prairies with prairies from other geographical areas (Table 3.3). The stands that represented examples of the Tallgrass Prairie had about a 50 percent similarity in species composition with the Wisconsin grasslands, while examples

TABLE 3.2 Total green forage of important species on a tallgrass pasture, Osage County, Oklahoma (Jones 1962).

| Species | Forage (percent) | | | | | | | | | | | |
| | 1960 | | | | | | | 1961 | | | | |
	June	July*	Aug.	Sept.	Oct.	Nov.	Dec.	Jan.	Feb.	Mar.	Apr.	May
Sedge (Carex sp.)	2.4	T	1.1	0.2	0.4	27.5	53.1	71.0	53.6	40.5	20.8	8.2
Common oxalis (Oxalis stricta)	0.1	0.3	0.2	0.2	T	2.1	6.7	3.0	3.9	4.2	2.5	1.7
Japanese brome (Bromus japonicus)	2.4	T				3.6	25.1	15.0	33.6	30.7	18.1	3.6
Wavyleaf thistle (Cirsium undulatum)	0.3	0.2				0.5	2.4	7.0	1.3	2.5	5.0	2.8
Carolina geranium (Geranium carolinianum)	0.1							1.0	0.6	2.0	1.6	0.7
Inland rush (Juncus interior)		1.2	1.1						3.2	9.6	8.2	2.6
Western ragweed (Ambrosia psilostachya)	7.0	8.6	8.4	10.3	11.1	1.0				3.3	22.6	13.4
Big bluestem (Andropogon gerardi)	24.4	26.9	21.7	28.8	20.0	0.6				0.4	4.8	17.2
Common goldstar (Hypoxis hirsuta)										0.2	2.9	1.1
Scribner panicum (Panicum scribnerianum)	3.1	2.1	2.4	3.0	2.4	7.2					3.0	3.2
Little bluestem (Schizachyrium scoparium)	41.6	46.5	52.4	37.1	46.6	46.5					1.0	33.1
Tall dropseed (Sporobolus asper)	0.3	1.7	1.1	2.5	5.6	1.5						1.6
Switchgrass (Panicum virgatum)	6.2	7.0	5.5	7.0	7.6							2.6
Total area of green forage available in cm²/1 m² plot	631.3	499.4*	557.5	477.3	464.5	111.0	25.9	19.8	30.8	55.0	116.1	430.3

*Pasture where half of the plots were located was heavily grazed this month.

TABLE 3.3   Geographical Relations of Wisconsin Prairies (From J. T. Curtis, Vegetation of
Wisconsin, Copyright 1959, University of Wisconsin Press, Madison).

Location of study	Species listed (No.)	Percent in common with Wisconsin stands	References
Northeastern Illinois	95	95.7	Vestal (1914)
Ann Arbor, Michigan	26	92.3	Gleason (1917)
Peoria, Illinois	92	90.2	Brendel (1887)
Wilton, Iowa	156	83.9	Shimek (1925)
Central Iowa	233	78.1	Conard (1952)
Clay Co., Kansas	48	70.8	Schaffner (1926)
Hancock, Illinois	247	68.4	Kibbe (1952)
Oklahoma	47	65.9	Smith (1940)
Southern Manitoba	131	58.0	Shimek (1925)
Ohio	157	52.3	Jones (1944)
Central Saskatchewan	85	41.2	Coupland and Brayshaw (1953)
Western North Dakota	57	36.8	Hanson and Whitman (1938)
Sand Hills, Nebraska	101	36.6	Pool (1914)
Northwestern Alberta	139	30.2	Moss (1952)
Eastern Colorado	89	22.5	Hanson (1955)
Southwestern Saskatchewan	94	20.2	Coupland (1950)
Southern Alberta	260	20.0	Moss and Campbell (1947)
Southern Montana	50	16.0	Wright and Wright (1948)
Western Idaho	77	10.4	Daubenmire (1942)

from the distant Shrub Steppe or Palouse Prairie in Idaho showed only a 10 percent similarity.

Plants of the Wisconsin lowland prairies generally have ranges that extend toward the southeastern part of the United States, while the dry prairie plants range to the southwest. The differences in affinities of the floras of these wet and dry prairies indicate that all of the grassland species could not have had the same geographical origin. The large Alleghenian meadow element in the wet prairies of Wisconsin indicates that this community migrated from the southeast along with the hardwood forest. Environmental conditions favorable for such a migration would undoubtedly be very different from conditions suitable for the entrance of xeric flora from the southwestern desert or plains. Thus, the homogeneous prairies in Wisconsin, and indeed throughout the True Prairie, probably did not develop as a single event. Rather a variety of elements arrived during separate climatic regimes that were favorable to each particular group. In terms of species originations, the extreme northern and southern portions of the True Prairie are more homogeneous than the major central portion (Curtis, 1959).

Certain similarities in environmental tolerance enable some species representative of various communities to diffuse from one element to another and impart a degree of uniformity to the whole. For example, the bluestem grasses and other broadly tolerant species have assumed prominent roles in many prairies regardless of their origin. The prairies of other midwest states have a similar history, but since they are located differently with respect to the source origins and since they have been affected differently by a post-glacial climatic change, the relative composition of their flora is correspondingly different. Thus, Gleason (1901) reported that 48 percent of the 415 species he listed for the prairies of Illinois were of southern origin. The prairies of Iowa have a higher percentage of southwestern species (Shimek, 1911) as do the prairies of Minnesota (Moyer, 1910), since they are closer to that source region and have been under the influence of dry climates more frequently and for longer periods than other regions to the east.

The Mesic Prairies of Wisconsin closely resemble prairies just to the west. Curtis (1956) studied a series of prairies in northeastern Iowa and southeastern Minnesota and an index of similarity between this group and the Wisconsin Mesic Prairies was 67.5 percent. Conard (1952), in his book on the vegetation of Iowa, listed the plants from prairies in seven central Iowa counties and 78.1 percent of the 233 species were also found in Wisconsin. Weaver and Fitzpatrick (1934) conducted an extensive study comparing 100 stands of Oklahoma Prairie with stands in Nebraska, Missouri, and Iowa. On the basis of the selected 142 important species, an index of a similarity of 49.8 percent was obtained between Oklahoma and Nebraska, Missouri, and Iowa.

A floristic analysis of the total flora of Wisconsin Tallgrass Prairies reveals that the Fabaceae comprise about 7 percent of all species. Four other families, in combination with the legumes, include 50 percent of the total species. These families are the Asteraceae (26 percent), Poaceae (10 percent), Lamiaceae (4 percent), and the Liliaceae (3 percent). As Curtis (1959) suggests, on the basis of taxonomy these areas should be called "daisylands"

rather than grasslands. However, forbs seem to become relatively less important farther south in the True Prairie.

Weaver (1954) studied the Flint Hills and Osage Hills in 1952 and 1955. During these investigations he sampled several prairies that were not grazed but were cut for prairie hay. A summary of these results show that big and little bluestem were by far the most important (Table 3.4). The eight grass species plus forbs composed 87 to 98 percent of the cover, so other grass species were relatively minor. In these mowed grasslands, because of the location of the apical meristem, the forbs are relatively unimportant, averaging only 2.2 percent of the cover. Site C on Table 3.4 is probably very near the Osage Site.

One of the fascinating aspects of the floristics of the True Prairie is that some dominant species are found over very large geographical areas. For example, little bluestem and sideoats grama are found essentially throughout the Tallgrass Prairie but are relatively more prominent in the South, while others like needlegrass are more important in the North. However, most of the True Prairie dominants span relatively large gradients of environmental factors such as moisture, temperature, soil, and day-length.

The dominant species apparently are adapted by both plasticity in response to a wide range of environmental factors and ecotypic variation within a species (Larsen, 1947). For example, when individuals from a needlegrass population were transplanted from widely distributed points into one location, the grasses showed almost identical behavior when grown under the same habitat influences. The populations from southern communities in their native habitats flowered earlier than individuals from northern habitats when conditions favored active early spring growth. However, in other transplant studies, clones of Canada wildrye (Elymus canadensis), which have displayed earlier flowering in southern communities and later flowering in the north and west, showed widely divergent behavior when transplanted and grown under the same conditions. Clones from western and northern communities required less time to reach flowering than southern and eastern clones. Clearly, the flowering time was genetically fixed and had arisen presumably because of the shorter growing season in the northern and western part of the True Prairie (McMillan, 1959; 1965). In early successional stages ranging from 2-40 years in New Jersey, where little bluestem is a successional species, populations from earlier stages flowered earlier and there was a greater amount of energy devoted to production. However, greenhouse experiments showed that these differences were largely the result of phenotypic plasticity (Roos and Quinn, 1977).

Habitat pressures provide a competitive framework on which the available genetic materials are sorted into communities of groups of species that conform to biotic, edaphic, and climatic constraints. The biotic manifestations may result from plasticity or ecotypic variation. The northern and western communities of the True Prairie have been shaped under pressure of late spring growing conditions, short periods of adequate moisture, and early frost. These communities contain early flowering variants of many species. In the eastern and southern communities, the requirements of all species are more likely to be met at some time during the long

TABLE 3.4   Percentage Composition of Grasses in Seven Locations in the Flint Hills-Osage Hills Region (Weaver, 1954).

Species	Location[a]							Average
	A	B	C	D	E	F	G	
Big bluestem	36.7	24.5	29.2	29.2	73.7	57.1	22.2	38.9
Little bluestem	45.8	38.0	41.3	22.0	7.5	32.8	53.6	34.4
Sideoats grama	5.0	0.9	3.3	23.5	1.0	--	3.6	5.3
Blue grama	0.6	16.1	1.1	8.5	--	0.8	2.2	4.2
Indiangrass	2.0	2.2	0.7	2.6	10.5	2.6	--	2.8
Switchgrass	2.9	8.8	3.3	--	0.1	2.1	2.5	2.8
Buffalograss	--	--	10.0	3.5	--	0.1	--	1.9
Tall dropseed	0.5	4.0	7.1	0.6	--	--	--	1.7
Forbs	2.2	1.0	1.6	2.8	3.8	0.9	3.0	2.2
Total	95.7	95.5	97.6	92.7	96.6	96.4	87.1	94.5

[a] A = Level to north sloping upland, 64 km east of Newton, Kansas.
   B = Level to rolling land between Augusta and Winfield, Kansas.
   C = Gentle slope near Grainola, Oklahoma (near Osage Site).
   D = Gently rolling land with outcropping limestone, near Foraker, Oklahoma.
   E = Same prairie on a north slope below the outcrop.
   F = Gently sloping lowland near Cedarvale, Kansas.
   G = Rolling land south of Cassoday, Kansas.

growing season, and these communities contain a sequence of later flowering species and types. In spite of its apparent species compositional uniformity, the True Prairie grassland presents an extensive display of ecotypic variation (McMillan, 1959).

Clearly, biological communities change along multifactor gradients such as the north-south and/or east-west axes of the True Prairie. These vegetational changes result from the ecological amplitudes of the individual species that result in species compositional variation and different species diversity patterns along these axes. Superimposed on the differential species responses is the intraspecific ecotypic variation. Grass species in the tallgrass region have been shown to possess ecotypes for photoperiod (McMillan, 1965), moisture (Porter, 1965), and nutrients

(Maschmeyer and Quinn, 1976). In addition to the definitional question regarding a species, an interesting and unexamined question revolves around the identification of situations that display ecotypic variation and situations that are manifest in species compositional variations.

Vegetation Types

A number of associations within the True Prairie have been recognized as distinct vegetational units. These associations all appear to relate to soil conditions and include the Cross Timbers and Oak Savanna, Blackland Prairie, and Black Belt Prairie. Each association will be described briefly, but the subsequent analytical aspects of the book will not include these vegetational units.

Cross Timbers and Oak Savanna

Cross Timbers, as described by Kennedy (1841), is ". . . a continuous series of forests, extending from the forested region at the sources of the central Texas Trinity River in a direct line north, across the apparently interminable prairies of northern Texas and the Ozark territory, to the southern bank of the Arkansas River" (Dyksterhuis, 1948). The actual width of the Cross Timbers varies depending on the author. Kennedy (1841) said that it was from 10-90 km wide but Küchler (1964) mapped it at about 180 km wide. Küchler also mapped the Cross Timbers as a large area in the Prairie Peninsula described by Borchert (1950), which is 500 km wide. At the southern end, the Cross Timbers split with the Texas Blackland Prairie that is located between the eastern and western Cross Timbers.

The origin of the term Cross Timbers is not known but probably arises from the fact that the forest extends north and south across, rather than along, the major eastwardly flowing rivers of the region, or because the westbound travelers, having gone through the eastern forests and some True Prairie, were confronted by yet another forest to cross (Dyksterhuis, 1948). Cross Timbers was mentioned frequently in the writings of early southwest travelers--as early as 1772. Since the prairie fires from the west swept into the Cross Timbers, and since these oak species have persistent branches made more formidable by fires, the area was considered quite inhospitable to travelers who also had to contend with the well-developed underbrush.

Rainfall in the Cross Timbers ranges from 50 cm in the west to 100 cm in the east. Drought conditions frequently prevail in the months of July and August. The soils are sandy and result from beaches left by the Cretaceous Sea. The western Cross Timbers are located in part of the upper Pennsylvanian while the eastern or main belt Cross Timbers are apparently all Cretaceous (Hill, 1901). Dyksterhuis (1948) referred to the southern soils as Immature Reddish Prairie Soils. As would be expected, the soils are more coarse than soils found under the adjacent prairies; within the Cross Timbers coarse-textured soils support more vigorous forests,

and fine-textured soils have fewer trees and a greater development of grasses.

In a study of the southern portion of the Cross Timbers, the overstory consisted of 63 percent post oak (Quercus stellata), 29 percent blackjack oak (Quercus marilandica), and 8 percent of ten other species, the most common of which were cedar elm (Ulmus crassifolia) and hackberry (Celtis spp.) (Dyksterhuis, 1948).

The information in the table below, as reported by Dyksterhuis (1948), provides an indication of the fire frequency in the western Cross Timbers (based on 269 sample plots, analysis conducted in 1942):

Fire History	Percent
Had not been burned for 10 years or more	16
Had not been burned for 6-9 years	37
Had not been burned for 1-5 years	44
Had been burned the preceding year	3

Although the vegetation has been modified with settling, grazing, and erosion, the major grass species is little bluestem. Table 3.5 makes a comparison of the original or relic vegetation with that sampled by Dyksterhuis in 1948. Two other tallgrasses, Indiangrass and big bluestem, were important in the original prairie but had virtually disappeared from the prairies by 1946. With grazing and disturbance, a compositional shift toward annual grasses, forbs, and buffalograss occurs.

The Oak Savanna occurs north of the Cross Timbers and east of the True Prairie in Wisconsin, Minnesota, and North Dakota. The understory is True Prairie grassland species and the overstory is composed of bur oak (Quercus macrocarpa) along with Hill's oak (Q. ellipsoidalis), white oak (Q. alba), and shagbark hickory (Carya ovata).

Blackland Prairie

The Blackland Prairie is an area of eastern Texas that extends from the mesquite-chaparral in south-central Texas northeastward to the Red River and then eastward to the eastern Cross Timbers. The northwestern area, which differs in substrate (Hill, 1901), has also been referred to as the Grand Prairie and the northeastern portion as the Fort Worth Prairie. Livingston and Shreve (1921) included the Blackland Prairie in their Grassland-Deciduous Forest Transition Area; Shantz and Zon (1924) placed it in the Prairie Grassland; Shelford (1926) called it part of the Oak Savanna; and both Clements and Shelford (1939) and Weaver and Clements (1938) included it in the Coastal Prairie Association (Dyksterhuis, 1946).

The soils of the Grand Prairie do not belong to the typical grassland mollisols but rather to the vertisols, which are immature profiles resting upon soft limestone materials. These calcareous clays have the dark color and high organic content characteristic of prairie soils, but because of their texture and the rainfall they have not been leached with the consequent development of normal profiles (Dyksterhuis, 1946). Although rainfall is less in the

TABLE 3.5 Species composition of the understory vegetation of the western Cross Timbers both as original vegetation and in 1948 (Dyksterhuis, 1948; copyright 1948 by the Ecological Society of America).

Principal species	Cover (percent)[a]	
	Original	Present (1948)
Little bluestem (Schizachyrium scoparius)	64.8	0.8
Indiangrass (Sorghastrum nutans)	5.7	--
Big bluestem (Andropogon gerardi)	3.3	--
Hairy grama (Bouteloua hirsutai)	3.5	4.6
Sideoats grama (B. curtipendula)	2.7	2.7
Tall dropseed (Sporobolus asper)	2.2	0.6
Oak species (Quercus spp.)	3.7	7.4
Annual forbs	--	18.8
Annual grasses (except Aristida)	--	10.2
Buffalograss (Buchloe dactyloides)	--	9.4
Aristida spp. (annual)	--	5.9
Aristida spp. (perennial)	--	4.0
Fringeleaf paspalum (Paspalum ciliatifolium)	--	4.0
Texas needlegrass (Stipa leucotricha)	--	4.0
Silver bluestem (Andropogon saccharoides)	--	2.8
Bermudagrass (Cynodon dactylon)	--	2.4
Western ragweed (Ambrosia psilostachya)	--	2.3
Tumblegrass (Schedonnardus paniculatus)	--	1.8
Tumble windmillgrass (Chloris verticillata)	--	1.3
Mesquite (Prosopis juliflora)	--	0.9
Percent of total	85.9	83.9

[a] Minimum values are 0.52 percent.

western Fort Worth Prairie, the Blackland Prairie falls largely in a climate that supports forest vegetation. In fact, the trees that occur in the city of Fort Worth are on ancient terraces of coarse materials that are now high above the Trinity River with its floodplain forest.

The most common grass throughout the Blackland Prairie is little bluestem. On the Fort Worth Prairie, Dyksterhuis (1946) found Texas needlegrass (Stipa leucotricha) along with several annual grasses such as little barley (Hordeum pusillum), Japanese brome, sixweeks fescue (Festuca octoflora), rescuegrass (Bromus catharticus), and common Ozarkgrass (Limnodea arkansana). He also noted the presence of some annual forbs, silver bluestem (Andropogon saccharoides), sideoats grama, several species of threeawn (Aristida), buffalograss, and little bluestem. Other important plant species include tall dropseed, hairy grama (Bouteloua hirsuta), Texas grama (B. rigidiseta), Indiangrass, hairy tridens (Tridens pilosus), eastern gama grass (Tripsacum dactyloides), Reverchon's muhly (Muhlenbergia reverchonii), blue threeawn (Aristida glauca), common broomweed (Gutierrezia dracunculoides), western ragweed (Ambrosia psilostachya), heath aster (Aster ericoides), branching noseburn (Tragia ramosa), yellow falsegarlic (Nothoscordum bivalve), and species of onion (Allium) and eveningprimrose (Oenothera) (Dyksterhuis, 1946; Gould, 1968; Hitchcock, 1971). In relict areas of the Fort Worth Prairie, the seven most important species were little bluestem, sideoats grama, Indiangrass, tall dropseed, hairy grama, big bluestem, and Texas needlegrass. Differences in percent composition depend on topographic position. For example, tumble windmillgrass (Chloris verticillata) and tumblegrass (Schedonnardus paniculatus) are more important on hilltops and ridges than elsewhere, and Texas needlegrass, tall dropseed, big bluestem, and Indiangrass are more common on the colluvial benches and alluvial bottoms (Dyksterhuis, 1946).

Although large portions of the prairie have been converted into cropland, significant prairie exists as grazing land. Overgrazing results in (1) the elimination of tallgrasses; (2) tallgrass replacement by buffalograss, broom snakeweed (Gutierrezia sarothrae), and common broomweed under grazing by horses and cows alone; or (3) the elimination of broomweeds if a few sheep are included in the grazing program (Gould, 1968).

Johnsongrass (Sorghum halepense) was introduced as an early meadow grass for improved pasture. This species, though it provides large amounts of forage, has proven to be very tenacious and difficult to eradicate. It grows best when plowed up each year but will not invade an established native grassland. Several other species and varieties (e.g., King Ranch bluestem [Bothriochloa ischaemum var. songaricus]) have now been introduced as grassland species (Gould 1968).

The Blackland Prairie, composed of species characteristic of the True Prairie and recognized as such by Weaver and Clements (1938), is apparently a function of soil conditions that exclude trees that would otherwise be supported under the climatic regime. Under such favorable conditions, the soils are rich and plant production is high. As a result, much of the original prairie has been converted into cropland.

Black Belt Prairie

Early notes of explorers and geographers, combined with plats of the original land surveys, all indicate the presence of the so-called Black Belt Prairie (Jones and Patton, 1966; Rankin and Davis, 1971). This crescent-shaped area extends from central Alabama northwestward through Mississippi and a short distance into western Tennessee. Several hundred kilometers long and about 40 km at its widest point (near the Alabama-Mississippi line [Harper, 1943]), the Black Belt Prairie has been both grazed and cultivated since its settlement before 1850 and has been a victim of considerable subsequent erosion.

On the Selma Chalk of the Late Cretaceous age grasslands have predominated in this area that is otherwise covered by forests. The variable alkaline soil ranges from 85 percent or more calcium carbonate to impure chalky clay and sand (Jones and Patton, 1966). Braun (1950) described red cedar (Juniperus virginiana) as an important species, but Rankin and Davis (1971) examined the surveyors' records of 1845-46 and found 52 percent oak, 9 percent pine, and 5 percent each of hickory, gum, and ash but did not mention red cedar as being important. The patches of grassland are typical True Prairie species dominated by the bluestems, Indiangrass, and switchgrass.

Like the Blackland Prairie in eastern Texas, the Black Belt Prairie represents a grassland located in climatic conditions that potentially would support forests. In both prairies the underlying soils are developed from Cretaceous alkaline clays, which occupy somewhat analogous positions on either side of the Mississippi Embayment. The controlling conditions are the fine-textured soils, which contribute to seasonal drought conditions, that preclude well-developed forests but are characteristic of grasslands.

Invertebrates

Insects are an old group and most of the present families were also present in the Permian. Insects have coevolved with other animals and plants of the prairie and have played a dominant role especially in areas of pollination strategy, predation and parasitism, decomposition, and as scavengers and disease vectors.

Major taxonomic groups of insects and other invertebrates of the True Prairie are found both above- and belowground, although many groups are difficult to categorize because they are found commonly in both habitats. For example, the immature stage of some phytophagous scarab beetles lives belowground and feeds on roots, while the free-living adult is aboveground. On the other hand, ants forage aboveground but live belowground.

The following descriptions of major groups found in the prairie are intended to acquaint the reader with the large number and diversity of invertebrate taxa. None of the orders or families is restricted to the prairie and all are widely distributed.

Protura. Protura, which feed on decomposing organic matter, are minute forms found in large numbers and considerable biomass; they represent the only group found exclusively belowground. A common order, Collembola or springtails, consists of very small

(5-6 mm) insects and most have a structure on their abdomens that allows them to jump commonly to a height of 75-100 mm. Though a few are herbivorous, most springtails are found just below the soil surface or deeper feeding on decaying plant material, fungi, bacteria, and arthropod feces. Four families of Collembola were found at the Osage Site, often in high numbers. A single family, Japygidae, represented Diplura at the Osage Site. This predaceous, numerous family was often found below a depth of 20 cm but rarely aboveground.

Orthoptera. Orthoptera (e.g., crickets, grasshoppers, mantids, and cockroaches) are diverse in habit and contain some of the major aboveground prairie herbivores. Acrididae (short-horned grasshoppers) are widespread and, because many are agricultural pests, are probably the most extensively studied prairie group. The prairie may serve as a reservoir for populations of grasshoppers and other insects that periodically move into cropland or vice versa. Of the nineteen dominant species of grasshoppers (of a total of fifty-three) found in Donaldson Research Pastures near Manhattan, Kansas (Table 3.6), nine are predominately graminivorous, five are forbivorous, and the remainder are mixed feeders. Only seven of these species were collected from the Osage Site; high populations of insects were never found at that site and a considerable difference in the Orthopteran fauna between the two sites seems to exist.

Thysanoptera. Thysanoptera (thrips) are numerous small insects that are mostly herbivorous although some are predaceous. The herbivores destroy plant cells with a rasping-sucking action and feed on plant sap.

Hemiptera. Hemiptera has numerous families that feed on plant sap. Some, like the assassin bugs, are predaceous; a few families are disease vectors. Major families in the southern True Prairie are Scutelleridae (shield-backed bugs), Lygaeidae (seed bugs), and Pentatomidae (stink bugs).

Homoptera. Homoptera (e.g., aphids, scale insects, and leafhoppers) are all herbivorous and feed on plant sap. Most Homoptera remove copious quantities of sap from the xylem, phloem, or intracellular leaf spaces. Much of the excess liquid is later excreted as a substance termed honeydew and is immediately returned to the system. The amount of honeydew produced may be as high as 90 percent of the amount of plant sap ingested. Important families of Homoptera are Cicadellidae (leafhoppers), Delphacidae (planthoppers), Coccoidea (scale insects) and Aphididae (plant lice). The major species of leafhoppers from the Donaldson Pastures and the Osage Site are given in Table 3.7; more than 80 total species were determined from the two sites, and about 60 percent of these occurred at both sites. The Osage Site was apparently near the northern range limits for high populations of five genera and the Donaldson Pastures were near the southern limit for high populations of four genera within the True Prairie.

Coleoptera. Coleoptera (beetles) have the largest number of species in the True Prairie. Scarab beetles are a dominant group in the process of dung removal from the prairie. Chrysomelidae (leaf beetles) and Curculionidae (weevils) are two dominant groups of herbivorous beetles and, surprisingly, weevils displaced leaf beetles as the most prevalent group during one year of the studies

TABLE 3.6 Major species of Acrididae found in Donaldson Pastures near Manhattan, Kansas (asterisk denotes principal species).[a]

Species[b]
<u>Ageneotettix</u> <u>deorum</u> (Scudder)
<u>Arphia</u> <u>simplex</u> (Scudder)*
<u>Campylacantha</u> <u>olivacea</u> <u>olivacea</u> (Scudder)
<u>Eritettix</u> <u>simplex</u> (Scudder)
<u>Hesperotettix</u> <u>viridis</u> <u>pratensis</u> (Scudder)*
<u>Hypochlora</u> <u>alba</u> (Dodge)
<u>Melanoplus</u> <u>bivittatus</u> (Say)*
<u>M</u>. <u>confusus</u> Scudder
<u>M</u>. <u>femurrubrum</u> <u>femurrubrum</u> (DeGeer)*
<u>M</u>. <u>keeleri</u> <u>luridus</u> (Dodge)
<u>M</u>. <u>sanguinipes</u> (Fabricius)
<u>M</u>. <u>scudderi</u> <u>latus</u> Morse
<u>Mermeria</u> <u>bivittatus</u> <u>maculipennis</u> Bruner
<u>M</u>. <u>picta</u> <u>neomexicana</u> (Thomas)
<u>Opeia</u> <u>obscura</u> (Thomas)
<u>Orphulella</u> <u>speciosa</u> (Scudder)*
<u>Pardalophora</u> <u>haldemanii</u> (Scudder)
<u>Phoetaliotes</u> <u>nebrascensis</u> (Thomas)*
<u>Syrbula</u> <u>admirabilis</u> (Uhler)*

[a] From Campbell et al. (1974).

[b] Eleven species were identified from the Osage Site; ten of these were also found in Donaldson Pastures. <u>Melanoplus</u> <u>admirabilis</u> (Uhler) was found at the Osage Site but not at Donaldson Pastures. The list from the Osage Site is not nearly as complete as the Donaldson Pasture list.

TABLE 3.7   Dominant leafhoppers found at Donaldson Pastures (Kansas) and the Osage Site (Oklahoma).

Taxon	Donaldson	Osage
Aceratagallia spp.	X	X
Athysanella spp.	X	
Balclutha neglecta (DeLong and Davidson)	X	X
Chlorotettix spatulatus (Osborn and Ball)	X	X
Draeculacephala mollipes (Say)	X	
Empoasca spp.	X	
Endria inimica (Say)	X	X
Exitianus exitiosus (Uhler)	X	X
Extrusanus ovatus (Sanders and DeLong)		X
Flexamia spp.	X	X
Graminella mohri (DeLong)	X	X
Kansendria kansiensis (Tuthill)		X
Lavicephalus spp.	X	X
Macrosteles fascifrons (Stal)	X	X
Mesamia coloradensis (Gillette and Baker)	X	
Parabolocratus spp.		X
Paraphlepsius spp.		X
Polyamia spp.	X	X
Prairiana spp.		X
Scaphytopius spp.	X	X
Stirellus spp.	X	X
Xestocephalus pulicarius (Van Duzee)	X	X

at both the Donaldson Pastures and the Osage Site. Greene (1970) identified fifty species of Chrysomelidae from the Donaldson Pastures. This family, which has chewing mouthparts and ingests most of the removed foilage, feeds primarily on forbs. Major species include Altica foliaceae Leconte, Chrysodina globosa (Olivier), Diabrotica undecimpunctata howardi Barber, Nodonota tristis (Olivier), Ophraella spp., Pachybrachis spp., Paria thoracica (Melsheimer), and Zygospila suturalis casta (Rogers).

Bertwell (1972) found thirty-seven families of beetles at the Donaldson Pastures including thirty genera and thirty-five species of Curculionidae. Only three species of weevils are known to be associated with grasses; the remainder are forb feeders. Reed (1972) found seventeen genera and twenty-one species of weevils at the Osage Site, thirteen of which were also present at the Donaldson Pastures. Many species differences, however, indicate considerable north and south faunal change. For example, four of the eight dominant taxa (indicated by an asterisk*) at the Donaldson Pastures were also found at the Osage Site. The major taxa at the Donaldson Pastures are Aulobaris nasutus (LeConte),* Apion spp.,* Ceutorhynchus sp.,* Chelonychus longipes Dietz, Odontocorynus sp.,* Pseudobaris farcta (LeConte), Smicronyx fulvus LeConte, and Smicronyx sordidus (LeConte).

Lepidoptera. Lepidoptera (moths and butterflies) are always prominent in the True Prairie. The immature forms (caterpillars) are chewing herbivores; adults feed on nectar or do not feed at all. Most caterpillers feed aboveground but some, such as cutworms, feed at or below the soil surface. Caterpillars were common at the Osage Site, especially in the spring and most were observed feeding on forbs.

Diptera. Diptera (flies) are also well represented in the True Prairie. Immature flies (maggots) feed on plants and generally live within a plant as leaf miners, gall insects, or stem and root borers. Many adults are predators or parasites, others feed on plant or animal juices, and many feed on blood. Groups commonly found at the Osage Site were Chloropidae, Tachinidae (flesh flies), Calliphoridae (blow flies), and Muscidae (house flies and face flies). Chloropid flies commonly feed in grass stems while immatures. This group is also the vector of pinkeye in cattle. House flies, face flies, and horn flies can be serious pests to adult cattle. Flesh flies and blow flies are commonly scavengers, but flesh flies contain many parasitic species.

Hymenoptera. Because of the large numbers of ants present, the Hymenoptera (bees, wasps, and ants) was the most abundant insect group collected at the Osage Site. Ants are variable in habit and are difficult to place in trophic categories. They have elaborated an intricate social organization, and the immatures are protected in the nest attended by adult workers. Adults can act alternately as scavengers and herbivores (seeds and foliage), and often feed on the exudate of sucking insects. Only seven taxa were found at the Osage Site: Crematogaster lineolata subopaca (Emery), Formica sp., F. pallidefulva (Latreille), F. neogagates (Emery), Leptothorax pergandei Emery, Momomorium viridum peninsulatum Gregg, and Tapinoma sessile (Say).

Other prominent families of Hymenoptera include Vespidae (paper wasps) and Scelionidae. Vespid wasps commonly feed on nectar and

sap as adults, while immatures feed on other insects and spiders. The scelionids are parasites of insect and spider eggs.

Most of the other orders of insects are present in the True Prairie, but their numbers are insignificant. Isoptera (termites) have been observed in True Prairie plots where they were detected feeding belowground on wood plot markers. Isoptera will be discussed in Chapter 7 as decomposers of cellulose.

<u>Noninsect Invertebrates</u>. The major soil insects are included in the groups discussed above. Families of belowground insects and other soil invertebrates from six collections at the Osage Site are given in Table 3.8 (Stepanich, 1975). Fifteen orders and forty-six families were identified. Major groups of insects include Protura, Collembola, Diplura, Hymenoptera (ants), and Thysanoptera (thrips). Important noninsect groups include Acarina (mites) and Araneida (spiders). Only mites were determined below the family level and these groups are shown in Table 3.9. While the role of mites is not well known, they can be herbivores, fungivores, and predators, and they constitute about half of the numbers and 10 percent of the belowground arthropod biomass. Other noninsects commonly collected include pseudoscorpions, which are predaceous, and symphylans, which are scavengers or herbivores.

Aboveground noninsect invertebrates also include mites and spiders. Spiders are predaceous and their numbers are determined by the numbers of prey species present, with high spider population increases invariably following population increases of insects. Larger numbers of mites were collected at the Osage Site in 1972 than any other group.

During 1972, other aboveground invertebrates recorded at the Osage Site included sowbugs (Isopoda), which feed primarily on organic debris but are occasionally herbivorous, predaceous millipedes (class Diplopoda), and daddy longlegs (Phalangida).

Soil nematodes (phylum Aschelminthes) were sampled at the Osage Site on 28 August and 5 December 1973. These nonarthropod invertebrates may be herbivorous, predaceous, or saprophagous (feeding on decaying organic matter), and some are known to be vectors of plant disease. The results reported here were compiled by James D. Smolik of South Dakota State University and show that nematodes constitute a significant proportion of the belowground biomass at the Osage Site, but numbers recovered were lower than expected. More than eighty species (some not determined below genus) were determined (Table 3.10). This group undoubtedly contributes significantly to energy flow at the Osage Site.

Smolik (1974) studied nematodes in a South Dakota mixed-grass Prairie and found 2-6 million phytophagous (herbivorous) forms m^{-2} to a depth of 60 cm in both grazed and ungrazed pastures. Predaceous forms nearly equaled phytophagous forms while saprophagous forms contributed only a small proportion of the total. He concluded that nematodes constituted a significant pathway of energy flow in grasslands and could consume more range vegetation than cattle.

TABLE 3.8   Arthropod orders and families determined from soil at
the Osage Site (Oklahoma) from 18 June 1971 through 21
November 1972.[a,b,c]

Order	Family	Trophic level
Protura		Scavenger
Diplura	Japygidae	Predator
Collembola	Entomobryidae	Herbivore, scavenger
	Onychiuridae	Scavenger
	Poduridae	Herbivore, scavenger
	Sminthuridae	Scavenger
Orthoptera	Blattidae	Scavenger
Psocoptera		Scavenger
Thysanoptera	Phloeothripidae	Herbivore, predator
	Thripidae	Herbivore
Hemiptera	Anthocoridae	Predator
	Miridae	Herbivore
	Pentatomidae	Herbivore, predator
	Scutelleridae	Herbivore
Homoptera	Aphididae	Herbivore
	Coccoidea	Herbivore
	Delphacidae	Herbivore
Coleoptera	Carabidae	Predator, herbivore
	Curculionidae	Herbivore
	Dermestidae	Scavenger
	Elateridae	Predator
	Nitidulidae	Herbivore
	Staphylinidae	Predator
Lepidoptera		Herbivore
Diptera	Cecidomyiidae	Herbivore
	Immatures	Undetermined
	Psychodidae	Undetermined
	Sciaridae	Herbivore
Hymenoptera	Formicidae	Scavenger, predator
Symphyla		Herbivore, scavenger

Table 3.8   Continued.

Order	Family	Trophic level
Acarina	Ameroseiidae	Fungivore
	Ascidae	Predator
	Parholaspididae	Predator
	Laelapidae	Predator
	Rhodacaridae	Predator
	Oplitidae	Myrmecophilore
	Trachyuropodidae	Myrmecophilore
	Pyemotidae	Unknown
	Cunzxidae	Predator
	Trombidiidae	Predator
	Eniochthoniidae	Undetermined
	Epilohmanniidae	Undetermined
	Euphthiracaridae	Undetermined
	Lohmanniidae	Undetermined
	Nanhermanniidae	Undetermined
	Nothridae	Undetermined
	Opiidae	Undetermined
	Eremobelbidae	Undetermined
	Oribatulidae	Undetermined
	Haplozetidae	Undetermined
	Ceratozetidae	Undetermined
	Galumnidae	Undetermined
Araneida		Predator
Pseudoscorpionida		Predator

[a] All orders were not determined to family.

[b] All immatures were determined to order.

[c] Acarina classification by Dr. Donald E. Johnston, Ohio State University.

Summary

Using the data from the Osage Site as a guide, the invertebrate fauna of the True Prairie is tremendously diverse. From 1970 through 1972, 16 orders and more than 131 families of aboveground insects were determined (Table 3.11). Differences between years indicate that all insects are not present every year, or collecting methods do not sample the entire fauna in any given year. Extensive leafhopper sampling over a four-year period in western Kansas (Blocker et al., 1972) indicated that some taxa are present in hot dry years and others in wetter years. Only 13 orders and 92 families were collected using the trapping method, while the

TABLE 3.9   Acarina collected from soil at the Osage Site, Oklahoma, August and November 1972.[a]

Suborder	Supercohort	Cohort	Genera
Mesostigmata		Gamasina	Ameroseius sp.
			Antennoseius sp.
			Cheiroseius sp.
			Holaspina sp.
			Hypoaspis sp. 1
			Hypoaspis sp. 2
			Hypoaspis sp. 3
			Rhodacarus sp.
		Uropodina	Oplitis sp.
			Trachyuropoda sp.
Cryptostigmata	Oribatei		Hypochthoniella sp.
			Epilohmannia sp.
			Thysotritia sp.
			Lohmannia sp.
			Masthermannia sp.
			Nothrus sp.
			Eremobelba sp.
			Scheloribates sp. 1
			Scheloribates sp. 2
			Scheloribates sp. 3
			Rostrozetes sp.
			Trichoribates sp.
			Galumna sp.
Prostigmata		Tarsonemina	Pygmephorus sp.
		Eleutherengona	Bonzia sp.
			Cunaxa sp.
			Cunaxoides sp.
	Parasitengona		Allothrombium sp.

[a] Acarina classification by Dr. Donald E. Johnston, Ohio State University.

remainder of the taxa in Table 3.11 were captured by sweeping foliage, by pitfall trapping, by hand-collecting, or simply by observing the research area. The IBP method involved dropping a 0.5 m^2 trap and suctioning out the contents; only invertebrates collected in this manner are considered in Chapter 7 of this volume and the method of collection will be amplified there. No single trapping method, however, is adequate for sampling the invertebrate

TABLE 3.10   Nematode taxa found at the Osage Site, northeastern
Oklahoma, 28 August 1973.

Species	Grazed	Ungrazed
Acrobeles sp.	X	X
A. ctenocephalus	X	X
Acrobeloides sp.	X	X
A. buetschli	X	X
A. minor	X	X
Aorolaimus sp.	X	X
Aphelenchoides sp.	X	
Aphelenchus sp.	X	X
Aporcelaimellus sp.	X	X
A. obscuroides	X	
A. obscurus	X	X
Axonchium sp.		X
Boleodorus sp.	X	X
Carcharolaimus sp.		X
Chiloplacus sp.	X	
C. contractus	X	X
Diphtherophora sp.	X	X
D. obesum		X
Discolaimium sp.	X	X
Discolaimus sp.		X
Ditylenchus sp.	X	X
Dorylaim sp.	X	X
Dorylaimellus sp.	X	X
Eucephalobus sp.	X	X
E. oxyuroides	X	X
Eudorylaimus sp. 1	X	X
Eudorylaimus sp. 2	X	X
E. angleus		X
E. aquilonorius		X
E. carteri		X
E. leptus		X
Mesodorylaimus sp.	X	X
Monhystera sp.		X
Mononchus sp.		X
Mylonchulus sp.		X
Nothotylenchus sp.	X	
Paraphelenchus sp.	X	X
Plectus sp.	X	X
Prismatolaimus sp.	X	X
Pungentus sp.	X	X
P. pungens	X	X
Rhabditid sp.	X	X
Thonus sp.	X	
Tripyla sp.		X

Table 3.10   Continued.

Species	Grazed	Ungrazed
Tylencholainellus sp.	X	
Tylencholaimus sp.	X	X
T. proximus		X
Tylenchorhynchus sp.	X	
T. acutus	X	
T. nudus	X	X
Tylenchus sp. 1	X	X
Tylenchus sp. 2	X	X
T. exiguus	X	X
T. fusiformis	X	X
T. plattensis	X	
Wilsonema sp.	X	
Xiphinema americanumi		X

population of the True Prairie, and quantitative data are difficult to gather.

The wide diversity found in insects, especially at the family level, makes a generalization concerning their function in any ecosystem extremely difficult. Several million species of insects abound worldwide, and an estimate of 3,000 at Osage is probably ultraconservative. More than 1,600 species have been identified from a Shortgrass Prairie in Colorado, and still this list is incomplete (Kumar et al., 1976). The True Prairie probably has a more diverse fauna because of higher rainfall and its proximity to the eastern forests. Blocker et al. (1972) have noted fewer species of leafhoppers in the Shortgrass Prairie, but greater numbers of individuals.

Birds

North American grasslands today are characterized by simple and meager avifaunas (Cody, 1966a; Mengel, 1970; Udvardy, 1963; Wiens, 1973; 1974a). The relative paucity of bird species in comparison with other habitat types is most readily explained in terms of the general simplicity of prairie habitats and the attendant limitations of opportunities for niche diversification (e.g., Cody, 1968), but such explanations are only proximate and rather superficial. Since history may contribute to current patterns, analysis of present grassland avifaunas should relate closely to past biogeographic patterns. The constraints on meaningful avian biogeographic analyses are formidable, however, and inferences must be based upon present distributional patterns of species and genera. The most comprehensive treatment of grassland avian biogeography is that of Mengel (1970), upon which much of the following information is based.

TABLE 3.11   Orders and families of aboveground insects determined from the Osage Site for 1970-72.

| Order | Family | Trophic level[c] | |
		Adult	Immature
Diplura[a,b]	Japygidae	5	5
Collembola	Entomobryidae	8	8
	Poduridae	8	8
	Sminthuridae	1	1
Ephemeroptera		9	1
Odonata	Coenagrionidae	5	5
	Libellulidae	5	5
Orthoptera	Acrididae	1	1
	Blattidae	8	8
	Gryllidae	7	7
	Mantidae	5	5
	Phasmidae	1	1
	Tettigoniidae	1	1
Dermaptera		8	8
Psocoptera		8	8
Thysanoptera		2	2
Hemiptera	Coreidae	2	2
	Corimelaenidae	2	2
	Corixidae	5	5
	Cydnidae	2	2
	Gerridae	5	5
	Lygaeidae	2	2
	Miridae	2	2
	Nabidae	5	5
	Neididae	2	2
	Pentatomidae	2	2
	Phymatidae	5	5
	Piesmidae	2	2
	Ploiariidae	5	5
	Reduviidae	5	5
	Scutelleridae	2	2
	Tingidae	2	2

Table 3.11 Continued.

| Order | Family | Trophic level[c] | |
		Adult	Immature
Homoptera	Aphididae	2	2
	Cercopidae	2	2
	Cicadellidae	2	2
	Cixiidae	2	2
	Coccoidea	2	2
	Delphacidae	2	2
	Dictyopharidae	2	2
	Fulgoridae	2	2
	Issidae	2	2
	Membracidae	2	2
	Psyllidae	2	2
Coleoptera	Cantharidae	3	5
	Carabidae	5	5
	Cerambycidae	3	1
	Chrysomelidae	1	1
	Cicindelidae	5	5
	Cisidae	8	8
	Cleridae	5	5
	Coccinellidae	5	5
	Cucujidae	5	5
	Curculionidae	1	1
	Dermestidae	8	8
	Elateridae	1	1
	Erotylidae	8	8
	Eucnemidae	8	8
	Histeridae	5	5
	Lampyridae	3	5
	Lathrididae	8	8
	Limnebiidae	8	8
	Malachiidae	5	5
	Meloidae	1	5
	Mordellidae	3	1
	Nitidulidae	3	8
	Phalacrididae	1	1
	Pselaphidae	5	5
	Ptilidae	8	8
	Scaphidiidae	8	8
	Scarabaeidae	1,8	1,8
	Scolytidae	1	1
	Scydmaenidae	8	8
	Silphidae	8	8
	Staphylinidae	5	5
	Tenebrionidae	8	1
	Throscidae	3	3

Table 3.11   Continued.

| Order | Family | Trophic level[c] | |
		Adult	Immature
Strepsiptera		6	6
Neuroptera	Chrysopidae	5	5
	Hemerobiidae	5	5
	Myrmeleontidae	5	5
Lepidoptera	Danaidae	3	1
	Geometridae	3	1
	Noctuidae	3	1
	Nymphalidae	3	1
	Pyralidae	3	1
	Satyridae	3	1
Diptera	Acroceridae	0	6
	Asilidae	5	5
	Bombyliidae	3	5
	Calliphoridae	3	8
	Cecidomyiidae	3	1
	Ceratopogonidae	6	8
	Chironomidae	3	1
	Chloropidae	3	1
	Culicidae	6	8
	Dolichopodidae	5	5
	Empididae	5	5
	Muscidae	3	8
	Otitidae	0	8
	Phoridae	3	8
	Piophilidae	0	8
	Pipunculidae	3	6
	Pyrgotidae	3	6
	Rhagionidae	5	5
	Sarcophagidae	3	6
	Scatopsidae	3	8
	Sciaridae	8	8
	Sciomyzidae	6	6
	Stratiomyiidae	3	8
	Syrphidae	3	5
	Tabanidae	6	5
	Tachinidae	3	6
	Tipulidae	3	1

Table 3.11 Continued.

| Order | Family | Trophic level[c] | |
		Adult	Immature
Hymenoptera	Apidae	3	3
	Braconidae	3	6
	Dryinidae	6	6
	Encyrtidae	3	6
	Eulopidae	6	6
	Figitidae	3	6
	Formicidae	7	7
	Halictidae	3	3
	Ichneumonidae	3	6
	Mutillidae	3	6
	Proctotrupidae	3	6
	Pteromalidae	3	6
	Scelionidae	3	6
	Sierolomorphidae	0	0
	Tenthredinidae	1	1
	Tiphiidae	3	6
	Thysanidae	3	6
	Trichogrammatidae	3	6
	Vespidae	5	5

[a] All orders were not determined to family.

[b] All immatures were not determined to family.

[c] Trophic level: 0 = unknown
                    1 = plant tissue
                    2 = plant sap
                    3 = plant pollen and nectar
                    4 = plant seed
                    5 = predators
                    6 = parasites
                    7 = omnivores
                    8 = saprophages
                    9 = nonfeeding stage

The absence of a crisply delimited grassland avifauna is apparent from the tabulations of Table 3.12. Mengel considered two groups of grassland species: a primary group of twelve endemic species and a larger set of twenty-five secondary species--that is, birds that have strong affinities for the central North American grasslands but that also live beyond the limits of this area. Five of the endemics are associated to varying degrees with aquatic or semiaquatic habitats within the grasslands, so the assemblage of strictly grassland endemics is quite small. Udvardy's analysis considered only passerine species, of which he recognized only seven

TABLE 3.12 Distribution of major grassland bird species in faunal groups, according to the analyses by Mengel (1970), Kendeigh (1974), and Udvardy (1963). Species occurrences in roadside counts conducted at the Osage Site are included for comparison.

Species	Mengel Endemic species	Mengel Secondary species	Kendeigh Grassland	Udvardy[a] Prairie	Udvardy[a] Eastern ecotone	Udvardy[a] Eastern deciduous forest	Udvardy[a] Western woodland edge	Udvardy[a] Great Basin	Udvardy[a] Edwards Plateau	Osage roadside counts
Ferrugineus hawk (Buteo regalis)	X									X
Swanson's hawk (Buteo swainsoni)		X								X
Mississippi kite (Ictinia mississippiensis)		X								
Marsh hawk[b] (Circus cyaneus)		X	X							
Prairie falcon (Falco mexicanus)		X								X
Greater prairie chicken[b] (Tympanuchus cupido)		X	X							
Lesser prairie chicken (Tympanuchus pallidicinctus)		X	X							X
Sharp-tailed grouse (Pedioecetes phasianellus)	X									
Mountain plover (Charadrius montanus)	X									
Long-billed curlew (Numenius americanus)	X		X							
Marbled godwit (Limosa fedoa)	X									
Upland plover[b] (Bartramia longicauda)	X		X							X
Wilson's phalarope (Steganopus tricolor)	X		X							
Franklin's gull (Larus pipixcan)			X							
Burrowing owl (Athene cunicularia)	X	X	X							
Short-eared owl (Asio flammeus)	X	X								
Horned lark[b] (Eremophila alpestris)	X	X								X
Sprague's pipit (Anthus spragueii)	X		X	X						
Sage thrasher (Oreoscoptes montanus)	X	X						X		
Eastern meadowlark[b] (Sturnella magna)		X								X

79

Table 3.12 Continued.

Species	Mengel Endemic species	Mengel Secondary species	Kendeigh Grassland	Prairie	Udvardy[a] Eastern ecotone	Udvardy[a] Eastern deciduous forest	Udvardy[a] Western woodland edge	Udvardy[a] Great Basin	Udvardy[a] Edwards Plateau	Osage roadside counts
Western meadowlark (Sturnella neglecta)		X	X							X
Dickcissel[b] (Spiza americana)		X	X			X				X
Lark bunting (Calamospiza melanocorys)	X			X						
Green-tailed towhee (Pipilo chlorurus)		X						X		
Savannah sparrow[b] (Passerculus sandwichensis)		X	X							X
Baird's sparrow[b] (Ammodramus bairdii)	X		X							
Grasshopper sparrow[b] (Ammodramus savannarum)		X	X							X
Henslow's sparrow[b] (Ammodramus henslowii)					X					
LeConte's sparrow[b] (Ammospiza leconteii)				X						
Sharp-tailed sparrow (Ammospiza caudacuta)				X						
Cassin's sparrow (Aimophila cassinii)	X		X						X	
Vesper sparrow[b] (Pooecetes gramineus)		X	X		X					X
Lark sparrow[b] (Chondestes grammacus)		X	X				X			X
Sage sparrow (Amphispiza belli)		X						X		
Brewer's sparrow (Spizella breweri)		X						X		
Clay-colored sparrow (Spizella pallida)		X								
McCown's longspur (Calcarius mccownii)	X		X	X						
Chestnut-collared longspur (Calcarius ornatus)	X		X	X						

[a] Nonpasserines are not included in Udvardy's analysis.
[b] Species that regularly occur in true prairies.

80

in a prairie fauna (two of which were not included in Mengel's listing). The remaining species from Mengel's groups were assigned to five distinctive faunas by Udvardy, which demonstrates the difficulty of biogeographic analyses in this region. Only twelve of the characteristic grassland species of Table 3.12 occur with any regularity in the True Prairie, and none of these species is a grassland endemic. Two species characteristic of eastern portions of the True Prairie, the bobolink (Dolichonyx oryzivorus) and brown-headed cowbird (Molothrus ater), were omitted from Mengel's analysis because they are most frequently associated with meadows or ecotonal areas.

The present avifauna of North American grasslands bears only faint affinities with its Eurasian Steppe counterpart, probably because of the substantial barriers to dispersal posed by intervening mountain ranges and vast areas of tundra (Mengel, 1970). The roles played by one species of lark, one species of pipit, various icterids, and a fairly substantial assemblage of emberizid finches in North America are in the Old World apparently assumed by a complex of emberizid finches (chiefly Emberiza) and a multitude of larks, pipits, and in some areas Old World warblers (Sylviidae). The shorebird and predator roles are generally played by the same families and genera but different species, while the plains grouse of North America is apparently replaced by a variety of bustards (Otididae), sand grouse (Pteroclidae), coursers (Glareolidae), and partridges and pheasants (Phasianidae). The avifaunal differences are thus substantial. Faunal relationships to South American grasslands and savannas are closer, as might be expected. Three of the so-called typical grassland passerine species range into at least northern South America (horned lark [Eremophila alpestris], eastern meadowlark [Sturnella magna], and grasshopper sparrow [Ammodramus savannarum]--all of which are broadly distributed in North American grasslands, including the True Prairie). The meadowlark group (genus Sturnella, Short, 1968) apparently originated in South America.

The central North American grasslands have not been a hotbed of avian speciation. This fact is evidenced by the rather small total species list, the general restriction of endemics, and the relatively broad distributions of the typical species. Mengel's analysis indicates that grassland avifaunas have a small number of species per genus (1.1), compared with the generic diversity characterizing adjacent habitat types (2.9-3.6 species/genus). The numbers strongly suggest that the prairie environment has rarely been fragmented for long enough periods to permit the specific differentiation of isolated parental stocks. Mengel was able to suggest only four cases of closely related congeners with central Great Plains distributions. To some degree the relative lack of speciation in the grasslands reflects the paleoecological history of the region: while the grasslands have shifted in distributional position with climatic shifts from the mid-Pliocene onward, they have apparently retained a relatively broad, unbroken distribution. Further, the climatic vicissitudes have undoubtedly tested the resiliency of species, eliminating highly specialized forms. But other, more immediate factors may also contribute to the relatively low avifaunal diversity of grasslands. Many of the species are migratory, presumably in adaptation to the seasonal extremes in

climate and food availability, and their migrations may promote distributional spread and genetic interchange among localized breeding populations. Grassland climates are also characterized by long-term irregularity and unpredictability. To the extent that occasional climatic extremes (e.g., droughts) influence production and food supplies, these extremes may exert substantial selective pressures on bird populations, excluding all but the best-adapted species (Wiens, 1974b). Climatically severe years are often widespread in grasslands (Borchert, 1950; Chapter 5), and since few "islands" of other habitat types occur, no suitable refugia exist in which poorly adapted species may persist during such conditions. These irregular but inevitable climatic pulses may act as a deterrent to specialization and speciation, with the result that the number of species exploiting grassland habitats may be less than might be expected given normal resource conditions (Wiens, 1974b; 1977a).

These comments apply most forcefully to the Shortgrass and Mixed-grass Prairies of the central Great Plains. True Prairie regions are characterized by more favorable and less variable climates, and, given the greater stability and the greater degree of local fragmentation of True Prairie in its interdigitation with eastern deciduous woodlands, speciation must be expected to proceed more rapidly. Rapid speciation has apparently not occurred, however, perhaps because of the distributional instability of the prairie-forest ecotone, or perhaps because of the graded continuity with increasingly arid grasslands to the west.

While North American grasslands have not formed a focus for avian speciation, Mengel and others believe that the grasslands have played a major role as an isolating agent in species formation in adjacent habitat types. Mengel (1970) and Selander (1966) detail a number of current distributional patterns of congeneric species that stem from repeated fractionation of eastern, northern, and western woodland habitats by glacial advances and retreats through the Pleistocene. During interglacial periods the expanded grassland has formed an effective barrier between eastern and western woodland or forest-edge species or populations, leading to varying degrees of specific identity.

These areas may be breeding and/or wintering habitats for several species of raptors and gallinaceous game birds (Table 3.12). Because individuals of these species occupy large home ranges, it is extremely difficult to obtain quantitative measures of their abundance or occurrence. Those species recorded on roadside censuses conducted over several years at the Osage Site are indicated in Table 3.12, but the species that are readily censused in small, intensive plot counts are emphasized in the following discussion. This emphasis should not be taken to suggest that raptors or large gallinaceous species are not important or characteristic elements of True Prairie avifaunas, but only that our studies have not provided quantitative information on them.

In a collection of 20 plot-based counts made during the breeding season over a range of True Prairie situations, a total of 13 species was recorded (Table 3.13). Three species, the eastern meadowlark, grasshopper sparrow, and dickcissel (Spiza americana), occurred in well over half of the censuses, and these species may thus be considered dominant species in these habitats, at least in

TABLE 3.13   Species recorded in plot censuses conducted in grassland habitats in North America. From Wiens and Dyer (1975).

Species	Relative frequency[a]			Residency[c]	Trophic group[d]	Body weight (g wet)[e]
	Mixed-grass (19)	Tallgrass (20)	Agricultural[b] (11)			
Killdeer (Charadrius vociferus)	0.05	--	0.36	S	I	125
Long-billed curlew (Numenius americanus)	0.21	--	--	S	I	587
Upland plover (Bartramia longicauda)	0.21	0.30	0.18	S	I	130
Mourning dove (Zenaidura macroura)	--	0.15	--	P	G	123
Common nighthawk (Chordeiles minor)	--	0.05	--	S	I	75
Horned lark (Eremophila alpestris)	0.89	--	0.36	P	O	32
Sprague's pipit (Anthus spragueii)	0.47	--	0.09	S	I	25
Eastern meadowlark (Sturnella magna)	--	0.80	0.55	P	I	98
Western meadowlark (Sturnella neglecta)	1.00	0.35	--	P	I	100
Bobolink (Dolichonyx oryzivorus)	--	0.20	0.27	S	O	30
Red-winged blackbird (Agelaius phoeniceus)	--	0.15	0.45	P	O	65
Lark bunting (Calamospiza melanocorys)	0.05	--	--	S	O	36
Savannah sparrow (Passerculus sandwichensis)	0.32	0.15	0.36	S[f]	O	18
Grasshopper sparrow (Ammodramus savannarum)	0.37	0.80	0.73	S[f]	O	17
Baird's sparrow (Ammodramus bairdii)	0.37	--	--	S	O	19
Henslow's sparrow (Ammodramus henslowii)	--	0.10	0.09	S	O	14
Dickcissel (Spiza americana)	--	0.65	0.09	S	O	27
Vesper sparrow (Pooecetes gramineus)	0.26	0.25	0.64	S[f]	O	24
Chipping sparrow (Spizella passerina)	--	0.05	--	S	O	13
Field sparrow (Spizella pusilla)	--	--	0.36	S[f]	O	12
Song sparrow (Melospiza melodia)	--	--	0.36	P	O	20
Chestnut-collared longspur (Calcarius ornatus)	0.68	--	0.09	S	O	20
McCown's longspur (Calcarius mccownii)	0.21	--	--	S	O	25

[a] Percent of censuses (in parentheses) in which species was recorded.

[b] Includes primarily pastures, fallow fields, hayfields, and small grain croplands.

[c] S = seasonal (migratory; species absent from plots in winter); P = permanent.

[d] I = insectivore (>75% of adult breeding season diet arthropods); O = omnivore (25% to 75% arthropod prey); G = granivore (>75% of diet seeds).

[e] Mean of ♂ and ♀ weights.

[f] Resident in southern U.S.

terms of their broad distributional patterns. Western meadowlarks (<u>Sturnella</u> <u>neglecta</u>), upland plovers (<u>Bartramia</u> <u>longicauda</u>), vesper sparrows (<u>Pooecetes</u> <u>gramineus</u>), and bobolinks occurred in 20% to 35% of the censuses, while the remaining six species were of limited distributional occurrence.   Roughly two-thirds of the species typically occurring in these True Prairie censuses are seasonal residents, residing in the census area only during the breeding season, and slightly over 60% of the species are essentially omnivorous in their diet (Table 3.14).

Another way of looking at the distributional spread of the more typical True Prairie birds is to consider it as an equivalent of "niche breadth" and employ appropriate indices. Such indices have been reviewed by Colwell and Futuyma (1971), and their equation 20 was employed with $\underline{k}$ = 10,000. This index adjusts the value to reflect the relative importance of different niche components (in this case, relative densities of species over the range of twenty True Prairie census locations, Chapter 7). The index ranges in value from 0 to 1, with high values indicating broad niches or, in this instance, the breadth of distribution of several species over the True Prairie census locations, adjusted to reflect their relative abundances where species do occur. The values obtained were:

Species	Distributional breadth index
Eastern meadowlark	0.69
Grasshopper sparrow	0.60
Western meadowlark	0.45
Bobolink	0.39
Dickcissel	0.35
Savannah sparrow	0.25
Upland plover	0.12
Average for all other species recorded	0.12

TABLE 3.14   Migratory and trophic characterizations of the species groups listed in Table 3.13. Values are sums of the species frequency values in each category for each habitat type.

Habitat type	Residency		Trophic group	
	Seasonal	Permanent	Insectivore	Omnivore
Mixed-grass	0.63	0.37	0.38	0.62
Tallgrass	0.64	0.36	0.38	0.62
Agricultural	0.65	0.35	0.24	0.76

Apparently, the two meadowlark species and the grasshopper sparrow have the broadest distributions with consistently high relative abundances among the birds occupying True Prairie locations.

Comparisons may be drawn with avifaunas of Mixed-grass Prairies to the west and with various grassland-like agricultural habitats to the east (Wiens and Dyer, 1975). These latter agricultural habitats (fallow fields, pastures, hayfields, old fields) frequently represent areas formerly occupied by the True Prairie. The number of species recorded in these other habitat types is quite similar to the number recorded in the True Prairie censuses (Tables 3.13 and 3.15), but the species compositions differ substantially. Roughly one-third of the species recorded in the Mixed-grass Prairie counts were not present in the more eastern areas, and all of those species were listed by Mengel (1970) as grassland endemics (Table 3.12). Half of the remaining species were shared with True Prairie sites, but nearly all of them were present in various agricultural situations. Thus, while Mixed-grass Prairies seem avifaunally more typical of North American grasslands than True Prairies, the component species disperse broadly into the wider range of agricultural habitats. A substantial share of the species recorded in the True Prairie also was present in agricultural habitats. This sampling considers only a limited range of agricultural environments, but agricultural-use patterns that retain a grassland-like physiognomy seem to be populated primarily by species that also form the nucleus of breeding species of native grassland habitats farther west.

During the breeding season both Mixed-grass Prairie and agricultural avifaunas are composed predominately of species that winter to the south (Table 3.14), as are species of the True Prairie. In contrast, more arid Shortgrass Prairies of the western Great Plains support a larger share of full-year resident species, while avifaunas of the Great Basin Shrub Steppe are almost totally dominated by migrants (Wiens, 1974b). Like the True Prairie, Mixed-grass Prairie samples are composed of 62 percent omnivorous species and the remainder are insectivorous. In agricultural habitats, a greater proportion of the species are omnivorous (Table 3.14).

These distributional patterns suggest that while the numerically dominant species in central North American grasslands are widespread, some differences in species affinities nonetheless occur. Using census values collected during 1970 at seventeen plots in IBP grassland sites over a broad range of grassland conditions, Wiens (1973) conducted tests of species associations, using Cole's (1949) index of interspecific association. In such tests, positive associations between two species may result from a common response to environmental factors, a social cohesiveness between the species, or a behavioral or ecological repulsion from other areas, forcing individuals to co-occupy the same areas. Negative associations, on the other hand, may stem from differences in habitat preferences, behavioral or ecological exclusion, or effects of past population histories. The association test does not unravel these various causal possibilities but only points to the existence of statistically significant deviations from chance co-occurrence between species pairs. Still, the results of this analysis are revealing and are a reinforcement of the more qualitative statements

TABLE 3.15 Avifaunal features of breeding bird communities in North American grassland types, calculated from Table 3.13.

Type	Number of species	Unshared	Species (percent) Shared with			Passerine	Faunal element [a]			
			Mixed-grass	Tall-grass	Agricultural		OW	SA-PA	NA	UN
Mixed-grass	13	31	--	38	62	77	15	8	54	23
Tallgrass	13	23	38	--	69	77	8	38	46	8
Agricultural	15	13	53	60	--	87	13	27	53	7

[a] From Mayr, 1946: OW = species representing families of Old World faunal origin; SA-PA = families of South American-Pan American origin; NA = families of North American origin; UN = families of undetermined origin.

made above. Over the series of IBP sites, western meadowlarks, lark buntings (<u>Calamospiza melanocorys</u>), and horned larks were positively associated in their distributions, with Brewer's sparrows (<u>Spizella breweri</u>) and McCown's longspurs (<u>Calcarius mccownii</u>) somewhat less so (Figure 3.19). Eastern meadowlarks, dickcissels, grasshopper sparrows, and upland plovers formed another loosely associated group that was negatively associated with the first group. The former group includes species generally characteristic of Shortgrass and Mixed-grass Prairies, while the second assemblage is rather closely associated with more mesic True Prairie conditions.

One additional general analysis of these species compositional patterns may be offered. Mayr (1946) suggested that bird faunas might fruitfully be analyzed in terms of the probable regions of

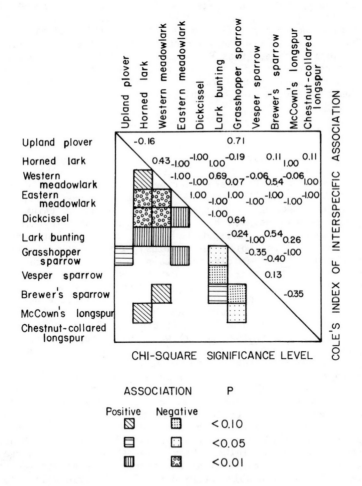

FIGURE 3.19 Bird association values from a wide range of grasslands (Redrawn from Wiens, 1973; copyright 1973 by The Ecological Society of America).

origin of the component families. He specifically drew attention to grasslands, noting that all of the species "usually listed as typical for the midwestern prairie" (e.g., prairie chicken [Tympanuchus sp.], upland plover, burrowing owl [Athene cunicularia], western meadowlark, bobolink, grasshopper sparrow, and savannah sparrow [Passerculus sandwichensis]) are of North American origin. "This may mean," Mayr (1946:39) continued, "that the great humidity of both the Bering and the Panama bridges prevented an influx of the faunas of the more arid habitat of Eurasia and South America. The ecological niche of the North American grasslands thus could be filled by the autochthonous North American element." The data of Table 3.13 provide a firmer base for consideration of faunal origins than Mayr's list of typical species. Grouping the species in terms of Mayr's assessment of the probable origins of their families (see Table 3.15) suggests that the grasslands and similar agricultural habitats are indeed dominated by North American elements, although the dominance is not so complete as Mayr thought. The South American and Pan-American elements contribute little to the avifauna of Mixed-grass Prairies, but comprise roughly one-third of the breeding species in True Prairies and agricultural habitats. Both Old World and South American elements thus do occur, casting some doubt on Mayr's biogeographic thesis.

Mammals

Major distributional changes occurred throughout the North American prairies during the Pleistocene (C. H. Hibbard, 1970). Further environmental changes during the last two centuries have caused and are still causing changes in the ecological and distributional status of certain species. We often think first of the range reductions, which we speak of in terms of this or that species having been extirpated from a given region. Mammalian examples such as red wolf (Canis rufus), elk (Cervus elaphus), and bison (Bison bison) vividly make this point for the True Prairie. In a few cases, species may be extirpated from an area and then become reestablished. Beaver (Castor canadensis) were overtrapped throughout most of their range within the True Prairie near the end of the nineteenth century, but now maintain self-sustaining populations throughout the region (Stains and Baker, 1958).

The swift fox (Vulpes velox) was considered extinct in much of its range during the first half of this century, but now occupies most of its former range. The present success of the species in areas adjacent to the True Prairie indicates it may become more common and widespread within the True Prairie than ever before. The other two species of foxes that occur in the True Prairie are currently expanding their distributions. Red foxes (Vulpes vulpes) are now common over much of the Great Plains, and have even come to occupy much of the Shortgrass Prairie and adjacent Desert Grasslands (Choate and Fleharty, 1975). Gray foxes (Urocyon cinereoargenteus) are invading certain mixed-grass and shortgrass habitats (Choate and Krause, 1976) but probably occupied the True Prairie well before its settlement by Europeans.

One rodent, the fox squirrel (Sciurus niger), is also expanding its distribution westward in the Great Plains, presumably by a

combination of introductions and natural or unaided range expansions (Armstrong, 1972; Hibbard, 1956). Again, however, the area of invasion is largely west of the True Prairie, where the species has long been resident.

Six species of mammals appear to be extending their distributions either into or within the True Prairie. One, the least weasel (Mustela nivalis) may be expanding in a southerly direction (Choate and Fleharty, 1975) whereas the other five species are all of Neotropical origin and appear to be expanding northward.

The opossum (Didelphis virginiana) was common over the southern two-thirds of the True Prairie (Hall and Kelson, 1959) by the mid-1950s. Subsequently, the species has been detected as far north as northern Minnesota (Hazard, 1963; Hibbard, 1970). These records probably represent range expansion rather than initial discovery of a long-term resident. Westward expansion of range by opossums in Kansas (Choate and Fleharty, 1975) and Nebraska (Choate and Genoways, 1967) is also indicative of a species successfully invading new territory, perhaps in response to human activities.

The distributional status of two Neotropical bats, Tadarida macrotis and T. brasiliensis, in the True Prairie is curious and may or may not indicate expansion of range into the area. Only three specimens of T. macrotis are known within the True Prairie, two from central Iowa captured in 1910 and 1914 (Bowles, 1975) and a third captured in southeastern Kansas in 1966 (Hays and Ireland, 1967). All three bats were taken in autumn and probably represent stragglers or wanderers. Poorly known anywhere in North America, the status of this bat in the True Prairie presents the sort of enigma that both demands and defies investigation.

The northeasternmost known breeding colony of T. brasiliensis is in north-central Oklahoma (Jones et al., 1967), but individuals are known from much of Kansas (Choate and Fleharty, 1975) and from eastern Nebraska (Jones, 1964). Birney and Rising (1968) suggested that such wandering individuals might have the advantage of reduced intraspecific competition for food and space and the opportunity to investigate potential maternity sites. Such wanderings in a highly mobile animal could lead to rapid invasion of new areas if environmental change resulted in suitable conditions. T. brasiliensis, however, probably is limited largely by the lack of appropriate caves and mine tunnels for usable maternity sites, and the kinds of changes taking place in the True Prairie are not forming such sites.

Spectacular northward advance of two mammalian species in the Great Plains, including the True Prairie, has been well documented by careful field effort during the past 65 years. In 1910 the armadillo (Dasypus novemcinctus) was evident no farther north than Stevens County, Texas, and not much east of the Brazos River in Texas (Taber, 1939). Armadillos since have occupied most of Oklahoma and Kansas and probably are established now as far north as the Platte River Valley in Nebraska (Choate and Fleharty, 1975). The biological basis of this advance has not been studied in any detail, but it may be the result of a combination of evolutionary adaptations of the expanding armadillo populations and environmental alterations attributable at least in part to human activities.

The northern advance of the hispid cotton rat (Sigmodon hispidus) is similar to that described for the armadillo but has

been studied in greater detail (Fleharty and Choate, 1973; Fleharty et al., 1972; Kilgore, 1970). The northward expansion through Kansas and well into Nebraska and Missouri has been discussed by numerous authors but is summarized by Genoways and Schlitter (1967). Two estimates of northward advance were 12 km per year in the True Prairie region of eastern Kansas (Cockrum, 1948) and 9 km per year in the Mixed-grass Prairie of central Kansas and adjacent Nebraska (Genoways and Schlitter, 1967).

Three species of mammals are known to have been introduced by humans and now have established populations within the True Prairie. These three, the Norway rat (Rattus norvegicus), house mouse (Mus musculus), and nutria (Myocaster coypus) are in addition to the domesticated mammals associated directly with people and their agricultural practices. The Norway rat and house mouse occur essentially wherever people live throughout the True Prairie and are also often found in fields, around ditches, and in other disturbed habitats well away from human dwellings.

The nutria is well established in such places as Louisiana (Lowery, 1974) and occurs in smaller but apparently permanent breeding populations farther north and west (Evans, 1970). Feral individuals have been captured in the True Prairie as far north as Minnesota (Gunderson, 1955), but these individuals apparently are escapees from fur farms rather than members of established feral populations.

Jones (1964) discussed instances of releases of European rabbit (Oryctolagus cuniculus) and fallow deer (Dama dama) in the True Prairie, and Cockrum (1952) reported a roof rat (Rattus rattus) taken in Wichita, Kansas, in 1950. No feral breeding populations of these three species are in existence in the True Prairie at present.

The area broadly outlined in Figure 2.1 as the True Prairie includes at least a portion of the distribution of 102 species of native mammals (Table 3.16), which may be classified taxonomically into 9 orders, 23 families, 65 genera, and the 3 introduced species, Norway rat, house mouse, and nutria, discussed above. Five native species, red wolf, grizzly (Ursus arctos), elk, caribou (Rangifer tarandus), and bison, have been extirpated from the True Prairie by post-Columbian humans and now occur there only (if at all) as a result of reintroduction. Gray wolves (Canis lupus) occur naturally in self-sustaining populations in sanctuaries such as the Beltrami Island State Forest of northwestern Minnesota. Individuals disperse into grassland pastures and other agricultural areas whenever the population within a sanctuary approaches saturation. Mountain lions (Felis concolor) are the subject of many sight records throughout the northern two-thirds of the Great Plains, but only a recent specimen from Manitoba (Nero, 1974) documents their continued existence anywhere near the True Prairie. Modern records of wolverine (Gulo gulo) from Iowa, Minnesota, and South Dakota--all clouded by uncertainties--and numerous sight records from Minnesota leave in question the status of this species where its original range overlapped the northern extent of the True Prairie (Birney, 1974b).

Taken collectively, the native mammals of the True Prairie include 45 species found primarily in forest-grassland ecotonal habitats, 18 species that are more or less restricted to the grasslands, and 39 species that occur in such ecologically diverse

TABLE 3.16 Mammals found in the True Prairie, with status of distribution, their
biogeographic affinity, habitat, and dietary category. (See text for
sources of information and discussion.)

Species	Margin in True Prairie[a]	Biogeographic affinity[b]	Habitat[c]	Dietary category[d]
Marsupicarnivora				
Didelphidae				
Virginia opossum (Didelphis virginiana)	Y	NT	ECOT	OMNI
Insectivora				
Soricidae				
Masked shrew (Sorex cinereus)	Y	CF	UBIQ	INSECT
Northern water shrew (Sorex palustris)	Y	CF	UBIQ	INSECT
Arctic shrew (Sorex arcticus)	Y	CF	ECOT	INSECT
Pygmy shrew (Microsorex hoyi)	Y	CF	ECOT	INSECT
Short-tailed shrew (Blarina brevicauda)	Y	DF	UBIQ	INSECT
Southern short-tailed shrew (Blarina carolinensis)	Y	DF	UBIQ	INSECT
Least shrew (Cryptotis parva)	Y	DF	UBIQ	INSECT
Desert shrew (Notiosorex crawfordi)	Y	DSW	ECOT	INSECT
Talpidae				
Eastern mole (Scalopus aquaticus)	Y	DF	UBIQ	INSECT
Star-nosed mole (Condylura cristata)	Y	CF	ECOT	INSECT
Chiroptera				
Vespertilionidae				
Little brown bat (Myotis lucifugus)	Y	WS	UBIQ	INSECT
Cave bat (Myotis velifer)	Y	DSW	UBIQ	INSECT

TABLE 3.16 Continued.

Species	Margin in True Prairie[a]	Biogeographic affinity[b]	Habitat[c]	Dietary category[d]
Chiroptera (cont.)				
Vespertilionidae (cont.)				
Gray myotis (Myotis grisescens)	Y	DF	UBIQ	INSECT
Keen's myotis (Myotis keenii)	Y	DF	UBIQ	INSECT
Indiana myotis (Myotis sodalis)	Y	DF	UBIQ	INSECT
Small-footed myotis (Myotis leibii)	Y	WS	UBIQ	INSECT
Silver-haired bat (Lasionycteris noctivagans)	N	WS	UBIQ	INSECT
Eastern pipistrelle (Pipistrellus subflavus)	Y	DF	UBIQ	INSECT
Big brown bat (Eptesicus fuscus)	Y	WS	UBIQ	INSECT
Red bat (Lasiurus borealis)	Y	DF	UBIQ	INSECT
Hoary bat (Lasiurus cinereus)	N	WS	UBIQ	INSECT
Evening bat (Nycticeius humeralis)	Y	DF	UBIQ	INSECT
Western big-eared bat (Plecotus townsendii)	Y	CF	UBIQ	INSECT
Molossidae				
Mexican free-tailed bat (Tadarida brasiliensis)	Y	NT	UBIQ	INSECT
Big free-tailed bat (Tadarida molossa)	Y	NT	UBIQ	INSECT
Primates				
Hominidae				
Man (Homo sapiens)	N	WS	UBIQ	OMNI

TABLE 3.16   Continued.

Species	Margin in True Prairie[a]	Biogeographic affinity[b]	Habitat[c]	Dietary category[d]
Edentata				
Dasypodidae				
Nine-banded armadillo (Dasypus novemcinctus)	Y	NT	ECOT	INSECT
Lagomorpha				
Leporidae				
Swamp rabbit (Sylvilagus aquaticus)	Y	DF	ECOT	HERB
Eastern cottontail (Sylvilagus floridanus)	Y	DF	ECOT	HERB
Snowshoe hare (Lepus americanus)	Y	CF	ECOT	HERB
White-tailed jackrabbit (Lepus townsendii)	Y	GL	GL	HERB
Black-tailed jackrabbit (Lepus californicus)	Y	DSW	GL	HERB
Rodentia				
Sciuridae				
Eastern chipmunk (Tamias striatus)	Y	DF	ECOT	OMNI
Least chipmunk (Eutamias minimus)	Y	CF	ECOT	OMNI
Woodchuck (Marmota monax)	Y	DF	ECOT	HERB
Richardson's ground squirrel (Spermophilus richardsonii)	Y	CF	GL	HERB
Thirteen-lined ground squirrel (Spermophilus tridecemlineatus)	N	GL	GL	OMNI
Franklin's ground squirrel (Spermophilus franklinii)	Y	GL	UBIQ	OMNI
Black-tailed prairie dog (Cynomys ludovicianus)	Y	GL	GL	HERB
Eastern gray squirrel (Sciurus carolinensis)	Y	DF	ECOT	OMNI
Eastern fox squirrel (Sciurus niger)	Y	DF	ECOT	OMNI

TABLE 3.16 Continued.

Species	Margin in True Prairie[a]	Biogeographic affinity[b]	Habitat[c]	Dietary category[d]
Rodentia (cont.)				
Sciuridae (cont.)				
Red squirrel (Tamiasciurus hudsonicus)	Y	CF	ECOT	HERB
Southern flying squirrel (Glaucomys volans)	Y	DF	ECOT	OMNI
Northern flying squirrel (Glaucomys sabrinus)	Y	CF	ECOT	OMNI
Geomyidae				
Northern pocket gopher (Thomomys talpoides)	Y	CF	GL	HERB
Plains pocket gopher (Geomys bursarius)	Y	GL	GL	HERB
Heteromyidae				
Olive-backed pocket mouse (Perognathus fasciatus)	Y	GL	GL	HERB
Plains pocket mouse (Perognathus flavescens)	Y	GL	GL	HERB
Hispid pocket mouse (Perognathus hispidus)	Y	GL	GL	HERB
Ord's kangaroo rat (Dipodomys ordii)	Y	DSW	GL	HERB
Castoridae				
Beaver (Castor canadensis)	N	WS	ECOT	HERB
Cricetidae				
Marsh rice rat (Oryzomys palustris)	Y	NT	ECOT	OMNI
Plains harvest mouse (Reithrodontomys montanus)	Y	GL	GL	OMNI
Western harvest mouse (Reithrodontomys megalotis)	Y	DSW	UBIQ	OMNI
Fulvous harvest mouse (Reithrodontomys fulvescens)	Y	NT	GL	OMNI

TABLE 3.16 Continued.

Species	Margin in True Prairie[a]	Biogeographic affinity[b]	Habitat[c]	Dietary category[d]
Cricetidae (cont.)				
Deer mouse (Peromyscus maniculatus)	N	WS	UBIQ	OMNI
White-footed mouse (Peromyscus leucopus)	Y	DF	ECOT	OMNI
Texas mouse (Peromyscus attwateri)	Y	D	ECOT	OMNI
Northern grasshopper mouse (Onychomys leucogaster)	Y	D	UBIQ	OMNI
Hispid cotton rat (Sigmodon hispidus)	Y	NT	GL	HERB
Eastern woodrat (Neotoma floridana)	Y	DF	ECOT	HERB
Southern red-backed vole (Clethrionomys gapperi)	Y	CF	ECOT	HERB
Meadow vole (Microtus pennsylvanicus)	Y	CF	ECOT	HERB
Prairie vole (Microtus ochrogaster)	Y	GL	GL	HERB
Pine vole (Microtus pinetorum)	Y	DF	ECOT	HERB
Muskrat (Ondatra zibethicus)	Y	WS	ECOT	HERB
Southern bog lemming (Synaptomys cooperi)	Y	CF	ECOT	HERB
Zapodidae				
Meadow jumping mouse (Zapus hudsonius)	Y	CF	UBIQ	OMNI
Western jumping mouse (Zapus princeps)	Y	CF	ECOT	OMNI
Erethizontidae				
Porcupine (Erethizon dorsatum)	Y	WS	ECOT	HERB

TABLE 3.16 Continued.

Species	Margin in True Prairie[a]	Biogeographic affinity[b]	Habitat[c]	Dietary category[d]
Carnivora				
Canidae				
Coyote (Canis latrans)	N	WS	UBIQ	CARN
Gray wolf (Canis lupus)	Y	WS	UBIQ	CARN
Red wolf (Canis rufus)	Y	DF	UBIQ	CARN
Red fox (Vulpes vulpes)	N	WS	ECOT	CARN
Swift fox (Vulpes velox)	Y	GL	GL	OMNI
Gray fox (Urocyon cinereoargenteus)	Y	WS	ECOT	OMNI
Ursidae				
Black bear (Ursus americanus)	N	WS	ECOT	OMNI
Grizzly (Ursus arctos)	Y	CF	ECOT	OMNI
Procyonidae				
Raccoon (Procyon lotor)	N	WS	ECOT	OMNI
Mustelidae				
Marten (Martes americana)	Y	CF	ECOT	CARN
Fisher (Martes pennanti)	Y	CF	ECOT	CARN
Ermine (Mustela erminea)	Y	CF	UBIQ	CARN
Least weasel (Mustela nivalis)	Y	CF	ECOT	CARN
Long-tailed weasel (Mustela frenata)	N	WS	UBIQ	CARN
Black-footed ferret (Mustela nigripes)	Y	GL	GL	CARN

TABLE 3.16 Continued.

Species	Margin in True Prairie[a]	Biogeographic affinity[b]	Habitat[c]	Dietary category[d]
Carnivora (cont.)				
Mustelidae (cont.)				
Mink (Mustela vison)	N	WS	ECOT	CARN
Wolverine (Gulo gulo)	Y	CF	ECOT	CARN
Badger (Taxidea taxus)	Y	WS	UBIQ	CARN
Spotted skunk (Spilogale putorius)	Y	DF	ECOT	OMNI
Striped skunk (Mephitis mephitis)	N	WS	UBIQ	OMNI
River otter (Lutra canadensis)	N	WS	ECOT	CARN
Felidae				
Mountain lion (Felis concolor)	Y	WS	UBIQ	CARN
Lynx (Lynx canadensis)	Y	CF	ECOT	CARN
Bobcat (Lynx rufus)	N	WS	UBIQ	CARN
Artiodactyla				
Cervidae				
Elk (Cervus elaphus)	Y	WS	UBIQ	HERB
Mule deer (Odocoileus hemionus)	Y	WS	UBIQ	HERB
White-tailed deer (Odocoileus virginianus)	N	WS	UBIQ	HERB
Moose (Alces alces)	Y	CF	ECOT	HERB
Caribou (Rangifer tarandus)	Y	CF	ECOT	HERB

TABLE 3.16 Continued.

Species	Margin in True Prairie[a]	Biogeographic affinity[b]	Habitat[c]	Dietary category[d]
Artiodactyla (cont.)				
Antilocapridae				
Pronghorn (Antilocapra americana)	Y	WS	GL	HERB
Bovidae				
Bison (Bison bison)	Y	WS	UBIQ	HERB

[a] Margin: N = range includes all of the True Prairie; Y = range termination within the True Prairie

[b] Biogeographic affinity: CF = coniferous forest; D = desert; DF = deciduous forest; DSW = desert southwest; GL = grassland; NT = neotropical; WS = widespread

[c] Habitat: ECOT = ecotonal; GL = grassland; UBIQ = ubiquitous

[d] Dietary category: CARN = carnivore; HERB = herbivore; INSECT = insectivore; OMNI = omnivore

habitats that they may be loosely considered ubiquitous. One species, Homo sapiens, has an essentially worldwide distribution, but not a single mammalian species has a distribution that falls entirely within the True Prairie. In other words, the True Prairie does not present a unique ecological opportunity for even a single species of mammal. Twelve species may be considered biogeographically as having a grassland origin, but these species occur also in Mixed-grass and/or Shortgrass and Desert Grasslands as well. So the mammalogist's viewpoint is apparently the same as that detected by study of invertebrates, the birds, and to a lesser degree the plants--namely the True Prairie appears not to be a unique ecological area, but rather appears as a transition zone between drier grasslands and either deciduous or coniferous forests.

Seven species having a Desert Southwest and/or Great Basin origin became established in the regions of shorter, drier grasslands and subsequently invaded part or all of the True Prairie to the east. An equal number of Neotropical species have been documented as occurring within the region, but of these species only the opossum, armadillo, fulvous harvest mouse (Reithrodontomys fulvescens), and hispid cotton rat are common and potentially significant ecological elements of True Prairie communities.

Nearly 50 percent of the mammalian species in the True Prairie fauna have arrived there from forest habitats, either the coniferous (26 species) or the deciduous (22 species) forest. Most of these mammalian species occupy mesic habitats (usually woodlands or savannas) within the True Prairie or occur there as ecological generalists in a wide variety of habitats. Richardson's ground squirrel (Spermophilus richardsonii) is an interesting exception. This so-called coniferous forest species is highly adapted to montane grasslands within the coniferous forest; within the prairie Richardson's ground squirrel lives in open pastures and rocky areas (Bailey, 1926).

The remaining 28 species of native mammals usually occupy large geographic areas and are classified in Table 3.16 as "widespread" (Armstrong, 1972). Most species (18) are not limited to specific habitat types but rather are found in several ecological settings. Another group of widespread species (beaver; muskrat, Ondatra zibethicus; mink, Mustela vison; and otter, Lutra canadensis) is partially aquatic and thus is limited by available water, but when not controlled or molested by people, these species successfully occupy aquatic habitats in a wide variety of vegetational communities. Similarly, most bat species are considered widespread, but in their case availability of caves and other roosting sites probably is locally more limiting than vegetation or other obvious features of the habitat.

Although more than 100 species of mammals occur within the region broadly defined as True Prairie, fewer species are found at any given locality. Simpson (1964) studied species density of North American mammals based on numbers of species occurring within quadrats of 57,600 km^2. Species density contours within the True Prairie were at about the 60-70 species level. Elsewhere in North America the density contours range from 163 species at one coastal area in Costa Rica to 15 species along one isogram north of Hudson Bay in northern Canada. Isograms showing 100-115 species per quadrat were typical of the western United States whereas isograms

showing 35-50 species per quadrat typify the southeastern United States.   An east-west transect throughout the mid-United States at about 35°N showed a precipitous drop in species density from the Rocky Mountains to the Great Plains, an additional slight decrease from the Great Plains into the Mississippi Valley, and then an increase into the Appalachians.

Simpson (1964) also discussed the well-known trend that diversity or species richness increases from the poles toward the equator.   According to his contours of species density, however, that trend is poorly defined or absent within the True Prairie.   A tally (Hall and Kelson, 1959) of mammal species occurring in northwestern Minnesota (69 species), eastern Nebraska (62 species), and northeastern Oklahoma (60 species) indicates slight reversal within the True Prairie of the pattern quoted so often as dogma. For North America as a whole, Wilson (1974) has shown that the increase in mammalian diversity on a north-south gradient is largely the result of increase in the number of tropical bats, whose numbers rise sharply in the absence of frosts where insects and fruits are available year-round.   Diversity of terrestrial mammals was shown to be only slightly greater in southern compared to northern areas (Wilson, 1974).   Over 85 percent of the mammals found in the True Prairie are terrestrial.   Furthermore, six of the fifteen bat species listed in Table 3.16 occur only on the periphery of the True Prairie or only as individual stragglers in the absence of known breeding colonies.

Hagmeier and Stults (1964) analyzed the distributional patterns of North American mammals, but they used a technique that minimizes the estimate of number of species at a given site; in effect, they considered only so-called major species by including only species that occur in over one-half of a given province, that have over one-half of their range within the province, or are endemic to the province.   Thusly treated, numbers of species in True Prairie provinces range from 39 to 44, but the trend was not distinctly north to south.   They also found that the number of species per mammalian genus is about average (1.3-1.5) for the True Prairie, and provinces that include part of the True Prairie have few endemics (0-2).

Perhaps most significantly, Hagmeier and Stults did not identify a province that corresponds closely to the True Prairie on the basis of mammalian comparisons.   Instead, they identified five or six provinces that lie largely in other habitat types and extend into parts of the True Prairie.   Figure 3.20, showing the margins of mammal distributions for species that occur in the True Prairie, reinforces their observation that no concentration of species boundaries is closely associated with the boundaries of the True Prairie.   A comparison of the mammals that occur in the Shortgrass and Mixed-grass Prairies, but that have not invaded the True Prairie, does little more to suggest a sharp line of demarcation around the True Prairie.   This boundary, which excludes only three western mammal species, corresponds closely in Kansas to the "Tatschl Line" (Küchler, 1971) designated on the basis of plant distributions.   Interestingly, environmental change along this north-south line seems to be more important to the distributions of plants than of mammals.   Two species of woodrats, Neotoma floridana, an eastern species, and N. micropus, a western species, form sharp

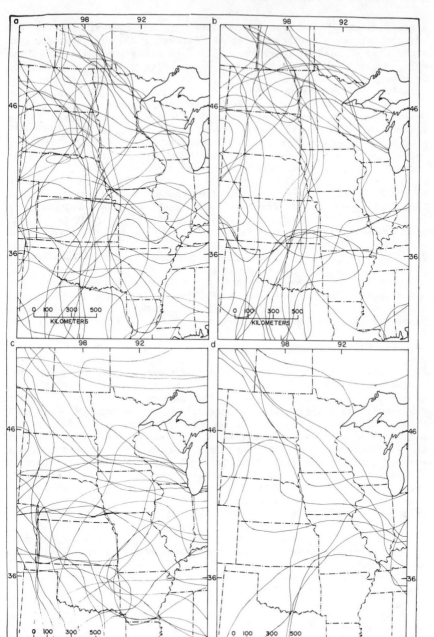

FIGURE 3.20 Margins of mammal distributions for species that occur in the True Prairie: (a) herbivores, (b) omnivores, (c) insectivores, and (d) carnivores.

distributional boundaries along the line in southern Kansas and adjacent parts of northern Oklahoma (Birney, 1973). A few mammalian subspecies boundaries (Cockrum, 1952) fall along the Tatschl Line as well, but viewed overall it seems of relatively little significance to mammals.

True Prairie mammals may be divided into four basic groups on the basis of food habits. Seventeen species may be regarded as primarily carnivores, twenty-six species feed largely on invertebrates (insectivores), twenty-nine species are for all practical purposes herbivores (including species that feed on seeds), and thirty species mix plant and animal material in their diet sufficiently to be termed omnivores.

Partial distributional margins are shown by dietary group in Figure 3.20 for the eighty-six species whose ranges terminate within the True Prairie. A trend of east-west margins is clear for insectivores (Figure 3.20c), indicating that their distributions tend to be limited in the True Prairie by factor or factors that form north-south gradients. Temperature is the most likely candidate for the ultimate cause, perhaps as it influences the general availability and seasonality of invertebrates.

Within the group regarded as omnivores (Figure 3.20b) an appreciable number of species reach their distributional limit along an east-west gradient that extends north and south from Texas to Minnesota and the Dakotas. Several herbivores (Figure 3.20a) and one carnivore (Figure 3.20d) show similar distributions. The north-south lines generally parallel the western edge of the True Prairie but most are somewhat east or west of it. A similar pattern of roughly parallel lines extends northwest-southeast along the boundary of the coniferous forest and grassland in Wisconsin, Minnesota, and Manitoba. Again the absence of an area in which distributional margins are concentrated is noteworthy. In both cases, this pattern appears to reflect an ecological buffer between mammals and their abiotic environment (i.e., a buffering influence of the biota itself). Whereas plant distributions tend to truncate sharply in response to edaphic factors, temperature, and rainfall, mammals respond to all of these as well as to the vegetation and to the associated fauna. Additionally, mammals respond to short-term environmental crises with decision making and mobility. Thus, relative to plants, mammals have a greater number of options for dealing with temporarily unsuitable environments.

Despite this absence of well-defined faunal associations with the True Prairie, several species occupying the area are distinctive and merit special mention. The following descriptions will familiarize the reader with some typical species of the True Prairie.

Least Shrew

The least shrew (<u>Cryptotis parva</u>) is one of the smallest mammals found in the True Prairie. These tiny insectivores occur as far north as Minnesota and south into Mexico (Figure 3.21), but most species of the genus occur in Mexico and Central America (Choate, 1970).

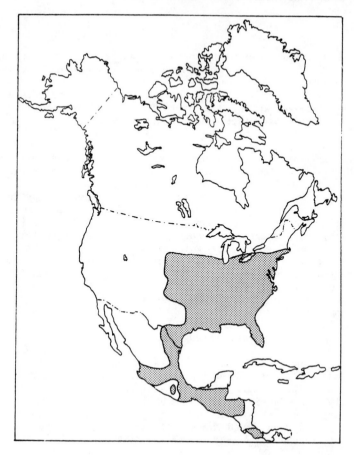

FIGURE 3.21 Distribution of the least shrew (<u>Cryptotis</u> <u>parva</u>).
Redrawn from Whitaker, 1974.

Although never contributing heavily to total mammal biomass in
any grassland habitat, least shrews may be found in the dense litter
of ungrazed tallgrass where they coexist with voles (<u>Microtus</u>) and
short-tailed shrews (<u>Blarina</u>), or in the open habitat of grazed
pastures where the voles and short-tailed shrews frequently are
excluded. Here least shrews occur most frequently with deer mice
(<u>Peromyscus</u> <u>maniculatus</u>) and pocket mice (<u>Perognathus</u> spp.).

Gray or tannish gray in color, about 70-80 mm in length, and
weighing 3-7 g, the least shrew has a long pointed snout, a short
tail, minute eyes, inconspicuous ears, and red pigmented teeth.
They feed almost exclusively on invertebrates but have been observed
preying upon small anurans.

Females have several litters per year and the young are born
after a brief gestation of about 14 days. Litter size is 3 to 6
young with each neonate weighing less than 1 g. By the eighth day
they may weigh about 3 g and by two weeks of age the young are adult
size (Walker, 1964). Nests are of shredded vegetation and usually
placed under a shelter or in a ground cavity (Whitaker, 1974).

Prairie Vole

The prairie vole (Microtus ochrogaster) is found throughout most of the True Prairie and Mixed-grass Prairie and extends into the Shortgrass Prairie in some places (Figure 3.22). Choate and Williams (1978) concluded that the species originated from an early Microtus lineage within the True Prairie, which would make the species unique biogeographically among mammals.

Prairie voles are brownish gray in color, flecked with a mixture of blackish and buffy-tipped guard hairs that give them a grizzled appearance. The cinnamon or ochraceous belly differs from most Microtus and is the source of the specific name, ochrogaster. Adults weigh 40-60 g, and have heavy, rounded faces, short legs, and short tails (30 to 40 mm). The species occurs exclusively in grassy habitats, usually where the vegetative cover is adequate, either standing or in the litter layer, to provide protection (Birney et al., 1976). The vegetative cover is characterized by a reticulum of runways, through which the voles can travel with amazing speed. Almost exclusively herbivorous, prairie voles may be important consumers of grass, especially at higher population densities. At such densities, they may exceed 100 individuals per ha, but high densities never persist for long, even though decay of the habitat is seldom detected.

Depending on latitude and local environmental conditions, prairie voles may breed throughout the year, but a decrease or cessation of reproductive activity usually is seen during the winter. Young females may mate as young as 30 days of age. Gestation is about 21 days, females normally breed during a postpartum estrus, and most litters contain 3 to 5 young. In captivity males assist with nest building and care of the young, suggesting a social system that perhaps tends more toward monogamy than is usually observed for small, short-lived rodents with high rates of reproduction and mortality (Thomas and Birney, 1979).

Meadow Jumping Mouse

The meadow jumping mouse (Zapus hudsonius) is widespread in northern and northeastern North America (Figure 3.23). The biology of the species recently was summarized by Whitaker (1972), and the following comments are condensed primarily from that source. The species was selected for inclusion as a representative of the True Prairie because of its broad distribution and because it differs from most grassland rodents in that it is an obligate hibernator.

Meadow jumping mice are yellow and brown, with long rear legs and a long tail that easily exceeds the combined length of the head and body. Total mouse length ranges from about 185-230 mm, and the tail ranges from about 110-140 mm. Weight of adults varies greatly depending on the season, with about 100 percent weight changes owing primarily to hibernation.

Grassy fields, dense vegetation along ponds, streams and marshes, and even wooded areas provide habitat for meadow jumping mice. Young are born between early June and early September. Litters vary in size geographically with slightly larger litters seen in the north, but litters usually contain between two and eight

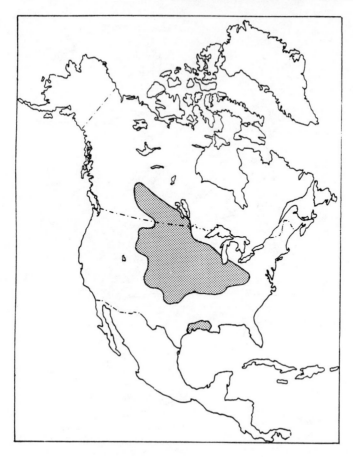

FIGURE 3.22 Distribution of the prairie vole (<u>Microtus</u>
<u>ochrogaster</u>). Redrawn from E. Raymond Hall and <u>Keith R.</u> Kelson,
The Mammals of North America, Vol. II, copyright 1959, The Ronald
Press Company, New York.

young, with four and five young often being modal litter size.
Seeds, especially seeds of grasses, are dietary staples, but
berries, nuts, fruits, and insects also are eaten. Hibernation is a
lengthy process, usually beginning in September or October.
Although an occasional individual may be recorded even during the
winter months, meadow jumping mice usually emerge from hibernation
late in April or early in May.

North American Badger

The single species in this genus of mustelid carnivore (North
American badger [<u>Taxidea</u> <u>taxus</u>]) is widely distributed in the
grasslands of North America (Figure 3.24). Badgers are fairly
large-sized predators (5-10 kg) of fossorial and burrow-inhabiting

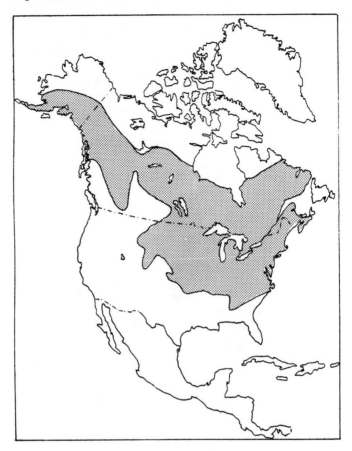

FIGURE 3.23 Distribution of the meadow jumping mouse (<u>Zapus hudsonius</u>). Redrawn from Whitaker (1972).

species. The tail is short and bushy, upper parts are grayish to reddish gray, and they have a white stripe that extends from the nose to at least the shoulders. Black patches are present on the sides of the face; the chin, throat, and belly are whitish.

Badgers normally are solitary, although family groups consisting of a female and her young, two or three yearling males, or an adult male with an estrous female may be seen together. Aboveground activity usually is greatest at night, but may occur at any time.

Mating takes place in late summer and young are born in March or April. Implantation does not take place until about February, however, and the actual period of embryonic development is only about six weeks.

An attempt at understanding the predatory and bioenergetic strategy of the North American badger was conducted recently by Lampe (1976). Morphological adaptations such as the powerful forelimbs and broad front claws are obvious for a fossorial

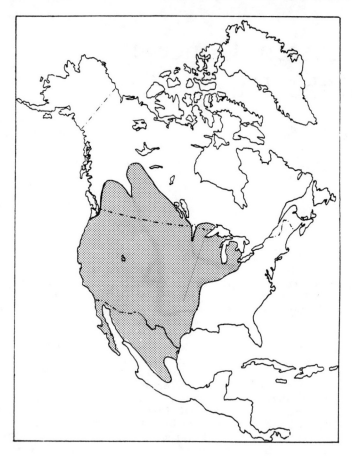

FIGURE 3.24 Distribution of the North American badger (<u>Taxidea</u>
<u>taxus</u>).

predator, but the highly selective foraging and digging behavior of
the species is less obvious as is the ability to learn effective
ways of making the predatory attempt. The dense medium of soil in
which the badger works is extremely costly to move, and badgers are
adept at capturing fossorial prey with a minimum of digging. Their
adeptness involves a good deal of surface investigation of the
subsurface tunnels of the prey and timely use of short bursts of
energy to dig rapidly and make a capture quickly. A foraging adult
badger must capture slightly less than the equivalent of two ground
squirrels or two pocket gophers, on the average, to break even
energetically for a night's foraging, but additional captures and
food items taken opportunistically (such as bumble bees and their
honey pots) contribute significantly to net energy gains. When prey
is readily available, huge energy stores in the form of fat are
furbished, but during winter when soils are snow covered and/or
frozen, badgers spend most of their time in the burrow and conserve
energy.

4

Ecosystem Concept
and the True Prairie

Early settlers of the True Prairie region knew that the grasses that grew on the rocky outcrops were not as tall, nor perhaps even the same kind, as the grasses that grew on the lower slopes. They also could see that whereas jackrabbits were common in the grasslands, they were less common in the forests, and these forests were largely restricted to the lowland areas. On the other hand, mosquitoes were constant pests in the wetter areas but were largely absent on the exposed uplands that had less moisture and more wind. In years blessed with plentiful rainfall, mosquitoes were more copious, and the grasses grew taller and sometimes had more flowers than the years with drought conditions. These early settlers also were aware that tree leaves that fell to the ground each year did not pile up indefinitely, but rather the material decayed in layers so that the leaves nearest the ground were no longer recognizable. Finally, if the grassland burned in the fall, in the next spring growth was rapid, the prairie appeared very green, and movement through the prairie was much easier without becoming entangled in the old grass.

If a person were to move 1500 miles to another location in the prairie, different species might be involved, but the same general pattern of relationships would be observable. In other words, the soils and climate would show a predictable relationship to the kinds and numbers of plants and animals. The early human inhabitants of the prairie were recognizing ecological principles that we have now formalized, defined, and described.

THE ECOSYSTEM CONCEPT

Ecology can be defined most simply as the relationship of living organisms to their environment. The primitive examples just described indicate a similarity of the relationships between organisms and the environment and among organisms through both time and space. This similarity suggests an organized system for the relationships (i.e., the relationships occur repeatedly in a predictable manner and one change at one point in the system may be

108

manifest at some other time or place). Thus, if the climatic conditions resulted in plentiful rainfall, the grass growth would be more vigorous the next year, and game would be in good condition, but perhaps more difficult to locate.

This example indicates that plant growth, which results from the capture of the sun's energy, is transferred to the animals, either in the form of cover or food. This concept is basic to our understanding of the prairie since plants, as primary producers by virtue of the photosynthetic process, capture solar energy and convert it into more plant material. The resulting material includes not only high energy organic compounds like sugars, fats, and proteins but also inorganic nutrients absorbed from the soil. When the plant material is eaten by consumers such as rodents, grasshoppers, or cows, the ingested plant material supplies energy and nutrients. The rodent may subsequently be consumed by a hawk. When the hawk dies, its body will be decayed by decomposer organisms such as bacteria and fungi, and in this process energy and nutrients are released. The nutrients can then be reabsorbed by plants. Some of the energy is lost as heat and some is retained by the microorganisms. Ultimately this energy is also lost and the process only continues because of incoming energy through the producers.

These concepts must be formalized so they can be utilized throughout the remainder of the volume. We have already noted that ecology is the study of the relationships of living organisms to their environment. This is a biotic-abiotic relationship where the biotic components are organic or living (i.e., plants and animals) and the abiotic components are nonliving (e.g., dissolved nutrients in the soil or precipitation as rainfall). Many relationships exist within these biotic and abiotic components. For example, plants are eaten by grasshoppers and grasshoppers are eaten by hawks. Similarly, whether or not dissolved nutrients are present in the soil depends on the kind of rock present and the amount of soil water, which in turn might be a function of the type of soil, topographic position, and amount of rainfall.

So, we are dealing with a complex system of relationships that involves the transfer of energy and nutrients among and between abiotic and biotic components. The abiotic components can be more specifically defined as inorganic substances, organic compounds, and climatic regime; the biotic components can be functionally recognized as producers, consumers, and decomposers. Finally, an ecosystem can be defined as a system in which living organisms interact with the physical environment, so that material and energy move along organized pathways (Odum, 1971). Each energy level is termed a trophic level. A sequence such as plants, rodents, hawks, in which energy and material move from one trophic level to another, is called a food chain.

Controlling Variables in the Ecosystem

Not all ecosystems are alike. We have already seen that all grassland ecosystems are not alike, and we will see differences even within the True Prairie. Ecosystems develop and exist under constraints of controlling variables. Some controlling variables are independent of the ecosystem while other variables are

controlled, at least partially, by the structure and function of the ecosystem.

External or independent controlling variables, constraints that are not a function of the ecosystem, can be classified into three broad categories: (1) geology, (2) macroclimate, and (3) available organisms. Geology is defined in a most general way to include such items as undifferentiated parent material or rocks, relief, and topography; for example, conditions just after the glaciers have retreated. Macroclimate is the large-scale climatic pattern that exists over a geographical area where the climate is determined by influences like oceanic currents, mountain ranges, and prevailing wind directions. Available organisms are organisms that are preadapted to live under a set of environmental conditions. These independent controlling variables define the possible components of an array of ecosystems but are not controlled by ecosystems themselves.

Dependent controlling variables result from ecosystem properties operating within the constraints of the independent variables. The dependent variables are: (1) soils, (2) microclimate, and (3) biological communities. The soil type that develops is a function of the prevailing climate, the parent materials, and the organisms that live on and in it. Microclimate refers to macroclimate that has been modified by elements such as the topography and plant canopy. Finally, the biological communities are composed of organisms that are capable of existing and reproducing under current soil and climatic conditions. Since soil and climate change through time and space, we would expect that biological communities would do likewise.

Our present perception of ecosystems, particularly the True Prairie ecosystem, comes from past observations. Early investigators noted the geographical ranges of species and their correspondence with environmental conditions. Many painstaking studies have shown both the requirements for some individual species and some interactions between species. From these past studies, we can now draw some generalities sufficient to construct the conceptual framework of the True Prairie ecosystem.

Based on their similarity of function in the ecosystem, species can be placed in one or more generalized components or compartments. Figure 4.1 shows examples of species that might typically be found in the True Prairie. Each species is placed in a functional compartment that demonstrates a generalized behavior in the system. The arrows depict processes that occur between compartments. These processes may refer to carbon flow or they may indicate water, nutrients, or even information flow. This book seeks to build this ecosystem concept, to organize information around it, and to examine interesting and pragmatic conclusions drawn from this information synthesis.

Studying the True Prairie Ecosystem

Seen in perspective, our research is simply the next step in a long sequence of investigative efforts aimed at the True Prairie. Some of the earliest ecological research in the United States was performed in the prairie-forest border, along the eastern edge of

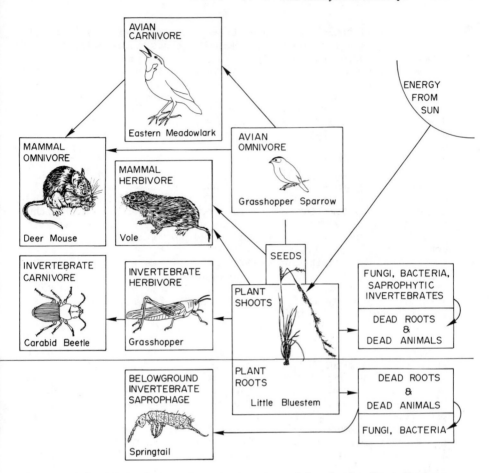

Figure 4.1   Pictorial and compartment model of the True Prairie ecosystem.

the True Prairie.   Around 1900, two major centers of plant ecology studies were in operation, one at the University of Chicago and the other at the University of Nebraska.   Several well-known ecologists received their training from the University of Chicago under Coulter: Cowles, Livingston, and Transeau (who completed his training from the University of Michigan).   At the University of Nebraska, Bessey had a student named Clements who went on to become perhaps the most dominant plant ecologist in North America during the first half of the twentieth century (Egler, 1951; McIntosh, 1974).

One of Clements's early concepts was that ontogeny, or the biological development of an individual organism, was similar to the developmental processes of vegetation or community types.   He also contended that biological communities underwent a successional sequence until a stable community was reached, and this final or terminal community was a function of the prevailing climatic conditions.   Beginning as early as 1904, he developed a whole school

of thought founded on these concepts, generating a set of terminology that described all community types in relation to the climatic climax. Subsequently ecologists have provided pragmatic arguments against this cumbersome system. Still other investigators argued against applying the organism connotation to biological communities, saying that the species present were chance occurrences of species with similar amplitudes (Curtis, 1959). The previous discussion (Chapter 3) concerning pre-Pleistocene and Pleistocene history has shown that members of the True Prairie flora and fauna were not always coexisting, which presumably precludes an organismal concept of community development.

Transeau (1905), in a study dealing with what is now called the Prairie Peninsula, described the extension of the Illinois, Indiana, and Ohio Prairie as a projection of the grassland between the northeastern conifer forest and the deciduous forest centers of the south and attempted to correlate this prairie with various climatic factors. No one single factor correlated particularly well with the occurrence of the Prairie Peninsula, so he used the idea of combining the precipitation-evaporation ratio as a single index for mapping vegetation regions. He noted that areas in which precipitation was only 20 to 60 percent of the potential evaporation were areas in which the prairies occurred. Further, the Tallgrass Prairie would occur in areas where the ratio of rainfall to evaporation was 60 to 80 percent, and the Oak Forest and Oak Savanna would occur in regions in which the ratio was between 80 and 100 percent. Both the deciduous forest and the coniferous forest required rainfall of 100 to 110 percent of the evaporation.

Briggs and Shantz (1911; 1916) published autecological studies describing the water requirements of the plants from the Great Plains. They first noted that the root systems were very large in proportion to the aboveground plant parts. As a result of this investigation, more attention was paid to the belowground plant parts and soil processes. The subsequent investigations were largely descriptive, and in spite of these early studies sixty years ago, we still have not adequately analyzed the belowground parts of the system (Coleman, 1976).

Livingston and Shreve (1921) did the first comprehensive study of the vegetation of the United States. The following years until about 1930 were concerned primarily with the drought adaptations exhibited by various plants. Some of the early work (Kearney and Shantz, 1911) emphasized the morphological adaptation of plants to either avoid or withstand drought conditions. Maksimov (1929) felt that drought resistance was a characteristic within individual cells. Subsequently, a long and perhaps still-continuing discussion ensued regarding the importance of stomates and their relation to drought resistance. The whole area of plant physiology was developing at this time and considerable interest was growing in fundamental processes like photosynthesis, vernalization, and the influence of the length of day or night on plant growth and development. This latter process, photoperiodism, was extensively studied in native grasses by Olmsted (1941; 1945), who indicated that a number of dominant prairie species show different photoperiodic responses that are correlated with their point of origin in the True Prairie. Clearly present studies dealing with

ecotypic differentiation, species specific gas exchange properties, and photosynthetic pathways follow from these earlier papers.

J. E. Weaver is well known for his detailed and meticulous work on the root systems of grassland plants. His paper on the ecological relations of roots (Weaver, 1919) was to be the first of numerous studies. From about 1920 to 1950 he and his students generated a tremendous body of information concerning the grasslands, and certainly the University of Nebraska and Weaver should be recognized as the center of early grassland plant ecology.

One of the earliest scientific descriptions of the American prairie fauna was done by Allen (1871). He felt that the grasslands were very monotonous and since animal diversity was tied to plant diversity, not much hope existed for a diverse fauna. Ruthven (1908) insisted that "The prairie is the most interesting biotic region of North America . . ." and emphasized that the vertebrate component had never been examined very carefully. He considered the True Prairie to be a transition zone between the eastern deciduous forest and the Great Plains, and concluded there was no particular fauna associated with this prairie. Shackleford (1929) studied the small animals of the dry and mesic prairies of Illinois, and noted that the animal population exhibited more uniformity on the high prairie than on the low prairie where it varied from place to place and from year to year. He also pointed out that topographic soil and vegetational variations influenced the fluctuations in animal populations.

By about 1940 numerous studies had been done concerning the distribution of small mammals (Blair and Hubbell, 1938; Clements and Shelford, 1939; Schaffner, 1926), and thereafter several attempts were made to correlate the plant communities with animal distributions.

The study of prairie insects was begun much later. Osborn (1939) published a book, Meadow and Pasture Insects, in which he stressed that little had been done with the grassland insects. Allen's (1871) general survey of the flora and fauna of the prairie included a brief mention of insects, especially commenting upon the number of grasshoppers. Allee (1927) pointed out that in some land communities the most abundant and important animals were insects, although in other land communities researchers generally held that mammals and birds were the most important (Shelford, 1931). Blair and Hubbell (1938) suggested that the area of Oklahoma, Kansas, and Nebraska in the True Prairie was the meeting ground for the insects both from the east and west, as well as the north and south. The fact that climatic conditions might affect insects in different ways was described by Shelford and Flint (1943). Grasshoppers received a considerable amount of early publicity, particularly in an attempt to correlate their numbers and activities to climatic conditions and grazing intensity.

The early inhabitants of the True Prairie recognized ecological patterns that we have now formalized into ecological principles. Early ecologists collected the data used in relating vegetation to climatic conditions and various consumer species to their prey. This information, along with that collected subsequently, can now be synthesized into an organized system of information. That synthesis is the purpose of this book. We will first discuss the abiotic controlling factors--namely, the climate and soils of the True

Prairie.   Next the biotic components will be described in detail, and then the abiotic and biotic components will be integrated into a comprehensive model.   Finally, the field, laboratory, and model results will be combined to discuss the responses of the ecosystem to various management approaches and to describe some whole system properties of the True Prairie ecosystem.

5

Controlling Variables
of the True Prairie

Generally, the type of vegetation occurring in a particular region is determined in large part by climate and soil. Walter (1973) suggested that the primary factors influencing vegetation are temperature, water, climate, and chemical and mechanical (soil) factors. Temperature and water are the major motive forces driving the ecosystem and determining its structure, but their influence is modified by soil and topography. Topographic factors such as slope and exposure, and soil factors such as water-holding capacity and infiltration rate, can have a great influence on vegetation by modifying soil temperature and available soil water for plant growth. The location of different vegetation zones or types is primarily determined by climate, but variations within a given type or zone may be the result of local differences in slope, exposure, or soil type.

Another dimension in the climate-soil-vegetation interaction is the influence of vegetation upon climate. Macrolevel effects of vegetation on climate have been hypothesized but are not well understood or verified (Sellers, 1965). Microclimate (the climate near the surface of the earth) is known to be under the influence of vegetation through the impact on the soil and air temperature, wind speed profile, and the radiation balance (Geiger, 1965).

CLIMATE

Trewartha (1954) defines weather as the sum total of the atmospheric condition (temperature, pressure, wind speed, moisture, and precipitation) over a short time period, while climate is defined as a composite of the variety of day-to-day weather conditions. The climatic description of a region deals largely with the composite state of the atmosphere and with distribution and characteristics of the climatic elements.

Climate Classification Schemes

Two approaches are generally used to describe climate: (1) an analysis of the average of the climatic elements, particularly temperature and precipitation or (2) a portrayal of the various types of weather that together comprise climate. The first approach has been used to develop climatic classification schemes that differentiate climate in different regions. Most frequently cited climate classification schemes are the schemes developed by Köppen (1936) and Thornthwaite (1933). The input variables required for these schemes include long-term average monthly temperature and precipitation.

Köppen classified the southern half of the True Prairie as a warm temperate rainy climate (Cfa climate) and the northern half as a cold snowy forest climate (Dfa climate). Cfa climates are characterized by an average temperature of the coldest month below 18°C but above -3°C, an average temperature of the warmest month greater than 22°C, and the driest month of summer receiving more than 3 cm of rainfall. The difference between the Cfa and Dfa climate is that the average temperature for the coldest month is less than -3°C for the Dfa climate. The significance of an average monthly temperature of -3°C for the coldest month is that this temperature supposedly coincides with the northern limit of frozen ground and a snow cover lasting for a month or more.

Thornthwaite classified the True Prairie as a subhumid climate with adequate rainfall. His scheme is primarily based upon the ratio of precipitation to potential evapotranspiration with growing season values of this ratio ranging from 0.60 to 0.90 within the True Prairie. This ratio is an index of water stress experienced by plants (increased water stress with decreasing values of the index). Both of these climatic classification schemes were developed with the assumption that boundaries between different climatic zones are associated with boundaries between different vegetation zones. Both authors perceive the plant community as a meteorological instrument that is capable of integrating all the climatic elements.

The climatic classification scheme presented by Trewartha (1961) divides the North American continent into seven regions with each region as a function of rainfall distribution. The True Prairie is in the eastern part of climatic type 3, which includes much of the central and northern interior of Anglo-America. As a general rule, annual rainfall decreases from over 76 cm in the eastern part to 25 cm in the western part of this region. The characteristic that defines the region is a single primary minimum in precipitation in winter and a maximum coinciding with the warm season (Figure 5.1). The eastern part of climatic type 3, which includes most of the True Prairie, is different from the rest of the region because it has two peak periods of rainfall, one in early summer (May, June) and one in September. The maximum monthly rainfall generally occurs in early summer while the secondary maximum generally occurs in September. West and north of the True Prairie the secondary maximum of rainfall disappears with a single maximum occurring in June for the western part and in July or August north of the True Prairie. East of the True Prairie the rainfall distribution is more uniform with the maximum still occurring during the warm half of the year; to the south of the True Prairie, more

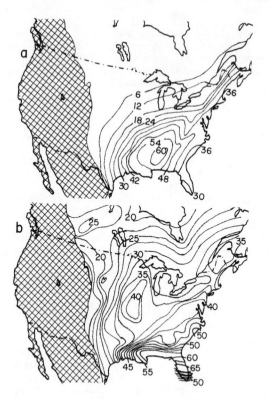

FIGURE 5.1 Average rainfall (cm) east of the Rocky Mountains,
November through March (a) and May through August (b). Isohyets are
based upon first-order Weather Bureau station for period of record
through 1930 (redrawn from Borchert, 1950).

precipitation falls during the winter months than during the summer.
South and east of the True Prairie, the annual rainfall increases
with all of the increase in precipitation coming during the colder
half of the year.

Important characteristics of rainfall patterns of the True
Prairie are the sharp northward decrease in precipitation during the
winter months, the fairly uniform distribution of rainfall during
the growing season, and the rapid decrease in precipitation west of
the True Prairie.

Types of Weather

Another way of looking at climate is to describe the various
types of weather that together comprise climate by introducing the
concepts of air masses and fronts and considering the sources of
moisture   for   precipitation,   the   mechanism   that   generates
precipitation,   and   the   general   circulation   patterns   of   the
atmosphere.   An air mass is a thick and extensive portion of the
atmosphere   where   temperature   and   humidity   characteristics   are

approximately homogenous in the horizontal direction. These air masses originate in regions where the atmosphere remains at rest over an extensive region for such a time that air is able to acquire the temperature and humidity characteristics of the underlying surface. Such a region is designated a source region. The types of air masses that influence the weather in the central part of the United States are:

1. Maritime tropical air (mT: warm, moist, and unstable) from the Gulf of Mexico and the Caribbean;
2. Continental tropical air (cT: hot, dry, and unstable) that originates over northern Mexico and the southwestern part of the United States;
3. Maritime polar air (mP: cool and moist, unstable in winter and stable in summer) that comes from the northern Pacific Ocean; and
4. Continental polar air (cP: cool, dry, and stable) that originates over the northern Canadian tundra.

These different types of air masses are separated by the more or less sharp boundaries referred to as frontal zones (10 to 100 km). Temperature discontinuity is the most significant characteristic of fronts and is responsible for maintaining differences in air mass densities (cold air masses have higher densities than warm air masses).

During the winter months the Great Plains are dominated by continental polar (cP) air masses more than 80 percent of the time (Figure 5.2). Just southeast of the True Prairie is a conflict region dominated by both cP and mT air masses. This conflict zone corresponds very well with the region of high winter precipitation in the southeastern United States, while the low precipitation in the Great Plains is the result of the dry continental air masses. The summer air mass distribution (Figure 5.3) shows that the True Prairie region is in a conflict zone between mT and cP air masses. Again this conflict zone corresponds very well with the region of maximum summer rainfall (see Figure 5.1). Source moisture for most of the precipitation in the United States comes from water evaporated from the oceans (Holzman, 1937). Of the total precipitation over the continents, 32 percent is returned to the ocean as runoff and 68 percent is returned to the atmosphere by evaporation. Most of the water evaporated from land surfaces is evaporated into air masses that have a continental source region. For the Great Plains and the True Prairie region, the primary sources of moisture are the tropical air masses that flow northward from the Gulf of Mexico. The primary mechanisms for generating precipitation include frontal activity (warm, cold, and stationary fronts) and large-scale extratropical cyclones. In summer the movement of fronts across the True Prairie is the primary mechanism for generating rainfall. These fronts separate air masses that have different characteristics, generally either mP and mT air masses or cP and mT air masses.

The rainfall during the summer months in the True Prairie is a result of the frequent change of air mass types (Figure 5.3) and the rainfall associated with the frontal activity. The decreasing rainfall westward across the Great Plains and the low precipitation

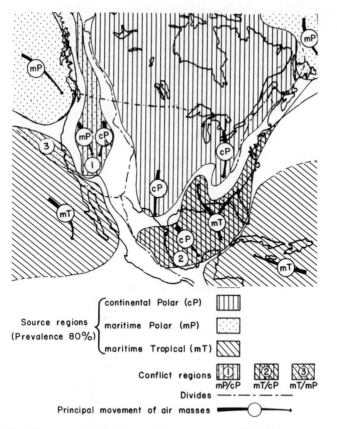

Source regions
(Prevalence 80%)
- continental Polar (cP) ▥
- maritime Polar (mP) ▦
- maritime Tropical (mT) ▨

Conflict regions ① mP/cP   ② mT/cP   ③ mT/mP

Divides — · — · — · —

Principal movement of air masses ━━━○━━━▶

FIGURE 5.2 Air mass distribution in winter. (Redrawn from Brunnschweiler, 1952.)

during the winter are a result of the high frequency of dry continental air masses and a reduction in frequency of tropical air masses.

The changes in the air mass distribution from winter to summer are caused by changes in the general circulation pattern of the atmosphere. The predominant feature of winter circulation patterns at the gradient level (.1 km above the ground) is the strong westerly wind across the entire United States (Figure 5.4). Across the Great Plains this strong westerly wind pattern prevails. To the south of the Great Plains the northerly flow of warm moist air from the Gulf of Mexico predominates, and to the north there is a southerly flow of continental polar air masses. From winter to summer the strength of the westerly wind decreases, while the flow of moist warm air from the Gulf of Mexico progresses northward (Figures 5.4, 5.5). During the winter a mean storm track occurs just south of the Great Plains, running from northern Texas to Ohio (Figure 5.4). In the spring and early summer the position of this storm track moves north- and westward until it is located from southern Colorado to Minnesota (Figure 5.5) in July. The movement of this storm track across the Great Plains and the northerly flow

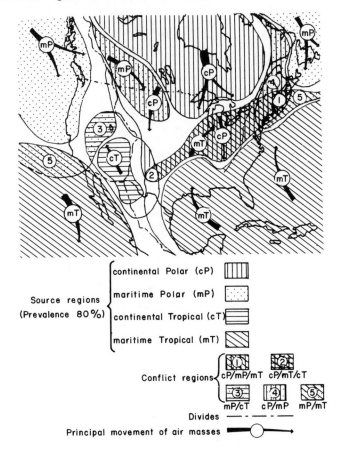

FIGURE 5.3 Air mass distribution in summer.   (Redrawn from Brunnschweiler, 1952.)

of warm maritime air from the Gulf of Mexico are responsible for the increased rainfall in the Great Plains during the spring and summer months.

Borchert (1950) indicates that the climate of the grasslands occupies a wedge-shaped region because of its unique position on the North American continent.   Between central Alberta and southern Texas, the midlatitude westerly wind descends from the Rocky Mountains upon the Great Plains. Maritime tropical air enters this circulation pattern to the south of the Great Plains while arctic air enters to the north. The stronger the mean westerly circulation in the midlatitude, the farther eastward the entering Arctic and tropical air streams are carried before they converge (compare Figures 5.2 and 5.3). The strength of the westerly circulation pattern influences the climate of the Great Plains by causing low rainfall and above-normal temperatures during periods of strong mean westerly circulation. The strong westerly circulation pushes the mean storm track to the south of the Great Plains and does not allow the northward advance of maritime tropical air. Borchert indicates

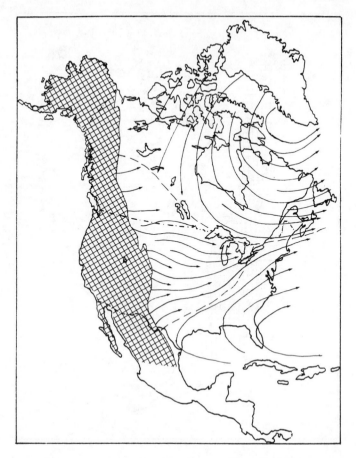

FIGURE 5.4   Streamline of resultant mean air flow at the gradient
level and axis of belts of maximum cycline frequency (dashed lines)
for January.   (Redrawn from Borchert, 1950.)

that the grassland climate of the Great Plains may be characterized
by:

1. Low snowfall and rainfall during the winter months, with
    the southern boundary of the True Prairie corresponding to
    the zone of steep rainfall gradient in the winter months
    (see Figure 5.1);
2. Larger risk of rainfall deficiency in summer within the
    grassland compared with bordering regions of forests;
3. The markedly low rainfall received in the shortgrass
    steppe during the summer compared with the rest of the
    Great Plains;
4. Fewer days with rain, more drought, and lower relative
    humidity during July and August than forest regions to the
    north;

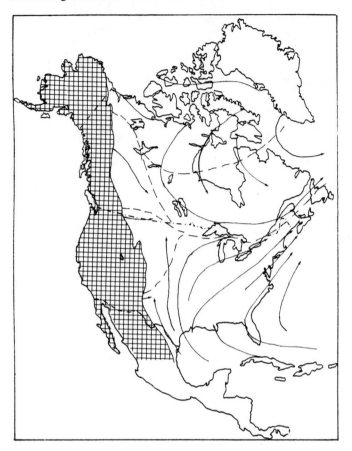

FIGURE 5.5 Streamlines of resultant mean airflow at the gradient
level and axes of belts of maximum cycline frequency (dashed lines)
for July. (Redrawn from Borchert, 1950.)

   5. The occasional major droughts in summer and the tendency
       for major summer drought to occur synchronously within the
       region;
   6. The continental source and trajectory of the mean air
       stream in the region during dry periods; and
   7. Large positive departure from average temperature and
       frequent hot wind during dry summers.

DETAILED ANALYSIS OF TRUE PRAIRIE CLIMATE

     The climate of the True Prairie has been studied in detail by
analyzing the monthly and seasonal distribution of a variety of
meteorological variables observed at several weather stations in the
eastern Great Plains. Pertinent features of the weather stations
used in this study are listed in Table 5.1. These stations (Figure
5.6) were chosen to study the change in climate along the

TABLE 5.1 Characteristics of the weather stations used in the detailed climate analysis of the True Prairie.

Station	Elevation (m)	Latitude	Longitude	Average yearly station pressure (mb)	Transmission coefficient[a]	Meteorological variable observed at the station
Foraker, Oklahoma	387.40	36°52'	96°34'	--	0.75	Precipitation
Pawhuska, Oklahoma	269.75	36°40'	96°21'	--	0.75	Precipitation and temperature
Tulsa, Oklahoma	205.74	36°09'	96°00'	991.1	0.75	Precipitation and temperature[b]
Manhattan, Kansas	327.05	39°11'	96°34'	--	0.75	Precipitation and temperature[b]
Topeka, Kansas	267.92	39°04'	95°37'	980.6	0.75	Precipitation and temperature[b]
Goodland, Kansas	1,112.82	39°22'	101°42'	887.0	0.81[c]	Precipitation and temperature[b]
Charles City, Iowa	308.76	43°04'	92°40'	--	0.75	Precipitation and temperature[b]
Des Moines, Iowa	290.78	41°32'	93°39'	985.5	0.75	Precipitation and temperature[b]
St. Cloud, Minnesota	315.16	45°35'	94°11'	977.3	0.75	Precipitation and temperature[b]
Columbia, Missouri	237.13	38°58'	92°22'	988.0	0.75	Precipitation and temperature[b]

[a] Transmission coefficient is set equal to transmission coefficient calculated for the Osage grassland site in northeastern Oklahoma.

[b] First-order weather station in which relative humidity, cloud cover, wind speed, and snowfall are also observed.

[c] The transmission coefficient is calculated by assuming that it decreases linearly with altitude between 304.8 and 1524 m with the transmission coefficient equal to 0.75 at 304.8 m and 0.84 at 1,524 m.

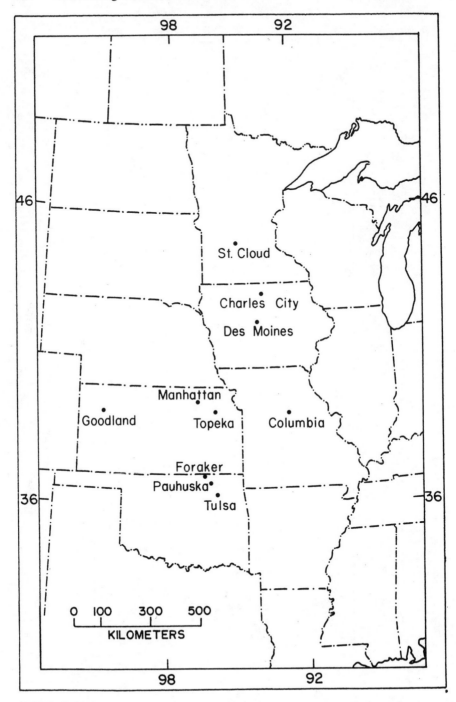

FIGURE 5.6 Location of the stations used in the detailed analysis
of the climate of the True Prairie.

north-south and east-west axes through the True Prairie. The analysis considered total incoming solar radiation, cloud cover, relative humidity, mixing ratio (ratio of the weight of atmospheric water vapor to the weight of the air), wind speed, snowfall, potential evapotranspiration, air temperature, and precipitation. Average values of the observed parameters are calculated using twenty years of daily and monthly average weather data (1953-1972 for all the stations except Goodland, Kansas, in which 1940-1959 were used).

Nine years of data (1950-1958) from each of the stations used in climate analysis were used to determine whether the climatic variables observed at each station were representative of the climate in the immediate area around each station. This analysis compared means and variances of monthly temperature and precipitation with means and variance of monthly temperature and precipitation determined from all of the weather stations within an 85 km radius of each site. An F-statistic was used to test the hypothesis that the variance of the mean monthly air temperature and precipitation for each station was the same as the variance for the circular area within an 85 km radius of each station. Variances were not significantly (p < 0.10) different except for the January mean monthly temperature at Columbia, Missouri (0.10 < p < 0.05).

The experimental design used to compare the mean monthly temperature and precipitation for each station with the means for the circular area within an 85 km radius is a randomized block analysis of variance where the nine years constitute the replicates or blocks.

Differences between months are expected because of seasonal changes in the abiotic variables. The site effect and the site-by-month interaction are the important aspects of the model. A significant site effect means that the average monthly precipitation or air temperature differed statistically between the site and its surrounding stations. A significant site-by-month interaction indicates that the site and the surrounding station show different patterns across the twelve months--that is, an increase or decrease in temperature or precipitation at the site is not followed by a comparable increase or decrease at the surrounding stations.

Precipitation data (Table 5.2) for all of the sites are truly representative of their surrounding stations (α = 0.10) and there are no significant site-by-month interactions (α = 0.10). Sites show the same trend across the twelve-month period as the surrounding stations and monthly rainfall at all of the sites is similar to the surrounding stations. Site differences in temperature are not significant (α = 0.10) for any of the stations except Goodland (α = 0.010), St. Cloud (α = 0.060), and Charles City (α = 0.058). There are no significant site-by-month interactions (α > 0.10) for the temperature data at any of the stations. Each site therefore shows the same temperature trend across the 12-month period as the surrounding stations, and all of the sites except Goodland, Charles City, and St. Cloud have the same mean monthly temperatures as the surrounding stations. Mean monthly temperatures at Goodland, St. Cloud, and Charles City are 1.8°C warmer, 1.0°C warmer, and 0.9°C colder, respectively, than the surrounding stations.

TABLE 5.2   Results of randomized block analysis of variance in which mean monthly precipitation and air temperatures are compared with mean monthly temperature and precipitation for the circular area within an 85-km radius of each station.

	Columbia--Temperature	Columbia--Precipitation	Goodland--Temperature	Goodland--Precipitation	St. Cloud--Temperature	St. Cloud--Precipitation	Manhattan--Temperature	Manhattan--Precipitation	Foraker--Precipitation	Pawhuska--Temperature	Charles City--Temperature	Charles City--Precipitation
Years	***	***	***	***	***		***	***	***	***	***	***
Site	***	***	***		**						**	
Month	***	***	***	***	***	***	***	***	***	***	***	***
Site by month												

*: Significant for α = 0.10

**: Significant for α = 0.05

***: Significant for α = 0.01

Blank: Non-significant for α = 0.10

North-south and east-west patterns for the climatic variables have been compared to climatic data presented by other authors (Borchert, 1950; Carpenter, 1940; Chow, 1964; Kincer, 1923; Rasmussen, 1971; Sellers, 1965; Trewartha, 1954; 1961; Wiens, 1972; 1974b) and showed that climatic patterns were in general agreement with these authors. Any significant discrepancy between the data presented in this chapter and data presented by other authors will be noted in the text.

Relative Humidity. Along the north-south cross-section through the True Prairie relative humidity is fairly homogenous, with a slight tendency for increasing relative humidity going to the north (Table 5.3). The increase in humidity is particularly evident during the nongrowing season. Maximum values of relative humidity occurred during the winter months, minimum values in March or April. The difference between the maximum and minimum values is from 10 percent to 15 percent relative humidity. Studying the east-west cross-section, apparently relative humidity changes little east of the True Prairie; west of the True Prairie humidity decreases, especially during the growing season.

Mixing Ratio. The mixing ratio is the ratio of the weight of atmospheric water vapor to the weight of the air (g H_2O:kg air) (Table 5.3) and is used as an indicator of the water content of the air. Maximum mixing ratios occur during July while minimum values occur in January. The mixing ratios decrease toward the north in all seasons, with the greatest decrease during the winter. The seasonal and north-south changes in the mixing ratio are primarily caused by the change in the air temperature, with the water-holding capacity of the air increasing as the air temperature increases. The air temperature increases from winter to summer and decreases to the north. Mixing ratios decrease to the west of the True Prairie and are unchanged to the east.

Cloud Cover. Cloud cover is highest during the winter months and lowest during the late summer months (July, August, and September). Cloud cover increases to the north, with the greatest difference during the growing season (Table 5.4). The greatest difference between the maximum and minimum mean monthly cloud cover is from 15 to 20 percent. East of the True Prairie the cloud cover is fairly homogenous during the summer and increases during the winter. Westward, the cloud cover decreases at all times of year, with the greatest difference during the summer.

Wind Speed. Mean monthly wind speed data show that the maximum wind speed occurs in the spring (March or April) and decreases to a minimum in August. Wind speed is fairly homogenous over the True Prairie, without any consistent north-south trends (Table 5.4). Wind speed decreases to the east and increases to the west of the True Prairie during the growing season and is fairly homogenous during the nongrowing season. The average wind speed for the growing season is consistently lower than the nongrowing season wind speed for all stations.

Snowfall. Total snowfall increases continuously toward the north and west (Table 5.4). Almost all of the snowfall is confined to the nongrowing season with the maximum monthly snowfall generally occurring during the winter months.

Solar Radiation. Values for solar radiation (Table 5.5) were calculated as a function of the station latitude, transmission

TABLE 5.3   Relative humidity and mixing ratio of the annual, growing season, and nongrowing season.

Station	Relative humidity (percent)			Mixing ratio (g H_2O:kg air)		
	Yearly average	Growing season[a]	Nongrowing season[b]	Yearly average	Growing season[a]	Nongrowing season[c]
North-south gradient						
Tulsa, Oklahoma	66	67	65	8.28	13.2	4.7
Topeka,[d] Kansas	68	69	68	7.76	12.8	4.2
Des Moines, Iowa	70	68	71	5.96	10.1	3.0
St. Cloud, Minnesota	70	69	71	5.20	9.0	2.5
East-west gradient						
Columbia, Missouri	67	67	67	7.35	11.9	4.1
Topeka,[d] Kansas	68	69	68	7.76	12.8	4.2
Goodland, Kansas	60	58	62	5.98	9.6	3.4

[a] May-September.

[b] October-April.

[c] April, October-December.

[d] North-south and East-west gradients intersect in Kansas.

TABLE 5.4 Mean annual, growing season (May-September), and nongrowing season (October-April) cloud cover, wind speed, and snowfall.

Station	Cloud cover (percent)			Wind speed (km/h)			Snowfall (cm)		
	Mean annual	Growing season	Nongrowing season	Mean annual	Growing season	Nongrowing season	Mean annual	Growing season	Nongrowing season
North-south gradient									
Tulsa, Oklahoma	55	52	58	16.7	15.6	17.5	27.2	0	27.2
Topeka, Kansas	58	55	60	17.1	16.1	17.9	57.9	0	57.9
Des Moines, Iowa	61	56	64	17.9	15.9	19.2	83.0	0	83.0
St. Cloud, Minnesota	62	57	66	15.6	14.6	16.3	100.8	.5	100.3
East-west gradient									
Columbia, Missouri	59	54	63	16.4	14.2	18.2	51.8	0	51.8
Topeka, Kansas	58	55	60	17.1	16.1	17.9	57.9	0	57.9
Goodland, Kansas	50	46	54	19.5	19.3	19.6	84.6	3.2	81.4

TABLE 5.5 Yearly average, growing season (May-September), and nongrowing season (October-April) total incoming solar radiation.[a]

Station	Solar radiation (Langleys/day)		
	Yearly average	Growing season	Nongrowing season
North-south gradient			
Tulsa, Oklahoma	387	512	298
Topeka, Kansas	369	495	274
Des Moines, Iowa	348	484	250
St. Cloud, Minnesota	326	472	223
East-west gradient			
Columbia, Missouri	365	498	270
Topeka, Kansas	369	495	274
Goodland, Kansas	418	570	309

[a] Solar radiation is calculated by using an equation presented by Sellers (1965) to determine the solar radiation outside the earth's atmosphere and then reducing this value as a function of the transmission coefficient and the cloud cover. Horwitz' equation (1941) is used to determine the reduction by cloud cover.

coefficient, and monthly mean cloud cover using the equations of Horwitz (1941) and Sellers (1965). Solar radiation was calculated since direct observations were not available. Solar radiation is at a minimum in December and a maximum in June or July. The solar radiation decreases to the north during all seasons; however, the maximum northward decrease is greater than 70 ly/day in January and less than 20 ly/day in July. The east-west cross-section shows that

solar radiation does not change going east in the True Prairie, but increases to the west. This increase is caused by a decrease in cloud cover and an increase in the transmission coefficient.

Potential Evapotranspiration. The potential evapotranspiration rates were estimated for each of the stations (Table 5.6) by assuming that the potential evapotranspiration rate is equal to 0.70 times the Class A pan evaporative rate (Chow, 1964). The potential evapotranspiration rate is assumed to be the maximum rate that water will be either evaporated or transpired into the earth's atmosphere. Minimum values of potential evapotranspiration rate occur in January and maximum rates in July. The rate decreases to the north, with the greatest north-south difference occurring during the nongrowing season. The potential evapotranspiration rate increases rapidly to the west of the True Prairie and decreases to the east of the True Prairie. The increase in the potential evapotranspiration rate south and west of the True Prairie is a result of increasing solar radiation and decreasing relative humidity.

Air Temperature. Average monthly maximum and minimum air temperatures are lowest in January and highest in July (Table 5.7). Along the north-south gradient, both maximum and minimum air temperatures decrease to the north in all seasons with the greatest north-south differential occurring during the winter months and the least difference during the summer. East of the True Prairie the maximum temperature is slightly lower during all seasons, while minimum air temperature does not change in the growing season but is warmer during the nongrowing season. To the west of the True Prairie the maximum air temperature is higher during the winter and lower during the spring and fall, while the minimum air temperature is 3° to 4°C lower during all seasons.

The variability of the maximum and minimum air temperature was assessed by examining the standard deviation of monthly mean values of maximum and minimum air temperatures. Maximum and minimum air temperatures are least variable in the summer months and most variable during the winter (see Table 5.7). In all cases the standard deviation of the minimum air temperature is less than that of the maximum air temperatures. Variability of maximum air temperature decreases going north in summer and increases during the winter. The variability of minimum air temperature does not show any clear trend on the north-south axis; however, the minimum air temperature tends to be more variable in the north during the winter. No clear pattern is evident between the variability of either maximum or minimum air temperature along the east-west axis, although maximum air temperature variability does tend to increase to the west of the True Prairie.

The air temperature data were used to calculate the mean frost-free period and the length of the growing season (Table 5.8). The growing season is defined as the length of time that the 14-day running mean air temperature exceeds 10°C, and the frost-free period is the number of days that the minimum air temperature is greater than -2.2°C. Both the frost-free period and the growing season decrease in length to the north. The growing season is reduced by a later start in the spring and an earlier end in the fall. The reduced frost-free period is caused by late spring frosts and early fall frosts; the reduction, however, is primarily caused by earlier frosts in the fall. The length of the growing season is not changed

TABLE 5.6   Yearly total, growing season (May-October), and nongrowing season (November-April) cumulative potential evapotranspiration.[a]

| Station | Potential evapotranspiration (cm of H_2O) | | |
	Yearly total	Growing season	Nongrowing season
North-south gradient			
Foraker, Oklahoma	132	94	38
Manhattan, Kansas	116	86	30
Charles City, Iowa	84	66	18
St. Cloud, Minnesota	78	63	15
East-west gradient			
Columbia, Missouri	89	69	20
Manhattan, Kansas	116	86	30
Goodland, Kansas	151	110	41

[a] Calculated by multiplying 0.70 times the Class A pan evaporation rates (Chow, 1964).

east of the True Prairie, but the frost-free period is increased because of later fall frosts. To the west, both the frost-free period and the growing season are reduced, with the frost-free period being shortened by late spring frosts, and the growing season being reduced by a late start and an early end. No systematic north-south changes in variability of these frost and growing season variables occur. However, the variability of the mean growing season and all of the frost variables tend to increase to the west across the True Prairie.

TABLE 5.7 Yearly, growing season, and nongrowing season averages of the maximum and minimum air temperatures and the standard deviation of the monthly average maximum and minimum air temperatures.

Station	Maximum air temperature (°C)			Minimum air temperature (°C)			Standard deviation of maximum air temperature			Standard deviation of minimum air temperature		
	Yearly	Growing season	Nongrowing season	Yearly	Growing season	Nongrowing season	Yearly	Growing season	Nongrowing season	Yearly	Growing season	Nongrowing season
North-south gradient												
Pawhuska, Oklahoma	22.2	31.1	15.6	8.3	17.8	1.7	2.24	2.00	2.44	1.73	1.44	1.94
Manhattan, Kansas	19.4	29.4	12.2	6.1	16.7	-1.1	2.22	1.89	2.44	1.79	1.50	2.00
Charles City, Iowa	13.9	26.1	5.6	2.8	13.3	-4.4	2.18	1.56	2.67	1.99	1.44	2.39
St. Cloud, Minnesota	11.1	23.9	2.2	-0.6	11.1	-8.3	2.28	1.50	2.83	2.16	1.39	2.72
East-west gradient												
Columbia, Missouri	18.3	28.3	11.1	7.2	16.7	0.6	2.17	1.72	2.44	1.73	1.33	2.00
Manhattan, Kansas	19.4	29.4	12.2	6.1	16.7	-1.1	2.22	1.88	2.44	1.79	1.50	2.00
Goodland, Kansas	18.3	28.3	11.7	2.8	11.1	-4.4	2.57	2.06	2.94	1.67	1.33	1.89

TABLE 5.8 Mean frost-free period and mean length of growing season.

| Station | Frost-free[a] | | | | | |
	Mean date of last frost day	Mean date of first frost day	Mean frost-free period (days)	Standard deviation of last frost in spring (days)	Standard deviation of first frost in fall (days)	Standard deviation (days)
North-south gradient						
Pawhuska, Oklahoma	28 March	30 October	216	13.3	12.9	14.7
Manhattan, Kansas	29 March	12 October	197	23.7	16.7	17.4
Charles City, Iowa	4 April	30 September	180	15.6	15.8	20.1
St. Cloud, Minnesota	10 April	15 September	158	22.2	23.6	17.2
East-west gradient						
Columbia, Missouri	30 March	29 October	213	14.4	14.1	13.3
Manhattan, Kansas	29 March	12 October	197	23.7	16.7	17.4
Goodland, Kansas	15 April	7 October	175	26.1	20.5	26.3

| Station | Growing season[b] | | | | | |
	Mean date for start of growing season	Mean date for end of growing season	Mean length of growing season (days)	Standard deviation of start (days)	Standard deviation of end (days)	Standard deviation (days)
North-south gradient						
Pawhuska, Oklahoma	20 March	15 November	240	13.1	10.1	12.9
Manhattan, Kansas	2 April	5 November	217	11.1	8.9	11.9
Charles City, Iowa	22 April	24 October	185	13.1	8.8	12.7
St. Cloud, Minnesota	26 April	3 October	160	16.3	12.8	15.8
East-west gradient						
Columbia, Missouri	3 April	5 November	217	13.5	11.1	9.3
Manhattan, Kansas	2 April	5 November	217	11.1	8.9	11.9
Goodland, Kansas	24 April	26 October	185	13.4	11.2	12.9

[a] Frost-free period is the number of days from spring to fall when the minimum air temperature is greater than -2.2°C.
[b] The growing season starts when 14-day running mean air temperature is greater than 10°C and ends when the running mean air temperature is less than 10°C.

Precipitation. Minimum rainfall occurs in January while maximum monthly rainfall generally occurs in June, with a secondary maximum in September and a secondary minimum in August (Table 5.9). The largest amounts of precipitation are received in summer with the least amounts received during the winter.   During the summer and spring total rainfall is fairly homogenous on a north-south axis, but during the winter precipitation decreases toward the north. To the east of the True Prairie, the total precipitation during the spring, summer, and fall is fairly similar, but during the winter precipitation is greater.   Precipitation decreases during all seasons to the west, with greatest precipitation decreases during the winter.

Trewartha (1961) suggested that one of the significant characteristics of the climate in the True Prairie region is the dual peak in rainfall during the summer. He also indicated that the September peak disappears to the west and north of the True Prairie region with a single maximum occurring during the summer. To the east the rainfall patterns become more uniform.   Our analyses support Trewartha's results and his climatic classification scheme.

Growing season rainfall is homogenous on a north-south axis in the True Prairie. Total precipitation is less, however, during the nongrowing season than the growing season, and rainfall in the north decreases (Table 5.10). The rain day and daily precipitation amount data show that the decreased precipitation during the nongrowing season is caused by a decreasing average daily precipitation amount. The north-south pattern for the average number of rain-days and average precipitation amounts reveals an increase in the number of rain-days and a decrease in rainfall amounts in the north for both the growing and nongrowing seasons.   This pattern suggests an increase in interception of rainfall by plants and a decrease in storm runoff to the north. A decrease in runoff is supported by rainfall intensity data that show a decrease in the number of days with excessive short duration rainfall to the north (Table 5.11). Growing season average rainfall amounts vary little to the east of the True Prairie, but an increase in nongrowing season rainfall is caused by an increase in the number of rain-days. To the west the rainfall, rainfall amounts, and rain-days decrease in both the growing and nongrowing seasons; however, the decrease is primarily caused by the decrease in the average rainfall amount.

The coefficient of variation (CV) of annual growing and nongrowing season characteristics (Table 5.10) is an indication of the variability of the precipitation variables.   Rainfall variability decreases to the north during the growing season, primarily because of a decrease in the variability of the number of rain-days, while during the nongrowing season the variability of precipitation does not follow any recognizable north-south pattern. The only distinct pattern along the east-west axis is an increase in variability of rainfall, rain-days, and rainfall amounts westward across the True Prairie in the nongrowing season. Annual rainfall is less variable to the north and more variable to the west across the True Prairie.

The ratio of precipitation to potential evapotranspiration indicates the water stress experienced by a plant, with high values of the ratio indicating lower water stress (Figures 5.7 and 5.8). Values of the ratio increase eastward across the True Prairie, are

TABLE 5.9 Mean monthly, seasonal, and yearly rainfall.

Station	Mean monthly rainfall (cm)												Mean seasonal rainfall[a]				Mean yearly rainfall
	Jan	Feb	Mar	Apr	May	June	July	Aug	Sept	Oct	Nov	Dec	Winter	Spring	Summer	Fall	
North-south gradient																	
Pawhuska, Oklahoma	2.46	2.79	4.83	7.14	11.33	11.81	8.76	8.05	11.07	7.62	3.89	4.06	9.35	23.29	28.63	22.58	83.85
Manhattan, Kansas	2.16	2.59	4.70	6.48	11.89	11.86	8.76	7.67	11.73	8.59	3.61	2.77	7.52	23.06	28.30	23.95	82.83
Charles City, Iowa	1.93	2.36	5.05	8.03	10.49	11.66	10.44	8.23	10.92	6.45	3.58	2.87	7.16	23.57	30.33	20.73	81.79
St. Cloud, Minnesota	1.65	1.68	2.84	5.92	9.12	10.72	8.43	11.20	6.76	4.93	3.20	2.08	5.41	17.88	30.35	14.88	68.53
East-west gradient																	
Columbia, Missouri	3.71	3.76	6.65	9.25	11.46	9.53	10.26	5.66	10.29	8.61	3.99	7.70	12.62	27.36	25.45	22.89	88.32
Manhattan, Kansas	2.16	2.59	4.70	6.48	11.89	11.86	8.76	7.67	11.73	8.59	3.61	2.77	7.52	23.06	28.30	23.95	82.83
Goodland, Kansas	1.12	1.02	2.59	3.94	6.30	7.16	7.75	5.94	3.10	1.83	1.52	1.07	3.23	12.85	20.85	6.48	43.36

[a] Winter: December-February; spring: March-May; summer: June-August; fall: September-November.

TABLE 5.10 Annual precipitation characteristics in the tallgrass prairie during growing season (May-September) and nongrowing season (October-April).

Station	Rainfall (cm)			Coefficient of variation			Rain days		Average precipitation amounts on rain days (cm)		Coefficient of variation on rain days		Coefficient of variation on rainfall amounts	
	Annual	Growing season	Nongrowing season	Annual rainfall	Growing season rain	Nongrowing season rain	Growing season	Nongrowing season	Growing season	Nongrowing season	Growing season	Nongrowing season	Growing season	Nongrowing season
North-south gradient														
Pawhuska, Oklahoma	83.87	51.05	32.82	27.84	35.6	27.8	38	39	1.32	0.84	18	21	127	144
Manhattan, Kansas	82.81	51.92	30.89	25.67	27.3	33.6	46	42	1.14	0.74	15	18	148	156
Charles City, Iowa	81.79	51.74	30.05	16.04	19.7	21.6	52	55	0.99	0.53	14	11	136	145
St. Cloud, Minnesota	68.56	46.23	22.33	18.99	21.1	32.1	50	56	0.94	0.41	13	13	138	159
East-west gradient														
Columbia, Missouri	88.29	47.17	41.12	21.94	29.5	28.4	45	59	1.07	0.71	14	17	130	141
Manhattan, Kansas	82.81	51.92	30.89	25.67	27.3	33.6	46	42	1.14	0.74	15	18	148	156
Goodland, Kansas	43.41	30.25	13.16	30.26	30.7	47.0	43	37	0.71	0.36	17	28	134	160

137

TABLE 5.11 Number of days per year that excess short duration
            rainfall is observed.

Location	Number of days with excessive rainfall[a,b]	Standard deviation	Coefficient of variation
North-south gradient			
Tulsa, Oklahoma	9.62	3.25	0.34
Topeka, Kansas	8.86	3.82	0.43
Des Moines, Iowa	6.71	2.99	0.45
St. Cloud, Minnesota	7.24	2.81	0.39
East-west gradient			
Goodland, Kansas	3.90	1.74	0.45
Topeka, Kansas	8.86	3.82	0.43
Columbia, Missouri	9.36	3.10	0.33

[a] Excessive short duration rainfall occurs on days when the rainfall
for any period from 5 to 180 minutes exceeds the rainfall amount
indicated by the following equation:

$$R = (t + 20)0.1$$

where R = rainfall amount (inches)
      t = time period (minutes)

[b] Based on climatological data, U.S. Department of Commerce
(1950-1970).

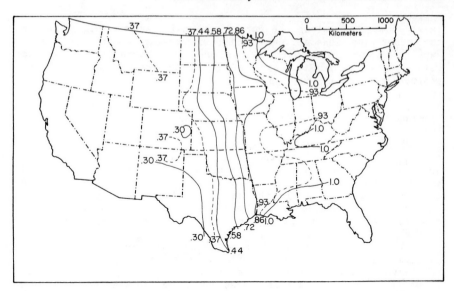

FIGURE 5.7 Ratio of growing season precipitation (May-October) to growing season potential evapotranspiration rate. The potential evapotranspiration rates were calculated by multiplying the growing season Class A pan evaporation rates (Data from Chow, 1964) by 0.7.

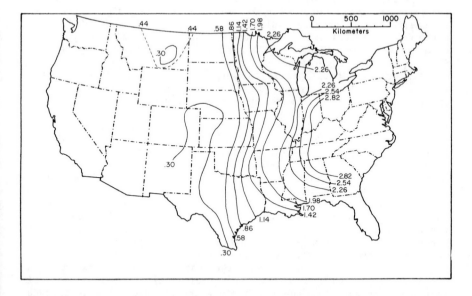

FIGURE 5.8 Ratio of nongrowing season precipitation (November-April) to nongrowing season potential evapotranspiration rate. The potential evapotranspiration rates were calculated by multiplying the nongrowing season Class A pan evaporation rates (Data from Chow, 1964) by 0.7.

fairly homogenous along the north-south axis, and are higher during the nongrowing season. During the nongrowing season the plants experience very little water stress, while during the growing season water stress is minimal east of the True Prairie and increases westward across the True Prairie.

Comparisons of Climatic Variability

Along the north-south axis in the True Prairie the climate is fairly homogenous during the growing season. The variables that show small north-south changes include relative humidity, cloud cover, wind speed, solar radiation, and total rainfall. A continuous drop occurs in the maximum and minimum air temperature that causes the average air temperature, frost-free days, length of growing season, mixing ratio, and potential evapotranspiration rate to decrease in the north.

During the nongrowing season large north-south changes take place in all the climatic variables except wind speed. The maximum north-south difference in climatic variables is generally two to three times larger than differences during the growing season. Some of the most significant changes include a drop in air temperature and precipitation and an increase in cloud cover and snowfall going northward. Thus, during the nongrowing season the climate of the northern True Prairie is greatly different from the climate of the southern part.

In the eastern part of the True Prairie most of the variables change very little during the growing season except for a decrease in the potential evapotranspiration rate, the ratio of rainfall to potential evapotranspiration, and wind speed; similarly, most climatic variables change little during the nongrowing season except for a significant decrease in snowfall and an increase in minimum air temperature, total precipitation, and rain days. These results indicate very little change in climate during the growing season except for a decrease in water stress experienced by plants.

Almost all of the climatic variables change fairly rapidly in the western part of the True Prairie during both the growing and nongrowing seasons. Some of the most significant changes include an increase in wind speed, solar radiation, and potential evapotranspiration and a decrease in relative humidity, cloud cover, minimum air temperature, length of growing season, total rainfall, average rainfall amounts, and the ratio of precipitation to potential evapotranspiration. During the growing season wind speed increases continuously and the number of frost-free days decreases westward across the True Prairie; during the nongrowing season the cloud cover, minimum air temperature, total precipitation, and rain-days increase to the east, while the snowfall decreases.

During the growing season the variability of maximum and average air temperatures, total precipitation, and number of rain-days decreases to the north along the north-south axis through the True Prairie, while other variables show no consistent north-south patterns. During the nongrowing season the variability of maximum, minimum, and average air temperatures increases to the north, while variability of the number of rain-days decreases. Annual rainfall variability increases to the south. Thus, during

the growing season the climate of the southern portions of the True Prairie is more variable; during the nongrowing season the northern areas are more variable.

Growing season precipitation and air temperature do not follow any systematic east-west pattern in variability. However, the variability of frost-free days and length of growing season increases from east to west across the True Prairie. Air temperature variables do not follow any east-west pattern during the nongrowing season, but the variability of total precipitation, rain-days, and average rainfall amounts increases. Thus, the data suggest the climate is more variable going westward across the prairie.

MICROCLIMATE

Almost all of the plants and animals in the True Prairie are found in narrow zones that extend from 1 m above to 1 m below ground surface. Within this zone the microclimatic variables like air temperature, wind speed, and relative humidity show very rapid changes along the vertical axis and show large diurnal variations.

The microclimate aboveground is greatly influenced by the structure of the plant canopy. The average canopy height is 25 cm at the beginning of the growing season (May) and increases to 50 cm by September. The leaf area at increasing height intervals is fairly homogenous up to the average height of the canopy and then decreases rapidly with height. A more complete description of the plant canopy is presented by Conant and Risser (1974).

Old (1969) presented profiles of the air temperature, light intensity, wind speed, and vapor pressure deficit (VPD) within the plant canopy of a tallgrass prairie in Illinois and soil temperature to the 200 cm depth. Diurnal temperature profiles within a tallgrass prairie in January and July 1967 (Figure 5.9) show that for July the maximum daytime air temperature occurred at the 40 to 60 cm level. This level corresponds to the plane of maximum leaf development with an average canopy height of 50 cm. Diurnal variation in temperature is greatest at the 50 cm level and decreases above and below this level. In January, minimum air temperature occurs at the 20 cm level and maximum air temperature at approximately 25 cm. During the winter the average plant canopy height is reduced to approximately half of the summer canopy height (Conant and Risser, 1974) as a result of the weathering of standing dead vegetation. In January isothermal conditions are observed between the 50 cm and 200 cm levels, and diurnal temperature variations are small at all levels compared to July. The diurnal heat pulse in the soil (zone in which soil temperature changes) goes down to -60 cm in July while the January heat pulse only reaches -25 cm, a result of the greater daytime heat input in summer. Old (1969) found a maximum soil surface temperature at the end of June and a maximum at 200 cm in early September. Minimum soil temperature at the surface occurred in January but did not occur until March at a depth of 200 cm. The lower diurnal variation of soil temperature compared to the air temperatures and the observed time lag in the occurrence of the maximum and minimum soil

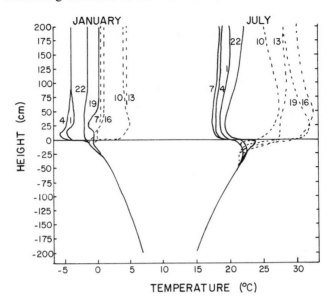

FIGURE 5.9 Diurnal temperature profiles in January and July 1967. Values for each height represent a weekly average for the hour specified. Solid lines represent night profiles, and dotted lines represent daylight profiles (Redrawn from Old, 1969; copyright 1969 by The Ecological Society of America).

temperature with increasing depth was a result of the high specific heat capacity of the soil.

Environmental profiles of air temperature, light intensity, wind speed, and VPD were taken at 0930 on 4 May 1968 by Old (1969) for three different pasture treatments as follows (Figure 5.10): (1) an ungrazed pasture that had been unburned for four years (Burn 4); (2) a pasture that had been burned recently (Burn 0); and (3) a pasture that had been clipped at ground level (clear-cut). The Burn 4 pasture had a maximum air temperature at 25 cm (average height of the plant canopy); the Burn 0 and clear-cut pastures had maximum air temperatures at soil surface. A comparison of soil surface temperatures of the Burn 4 treatment with the other two treatments showed that soil surface temperature is 5°C colder in the ungrazed and unburned pasture. Apparently, removal of the vegetation canopy by grazing or burning will cause an increase in soil temperature. Increase in soil temperature as a result of burning has been reported by Ehrenreich (1959) and Hensel (1923) while May and Risser (1973) have shown that the soil temperature at -10 cm is 2° to 3°C greater in grazed pastures than in an ungrazed pasture during the growing season. In Burn 4 light intensity was decreased by one-half of its maximum value at 35 cm and is less than 10 percent of the maximum value at heights less than 20 cm. Light intensity at the soil surface is reduced by 17 percent for the clear-cut pasture and less than 5 percent in burned pastures.

Wind speed decreased linearly from the 200 cm level to soil surface for the Burn 4 treatment, while wind speed was constant

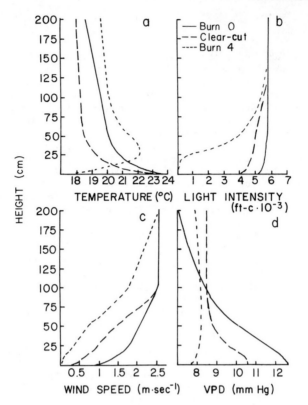

FIGURE 5.10 Environmental profiles taken on 4 May 1968 at 0930; (a) temperature profiles, (b) light intensity profiles, (c) wind speed profiles, and (d) VPD profiles (Redrawn from Old, 1969; copyright 1969 by The Ecological Society of America).

above the 100-cm level and decreased from 100 cm to ground surface in a linear manner for the clear-cut pasture and in a curvilinear manner in Burn 0. The wind speed and gradient of wind speed (change of wind speed with height) near the ground were much greater in clear-cut and burned pastures than in the unburned and ungrazed pasture.

For the Burn 4 treatment the VPD profiles showed minimum values at the soil surface and are fairly homogenous from the surface to 200 cm. For the clear-cut and Burn 0 treatments the maximum VPD was at the ground surface and decreased with increasing height. The VPD below the 100 cm level is much greater for the clear-cut and Burn 0 treatments than for the Burn 4 treatment.

Thus, Old's (1969) microclimatic data show that grazing and burning of the plant canopy in a Tallgrass Prairie cause the soil temperature, light intensity at ground surface, wind speed, wind speed gradient near the ground, and VPD below the 50 cm level to increase.

Osage Site Data

Temperature profiles from the Osage Site (Figure 5.11) are generally consistent with profiles presented by Old (1969) with the exception that Old's data indicate that average maximum air temperature at 30 cm should be greater than the maximum air temperature at 150 cm, while at Osage maximum air temperature was the same at both levels. Old observed the air temperature using thermocouple sensors placed in the plant canopy (the preferred technique), while a hydrothermograph enclosed in a shelter was used at the Osage Site. The discrepancy is probably related to the different techniques used to observe air temperature. The diurnal profile of relative humidity showed that the maximum, minimum, and average relative humidity was greater at 30 cm than at 150 cm; similarly Old showed that the VPD at 50 cm was 14 percent lower than the VPD at 200 cm.

Soil water data for three years (1970-72) are available for the 0-45 cm layer at the Osage Site for both a grazed and ungrazed pasture. The soil water content was approximately equal to field capacity (-0.3 bars) in the spring and decreased to a minimum value in August with the soil water tension less than -15 bars for the 0-45 cm soil layers. The major period of soil water recharge is during the fall and winter months. These data are consistent with the data presented by Anderson et al. (1970) that showed more water use than rainfall during early summer, thus indicating that the winter months were storage periods. These results differ from Old's data, in which soil water tension was never less than -6 bars at any time during the season. The lower soil water content at the Osage Site is consistent with the lower ratio of rainfall to potential evapotranspiration, indicating higher water stress.

Spring soil water content (0-45 cm layer) of the grazed site at Osage was 15 percent lower than the ungrazed site; at the end of the growing season the content of the grazed site was only 5 percent lower. The lower soil water content in the grazed site might have resulted from a larger amount of standing dead vegetation in the ungrazed site, which reduced the evaporation loss from the soil; or the warmer soil temperature in the grazed site (May and Risser, 1973) might have produced an earlier initiation of growth and transpired water loss. In an ungrazed pasture the soil temperature, wind speed, and gradient of wind speed near the ground surface are lower than in a grazed pasture. Lower values for these microclimatic variables reduce the evaporation water loss from the soil. Osage Site data showed that the growing season starts earlier in the grazed pasture and that larger amounts of live shoot biomass are in the grazed pasture during the early part of the growing season (April-June). Although the bulk density of the soil in the grazed treatment is slightly higher than the ungrazed treatment, both are essentially level and surface runoff is minimal.

Studies at Manhattan, Kansas indicate that grazing during the growing season tends to increase the soil water. Owensby et al. (1970) showed that clipping during the growing season tended to increase the soil water and postulated that the increase was a result of removal of plant material that would be transpiring over the growing season. The apparent discrepancy between the results of

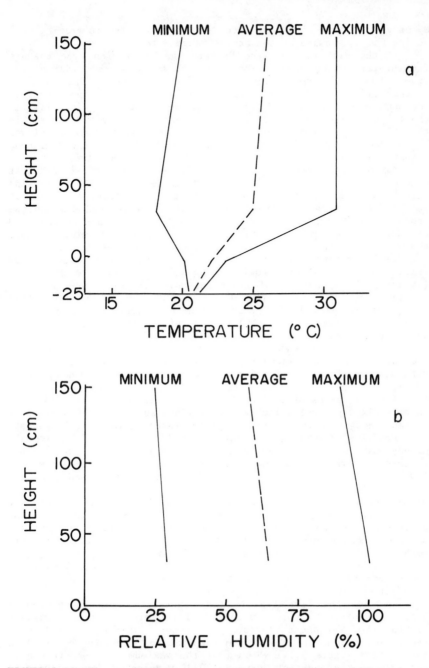

FIGURE 5.11   The average maximum, minimum, and average daily
temperature (a) and relative humidity (b) for the growing season
(May-September) in an ungrazed pasture at the Osage Site.   Data from
1971 were used for the temperature data; data from 1970 were used
for the humidity data.

Owensby et al. and the Osage Site data is attributed to grazing only during the nongrowing season at the Osage Site.

To the east of the True Prairie the vegetation type changes from prairie to deciduous forest, and on the western border of the True Prairie the vegetation changes to Mixed-grass Prairie. As Sprague (1959) noted, the microclimate depends to a large extent on the nature of the plant cover; thus we anticipate microclimatic changes on the eastern and western borders of the True Prairie. In the Mixed-grass Prairie of the Great Plains the average foliage height does not exceed 30 cm (Whitman, 1969), while in the deciduous forest ecosystem the height of the canopy can exceed 20 m.

Maximum air temperature in the mixed-grass prairie is greatest within the plant canopy at 3 cm height and decreases with increasing height up to 120 cm (Figure 5.12). Mean air temperature is lowest at 8 cm and increases upward in the plant canopy and down to the ground surface. Diurnal variation of soil temperature reaches -30 cm during the growing season, and maximum and minimum soil temperatures occur at ground surface level. Wind speed decreases in a curvilinear manner below 100 cm and increases slowly above 100 cm (Figure 5.13). Relative humidity decreases slowly from 15-150 cm (Figure 5.13). Profiles of light intensity in the canopy were not available for the Mixed-grass Prairie, but light intensity within the plant canopy will be greater than the light intensity in the True Prairie (see Figure 5.10) because of the lower leaf area index and lower canopy height in the Mixed-grass Prairie. The profile of light intensity in a Mixed-grass Prairie probably has a profile similar to the light intensity profile for a True Prairie treatment with the vegetation clipped to 5 cm and removed by raking (Old, 1969).

Typical curves of available soil water at the 0 to 120 cm depth under a Mixed-grass Prairie (Whitman, 1969) show that the soil water content is greatest in spring and decreases to minimum in August.

Profiles of the microclimatic variables in the Tallgrass and Mixed-grass Prairies have a similar shape. In a Mixed-grass Prairie, however, the maximum and minimum air temperatures occur at lower levels in the plant canopy (4 versus 50 cm), and the wind speed and light intensity are higher, while the decrease with height of the relative humidity is less. In the Mixed-grass Prairie the soil temperatures are greater during the summer months, and the difference between maximum air temperature in plant canopy and maximum air temperature above the plant canopy is much greater. Both the True Prairie and Mixed-grass Prairie have similar patterns in the soil water content through the growing season, although in the Mixed-grass Prairie the growing season rainfall is one-half the rainfall observed in the True Prairie region.

East of the True Prairie the vegetation type changes to a deciduous forest. Typical profiles of the microclimatic variables within the deciduous forest are shown in Figures 5.14-5.16 (Hutchison and Mott, 1976). The temperature regime in the leafless forest (see Figure 5.14) shows that the maximum air temperature is observed in the overstory canopy (20-25 m height). Temperatures in the overstory canopy are warmer during the day and most of the night and colder in the very early morning than in higher or lower strata. This same type of pattern occurs with the fully leafed canopy; however, the temperature gradient between the ground surface and

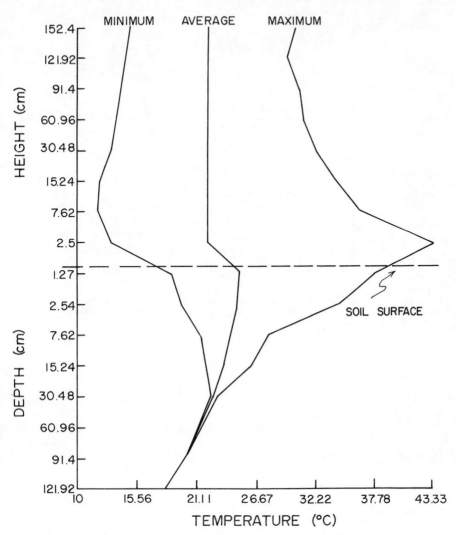

FIGURE 5.12 Average daily maximum, average, and minimum air and
soil temperature for a Mixed-grass Prairie (Redrawn from Whitman,
1969).

overstory canopy is much greater with the fully leafed canopy.
Relative humidity within a deciduous forest was not available, but
data from a dense fir (Abies) forest (Geiger, 1965) show that
relative humidity is greatest near the ground and decreases with
increasing height in the canopy. Light intensity within the forest
canopy (Figure 5.15) decreases very rapidly from the 30 m level to
the 15 m level, with the light intensity being reduced by 80 and 50
percent, respectively, for a fully leafed canopy and a leafless
canopy. Light intensity at ground surface is 6 and 30 percent,
respectively, for a fully leafed canopy and a leafless canopy.

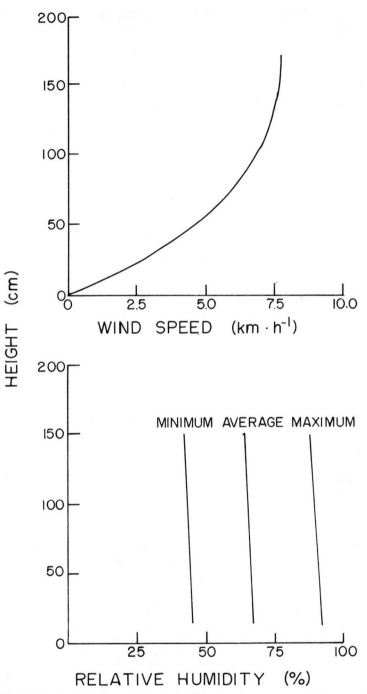

FIGURE 5.13 Gradients of wind speed and relative humidity with height in a Mixed-grass Prairie (Redrawn from Whitman, 1969).

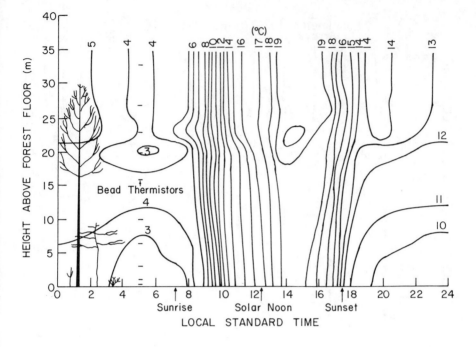

FIGURE 5.14 Average air temperature above and within a leafless deciduous forest (Redrawn from Hutchison and Mott, 1976).

Wind speed (Figure 5.16) has its maximum value above the canopy, decreases very rapidly from 33 to 25 m, and is fairly uniform below 25 m. The wind speed is greater on the leafless canopy than on the fully leafed canopy.

    A short transition zone exists between the grassland and forest in which the grassland microclimate changes to a forest microclimate within a few 100 m. The climate in this transition zone is primarily influenced by the edge of the forest, which forms a high step in topography, that can either increase or decrease sunshine, wind speed, and rainfall and alter long-wave radiation balance during the night. The change in rainfall and wind speed in the transition zone is determined by the orientation of the transition zone relative to wind direction, while shading of sunshine is dependent upon the orientation of stand edge and the time of day. A complete description of microclimate in this transition zone and the influence of orientation of the transition zone upon the microclimate is presented by Geiger (1965).

    A comparison of microclimatic variables east of the Mixed-grass Prairie to the forest shows that the profile of the microclimatic variables is similar within the plant canopies for all these systems; however, the average height of the plant canopies goes from 25 cm in the Mixed-grass Prairie to over 20 m in the deciduous forest. This increase in the average canopy height from Mixed-grass Prairie to forest results in a decrease in wind speed and light intensity near ground surface, a decrease in the soil temperature during the growing season, a decrease in the maximum air temperature

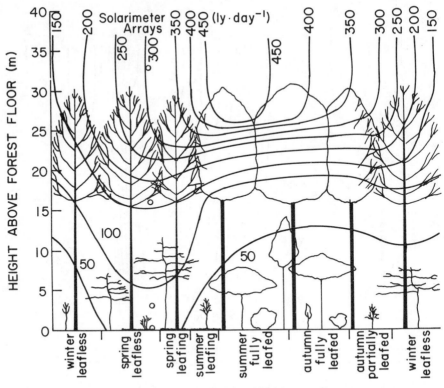

FIGURE 5.15 Annual course of average daily total solar radiation received within and above a deciduous forest (Redrawn from Hutchison and Mott, 1976).

within the plant canopy, and a decrease in the difference between the maximum air temperature in the plant canopy and the maximum air temperature above the plant canopy.

SOILS

Soils of the True Prairie are primarily Mollisols in the suborders Udolls and Ustolls (Soil Survey Staff, 1975). Some of the northern prairie soils belong to the suborders Aquolls and Borolls, and the tallgrasses along the eastern and southern borders of the True Prairie are found on soils in the order Alfisols.

In the northern portions of the True Prairie the soils are deposits of windblown loess or glacial drift. The loess has a mineralogical composition of quartz, feldspar, and lime and is usually silty in texture and unconsolidated in structure. In places the loess is as great as 20 m in depth. Glacial drift is composed of clay minerals, quartz, feldspar, and lime and is unconsolidated in structure and moderately fine in texture. The parent material

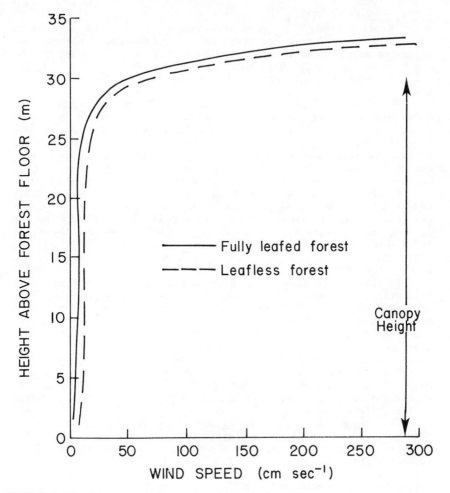

FIGURE 5.16 Average wind speeds above and within a deciduous forest (solar noon ± 3 h) (Redrawn from Hutchison and Mott, 1976).

for the southern sections of the True Prairie in Kansas and Oklahoma is a mixture of sandstone, limestone, and shale (Jenny, 1941).

Mollisols include soils previously called Brunizems, Chernozems, and Chestnut and Reddish Prairie soils. They are grassland soils developed under relatively high precipitation, 60 cm or more annually. Mollisols are found on somewhat poorly drained to well-drained sites under mid- and tallgrasses, chiefly bluestems. Leaching occurs to the extent that there is no horizon of calcium carbonate, but Mollisols do have a high degree of base saturation. Osage Site soils show a base saturation slightly over 90 percent in the A horizon and 80 percent in the B horizon. Especially in the upper portion Mollisols are rich in organic matter (~0.18 percent organic carbon and 5.5 percent organic matter from 0-15 cm) and slightly acid (pH 5.9 from 0-5 cm). These soils are well supplied with nutrients necessary for growth of grasses and forbs. Ungrazed

Osage Site soil has 0.17 percent nitrogen and 313 ppm phosphorus from 0-20 cm. Values for exchangeable potassium, calcium, and magnesium are 0.55, 8.65, and 1.56 meq/100 g, respectively, in the top 20 cm of soil. The high fertility and variable climate make the Mollisols the most productive in the world for grain and grass crops.

The profile of the Mollisols consists of a dark, grayish brown A horizon 20-40 cm thick, a grayish brown to brown or yellowish brown B horizon 30-60 cm thick, and a light yellowish brown to pale brown C horizon of parent material. Principal textures are loam and silty clay loam in the A horizon, clay loam to silty clay loam in the B horizon, and loam or silt loam in the C horizon. The B horizon is more finely textured than either the A or the C horizons and is distinctly blocky in structure. Soil from Tama County, Iowa, demonstrates these textural differences in the horizons of a typical Mollisol. Average values for sand in the A, B, and C horizons are 2, 1.6, and 3.4 percent, respectively. Silt values for the A, B, and C horizons are 66, 63, and 69 percent, respectively, and the percentages of clay in the A, B, and C horizon are 32, 35.5, and 27.6, respectively (Soil Survey Staff, 1975).

In the B horizon root penetration is more difficult and branching is usually less pronounced. Many rootlets are found appressed to the soil aggregates, between which water penetrates more readily. Air is abundant and nutrients are more concentrated, but relatively few roots penetrate the interior of the aggregates. Many grass roots occur in the C horizon and roots of forbs may extend several meters into the parent material.

Soil formation is a slow and complex process involving both physical and biological factors. The topography, climate, and vegetation are all important in reshaping the original parent materials into Mollisols.

The topography is characterized by a gently rolling plain. This relief provides the True Prairie with slow drainage, which aids in soil development and maturity but prevents serious erosion of the topsoil (Gibson and Batten, 1970).

Climate influences the development of the soil both directly by acting on the parent material and indirectly by influencing the vegetation. Physical weathering as a result of freezing and thawing, wind and water erosion, and root action is accompanied by chemical corrosion or decomposition. The chemical change includes specifically hydrolysis, hydration, carbonation, oxidation, and solution. These processes transform the parent material into the basic mineral components of the soil. In contrast to these destructive physical and chemical activities are the constructive processes that convert the modified parent materials into soil. The gradual introduction of living material, which may accompany or follow the accumulation of parent materials, provides the biological force that is responsible for the development of a mature soil.

Temperature and moisture contribute more to soil formation than do other climatic factors (Gibson and Batten, 1970). Although variations in temperature occur along the north-south axis of the True Prairie, the total precipitation and average air temperature are fairly uniform during the growing season.

The major process of soil formation of Mollisols is melanization--the darkening of soil by the addition of organic

matter. By this process the mollic epipedon is extended down into the profile (Buol et al., 1973). Factors involved in melanization include: (1) growth of the grass roots into the profile; (2) partial decay of organic materials; (3) eluviation and illuviation of organic and mineral colloids; (4) mixing of the soil and organic material by prairie fauna; and (5) the gradual accumulation of lignin compounds (Hole and Nielson, 1970).

The influence of temperature and moisture on decomposition of litter enhances the accumulation of the organic material necessary for melanization. The low moisture typical of this region in late summer slows the decomposition process while temperatures are still high and adds to the accumulation of humus (Bridges, 1970). As much as 33.6 to 123.5 g m^{-2} of raw organic material may be added to the soil in a year (Bridges, 1970), while Arnold and Riecken (1964) estimate that an A horizon can form in a prairie soil in 500 years but a textural B horizon requires 4000 years. In considering the reconstituted Curtis Prairie in the University of Wisconsin Arboretum, researchers showed that after cutting, the forest soil lost about 230 g m^{-2} of organic matter during the 90 years between 1850 and 1940. This amount of organic matter comprised about 25 percent of the organic matter contained in the undisturbed forest soil. However, during the 19 years from 1940 to 1959, the prairie was restored and regenerated about 140 g m^{-2} or about 60 percent of the organic matter that was lost (Nielson and Hole, 1963).

The native fertility of a soil is ultimately a function of the chemical composition of the parent materials. These materials may be residual or they may have been transported by glacial, alluvial, or aeolian forces. In the process of weathering, minerals from these materials are released and become a part of the available nutrient pool.

The biotic components of the ecosystem contribute significantly to the status of the available nutrient pools. Added organic material causes fundamental changes in the soil development. Throughout the soil, wastes are converted by the activities of microorganisms into dark-colored organic matter of high carbon content. Organic carbon values in the first 15 cm of soil vary from 2.4 percent in Tama County, Iowa, to 0.90 percent in the Oklahoma prairie (Soil Survey Staff, 1975). The total organic matter in the True Prairie soil constitutes between 1 and 6 percent of the dry weight of prairie soil. Roots are very important in supplying organic matter to the soil (Weaver, 1961; 1968) and in transporting nutrients from the lower layers to the upper layers.

Both microorganisms and higher plants absorb and immobilize nutrients from the equilibrium soil solution. The microflora mobilize minerals from the inorganic soil material, primarily as a result of acid formation. Mobilization of inorganic phosphorus by lactic acid-producing bacteria can increase the phosphorus uptake by vegetation (Witkamp, 1971).

Immobilization of nutrients in the litter, soil organic matter, and roots is a continuous process in maintaining the True Prairie. Biological retention is particularly effective for essential elements such as nitrogen and phosphorus. Since the catabolism processes release carbon dioxide, both nitrogen and phosphorus increase relative to carbon during decomposition. Leaching of soluble materials from the standing vegetation and litter is also

important in the supply of nutrient pools. Nitrogen, phosphorus, magnesium, and potassium are all added to the prairie soil by leaching from the plant canopy (Koelling and Kucera, 1965).

The soil fauna also play an important role in soil development in the True Prairie. The activity of earthworms is high in prairie soils (Buol et al., 1973). These animals aid in the mixing of organic matter and increase the percolation of water and movement of air through the soil. Baxter and Hole (1967) reported the common prairie ant (<u>Formica</u> <u>cinerea</u>) brought clay from the B horizon to the surface. Rodents may add 44,927 to 89,854 kg ha^{-1} subsoil to the surface annually (Buol et al., 1973). This combined activity of ants, worms, and rodents will turn over the top 0.6 m of the prairie every 100 years (Curtis, 1959).

6

Producers

Essentially all fixed carbon in the ecosystem originates from the biochemical steps of photosynthesis. However, the photosynthetic process for a particular species of plant may occur by one of several biochemical pathways, and the pathways vary among plant species depending on environmental factors or abiotic influences. Even though they may respond differently to external stimuli, all autotrophic plant species--that is, plants that make their own food--in an ecosystem fix carbon, which then becomes the energy source not only for the producers but also for all other trophic levels. Thus, photosynthesis is the basis for the energy-transfer function of ecosystems.

PHYSIOLOGICAL AND MORPHOLOGICAL PROCESSES

Morphology

Although all autotrophic plants function as producers in the grassland community, some morphological and physiological variations are important. A warm-season plant begins to develop in midspring and continues to develop until flowering occurs in late summer or early autumn. Most warm-season grasses are probably of tropical origin and include such True Prairie dominants as the bluestems (Andropogon spp.), Indiangrass (Sorghastrum nutans), and switchgrass (Panicum virgatum). Cool-season grass species mature rapidly, usually beginning vegetative growth early in the spring and completing it by early summer, although many exhibit a second period of growth in late fall, and other plants do not follow this seasonal cycle at all (Dickinson and Dodd, 1976). The cool-season plants are more predominant in northern than in other parts of the True Prairie but occur throughout, frequently increasing when the prairie is disturbed. Both bluegrasses (Poa spp.) and bromes (Bromus spp.) are cool-season grasses but have different photosynthetic pathways that affect the rate and time intervals of energy fixation.

Grasses that produce a thick mat (such as Bermuda grass [Cynodon dactylon]) are called sod grasses; grasses that grow in distinct clumps are called bunchgrasses. Sod grasses usually grow vegetatively with short rhizomes, stolons, or runners and are very effective in resisting water or wind erosion. Bunchgrasses reproduce vegetatively by tillers (shoots arising from the crown), which may number one hundred or more in one clump of some species (e.g., little bluestem [Schizachyrium scoparium]). Growth characteristics, however, may vary with climate. For example, big bluestem (Andropogon gerardi) forms a sod under moist conditions but frequently grows as a bunchgrass in the drier habitats.

Most perennial bunchgrasses of the True Prairie reproduce vegetatively. In a nineteen-month study in Kansas, McKendrick et al. (1975) found that most big bluestem tillers arose in late April and May, each tiller lived only one growing season, and reproductive tillers averaged two tiller offspring in the next growing season. About half of these second-year tillers were vegetatively reproductive but none produced seeds. Indiangrass tillers arose in late spring, early summer, and autumn, and the tillers were biennial.

Although the degree of reproduction by seeds is probably low (Canode and Law, 1975; Weaver, 1954), most species produce many seeds. Seed dormancy and delayed germination are general characteristics of native tallgrasses. Blake (1935) concluded that native prairie seeds germinated most abundantly in the spring with a second peak in the fall. Stratification of grass seeds through the winter generally resulted in higher germination; although after thorough ripening in dry storage, seedling production was found to be as high as the production that followed stratification.

Robocker et al. (1953) conducted greenhouse experiments over three winter seasons to determine the percent emergence for ten True Prairie species. Peak germination was highest the first year after seed harvest for little bluestem and sideoats grama (Bouteloua curtipendula); highest the second year for big bluestem, Canada wildrye (Elymus canadensis), Virginia wildrye (E. virginicus), and switchgrass; and highest the third year for green needlegrass (Stipa viridula) and sand dropseed (Sporobolus cryptandrus). Seeds of Indiangrass germinated equally well when planted the first, second, or third year after harvest. For all species, the average number of seedlings that emerged under greenhouse conditions was less than 50 percent of the sound seed planted.

Physiology

Although the term photosynthesis implies the necessity of light, several steps of the process can occur in the dark. The series of steps composing the photosynthetic process is usually separated into light and dark reactions, but these reactions are interdependent. The fixation of CO_2 in grasses and forbs occurs by one of two major routes--the C_4 (Hatch and Slack) or C_3 (Calvin-Benson) pathway. In the C_4-dicarboxylic acid pathway, which predominates in tropical and warm-season grasses, CO_2 is bound initially to phosphoenolpyruvate (PEP); but in the C_3 pathway, CO_2 is attached to ribulose 1, 5-diphosphate (RuDP). The formation of

two molecules of 3-carbon phosphoglyceric acid is common to both pathways. The C_4 cycle enhances the C_3 cycle by concentrating CO_2 for further processing by the C_3 pathway. In the initial C_4 reaction, which occurs in the mesophyll cells, phosphoenolpyruvic acid accepts CO_2 to form an intermediate compound, oxaloacetate, which is converted immediately to malate or aspartate. Both malate and aspartate form CO_2 in subsequent reactions in the thick-walled bundle-sheath cells of C_4 plants. This CO_2 is retained in the thick cell walls, which act as physical barriers to CO_2 diffusion (Burris and Black, 1976; Laetsch, 1974), thereby concentrating CO_2 for the initial reactions of the C_3 cycle. Therefore, C_4 plants have high rates of CO_2 fixation, especially at high light intensities and temperatures and at low ambient CO_2 concentrations. (Table 6.1 compares some physiological characteristics of C_3 and C_4 plants.) In C_4 plants, the detrimental effects of high plant-water stress often are reduced (Bjorkman, 1971; Black, 1971; Hsiao and Aceredo, 1974; Sesták et al., 1971), Na requirements are changed (Brownell and Crossland, 1972; Laetsch, 1974), and the rate of photosynthesis per nitrogen content is high (Wilson, 1975). Also, C_4 plants have relatively inferior nutritional quality because bundle-sheath cells are not readily broken down, at least in the gut of grasshoppers (Caswell and Reed, 1975; Caswell et al., 1973).

Some evidence indicates that C_4 plants evolved from C_3 plants. The C_4 species survive and may adapt better than many C_3 species under conditions of high water stress, high temperature, high oxygen concentration, low carbon dioxide concentrations, and high

TABLE 6.1   Pertinent physiological characteristics of C_3 and C_4 plants (from Kanai and Black, 1972).

Characteristic	C_3 plants	C_4 plants
Light saturation (lux)	10,764-43,056	None
Maximum rate of net photosynthesis (mg CO_2 dm^{-2} leaf area h^{-1})	15-35	40-80
Optimum temperature for net photosynthesis (°C)	15-25	30-45
CO_2 compensation concentration (ppm CO_2)	30-70	0-10
Transpiration rate (g H_2O g^{-1} dry wt)	490-950	250-350
Growth rate (g dry wt dm^{-2} leaf area day^{-1})	0.5-2.0	3.0-5.0

irradiances. These conditions probably were not typical during the evolution of higher plant life on this planet, so the C_3 plants must have been the first higher plants to evolve in the low-oxygen, high-CO_2 atmosphere of the earth at that time. The C_4 pathway probably evolved as an adaptive mechanism of plants that were originally native to moist tropical climates but then immigrated to temperate regions and the more temperate climates within tropical regions (Bjorkman, 1971; Downton, 1971).

Respiration rates in both C_3 and C_4 species are usually greater in light than in darkness. Photorespiration is not measurable under normal conditions in C_4 plants. The biochemical source of photorespiratory CO_2 is glycolate, which is probably synthesized from ribulose 1, 5-diphosphate (RuDP). Because light is necessary for regenerating RuDP in all plants, it controls the production of photorespiratory CO_2. The lack of photorespiration in C_4 plants may be caused partly by low mesophyll resistance to CO_2 diffusion (El-Sharkawy and Hesketh, 1965), which results in greater amounts of energy available for CO_2 fixation. Also, the C_4 plants readily reassimilate any photorespiratory CO_2. The lack of apparent photorespiration in C_4 plants might account for their greater net photosynthetic rates (Samish and Coller, 1968). With one exception (Syvertsen et al., 1976), whole communities of plants have not been characterized by carbon-reduction pathways.

Plant productivity is affected by (1) the soil, including soil water, nutrient availability, and CO_2 release during soil respiration; (2) the properties of leaves, such as size, shape and angle; stomatal numbers and behavior; response of mesophyll to irradiance; reflectance and transmission; effect of temperature on dark respiration and photorespiration; and physical resistances and carboxylation; (3) the architecture of the plant canopy, including the total leaf area, leaf distribution along the stem, and angle of the leaf orientation from the horizontal; (4) ambient climatic factors, such as air temperature, wind speed, CO_2 concentration, relative humidity, angle of the sun, and whether irradiation is diffuse or direct; and (5) the duration of photosynthesis, changes in photosynthesis with leaf size and age, efficiency of CO_2-assimilation rate, leaf-area enlargement, plant height, stand-species composition, photosynthesis by organs other than leaves, and efficiency of photosynthate transport to storage tissues (Zelitch, 1971).

Photosynthetically efficient plants have high net productivity rates, about 50 g dry wt m^{-2} leaf area day^{-1}, which is equivalent to a net CO_2 uptake of about 73.5 g CO_2 m^{-2} ground area day^{-1}. Most crop species, however, have maximum growth rates of 20 to 30 g m^{-2} leaf area day^{-1} (Zelitch, 1971).

Few data are available on the gas-exchange rates of True Prairie species. Risser and Johnson (1973) examined CO_2 exchange rates and CO_2 compensation points for seedlings of several True Prairie species and found that the rate of photosynthesis decreases with seedling age. Based on CO_2 compensation points, all the tested warm-season species (little bluestem, tall dropseed [Sporobolus asper], and western ragweed [Ambrosia psilostachya]) were C_4 species. Photosynthesis of prairie dropseed (Sporobolus heterolepis) increased with increasing light (Redmann, 1971); the net photosynthesis at 60,000 lux was about 13 mg CO_2 g^{-1} dry wt h^{-1}.

If the soil water was reduced to -15 bars, the photosynthetic rate dropped to zero.

PRIMARY PRODUCTION

The term _primary production_ refers to the amount of energy fixed by plants or autotrophs over some period of time (Woodwell and Whittaker, 1968). Energy input, or the total amount of energy fixed by a plant or plant community, is called gross primary production (GPP). However, plants use part of the GPP in respiration (R) for maintenance. The remainder appears, over a period of time, as a change in stored energy or new biomass, net primary production (NPP). Thus, for a single plant or plant community:

$$NPP = GPP - R.$$

If GPP equals R, there is no NPP; if GPP is greater than R, there is NPP; if R is greater than GPP, the plants lose energy. These relationships apply to whole ecosystems, except that the biomass and respiration of the heterotrophs must be included. Net ecosystem production (NEP) of all organisms is the gross ecosystem production (GEP) plus the increase in total biomass with time; the respiration term must include the total respiration (ER) of green plants as well as the heterotrophic animal and decomposer organisms. Thus, for an ecosystem,

$$NEP = GEP - ER.$$

At a single trophic level, the gross production may be lost by predation, death, and respiration.

Aboveground Primary Production

The amount of net energy captured by a grassland can be calculated indirectly by measuring the amount of biomass that accumulates during the growing season. The specific amount of NPP can be measured by harvesting biomass at intervals over the growing season, then variously manipulating data. One measure is the peak, or greatest value of live biomass obtained during the growing season. On the Osage Site, the aboveground plant biomass was harvested within 0.5 m^2 circular quadrats and sorted into major species and categories, which were then separated into live, current-year dead, and old dead. The material was oven-dried, weighed, corrected for mineral content, and expressed as g m^{-2}. The ungrazed peak live aboveground values for 1970, 1971, and 1972 were 270, 335, and 254 g m^{-2}, respectively.

The results from live herbage biomass sampling studies from the True Prairie are summarized in Table 6.2. Most sites are ungrazed, and the tabular value is simply the sample that has the highest live biomass. Only the Michigan sample is from an early successional-stage prairie, but some other samples are from previously grazed sites. The sites are in various topographic positions and are subtended by different soil types. Therefore,

TABLE 6.2 Summary of aboveground peak live standing crop values from several True Prairie sites.

State	Location	Production (g m^{-2})	Reference
Minnesota	Polk County	447	Smeins and Olsen (1970)
Michigan	Livingston County	238	Wiegert and Evans (1964)
North Dakota	Oakville Prairie	456	Wali et al. (1973)
North Dakota	Oakville Prairie	430	Hadley (1970)
South Dakota	Union County	500-566	Beebe and Hoffman (1968)
Nebraska	Lincoln County	344-432	Weaver and Tomanek (1951)
Iowa	Hayden Prairie	364	Ehrenreich (1959)
Iowa	Hayden Prairie	369	Ehrenreich and Aikman (1963)
Iowa	Hayden Prairie	390	Koelling and Kucera (1965)
Illinois	Trelease Prairie	302-489	Hadley and Kieckhefer (1963)
Illinois	Trelease Prairie	328	Old (1969)
Illinois	Mason County	280	Baier et al. (1972)
Missouri	Taberville Prairie	544	Koelling and Kucera (1965)
Missouri	Tucker Prairie	508	Koelling and Kucera (1965)
Missouri	Tucker Prairie	482-570	Kucera et al. (1967)
Kansas	Junction City	180	Hulbert (1969)
Kansas	Donaldson Pastures	325-473	Anderson et al. (1970)
Oklahoma	McClain County	316	Kelting (1954)
Oklahoma	Osage County	414	Hazel (1967)
Oklahoma	Wagoner County	348	Risser (1975)
Oklahoma	Pawnee County	402	Risser (1976)
Oklahoma	Oklahoma County	592	R. C. Anderson (personal communication, 1976)
Oklahoma	Osage Site	254-335	Risser and Kennedy (1975)

strict comparisons are impossible, but the average peak standing crop for the True Prairie sites is about 400 g m^{-2}. These annual biomass values do not show any apparent geographical gradient in any direction (Figure 6.1) and are remarkably uniform (CV = 25.6) considering the diverse conditions under which the samples were taken. Harlan (1960b) compared some earlier biomass values and also noted that, even though the comparable value was somewhat lower, consistency of average yield is one of the most characteristic features of native-range production. Variation in biomass from year to year can be significant, however, and in a 20-year study in Oklahoma (Harper, 1957), the range was from 127 to 522 g m^{-2}, or 35 to 167 percent of the mean.

All production estimates based on biomass are underestimates (Bradbury and Hofstra, 1976; Dalgarn and Wilson, 1975; Kelly et al., 1974; Kennedy, 1972; Singh et al., 1975). The peak standing-crop estimate is an underestimate because some species in the community may not have reached peak biomass by, and some may have senesced before, the time of the peak sample. Furthermore, all harvest methods ignore consumption by various herbivores.

To account for species that reach maximum biomass after the date of peak community standing crop, all the live and dead biomass of the current year can be harvested at the end of the growing season. These estimates for the Osage Site are 431, 383, and 164 for 1970-72, respectively. The accuracy of this method suffers because some biomass decomposes or changes into mulch or litter throughout the year and, consequently, is not measured in a single sample of the current year's standing dead taken at the end of the growing season. Furthermore, separation of the dead perennial-grass material of the current year from material remaining from past years is difficult.

One method that accounts for these phenological differences is the summing of peak live weights or of live weights plus the current year's dead weight of individual species. However, either version of this method still yields a conservative estimate because neither method accounts for material that is senesced or detached before the peak live weight of the individual species occurs. Species that respond rapidly and markedly to moisture events such as blue grama (Bouteloua gracilis), and cool-season grasses that grow both in spring and fall such as Japanese brome (Bromus japonicus) may have more than one peak in one growing season. This method confuses spatial and temporal variability. If samples are taken over a period of time and where spatial variation occurs in species composition, greater numbers of species will be encountered in subsequent samples, and the estimate will increase. Nevertheless, summing peak live weights is considered to be the appropriate method, and the ungrazed primary production values for the Osage Site are 331, 416, and 290 g m^{-2} for 1970, 1971, and 1972, respectively.

Aboveground NPP was estimated by summing the peak live plus recent dead values for all species combined and by functional groups (e.g., warm-season perennial). Peak live plus recent dead values were estimated by placing confidence intervals at ±1 SE of the biomass mean for each sample date and testing for overlap of the intervals. Observations having confidence intervals that overlapped the peak live value plus recent dead value were averaged to obtain

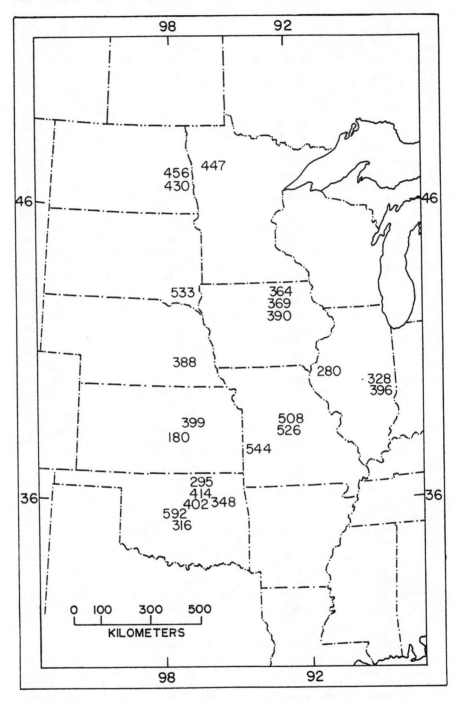

FIGURE 6.1 Aboveground peak live standing-crop values from several
True Prairie sites.

the estimate for primary production. This method differs from the original one of Singh et al. (1975) in that they used no statistical method for establishing the confidence-interval overlap.

The five computational schemes for estimating net annual aboveground production result in quite different estimates (Table 6.3). The summation of species peak live biomass values (Method 3) is about 20 percent higher than the peak live standing-crop value. Therefore, the average value of aboveground peak live biomass of about 400 g m^{-2} (Table 6.2) from ungrazed areas throughout the True Prairie should be about 480 g m^{-2}. If the summation of peak live plus recent dead values by functional groups (Method 5) is used, the estimate is about 40 percent greater than the simple peak live standing crop value. Accordingly, the estimate of aboveground NPP in the True Prairie is about 560 g m^{-2}. The last estimate, which includes the recent standing dead (e.g., current year), is probably the best estimate for ungrazed areas. However, in grazed areas, the current year's standing dead is minimal and its inclusion has much less influence on the aboveground NPP estimate than in the ungrazed areas (Table 6.3).

At the Osage Site, most of the dynamics of the plant biomass are attributable to the species or categories of little bluestem, big bluestem, Indiangrass, switchgrass, tall dropseed, miscellaneous grasses, and miscellaneous forbs. The total aboveground, live aboveground, recent dead, and old dead biomass of these seven categories were used in a multiple linear regression to predict the total aboveground biomass of the ungrazed and grazed treatments. Regressions were calculated for each year separately, then for all three years (1970-72) together; the lowest coefficient of

TABLE 6.3 Five computational schemes for estimating net annual aboveground production based on data from the Osage Site.

Methods	Ungrazed (g m^{-2})			Grazed (g m^{-2})		
	1970	1971	1972	1970	1971	1972
1. Peak live standing crop	270	335	254	286	314	311
2. End of season total biomass of current year	431	383	164	271	380	151
3. Summation of species peak live biomass	331	416	290	434	523	370
4. Peak live plus recent dead of all species combined	458	750	468	323	481	353
5. Summation of peak live plus recent dead by functional groups	490	800	412	424	461	367

determination (R^2) was 0.985 (level of significance was 0.001 percent or better in all cases). High coefficients of determination were evident because seven individual species or categories comprise the bulk of the total aboveground biomass and explain virtually all intraseasonal dynamics of the total biomass weights.

Multiple linear regressions were used to predict five measures of NPP from selected abiotic variables:

Independent Variables

Cumulative solar radiation (SOLR)
Cumulative potential evapotranspiration (PEV)
Cumulative precipitation (PPT)
Cumulative number of days above 4.4°C (TEMP)
Cumulative soil water (SWAT)

Dependent Variables

Cumulative standing crop of current live (SCCL)
Cumulative current live plus recent dead (CLRD)
Cumulative positive increments in total standing crop (INCC)
Cumulative positive increments in current live plus positive
    increments in recent dead (INCR)
All cumulative positive increments in current live plus
    positive increments in recent dead; values significant at
    the 90 percent confidence level (DVAL)

Eleven, ten, and seven observations came from the Osage Site for 1970, 1971, and 1972, respectively, from both the ungrazed and grazed treatments. Soil water was analyzed only for 1972, and all regressions were calculated separately for each year and for each treatment.

Only minor differences were evident between the two treatments and, with few exceptions, between years. Table 6.4 shows the coefficients of determination for the ungrazed treatment in 1970; 1971 and 1972 had even higher values. The best single predictors of each of the five dependent variables during the three years were:

1. Cumulative standing crop of current live was best predicted by cumulative potential evapotranspiration.
2. Cumulative current live plus recent dead was best predicted by cumulative numbers of days above 4.4°C.
3. Cumulative positive increments in total standing crop were best predicted by cumulative solar radiation.
4. Cumulative positive increments in current live and positive increments in recent dead were best predicted by cumulative solar radiation in 1970 and cumulative potential evapotranspiration in 1971 and 1972.
5. All cumulative positive increments in current live plus positive increments in recent dead (values significant at the 90 percent confidence level) were best predicted by cumulative solar radiation in 1970 and 1971 and a combination of cumulative potential evapotranspiration and cumulative number of days above 4.4°C in 1972.

TABLE 6.4   Coefficients of determination between five measurements
of net primary production (NPP) and selected abiotic
variables based on eleven observations in 1970 from the
ungrazed treatment on the Osage Site.

Independent variables, abiotic[a]	Dependent variables, measures of NPP[b]				
	SCCL	CLRD	INCC	INCR	DVAL
SOLR	.96	.94	.94	.68	.64
PEV	.97	.96	.90	.62	.56
PPT	.86	.88	.87	.63	.59
TEMP	.96	.96	.91	.61	.56
SOLR/PEV	.97	.96	.95	.81	.80
SOLR/PPT	.98	.94	.94	.69	.64
SOLR/TEMP	.96	.97	.95	.89	.88
PEV/PPT	.98	.96	.91	.63	.59
PEV/TEMP	.99	.96	.91	.64	.58
PPT/TEMP	.98	.97	.91	.63	.59
SOLR/PEV/PPT	.99	.96	.96	.84	.82
SOLR/PEV/TEMP	.99	.97	.95	.94	.92
SOLR/PPT/TEMP	.98	.97	.96	.90	.88
PEV/PPT/TEMP	.99	.98	.91	.76	.73
SOLR/PEV/PPT/TEMP	.99	.98	.96	.95	.94

[a]
SOLR = Cumulative solar radiation
PEV = Cumulative potential evapotranspiration
PPT = Cumulative precipitation
TEMP = Cumulative number of days above 4.4°C

[b]
SCCL = Cumulative standing crop of current live
CLRD = Cumulative current live plus recent dead
INCC = Cumulative positive increments in total standing crop
INCR = Cumulative positive increments in current live plus
       positive increments in recent dead
DVAL = All cumulative positive increments in current live plus
       positive increments in recent dead; values significant
       at the 90 percent confidence level

Even though these pairs represent the highest coefficients of determination, as shown in Table 6.4, other predictors had high values. Soil water was not as good a predictor as the other four abiotic variables.

For predicting cumulative standing crop of current live, cumulative current live plus recent dead, and cumulative positive increments in current live plus cumulative positive increments in recent dead, any single independent variable had a coefficient of determination greater than 0.9 in 1971 and 1972 but a low value in 1970. In 1970, combining solar radiation and temperature brought the prediction to 0.89. For predicting cumulative positive increments in current live plus positive increments in recent dead (significant at the 90 percent confidence level), any single 1972 independent variable had a high coefficient of determination. In 1970 and 1971, combining solar radiation and temperature brought the coefficient of determination values to near 0.9.

In most of these regressions, single independent variables were good predictors. Based on the data from 1972, however, cumulative soil water was not a good predictor unless combined with another variable. Predicting positive increments in current live plus recent dead estimates required two independent variables, and solar radiation plus temperature was the best of the pairs. The cumulative positive increases in total standing crop were a good example of a relationship that is best described using a nonlinear equation such as a logistic or Gompertz function. The equations that represent the best fit for each dependent variable are:

Equation	R^2
SCCL = -107.2517 + 21.3541 PEV	0.88
CLRD = -468.6940 + 0.7398 TEMP	0.96
INCC = -98.6210 + 0.0047 SOLR	0.94
INCR = -290.0191 + 0.0171 SOLR - 0.3775 TEMP	0.88
DVAL = -269.0079 + 0.0162 SOLR - 0.3627 TEMP	0.87

Belowground Primary Production

Studying belowground components--roots and rhizomes--has always been arduous, and the difficulty is increased because we cannot see the structure of the belowground system. As a result, the belowground system has been neglected even though, in the True Prairie, from one to four times as much plant material is belowground as aboveground, and root structural material provides most of the organic inputs in many biomes (Coleman, 1976). Furthermore, most water and nutrient uptake processes occur belowground and depend on the structure and function of the belowground systems.

John Weaver, the early grassland ecologist from Nebraska, did not overlook the importance of the belowground compartment. Indeed, Weaver and his students have contributed the most to our

understanding of the morphology of prairie root systems. Weaver
(1954) describes his efforts:

> There is no easy method of uncovering the roots, and it can
> be done successfully only at the expense of considerable
> time and energy and by exercising a great deal of patience.
> But once started, the work is not only interesting but even
> fascinating. A small hand pick with cutting edge, an ice
> pick, a tape measure and a meter stick, note book, and
> drawing paper are the only equipment needed.

The story goes that Weaver would allow his students to do the
initial separations, but in the final stages, when the greatest
concentration and painstaking patience were required, he worked
alone (in suit and necktie). During the first part of his work, he
used the trench, or direct, method--digging a trench adjacent to the
plant to be studied ("One 6 to 12 feet long, 3 feet wide, and 5 to 7
feet deep is most convenient"), then slowly extricating the root
system, sketching it as he went. Later, he used the monolith, or
indirect, method--extracting a block of soil with plants and roots
intact, taking the work to the laboratory, and washing the roots on
a specially designed screen. Figures 6.2 and 6.3 illustrate some
results of this work.

As summarized in Table 6.5, Weaver's descriptions of the root
systems of the major True Prairie species (Weaver and Darland, 1949)
show that native grasses routinely produce roots to a depth of
nearly 2 m and switchgrass to 3.7 m. Forb roots frequently extend
even deeper. Roots of leadplant (Amorpha canescens) and the tap
root of dotted gayfeather (Liatris punctata) extend to over 5 m.
Pavlychenko (1942) measured lengths of roots in the top 10 cm of a
0.5 m^2 section of prairie sod near Lincoln, Nebraska and found: big
bluestem, 21.5 km; little bluestem, 38.7 km; needlegrass, 18.3 km;
and Kentucky bluegrass, 176.7 km.

Some early studies of True Prairie plants in Illinois indicated
that the roots were not as deep as the roots Weaver recorded in
Nebraska (Sperry, 1935). At the western border of the True Prairie
in Nebraska, however, the same species had roots at even greater
depths than the roots listed in Table 6.5. With sufficient water
for growth, apparently the drier conditions are conducive to longer
root development (Weaver, 1954).

Using root excavations and observation windows in the
shortgrass prairie, Ares (1976) found that roots began to grow and
differentiate a short time before leaf growth was apparent. Soil
desiccation in mid-growing season resulted in the death and
subsequent decomposition of 30 to 60 percent of the newly formed
roots. We do not know whether midseason mortality is high in the
True Prairie.

Several recent studies show that clipping and grazing affect
root growth and the accompanying physiological processes. Clipping
aboveground parts of western wheatgrass (Agropyron smithii)
stimulated root exudation (Bokhari and Singh, 1974). Grazing of
blue grama by grasshoppers stimulated root respiration and enhanced
the excretion of organic acids (Dyer and Bokhari, 1976). The effect
of clipping on root growth and production is discussed in
Chapter 10.

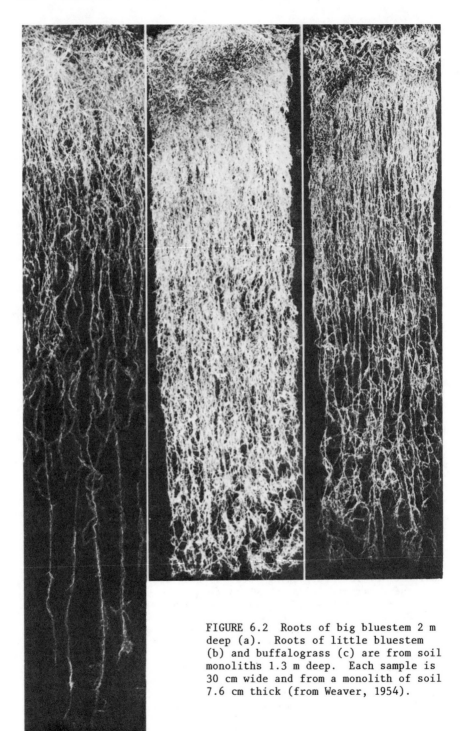

FIGURE 6.2 Roots of big bluestem 2 m
deep (a). Roots of little bluestem
(b) and buffalograss (c) are from soil
monoliths 1.3 m deep. Each sample is
30 cm wide and from a monolith of soil
7.6 cm thick (from Weaver, 1954).

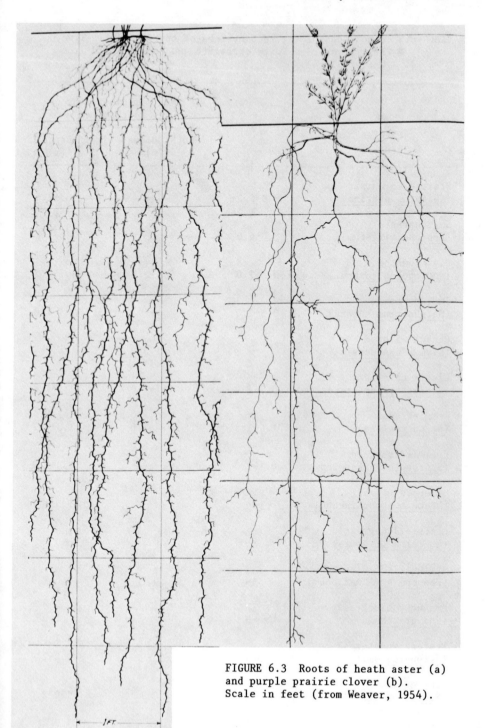

FIGURE 6.3 Roots of heath aster (a)
and purple prairie clover (b).
Scale in feet (from Weaver, 1954).

TABLE 6.5   Dimensions of the root systems of characteristic True
            Prairie species (data extracted and recalculated from
            Weaver, 1954).

Species	Root diameter (mm)	Maximum dimensions of root system	
		Width (diameter) (m)	Depth (m)
Prairie cordgrass (Spartina pectinata)	3.0-5.0	--	2.3-2.6
Switchgrass (Panicum virgatum)	3.0-4.0	--	2.6-3.7
Big bluestem (Andropogon gerardi)	0.5-3.0	0.7-1.0	1.7-2.3
Indiangrass (Sorghastrum nutans)	0.5-2.0	--	1.7-1.8
Canada wildrye (Elymus canadensis)	--	1.3	0.7
Little bluestem (Schizachyrium scoparium)	0.1-1.0	0.7-1.0	1.5-1.8
Needlegrass (Stipa spartea)	1.0-1.5	0.7-1.0	1.3-1.7
Prairie dropseed (Sporobolus heterolepis)	0.1-1.0	--	1.3-1.7
Sideoats grama (Bouteloua curtipendula	--	--	1.3-1.7
Prairie Junegrass (Koeleria cristata)	--	0.7	0.5
Scribner's panicum (Panicum scribnerianum)	--	0.7	0.8-1.2
Kentucky bluegrass (Poa pratensis)	--	--	1.0
Western wheatgrass (Agropyron smithii)	0.5-1.5	--	2.0-2.7
Blue grama (Bouteloua gracilis)	0.1-0.5	0.7	1.0-2.0

TABLE 6.5 Continued.

Species	Root diameter (mm)	Maximum dimensions of root system	
		Width (diameter) (m)	Depth (m)
Buffalograss (Buchloe dactyloides)	0.1-0.5	0.7	1.0-2.0
Stiff sunflower (Helianthus rigidus)	0.5-7.0	1.3	2.3-2.7
Heath aster (Aster ericoides)	1.5-2.0	1.0-1.3	2.3-2.7
Stiff goldenrod (Solidago rigida)	0.5-1.0	0.7-1.0	1.7-2.0
Missouri goldenrod (S. missouriensis)	0.5-3.0	1.0-1.3	1.8-2.7
Dotted gayfeather (Liatris punctata)	--	2.7-3.3	3.8-5.3
Compassplant (Silphium laciniatum)	--	2.0-2.7	3.0-4.7
Sunshine rose (Rosa suffulta)	--	3.3-4.0	5.0-7.0
Pale echinacea (Echinacea pallida)	1.3-2.5	--	2.0-2.7
Baldwin ironweed (Vernonia baldwini)	3.0-9.0	1.3	3.0-3.7
Purple prairieclover (Petalostemum purpureum)	--	1.0-1.3	1.8-2.2
Wild-alfalfa (Psoralea tenuiflora)	--	--	2.7-3.0
Groundplum milkvetch (Astragalus crassicarpus)	--	0.7	2.0-2.7
Plains wildindigo (Baptisia leucophaea)	--	--	2.3-2.7
Leadplant (Amorpha canescens)	8.0-14.0	2.7-3.3	5.5

In addition to these massive root systems, many grasses and sedges have rhizomes, or underground horizontal stems. Rhizomes anchor the plant, take up some water and nutrients, store food, and produce aerial shoots. The rhizomes of most prairie species are usually within the top 10 to 20 cm of the soil and may be 5 to 10 mm in diameter. They may be only 4 to 5 cm long (e.g., sideoats grama) or may grow more than 2 m in a growing season (e.g., western wheatgrass). Switchgrass roots have been known to grow at the rate of 2.0 cm per day, indicating that both roots and rhizomes grow rapidly. Rhizome growth usually begins at about the same time as shoot growth, and rhizomes frequently contain the food for the initial growth of the roots and shoots. In one experiment, big bluestem produced 100 m of rhizomes in a two-year period and western wheatgrass produced 200 m (Mueller, 1941). Tubers, corms, and bulbs, which are found on many True Prairie plants, store food for and protect the bud but do not supply the aggressive means for the vegetative propagation provided by rhizomes.

Horizontal distribution of roots is definitely species specific, and many of the forbs have much deeper roots than the grasses (Dalrymple and Dwyer, 1967). Of 43 species examined by Weaver in 1954, 14 percent were shallow-rooted--rarely producing roots at depths greater than 0.7 m. The second group, whose major rooting zone was 0.7 to 1.7 m, included 21 percent of the species. The last and largest group (65 percent) had roots that extended from 1.7 m to greater than 3 m. Except for roots of seedlings the deeper roots did not absorb nutrients and water from the top soil layers. Apparently, root morphology of True Prairie species is adapted for deep soils having uniform water supply and emphasizes deep penetration and widely spreading deep laterals (Weaver, 1954).

Measures of NPP also must include estimates of seasonal belowground biomass increments. Many difficulties complicate making these measurements, however, including the inability to collect and process large numbers of samples, the intraseasonal loss of musilage, and the decomposition of roots (Dahlman and Kucera, 1965; Lieth, 1968; Newbould, 1968; Samtsevich, 1965; Sims et al., 1978).

Most root biomass is in the upper soil horizons. Dahlman and Kucera (1965) found that over 80 percent of root mass was in the A horizon (top 25 cm) of the soil profile where the annual increment was 429 g m^{-2}. The A horizon contained about 80 percent of the total root system in a little bluestem grassland in Brazos County, Texas (Van Amburg and Dodd, 1970). In Illinois Old (1969) measured root biomass as 70, 15, 9, and 5 percent for soil depths of 0-25, 26-50, 51-75, and 76-100 cm, respectively.

Weaver and Zink (1946) grew native grasses in oil drums and measured the growth of the belowground plant parts. After three years, the total biomass in the top 10 and 30 cm of the soil was:

Species	Biomass (percent)	
	Top 10 cm	Top 30 cm
Big bluestem	43	78
Little bluestem	36	69
Blue grama	44	80

Weaver (1954) also measured the root depth distributions in native Nebraska grasslands (Table 6.6). In many grasses, particularly big bluestem, little bluestem, sideoats grama, and bluegrass (Poa sp.), the percentage of roots remains fairly constant even though the depth of the soil horizon varies. For example, 78 percent of big bluestem's belowground biomass was in an 18 cm A horizon and 90 percent was in a 50 cm A horizon. Similarly, in other cases 82, 92, and 97 percent of the belowground biomass of big bluestem grew in A horizons 18, 36, and 50 cm deep, respectively.

Total belowground biomass in the True Prairie is usually from two to four times greater than aboveground live biomass. Table 6.7 summarizes some belowground biomass data from several studies in the True Prairie. Similar belowground biomass values from Shortgrass and Mixed-grass Prairies range from 100 to 1,000 g m^{-2}, which includes both live and dead roots. From unpublished data using $^{14}CO_2$-labeled roots from the Osage Site, apparently 26 to 53 percent of the root biomass is functional.

Four methods to calculate net root production were used to process the belowground data from the Osage Site (Table 6.8). The first method is the difference between the maximum total and the preceding total minimum biomass (Dahlman and Kucera, 1965). This estimate, which corresponds to aboveground peak standing crop and assumes that the minimum is the starting point for the growing season, has the same limitations as the aboveground calculation compounded by the difficulty of obtaining an adequate sample. The second method is like the first except that the calculations are made individually for each depth increment. This method is intended to account for different growth rates in various layers of the root system. The third method is a summation of positive increases in total root biomass, and the fourth method is the same as the third except that the calculations are made within each depth intervals. The level of statistical significance for positive increase in the fourth method conforms with the projected initial sampling precision for the field samples--namely, the number of quadrats required to

TABLE 6.6   Belowground biomass from soil monoliths collected in Nebraska (Weaver, 1954).

Species	Depth (cm)						Total weight (g)
	0-15	15-30	30-60	60-90	90-120	120-150	
Big bluestem	68.5	13.9	10.5	4.4	1.7	1.0	49.3
Little bluestem	85.8	5.3	5.0	2.3	1.4	0.2	52.0
Sideoats grama	85.0	7.8	5.8	1.4	--	--	39.0
Blue grama	87.7	6.7	3.9	1.3	0.4	--	35.7
Western wheatgrass	50.9	14.1	13.5	12.1	9.4	--	19.3

TABLE 6.7   Peak values for belowground biomass at sites in the True
Prairie to a depth of 90 cm.

Site	Belowground biomass (g m^{-2})	Reference
Tucker Prairie, MO	1901	Dahlman and Kucera (1965)
Tucker Prairie, MO	1155	Kucera et al. (1967)
Livingston Co., MI	685	Wiegert and Evans (1964)
Bath Co., IL	1500	Baier et al. (1972)
Osage Site, OK	904-1618	Risser and Kennedy (1975)

TABLE 6.8   Four computational schemes for estimating net annual
root production based on data from the Osage Site
(calculations performed by Sims et al., 1978).

Methods	Ungrazed			Grazed		
	1970	1971	1972	1970	1971	1972
			(g m^{-2} yr^{-1})			
1. Maximum total biomass minus preceding total minimum biomass	169	0	159	84	0	366
2. Subtraction of the preceding minimum biomass from peak biomass by depth increments	215	0	250	244	57	593
3. Summation of positive increases in total biomass	215	20	244	209	112	366
4. Summation of statistically significant positive increases in total biomass	222	361	185	336	281	593

estimate the root biomass within 20 percent of the mean with 80
percent probability. No level of significance was applied in the
first three methods.
     The first two methods yielded a zero estimate for 1971. Had
any reasonable level of significance been applied in the third

method (the summation of positive increases in total root biomass), virtually all estimates of root production for both treatments in all three years would approach zero. After examining data from a number of grasslands, Sims et al. (1978) concluded that the fourth method seemed to give the most reliable results. However, determination of whether this method gives the best estimate is impossible because if root growth occurs in waves as carbohydrates are translocated through the root system, the use of positive charges will overestimate root production because increments may be counted more than once (Lauenroth, personal communication 1978).

Root biomass samples at the Osage Site were taken to a depth of 90 cm with a 5 cm-diameter hydraulic corer. The root material was washed, dried, weighed, corrected for mineral content, and referred to as total belowground biomass even though a few roots penetrated deeper than 90 cm. An attempt was made to predict the total belowground biomass using the amount of biomass in core samples of 0-5, 0-10, and/or 0-20 cm. The multiple linear regressions were calculated for each year separately (1970-72), then for all three years (Table 6.9). Not surprisingly, the 0-20 cm segment provided the most information and appeared in all the significant equations. Except for the ungrazed treatment in 1970, the total belowground biomass could be predicted with a 1 percent or less error by using only the top 20 cm.

TABLE 6.9   Multiple linear regression equations for total belowground biomass (TBB).

Year	Regression equation	R^2	α (percent)
	Ungrazed		
1970	No significant regression	--	--
1971	TBB = -23.37 + 1.33 (0 to 20 cm)	0.9064	0.1
1972	TBB = 25.85 - 0.59 (0 to 5 cm) + 1.54 (0 to 20 cm)	0.9742	0.1
All	TBB = -5.64 - 1.87 (0 to 5 cm) + 2.48 (0 to 20 cm)	0.7386	0.001
	Grazed		
1970	TBB = 1672.90 - 0.74 (0 to 5 cm) + 13.36 (0 to 10 cm) - 10.93 (0 to 20 cm)	0.7865	5.0
1971	TBB = -30.48 + 1.37 (0 to 20 cm)	0.7837	1.0
1972	TBB = 180.96 - 0.93 (0 to 10 cm) + 1.69 (0 to 20 cm)	0.9701	0.1
All	TBB = 102.88 + 1.19 (0 to 20 cm)	0.4710	0.1

Crown Primary Production

Crown primary production is difficult to define because the
amount of plant tissue that is neither aboveground shoot nor
belowground root or rhizome varies with the species. Annuals such
as Japanese brome have essentially no crown material, but the crown
of such perennial bunchgrasses as little bluestem can compose from
10 to 75 percent of the biomass. This wide range among different
plants and species is partly caused by the differences between
perennials and annuals, differences between young and old clumps,
and partly because collection of only crown material is difficult.
On the Osage Site, the crowns were collected after the aboveground
biomass had been harvested. In 1972, a small shovel was used to cut
the crown material from the roots, but in 1970 and 1971, the crowns
were separated from the root cores, which produced a much smaller
total sample. Crown and root material were treated alike in the
laboratory. In the samples taken over the three-year period, peak
crown biomass ranged from 56 to 555 g m^{-2}. Sims et al. (1978),
using the data from the Osage Site, calculated net crown production
by summing the statistically significant positive increase in crown
biomass within a growing season (Table 6.10).

Net Primary Production Estimate

A summary of NPP by absolute amounts, obtained by using the
best estimation techniques, for grazed and ungrazed treatments for
each year is shown in Figure 6.4. The average total production over
three years was 887 and 1,077 g m^{-2} in the ungrazed and grazed
treatments, respectively. Although other studies in the True
Prairie have demonstrated that grazing increased primary production
(e.g., Kelting, 1954), we should remember that the grazed treatment
at the Osage Site was grazed only during the nongrowing season.

TABLE 6.10   Net crown, root, and total belowground prairie
             production at the Osage Site, Oklahoma. Values are
             calculated by summing the positive increases in biomass.

Production	Treatment	1970	1971	1972
		(g m^{-2})		
Crown	Ungrazed	380	222	602
	Grazed	166	332	502
Root	Ungrazed	70	361	431
	Grazed	140	281	421
Total belowground	Ungrazed	407	185	592
	Grazed	309	593	983

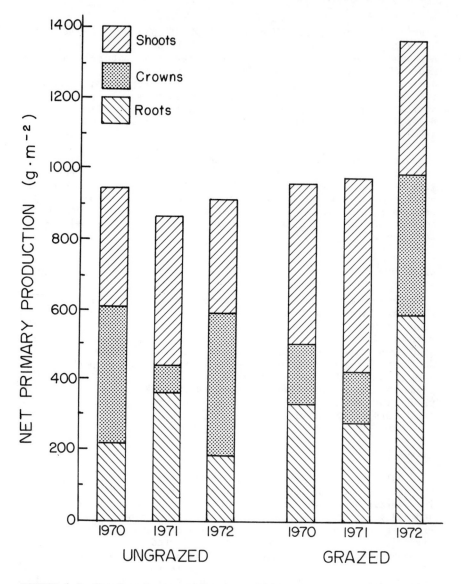

FIGURE 6.4 Total primary production of shoots, crowns, and roots under ungrazed and grazed treatments, 1970-72, at the Osage Site.

The proportion of production by shoots, crowns, and roots varied from year to year. Part of this variability is undoubtedly real, but, as noted, the method for separating roots from crowns changed after the first two years. The shoots averaged about 40 percent of the total primary production, 39 and 43 percent in the ungrazed and grazed treatments, respectively. The crowns, though quite variable (CV = 50 percent), averaged 32 percent of the total primary production in the ungrazed treatment and 21 percent in the

grazed treatment. The roots averaged 29 percent of the total primary production in the ungrazed and 37 percent in the grazed. Although some workers have reported increased root production at moderate grazing levels (Crider, 1955; Dahlman and Kucera, 1967; Jameson and Huss, 1959; Youngner, 1972), no other workers separated the crowns. Neiland and Curtis (1956) showed that the total available carbohydrate, much of it contained in the leaf and stem bases, decreased as the intensity of clipping increased in sideoats grama, Indiangrass, big bluestem, and little bluestem. Considering the trampling caused by grazing, it is not surprising that the proportion of crown production on grazed sites decreases. In general, about 40 percent of the total annual primary production goes into shoots and nearly 40 percent goes to roots. If crowns are considered as part of the total belowground production (which might be, inasmuch as photosynthesis in the crowns is minimal), about 60 percent of the total NPP at the Osage Site is in the belowground components.

Seeds

Seed biomass was collected during the 1972 growing season at the Osage Site on both grazed and ungrazed treatments. Early-season seed production was greater on the grazed treatment, where the maximum was 16.5 g m^{-2} on 6 June (Table 6.11). During the fall, seed production was again greater on the grazed treatment, but the dominant grasses contributed a relatively small proportion. If spring and fall are considered as discrete periods of seed production and production is estimated by peak values, annual seed production was 4.8 and 30.3 g m^{-2} on the ungrazed and grazed treatments, respectively.

Litter

Mulch or litter is the dead and detached plant biomass lying on the soil surface and may be several years old near the soil surface. The average amount of litter was 214 g m^{-2} on the ungrazed treatment and 353 g m^{-2} on the grazed treatment. Litter amounts in Shortgrass Prairies were about 100 g m^{-2} and ranged from 400 to 1,000 g m^{-2} on Mixed-grass Prairies. Samples were collected by hand during 1970, as were the first two samples in 1971; the remaining samples were collected with a vacuum. The samples were quite variable (CV = 50 percent). Although, in general, higher values might be expected on the ungrazed site, the fact that higher values were not found in the Osage Site data is probably attributable to the mowing history of the ungrazed treatment.

BIOMASS TURNOVER RATES

In steady-state conditions, turnover rates can be calculated by determining the amount of input or output during an interval of time as a fraction or percentage of the compartment. That is, if 100 g are in a compartment and 10 g go into the compartment each year,

TABLE 6.11   Seed biomass on the ungrazed and grazed treatments of
the Osage Site during the 1972 growing season.   Sample
varied but included a minimum of six, 0.5 m^2 quadrats
per size treatment.

| Sample date | Species or category | Biomass (g m^{-2}) | |
		Ungrazed treatment	Grazed treatment
15 May	Miscellaneous grasses	0.16	0.78
	Miscellaneous forbs	--	0.07
6 June	Miscellaneous grasses	1.44	16.49
9 July	Miscellaneous grasses	0.21	7.09
	Miscellaneous forbs	0.03	--
2 September	Miscellaneous grasses	0.25	9.83
	Miscellaneous forbs	0.51	1.06
	Big bluestem	0.26	0.08
	Little bluestem	1.80	1.09
	Indiangrass	0.21	0.43
	Switchgrass	0.30	1.36

then the turnover rate is 0.10, or 10 percent.   The turnover time is
the reciprocal of the rate; it would take 10 years to completely
turn over the compartment, so the turnover time is 10 years.

Turnover rates for the total belowground biomass in the True
Prairie were calculated by Dahlman and Kucera (1965) by dividing the
annual increment by the peak belowground biomass.   Sims et al.
(1978) performed the same calculations on root and crown data from
the Osage Site.   Averaged over the three years, root turnover in the
ungrazed and grazed treatments was 0.25 and 0.36 years,
respectively.   These rates compare favorably with the rates obtained
by Dahlman and Kucera (1965) on a prairie in Missouri.

Crown turnover rates for the three years averaged 0.83 percent
in the ungrazed treatment and 0.81 percent in the grazed treatment.
Clearly, crowns are active components of the intraseasonal energy
dynamics.

Shoots have a turnover rate greater than 1.0, even though all
shoots die at the end of the growing season and, by definition, have
a turnover rate of 1.0 or 100 percent.   But because some shoots grow
and die at different times during the growing season, the annual
increment is greater than the peak standing crop measured at any one
time.   The turnover rate is presumably close to 1.0 if the annual
increment is divided by the total of the current year dead material

at the end of the growing season; but because some material would have decomposed or fallen to the mulch, the turnover rate still would be greater than 1.0.

Aboveground turnover rates were calculated on the Osage Site data by dividing the aboveground NPP by the peak live biomass values; these values averaged 1.18 and 1.48 for the ungrazed and grazed treatments, respectively. Because one response to grazing is the introduction of cool-season grasses, which peak early in the growing season, the peak live biomass is greatly underestimated, which increases the turnover rate.

Litter turnover rates are difficult to calculate from the Osage Site data because neither actual input before the first growing season nor output after the last growing season was measured, the litter biomass on both treatments increased during the three field seasons, and the actual field collection method was changed after the first year. However, by using the litter dynamics information from the middle year (1971) and assuming that the throughput is the average of input and output of intervals over the season, we can calculate turnover rates: the litter turnover time would be once every 4.4 years on the ungrazed treatment and once every 3.7 years on the grazed treatment.

Hadley and Kieckhefer (1963) found that, after burning, close to two growing seasons were needed for the litter to accumulate to essentially the amount of the unburned control. They estimated the litter turnover to be two to three years on this site that is dominated by big bluestem and Indiangrass. Ehrenreich and Aikman (1963) estimated that turnover took five years in Iowa, and in Minnesota the litter recovery rate was two to three years (Tester and Marshall, 1961).

In summary, shoot material turns over once or slightly more each year, crown material turns over every 1.25 years, and root and litter material turns over about every 3 to 4 years. Grazing, however, increases these turnover rates by about 25 percent.

ENERGY EFFICIENCY OF PRIMARY PRODUCERS

Usable solar radiation can be calculated by multiplying the total growing season solar radiation energy value by 0.45, because only wavelengths from about 400 to 700 microns are used by higher plants in photosynthesis, and the energy within this interval contains about 45 percent of the energy of the measured total solar radiation. Efficiency can be calculated as the percentage of usable solar radiation captured in primary production (Botkin and Malone, 1968). Not all the light energy is intercepted by the aerial plant parts, even though in an ungrazed condition, the canopy coverage is essentially complete at midsummer (Conant and Risser, 1974). Some solar energy is reflected from the leaf surfaces, but the amount varies with the leaf color, orientation, amount of pubescence, and physiological condition.

Because previous analysis has permitted the calculation of aboveground, crown, and root production, calculation of the efficiencies of these processes during the growing season is possible (Table 6.12). The efficiency of the aboveground production was 0.26 percent on the ungrazed treatment and 0.33 percent on the

TABLE 6.12   Percent efficiency of energy capture in growing season:[a] total net primary production (TNP), aboveground net primary production (ANP), net crown production (CP), net root production (RP), and net root plus crown production (BNP) in a True Prairie grassland. The data for this table were collected from the Osage Site, and some of these calculations were compiled by Sims et al. (1978).

Production	Ungrazed				Grazed			
	1970	1971	1972	Mean	1970	1971	1972	Mean
TNP	0.77	0.70	0.70	0.72	0.75	0.77	1.09	0.87
ANP	0.25	0.32	0.21	0.26	0.33	0.40	0.27	0.33
CP	0.32	0.06	0.33	0.23	0.14	0.12	0.31	0.19
RP	0.20	0.32	0.16	0.23	0.30	0.25	0.51	0.35
BNP	0.52	0.38	0.49	0.46	0.44	0.37	0.82	0.51

[a] The calculation of efficiency is obviously a function of the growing season, because an increase in the designation of the growing season decreases efficiency of production. The dates used in these calculations are:

Year	Beginning date	Ending date	Length in days
1970	87 (March 28)	358 (December 24)	272
1971	69 (March 10)	338 (December 3)	270
1972	52 (February 22)	327 (November 23)	275

These seasons were determined by the number of consecutive days with a 15-day running mean air temperature greater than or equal to 4.4°C. Insignificant amounts of growth occur in February and December.

grazed. The increased efficiency under grazing was the result of the higher net aboveground production there. The efficiency of crown production was greater in the ungrazed treatment (0.23 percent) than in the grazed (0.19 percent). Efficiency of energy storage in roots or belowground parts was greater in the grazed treatment, perhaps partly because it was not grazed during the growing season. A striking feature in the ungrazed system is that although interseasonal differences exist, the aboveground, crown, and root efficiencies are approximately equal. The efficiency of this system is 0.72 percent, which is an underestimate of plant

energy capture because primary production estimates are underestimated.

Other literature values for efficiency of energy capture are somewhat higher (Botkin and Malone, 1968; Dalgarn and Wilson, 1975; Golley 1960; Kucera et al., 1967; Moir, 1969). Cooper (1970) reviewed studies of the potential production and energy conversion in temperate grasses and found that the efficiencies of temperate grasses range from 1.3 to 6.4 percent. However, some of these species are cultivated, and Cooper suggested that grasses growing without water and nutrient limitation should capture up to 3.0 percent of the incoming light energy.

The ratio of aboveground NPP and total NPP to growing season actual evapotranspiration is a relative measure of water-use efficiency. The water-use efficiency for both aboveground and total production at the Osage Site was higher under the grazed than the ungrazed regime (Table 6.13). This increase in efficiency is caused by the increase in aboveground production on the grazed treatment.

Rosenzweig (1968) showed a significant positive relationship between aboveground NPP of terrestrial communities and actual evapotranspiration. Black (1971) suggested that the warm-season grasses (C^4 plants) use water more efficiently than cool-season grasses. Thus, the change in species composition toward cool-season grasses (e.g., bromes and bluegrasses) under grazing should decrease the efficiency. However, interpretation of the ratio of total NPP to actual evapotranspiration is complicated because the relation between evaporation and transpiration is reciprocal. Both processes represent water losses to the system, but evaporation from the soil surface is high when the amount of aboveground biomass is low. On the other hand, large amounts of plant canopy increase transpiration and decrease rates of evaporation.

On the Osage Site and elsewhere in the True Prairie, the ratio of amount of biomass produced to amount of precipitation shows that

TABLE 6.13   Water-use efficiency of aboveground net primary production (ANP) and total net primary production (TNP) at the Osage Site. Values are the ratio of ANP and TNP to growing season actual evapotranspiration (calculations by Sims et al., 1978).

Year	Ungrazed		Grazed	
	ANP	TNP	ANP	TNP
1970	0.99	2.57	1.30	2.79
1971	0.69	1.41	0.87	1.57
1972	0.47	1.42	0.60	2.18
Mean	0.72	1.80	0.92	2.18

production of biomass is fairly insensitive to the amount of rainfall (Owensby et al., 1970) (Table 6.14). Rainfall of only 44 cm (about half the average annual rainfall) in 1970 produced essentially the same amount of biomass as in 1971 and 1972. The rainfall in 1970, even though considerably below normal, occurred in spring and late summer and undoubtedly contributed to the relatively high primary production. About 1 g of biomass is produced for every 1,000 g of precipitation.

COMPARTMENTAL ANALYSIS

The dynamics of the producer component of the True Prairie can be characterized by describing compartments and the transfer among compartments. The rate of net accumulation of biomass can be calculated by dividing the NPP by the number of days in the sampling season (233, 221, and 207 days in 1970, 1971, and 1972, respectively). The rate of total net accumulation ranged from 3.83 to 4.26 and from 4.02 to 6.54 g m^{-2} day^{-1} on the ungrazed and grazed treatments, respectively. Because these calculations are simply manipulations of total production based on the sampling season, the rate of accumulation of live and root biomass was greater on the grazed than on the ungrazed treatment at the Osage Site.

The amount of biomass transferred during the season to the standing-dead compartment can be evaluated by examining the amount of biomass that accumulates. Input to the standing-dead compartment from the live compartment for each growing season was considered to be equal to the aboveground NPP plus the initial live biomass minus the final live biomass. Actually, the initial live biomass should not be included if the estimate of aboveground NPP is based on species peak biomass values or positive increments in biomass. In these data, little error is introduced because initial samples in 1970 and 1971 were taken before any live material was present and only about 15 g m^{-2} were present in the first sample of 1972. The amount and rate of transfer to the standing dead were greater under the grazed treatment. Because no live biomass was found at the end of the growing season on the ungrazed treatment, the amount of accumulation in the standing-dead compartment equalled the season-long accumulation in live biomass.

TABLE 6.14   Ratio (× 1000) of total net production to total precipitation in grams during 1970-72 on the Osage Site in grams.

Year	Ungrazed	Grazed
1970	1.43	1.44
1971	0.89	1.00
1972	1.02	1.56

Input into the litter compartment from the standing-dead compartment for the grazing season was considered to be equal to the input into the standing-dead compartment plus the initial standing-dead biomass minus the final standing-dead biomass. The relatively small initial amount of litter on the ungrazed treatment and the trend of increasing accumulation were probably because the ungrazed treatment had been mowed for prairie hay perhaps as recently as three years before the initiation of the study.

Disappearance of litter from the litter compartment was assumed to be equal to the input into the litter compartment plus the initial litter biomass minus final litter biomass. Similarly, crown disappearance was calculated as the crown production plus the initial crown biomass minus the final crown biomass. Root disappearance was equal to the root production plus the initial biomass minus the final root biomass.

During the growing season, about 0.72 percent of the usable solar radiation was captured in the ungrazed treatment (3,843 kcal m^{-2}/532,000 kcal m^{-2}) and 0.88 percent was captured in the grazed treatment (4,721 kcal m^{-2}/532,000 kcal m^{-2}). Energy was captured at a rate of 17.4 and 21.5 kcal m^{-2} day^{-1} for the ungrazed and grazed treatments, respectively. About 64 percent of the captured energy was translocated to the belowground parts under both treatments; more energy goes to the roots in the grazed treatment (41 percent) than in the ungrazed treatment (31 percent). In the ungrazed treatment, the energy input to crown and roots is about equal, but the standing crop is much lower in the crowns, which indicates that the crown compartment is the more dynamic. Input to the standing-dead compartment was higher than output, indicating that an increase occurred in this compartment during the study period. Finally, the energy dissipated essentially equaled the amount captured in the ungrazed treatment (3,859 to 3,843 kcal m^{-2}), but only 77 percent of the energy captured was dissipated in the grazed treatment (3,654 to 4,721 kcal m^{-2}). These figures indicate that the ungrazed is a balanced system but that the grazed system accumulated energy in the producer compartments.

7

Consumers

INVERTEBRATE CONSUMERS

Insects and other invertebrates are found in all consumer trophic levels, both aboveground and belowground and may perform different consumer roles at various times during their life cycles (see Table 3.8). Since invertebrates have a wide diversity in feeding habits, assigning their trophic position to any taxonomic group is difficult. Herbivory, or the consumption of plant primary biomass, is a major activity of aboveground invertebrates, and, based on biomass estimates shown in Table 7.1, about 65 percent of the fauna at the Osage Site is herbivorous (Reed, 1972). This group can be further divided between herbivores that feed on plant tissue and herbivores that feed on plant fluid, but pollen and nectar feeders as well as seed parasites and predators belong in both of these categories. The remaining 35 percent of the invertebrates in the True Prairie are parasites, predators on other animals, and saprophages (terminology from Wiegert and Owen, 1971).

Wiegert and Evans (1967) suggest that only stable, natural ecosystems that contain populations of large grazing mammals can approach maximum utilization as determined for steady-state, managed rangeland. Ecosystems containing only consumer populations of invertebrates or small vertebrate grazers and granivores seldom achieve such utilization levels except under temporary departure from steady-state conditions such as rodent and insect plagues. Subsequent data will show that in the True Prairie insignificant amounts of herbivorous-invertebrate biomass are present compared with the large amount of primary productivity available.

Knutson and Campbell (1976) suggested that grasshoppers and many other insects have not multiplied in order to take advantage of the abundant food supply in the True Prairie. In fact, host plants are already well along in growth before grasshopper eggs hatch in the spring, and the predominant grasses, big bluestem (Andropogon gerardi) and little bluestem (Schizachyrium scoparium), are among the less preferred host plants. Evolution has resulted in a balance between grasshopper and host plant; the herbivore is rarely sufficiently abundant to depress significantly the long-term

TABLE 7.1 Estimate of herbivory Compared to Total Invertebrate (Total Insect and Arachnid) Biomass at the Osage Site.

	Ungrazed			Grazed		
Date	Herbivory (percent)	Herbivory biomass (mg m^{-2})	Total biomass (mg m^{-2})	Herbivory (percent)	Herbivory biomass (mg m^{-2})	Total biomass (mg m^{-2})
1971						
24 April	46	15	33	68	43	63
13 May	57	34	59	65	53	81
3 June	49	42	85	72	85	118
19 June	65	64	114	66	85	129
11 July	55	17	195	66	60	91
25 July	70	139	198	81	68	88
6 August	45	37	82	59	54	91
20 August	73	68	93	72	138	192
19 September	67	37	55	48	30	62
11 October	52	49	95	61	79	129
7 November	64	25	39	67	43	64
Average	59	49	83	65	61	94
1972						
23 April	41[a]	32	79	80	49	61
17 May	60	78	129	58	127	218
7 June	57	47	82	76	41	54
6 July	61	135	222	65	79	122
9 August	60	82	137	72	68	94
3 September	67	45	67	31[b]	15	49
29 September	51	88	174	56	150	270
15 November	56	35	63	67	35	52
Average	57	68	119	62	71	115

[a] A high proportion of predators (Araneida) and parasitoids (Tachinidae).

[b] A high proportion of predators (Carabidae).

productivity of its host, even though grasshoppers occur in highest densities in dry years when herbage production is low. Consumption of herbage by the grasshopper-spider segment of Tucker Prairie in central Missouri accounts for less than 2 percent of NPP (C. L. Kucera, personal communication, 1977). Besides consuming quantities of primary production, grasshoppers sever grass stems (up to four times as much as they consume), and sucking insects can affect water relationships of plants by removing large amounts of plant sap. Chew (1974) suggested that consumers are beneficial to ecosystems as regulators rather than as energy movers.

Adaptive strategies of grassland invertebrates are usually much the same on nongrassland habitats and on the True Prairie. Pollination, manure removal, litter decomposition, and avoidance of predation and parasitism are most important. Some general examples of these strategies will be discussed in the next sections.

Plant Tissue Feeders

Grasshoppers (Acrididae, Orthoptera) are some of the major plant tissue feeders and represent the greatest aboveground biomass of invertebrates in most range habitats. At a density of 10 m^{-2} and an estimated individual weight of 0.05 g, grasshoppers would more than double the highest previously recorded insect biomass at the Osage Site and would be five times as great as the average biomass found in our studies at the Osage Site. Densities as high as 10 m^{-2} are seldom observed but certainly are possible in the True Prairie, especially in burned and overgrazed areas (Smith, 1954).

Grasshopper density of 10 m^{-2} of two species in the Shortgrass Prairie can measurably control the pathway of primary production (Mitchell and Pfadt, 1974). Because grasshoppers tend to assimilate so small a proportion of what they chew, they are unlikely to have a major influence on ecosystem structure or a direct affect on either energy or nutrient storage. On the other hand, their indirect effect of clipping off plant parts ("litter making") is probably the primary role of grazing insects, at least in the Shortgrass Prairie.

Bailey and Riegert (1973), studying one species of grasshopper on the Canadian Grassland in Saskatchewan, found that the grasshoppers ingested 2 percent of the green shoot primary production and cut and dropped an additional 8 percent. Some 25 percent of the ingested food was metabolized, 51 percent of which was used for respiration. Exuviae (cast exoskeletons of immatures) made up 6 to 11 percent of secondary production. Some 97 percent of the energy of the green vegetation removed by grasshoppers was returned to the decomposer portion of the ecosystem, and the remaining 3 percent was removed or lost from the system (Figure 7.1).

According to Knutson and Campbell (1976), grasshoppers have done little noticeable damage in the True Prairie near Manhattan, Kansas during the past sixteen years, although earlier observers noted that row crops, but not the prairie, suffered heavy damage in the 1930s.

Beetles. Three groups of beetles (Coleoptera) are important plant tissue feeders in the True Prairie. Coleoptera, primarily weevils (Curculionidae) and leaf-feeding beetles (Chrysomelidae),

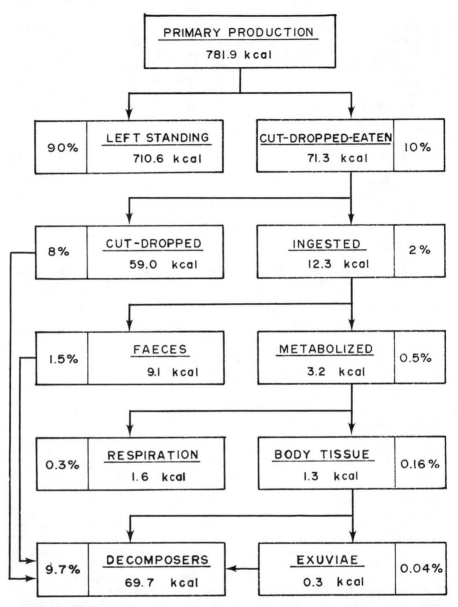

FIGURE 7.1 Annual energetics of grasshoppers in a Canadian grassland (redrawn from Bailey and Riegert, 1973). Reproduced by permission of the National Research Council of Canada from the Canadian Journal of Zoology, Vol. 51, pp. 91-100, 1973.

were second in aboveground biomass at the Osage Site in 1972. Weevil larvae attack plant tissue in nearly every way: root feeding, gall making, leaf rolling, or leaf mining. Adults feed on green tissue (not always the same plants attacked by their larvae),

pollen, or developing seeds. Chrysomelids are phytophagous as larvae and adults; they feed on roots, stems, and leaves of herbaceous plants, primarily forbs (Bertwell, 1972).

Another important group, grass-feeding scarab beetles (Scarabaeidae), removes dung, although some feed on grass roots as immatures (white grubs). Schumacher (1959) reported that these larvae destroyed 40 percent of the grasses in areas of four Kansas counties. Because damage to roots usually occurs during the summer and fall, it usually is not detected until the following spring when the affected grass fails to grow.

Caterpillars. Caperpillars (Lepidoptera larvae), commonly collected at the Osage Site, accounted for less than 3 percent of the biomass in 1972 but do feed voraciously on plant tissue. Every year, the range caterpillar significantly damages New Mexico and Texas rangelands; armyworms and other groups often become local problems in the True Prairie. Most caterpillars feed aboveground, but some (cutworms) are subterranean feeders.

Ants. The major herbivorous effect of ants (Formicidae), another dominant group, is probably seed removal rather than grazing (Reed, 1972). They undoubtedly harvest some plant parts both aboveground and belowground and often accompany populations of plant sap feeders (e.g., aphids and scale insects), sometimes actually transporting the sap feeders to and from host plants and ant nests.

Plant tissue feeders can increase productivity by increasing the production:respiration ratio of plants (Chew, 1974). The "mower" effect, or midseason cutting of such grasses as blue grama (Bouteloua gracilis), may delay senescence and thus increase yield. Likewise, rhizome growth of species of bluestem increased when senescence was delayed by removal of terminal buds. Also, invertebrate consumers may prolong plant growth and increase productivity by delaying plant senescence, which affects the quality of forage because the protein concentration of leaves decreases with maturity. Like cutting, feeding by consumers may increase total protein yield, but the consequences to plants are uncertain. Because nitrogen reserves are translocated to the aboveground plant parts when regrowth occurs, a reduction in nitrogen available for the next season's regrowth is likely (C. E. Owensby, personal communication, 1977).

Differential effects of feeding insects can occur. Beetle larvae may feed on older leaves but kill more tissue than they ingest by rasping leaf surfaces and severing veins. Because moth larvae (caterpillars) feed on leaves of all ages, eating them cleanly and leaving veins intact, they have less effect on leaf area.

Plant Sap Feeders

Hemiptera (true bugs) and Homoptera (aphids, leafhoppers, and scale insects) accounted for about 17 percent of the total insect biomass and 14 percent of the numbers at the Osage Site in 1972. The Hemiptera families Scutelleridae and Lygaeidae were highest in biomass at the Osage Site in 1972, and both are plant sap feeders. Of the Homoptera, leafhoppers (Cicadellidae) are usually dominant in biomass, and scale insects (Coccoidea) usually occur in greatest

numbers.   Aphids (Aphididae) ranked fifteenth in number of insects collected at the Osage Site in 1972.

The specific impact of plant sap feeders has not been measured on the True Prairie, but their large numbers and high metabolic rate make them important consumers (Menhenick, 1967). For example, in a South Carolina lespedeza field sucking insects accounted for 25 percent of all herbivore assimilation. Removing plant juices may affect the plant more seriously than removing a corresponding percentage of foliage. Wiegert (1964) studied spittlebugs in alfalfa and in old fields in Michigan and found that the removal of amino acids from the xylem of the host plant caused a decrease in the potential photosynthetic fixation of energy about five times the total energy ingested by the insects. If this is true for all Hemiptera-Homoptera (and some subsequent data indicate that probability), the depressing effect of this group might be greater than previously estimated. Many of these insects can ingest large quantities of liquid, concentrate it in a "filter chamber," and evacuate the excess as honeydew.

Scale Insects. Scale insects, which can remove large amounts of plant sap, were low in biomass but ranked fifth in total numbers at the Osage Site in 1972. Schuster (1967) estimated that Rhodesgrass scale insects reduced primary production by 30 percent in some south Texas ranges. In greenhouse tests, the scale insects killed up to 85 percent of some grasses and reduced yield by up to 88 percent. When an experimental biological control program of field populations of scale insects was initiated, primary productivity was increased by about 30 percent in some areas. No quantitative data on scale insect populations are available from this study to compare with those on populations found in the True Prairie, nor is Rhodesgrass scale recorded for the Osage Site.

Thrips. Thrips (Thysanoptera) and other small sap feeders are present but, because of their small size, are often overlooked when collections of prairie insects are made by sweep nets. However, thrips ranked second in insect numbers (third for all invertebrates) at the Osage Site in 1972. Many are herbivorous, since they rasp leaf tissue and suck juices. Thrips greatly damage seed production in New Mexico prairies (Watts, 1963), and commercial seed producers in all parts of the country often apply control measures.

Mites. Mites (Acarina) ranked first in numbers of small noninsect invertebrates at Osage in 1972 and are present in virtually every ecosystem. They may be predaceous or herbivorous, but no information is available on this ratio in True Prairie populations. Mites are categorized with sap feeders primarily because their feeding habits are poorly understood.

Nematodes. The impact of nematodes (Phylum Aschelminthes) is as enigmatic in the True Prairie as in most other ecosystems. Nematodes are commonly classified as agricultural pests, primarily because they damage plant roots, but others are predaceous or saprophytic. Population estimates for the True Prairie are based on a small number of samples but indicate that high numbers can occur (J. D. Smolik, personal communication, 1973). Andrews et al. (1974) cite Smolik's studies from a mixed-grass prairie where application of a nematicide was followed by 30 to 60 percent increased production in aboveground plants, suggesting that nematodes are the

major competitors with man for the available resources of such ecosystems.

Like leaf-feeding invertebrates, sap feeders show little evidence of an adaptive strategy that is limited to the True Prairie. For example, introducing anticoagulant and/or toxic material into the saliva of consumers occurs in all ecosystems. Likewise, several leafhopper genera that are found predominately in the Shortgrass Prairie have developed brachypterous (short-winged) forms, although macropterous (long-winged) forms are also still present at times. These genera are grass feeders, and the shorter wing length apparently represents an adaptation for a sedentary existence in a windy habitat. According to Kramer (1967), long-winged forms represent a migratory phase in the life cycle. Short-winged forms are not as prevalent in the True Prairie as in Mixed-grass and Shortgrass Prairies. Little published data are available on the evolution of these genera.

Pollen, Nectar, and Seed Feeders

Pollination by animals involves a coevolutionary process that has been proceeding perhaps for 225 million years, although the highly evolved pollinators such as butterflies, moths, and bees first appeared in the Oligocene (28 million years ago) and after the angiosperms (Price, 1975).

On the prairie, many insect species pollinate while seeking pollen, sucking nectar, or hunting other insects on prairie flowers (Costello, 1969). Insect pollinators include beetles, butterflies, moths, bugs, flies, and leafhoppers. Soldier beetles (Cantharidae) commonly feed on and pollinate goldenrod flowers, but their larvae live in the soil and are carnivorous. Moths and butterflies pollinate many flowers whose plants may be food for their caterpillars. Many adult flowerflies (Syrphidae) benefit plants as pollinators, and the flowerfly larvae can benefit plants further by preying on aphids, which can be important plant pests.

Bees are probably the most important pollinators of prairie plants (Costello, 1969). Bees, flies, and wasps pollinate many rose-family plants, many forbs with conspicuous and strongly scented flowers, and most legumes. No honeybees were in existence on the original prairie, because they were introduced from the Old World after the discovery of America. The native bees, however, were numerous.

Chew (1974) reviewed effects of phytophagy on seed production. Often plants can compensate for the removal, if not too severe, of leaf and/or flower tissue. When consumers eat mature seeds, however, parent plants cannot sense and compensate for the loss, and the consequences can be worked out only through natural selection. Coevolution of seed production and seed predation strategies can affect major parts of an ecosystem through high loss of seeds to predators. However, this loss could be a worthwhile tax in exchange for the dispersal of seeds to germination sites (Chew, 1974).

Seed production is not an important method of reproduction for some of the major native grasses in the True Prairie (Weaver, 1954). Less than 0.4 percent of the big bluestem produce flowers in any given year in the Donaldson Research Pastures (C. E. Owensby,

personal communication, 1977). In 1976, none of the big bluestem at Tucker Prairie in central Missouri produced seed (C. L. Kucera, personal communication, 1977). Other grasses, such as Japanese brome (Bromus japonicus) and annual bluegrass (Poa annuus), do produce seeds, as do the forbs.

The invertebrates so far discussed are primary consumers (herbivores), which might affect vegetation diversity (Chew, 1974) as well as primary production. Grass-dominated communities may endure a heavy impact by consumers; yet most dominants are palatable species. The disturbance of grazing animals makes the environment heterogeneous by encouraging a more complex flora (Harper, 1969). This floral diversity could be critical in allowing a grassland to maintain functional integrity during prolonged drought.

Parasitoids and Predators

One time-worn but correct precept is that if a pair of houseflies mated today and all their descendants lived and reproduced normally, their offspring would cover the earth several feet deep in four months. Several naturally occurring abiotic and biotic phenomena prevent such population accumulations. Parasitoids and predators are significant secondary consumers because they can be responsible for mortality in any given animal population in any stage of the life cycle, from egg to immature to adult.

A predator is an organism that kills and consumes many animals during its life-span. A parasitoid requires only one animal during its life-span (Price, 1975). The egg of a parasitoid is laid on or in its host; an immature hatches and develops on or in the host (often consuming it completely) and becomes a free-living adult. Because a female adult may deposit eggs on or in many hosts, females can be responsible for parasitizing many hosts. Both groups emerged before the Mesozoic (over 225 million years ago) and have coevolved with their hosts since that time. The importance of predation as an exploitation strategy has been divided into the following four categories (Price, 1975):

1. Predators play a prominent role in the energy flow through the community.
2. Predators and parasitoids have been singled out repeatedly as regulators of the populations of their hosts. This fact is important in biological control strategy and is one of the most visible aspects of mortality.
3. Predators are important in maintaining fitness of the prey population; many vulnerable individuals of the population are eaten by predators.
4. Predators act as selective agents in the evolution of their prey because any major mortality factor is likely to change a population permanently. At some stage in its development, every invertebrate species has been exposed to either heavy or prolonged predation. Perhaps predation pressure has produced one of the most visible evolutionary forces in the animal kingdom, and populations of invertebrates in the True Prairie are certainly regulated at least partly by predation.

Dominant parasitoids and predators found at the Osage Site include representatives of most orders of insects and of most other groups of invertebrates. Spiders (Araneida) were third in aboveground biomass and sixth in numbers at Osage in 1972. This group is entirely predaceous and generally considered to be highly beneficial. Major insect predators include vespid wasps (Hymenoptera) and carabid beetles (Coleoptera). Mites (Acarina), thrips (Thysanoptera), nematodes (Phylum Aschelminthes), and many other invertebrate groups contain large numbers of predaceous species. Major groups at Osage containing parasitoids include tachinid flies (Diptera) and scelionid wasps (Hymenoptera). Members of many of the parasitic groups are very small and probably were not recovered from the field samples.

Defensive strategies by invertebrates against predators are evident in the True Prairie. For example, the treehopper (Membracidae) blends perfectly with its background, the leafhopper automatically moves to the other side of a stem or leaf at the slightest disturbance, and the grasshopper may change color seasonally to match the vegetation. Mimicry, protective coloration, and chemical defenses are only three of the strategies in the invertebrate repertoire.

Saprophages and Fungivores

One major finding during the past decade has been the importance of insects and other invertebrates as saprophages (feeders on dead organic matter and/or fungi). Soil animals increase the speed of litter breakdown by disintegrating tissue, increasing surface area, and changing the physical and chemical nature of litter and mixing it with inorganic matter. All these activities facilitate metabolism by microorganisms. Reportedly, oak leaves that are mechanically ground up to the size of insect feces decompose as fast as feces and seven times faster than whole leaves (Chew, 1974).

Dung beetles are reportedly almost as effective in enabling plants to benefit from animal dung as dung that is mechanically mixed with the soil; their rolling, burying, digging, and sanitizing are important ecological functions on the prairie. Bacteria quickly act on and transform the manure carried underground into substances useful to plants. Moreover, removal of manure from the surface reduces the habitat of flies, which pester modern livestock as they once pestered bison.

Ants and other invertebrates move large quantities of soil. Some ants (Anta spp.) may reduce GPP by reducing leaf area, but the reduction is compensated for somewhat by minerals returned to the surface of the soil (Chew, 1974). Termites (Isoptera) are not major inhabitants of the True Prairie but are present wherever woody material is available for food. Their role in the True Prairie, assisted by flagellated Protozoa in their digestive systems, centers on breaking down cellulose in woody material to reusable components. When numerous wood marking stakes at our True Prairie research plots at Donaldson Pastures were removed in 1972 after being in the soil for two years, more than half were infested with termites.

The litter subsystem benefits from feces deposited by aboveground invertebrate herbivores. This fall of frass allows recycling of nutrients within the same growing season. In nature, the rate and pathways of recyling can be altered greatly during insect outbreaks (Chew, 1974). Chewing herbivores harvest and quickly return much of the primary spring growth and any subsequent regrowth to the soil in small particles, and sucking herbivores return a large proportion of their diets immediately to the soil as urine and feces (honeydew).

Some of the most important insects on the prairie are the scavengers found on carcasses of other animals. The effectiveness of insects in reducing dead bodies has been shown in the eastern United States. Flies arrive within minutes to feed and deposit eggs; ants begin feeding on the first day. Scarab beetles may arrive the second day, then fruit flies, coreid bugs, soldier flies, hister beetles, carrion beetles, and skin beetles. An animal carcass can be reduced to dry skin, cartilage, and bones in slightly more than a week. Insects and other small animals continue to feed. Centipedes, millipedes, snails, and roaches later appear and feed on anything remaining. These kinds of invertebrates reduced the bodies of countless bison and other prairie animals that died of starvation or natural disaster.

The presence of fungivores in soil and litter may prevent fungal overgrowth and senescence, thereby promoting vigorous fungal growth (Chew, 1974). By inoculating soil with spores, small arthropods such as mites and collembolans also can be disseminating agents for fungi.

Major groups of saprophages found in the True Prairie include some mites and ants, isopods, and most of the Collembola. Data presented in this volume may show a smaller ratio of saprophages to herbivores than would occur if samples had been taken regularly during the nongrowing season. Most of our samples were taken when herbivores, parasitoids, and predators were most active, but many soil saprophages remain active during winter months.

Adaptations by saprophages are much the same in the True Prairie as in other ecosystems. Evidence exists of vertical movement of some groups in response to soil water, the mouthparts of many soil invertebrates are modified and protectively enclosed in pouches, and in many groups external sensory structures such as antennae and eyes are often reduced or lost as an adaptation to their soil habitat.

Such are the major roles of the invertebrates, one of the most spectacular and dominant groups in any ecosystem, as consumers in the True Prairie. The reader is cautioned again against accepting generalizations about invertebrate impact. Very few groups of invertebrates can be placed into any single functional category. Diversity in trophic levels includes differences in the habits of matures and immatures, differences in habits between sexes in the adults of many species, seasonal differences, and feeding habits.

Site Specific Quantitative Patterns

Aboveground Invertebrates

From 1970 through 1972, invertebrate numbers and biomass at the Osage Site were quantitatively measured. For these 3 years, 13 orders and 92 families of insects plus mites and spiders were identified; specimens of 16 orders and 131 families plus mites and spiders were gathered by all collection methods. Numbers and biomass were recorded for insects, mites, and spiders (i.e., all invertebrates, Figure 7.2), for all insects (Figure 7.3), and for leafhoppers (Figure 7.4). In 1972, eight collections were made and fifteen invertebrate taxa were identified (ten orders of insects, mites, and spiders and three groups of invertebrates not collected in 1970 and 1971). Total numbers and biomass for the eight collections are given in Table 7.2. These data provide comparisons of the ratios of the groups to one another as well as the basis for subsequent discussion of the quantitative patterns of invertebrates.

Collection methods correspond closely to methods described by French (1971). Ten samples (a total of 5 m²) per treatment were taken during each collecting period in 1970 and 1971 and eight (a total of 4 m²) in 1972. Collections always were taken in the same area as and simultaneously with vegetative samples.

Free-fall traps were suspended on an aluminum tripod and held aloft by nylon rope (modified from Turnbull and Nicholls, 1966). All traps were set the afternoon before sampling, dropped at 10 A.M. the next day, and samples were taken as quickly as possible by clipping the enclosed vegetation and placing it in a paper sack. The interior of the trap was vacuumed with a "D-vac" collector, and samples were taken to the laboratory in ice chests. Laboratory separations were processed by: (1) extracting the D-vac collections in 70 percent isopropyl alcohol from Tullgren-type separatory funnels for 48 hours, (2) hand-sorting collections after extraction, (3) extracting vegetation clippings as with D-vac samples, and (4) handsorting vegetation clippings after extraction.

Specimens were counted, oven-dried at 60°C for 24 h, and weighed. Biomass and numbers data per m² were calculated, and the standard error was determined. Standard error quite often exceeded 20 percent of the mean, and the calculated number of samples necessary to reduce the error was often very high. Results of statistical analyses indicated that because of high standard errors, faunal differences between grazed and ungrazed treatments were not significant.

The shortcomings of this objective collecting method are obvious. It collected less than 80 percent of the total taxa recovered by all collecting methods. (Sweeping, pitfall traps, and special hand-collection methods were also used but not quantified.) The free-fall traps were more efficient than other methods in obtaining small parasitic invertebrates and litter inhabitants, but it did not efficiently trap large saprophagous and predaceous beetles, flies, and, apparently, grasshoppers. Grasshoppers and some other herbivorous Orthoptera seem to have avoided the traps.

Total herbivore biomass (Table 7.1) was estimated by calculating the weight of known herbivores and by attributing 50

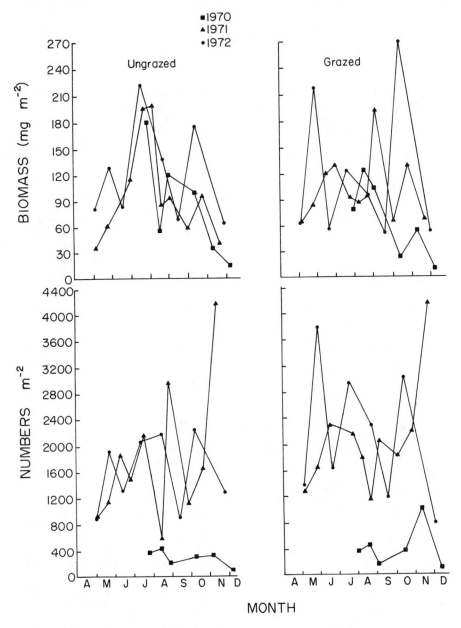

FIGURE 7.2   Invertebrate biomass and numbers.

percent of the biomass of omnivores and unknowns to the herbivore category. This value is probably conservative but is considered the best estimate in the absence of evidence of the exact role of some groups.

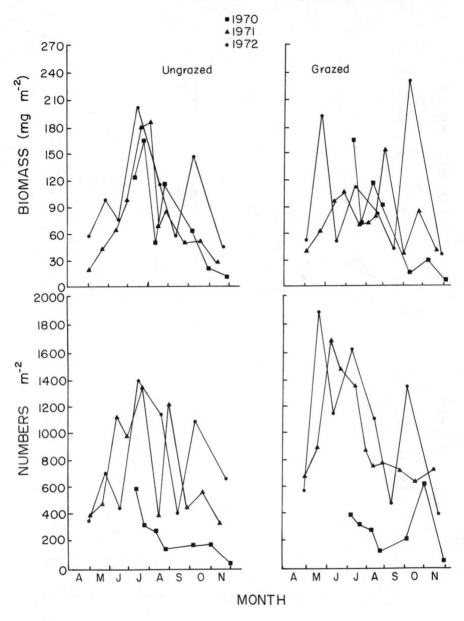

FIGURE 7.3 Insect biomass and numbers.

Major groups of invertebrates ranked by number were mites, ants, thrips, and springtails. Also found in high numbers were scale insects (Coccoidea), sap beetles (Nitidulidae), leafhoppers (Cicadellidae), spiders (Araneida), planthoppers (Delphacidae), seed bugs (Lygaeidae), and ground beetles (Carabidae). Ants, thrips, and

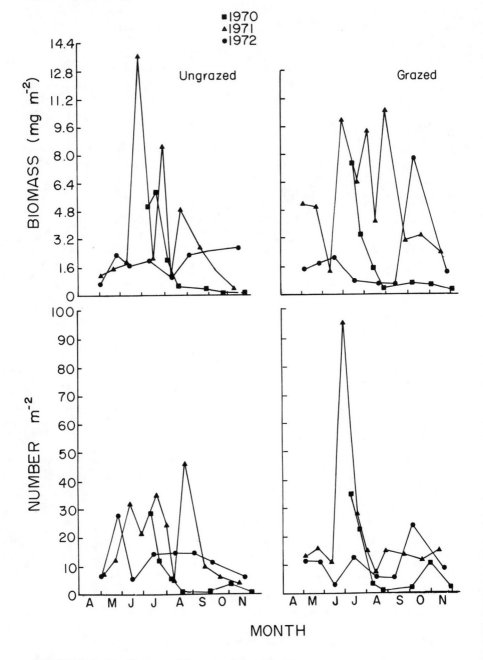

FIGURE 7.4 Leafhopper biomass and numbers.

mites   are   omnivorous;   many   are   herbivorous,   but   others   are
saprophagous   or   predaceous.   Springtails   and   sap   beetles   are
predominately   saprophagous;   spiders   and   ground   beetles   are

TABLE 7.2 Total Number and Biomass of Aboveground Invertebrate
Taxa Collected on the Osage Site, 1972.

Taxa	Biomass		Number	
	Total (mg)	Rank	Total	Rank
Collembola	50	13	5,499	5
Orthoptera	630	5	37	12
Acrididae	(380)		(11)	
Tettigoniidae	(210)		(8)	
Blattidae	(40)		(18)	
Thysanoptera	100	12	17,066	3
Hemiptera	800	4	1,223	8
Scutelleridae	(580)		(287)	
Lygaeidae	(150)		(911)	
Pentatomidae	(50)		(1)	
Homoptera	270	9	6,905	4
Cicadellidae	(120)		(705)	
Delphacidae	(50)		(699)	
Coccoidea	(20)		(5,018)	
Aphididae	(10)		(397)	
Coleoptera	1,480	2	4,226	6
Curculionidae	(510)		(214)	
Chrysomelidae	(260)		(328)	
(immatures)	(120)		(967)	
Scarabaeidae	(70)		(2)	
Nitidulidae	(20)		(810)	
Neuroptera	100	15	1	15
Lepidoptera	260	10	50	11
(immature)	(170)		(45)	
Diptera	280	8	1,119	9
Tachinidae	(170)		(23)	
(immature)	(20)		(862)	
Chloropidae	(20)		(173)	
Hymenoptera	2,240	1	18,561	2
Formicidae	(1,920)		(18,067)	
Vespidae	(280)		(9)	
(immature)	(10)		(340)	
Scelionidae	(10)		(97)	

TABLE 7.2 Continued.

Taxa	Biomass		Number	
	Total (mg)	Rank	Total	Rank
Acarina	550	7	55,813	1
Araneida	610	6	1,284	7
Diplopoda	160	11	7	13
Phalangida	10	14	2	14
Isopoda	1,310	3	531	10
Insect	6,110 (70 percent)		54,684 (49 percent)	
Noninsect	2,630 (30 percent)		57,637 (51 percent)	
Total	8,740		112,321	

predaceous; and scale insects, leafhoppers, planthoppers, and seed bugs are herbivorous.

Numbers of major groups fluctuated widely during a season. Ants, for example, ranked first in numbers for insects but numbered only 9.8 m^{-2} on 24 April 1971, compared with 258.4 Entomobryidae (Collembola). Only 2.6 ants m^{-2} were found in the ungrazed treatment on 7 November 1971, which indicates that ants move underground during the winter season.

According to biomass, major groups of invertebrates were Formicidae, Araneida, Acarina, Cicadellidae, Curculionidae, and Acrididae. Secondary contributors were Gryllidae, Chrysomelidae, immature Lepidoptera, and Scutelleridae. Of these groups, spiders are predaceous, ants and mites are omnivorous, and the remainder are herbivorous.

A higher number of invertebrates was evident in 1971 and 1972 than in 1970, but that difference was not reflected in total biomass, which was similar for all years. Environmental conditions were probably responsible for the difference, since rainfall was below average and temperatures were above average in 1970. Collembola decrease in dry habitats and increase under moist conditions (MacNamara, 1924). The same fluctuation is true for many of the small insects, which shows that environmental conditions may have a profound effect on numbers but not on biomass.

Different groups of invertebrates were responsible for number and biomass peaks during different years of the study (Figure 7.2). Neither differences between grazed and ungrazed treatments nor differences between taxa responsible for these peaks were statistically significant because of the high standard error encountered. In 1970, no invertebrate group was dominate in number,

but no early collections were made that year. Biomass peaks were primarily caused by acridid grasshoppers and mites in the ungrazed treatment and by tettigoniid grasshoppers in the grazed treatment.

Thrips, ants, and springtails were responsible for population peaks in both treatments in 1971 until late in the season, when mites became numerous. Ants and acridid grasshoppers were responsible for midseason biomass peaks in the ungrazed treatment, while Lepidoptera larvae, ants, crickets, and spiders were responsible for biomass peaks in the grazed treatment.

In 1972, large numbers of mites were recorded early in the season. Thrips, ants, and scale insects were dominant later in the ungrazed treatment, while thrips, ants, and springtails were responsible for numbers peaks in the grazed treatment. Early biomass peaks in both treatments were caused by ants and beetles. Later biomass peaks were caused by vespid wasps and true bugs in the ungrazed treatment and by acridid grasshoppers, vespid wasps, and ants in the grazed treatment.

Leafhopper numbers were lowest in 1970, but their density was not as depressed as that of total insects (Figure 7.4). Peaks in leafhopper densities were highest during 1971 in both treatments, suggesting that no single environmental or biological factor associated with the treatment was responsible. Numbers and biomass of leafhoppers were not significantly related, providing further evidence that both types of data must be gathered if inferences are to be made (Blocker and Reed, 1976).

The multivoltine (multiple annual life cycles) nature of many leafhopper species and the responses of different groups to changing environmental and biological factors make predicting population dynamics difficult (Blocker et al., 1972). However, univoltine (single annual life cycle) groups such as grasshoppers progressively decrease in numbers and increase in individual biomass during the season (at least until eggs are deposited).

Comparison of invertebrate (Figure 7.2) and insect data (Figure 7.3) shows that noninsects, especially mites, contribute significantly to numbers of invertebrates, but spiders are the only noninsects that make consequential contributions to invertebrate biomass.

Herbivore biomass and total invertebrate biomass are compared for 1971 and 1972 in Table 7.1. In the ungrazed treatment, biomass varied from 41 percent of the total on 23 April 1972 to 73 percent on 20 August 1971. Herbivore biomass in the grazed treatment ranged from 31 percent of the total on 3 September 1972 to 81 percent on 25 July 1971. The 1971 averages in the ungrazed and grazed treatments were 59 percent and 65 percent, respectively; biomass averages for 1972 were 57 percent and 62 percent for ungrazed and grazed, respectively.

Slightly more plant species and a higher frequency of forbs were recorded in the grazed than in the ungrazed area. Frequency and conspicuousness of forbs are highest during the early (May and June) and late (September and October) portions of the growing season. Invertebrate numbers appeared to be associated with forbs, so the higher frequency of forbs on the grazed area could account for high numbers of certain groups of invertebrates. The percentage of herbivores in the grazed treatment was also highest when forbs were most abundant. Thus, vegetation differences between the two

treatments probably accounted for the difference in invertebrate numbers and biomass.

Spiders and mites were collected and recorded during the entire study. Sowbugs (Isopoda), millipedes (Diplopoda), and daddy longlegs (Phalangida) were present but not recorded until 1972 (Table 7.2). They are not included in the numbers and biomass data discussed elsewhere. Isopods contributed most to numbers and biomass, and although either the density or biomass rank of most of these groups was similar, the biomass rank of Isopoda (sowbugs) and Orthoptera was much lower than the density rank (Figure 7.5). On the other hand, the biomass rank of mites, thrips, springtails, and Homoptera was much higher than the density rank.

After three years of collecting, we have reached the following conclusions about aboveground invertebrate populations:

1.   The high number of samples needed to reduce standard error to below 20 percent of the mean at the 80 percent confidence level is prohibitive in terms of cost and time. However, the density numbers and biomass measurements of invertebrates in the True Prairie that were actually determined can be valuable in estimating at least the minimal impact of invertebrates in an ecosystem.

2.   No single sampling method adequately assesses the invertebrate fauna of the True Prairie. Quick trap collecting was the most efficient method but should be supplemented with other methods.

3.   Estimates of invertebrates in the True Prairie based on numbers or biomass alone are inadequate because of wide variability within and among groups. Biomass in 1970 and 1971 was similar, but numbers varied greatly.

4.   Habitat conditions in the True Prairie affect invertebrate populations. Biomass at the Osage Site was negligible compared to the biomass of primary productivity, but this is partly because of the good to excellent range conditions at this particular site. Invertebrate numbers apparently were associated with the number of forbs, especially in the grazed treatment. Visual inspection of adjacent heavily grazed areas suggested higher populations of many invertebrates, including grasshoppers and leafhoppers.

Belowground Invertebrates

Soil arthropod data at the Osage Site are based on six collections taken between 18 June 1971 and 21 November 1972 in the grazed and ungrazed treatments (Stepanich, 1975). Three 5 cm diameter core samples in the two replicates of each treatment were collected at depths of 0-5, 5-10, 10-20, and 20-50 cm, with a hydraulic corer. Specimens were extracted in modified Tullgren separatory funnels and stored in 70 percent isopropyl alcohol until identification and weighing. The variation in numbers per sample was mostly high, indicating the need for more observations.

Fifteen orders and forty-six families of arthropods were identified (see Table 3.8). Major groups in numbers were Acarina, Formicidae, Collembola, Diplura, Thysanoptera, Protura, and Symphyla, respectively (Table 7.3), and major groups in biomass were

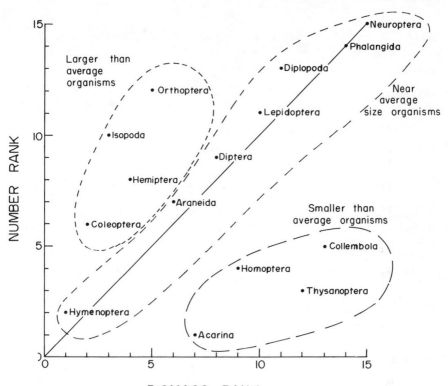

FIGURE 7.5 Relationship between the total number and biomass of the aboveground invertebrate taxa collected on the Osage Site in 1972.

Hymenoptera, Collembola, Thysanoptera, Diplura, Acarina, and Araneida. Only mites were taxonomically separated below the family level. Variation among and within groups made trophic analysis difficult, but, for analysis, partitioning was allocated as 40 percent herbivores, 30 percent predators, and 30 percent saprophages.

Large numbers of microarthropods were found below 20 cm (Table 7.4), and the occurrence of substantial proportions of the fauna below the humus layer is unusual, particularly for the Acarina and Collembola (Price, 1973). In England, 92.6 percent of the Collembola in grassland soil was found in the upper 4 cm (Dhillon and Gibson, 1962), and Wood (1967) found over 76 percent of the Collembola and Acarina in the upper 4 cm and more than 90 percent of the total fauna in the upper 6 cm in four English grassland sites.

Part of the population of every major category of soil microarthropods occurred at depths below 25 cm in the subsoil of a California pine forest (Price, 1973). Similarly, soil fauna of the Oklahoma Tallgrass Prairie can be found at greater depths than previously reported for grassland soil in England (Dhillon and Gibson, 1962; Wood, 1967).

TABLE 7.3   Estimated Numbers and Biomass of Soil Invertebrates
             Collected at the Osage Site in 1972.

Taxa	Biomass		Number	
	Total (mg)	Rank	Total	Rank
Protura	180	9	3,565	7
Collembola	890	2	12,478	3
Diplura Japygidae	590	4	11,714	4
Orthoptera Immature Blattidae	T[a]	T	905	10
Thysanoptera	710	3	7,130	5
Hemiptera	T	T	329	14
Homoptera	260	7	3,642	6
Coleoptera	T	T	822	11[b]
Lepidoptera Immatures	T	T	82	15
Diptera	T	T	1,233	9
Hymenoptera Formicidae	19,300	1	41,444	2
Acarina	520	5	94,305	1
Araneida	490	6	822	11
Symphyla	200	8	2,122	8
Pseudoscorpionida	100	10	748	12
Psocoptera	T	T	676	13

[a] T = Trace.

[b] Coleoptera and Araneida tied.

The function and impact of soil arthropods are poorly
understood, but as far as we know, trophic levels are indicated (see
Table 3.8) for the major groups. An analysis of covariance showed

TABLE 7.4 Major Soil Microarthropods Collected at Different Depths (0-5, 5-10, 10-20, and 20-50 cm) at the Osage Site, Northeastern Oklahoma, from 18 June 1971 Through 21 November 1972.

Taxa	Grazed					Ungrazed				
	Total (percent)				No. m^{-2} (0-50 cm) (×1000)	Total (percent)				No. m^{-2} (0-50 cm) (×1000)
	0-5	5-10	10-20	20-50		0-5	5-10	10-20	20-50	
18 June 1971										
Acarina	69	13	18		15.8	52	36	12		24.7
Collembola	79	15	6		4.1	80	10	10		1.7
Diplura	17	58	25		1.0	33	33	33		1.0
Thysanoptera	100				0.5	21	64	14		1.2
Total	69	15	16		21.4	52	36	13		28.6
10 October 1971										
Acarina	22	17	17	45	38.6	44	12	24	20	118.5
Collembola	76	3	7	84	8.8	10	2	23	65	11.4
Diplura		3	14	83	4.9			35	65	2.0
Protura	33		11	56	0.8	27		36	36	0.9
Symphyla			33	67	1.0	7	13	47	33	1.3
Thysanoptera	56		11	33	0.8	33	17	17	17	0.6
Total	18	12	15	55	54.9	40	11	25	25	134.7
16 May 1972										
Acarina	26	27	17	30	15.8	57	32	8	4	6.5
Collembola	33	25	17	25	1.0	50	25		25	0.3
Diplura		10	17	73	2.5	50		50		0.2
Protura				100	0.1	100				0.1
Thysanoptera	20			80	0.4	33	33	33		0.3
Total	23	25	16	36	19.8	53	29	9	4	7.4
5 July 1972										
Acarina	37	24	18	21	19.9	39	23	27	12	9.6
Collembola	47	3	6	44	2.7	64		7	29	1.2
Diplura	11	11	18	61	2.4		25	75		0.7
Protura	27		7	67	1.3	29	14	14	43	0.5
Symphyla		50		50	0.2	100				0.2
Thysanoptera	67	33			0.3		100			0.3
Total	35	20	16	29	26.8	39	22	26	14	12.5

TABLE 7.4.   Continued.

Taxa	Grazed					Ungrazed				
	Total (percent)				No. m^{-2} (0-50 cm) (×1000)	Total (percent)				No. m^{-2} (0-50 cm) (×1000)
	0-5	5-10	10-20	20-50		0-5	5-10	10-20	20-50	
8 August 1972										
Acarina	42	11	32	16	8.1	37	29	16	18	11.0
Collembola	44	22	25	9	2.8	55	9	9	27	1.9
Diplura	5	2	10	85	1.7	20		40	40	0.4
Protura			40	60	0.4	55		27	18	0.9
Symphyla			20	80	0.4	63	13	13	13	0.7
Thysanoptera	54	15	15	15	1.1			100		0.2
Total	37	11	27	26	14.4	40	23	18	19	15.1
21 November 1972										
Acarina	51	15	22	12	9.6	43	31	16	11	13.6
Collembola	42	33	8	17	1.0	53	26	16	5	1.6
Diplura			37	63	2.0	6		19	75	1.4
Protura	50	50			0.2					
Symphyla	17	17	17	50	0.5		25	38	38	0.7
Thysanoptera	8	77		15	1.1	86	2	5	7	3.6
Total	37	18	22	23	15.4	47	23	15	15	20.8

that neither root biomass nor soil water was related to the number of mites, springtails, or total arthropods at the Osage Site.

Nematode populations were examined at the Osage Site from collections made on 28 August and 5 December 1972 (Table 7.5). Numbers were lower than expected, but a more extensive sampling would be required to properly evaluate the extraction techniques. Samples were taken in late summer and late fall, seasons that do not include the peak growing season for most other organisms.

The nematode group with the largest numbers was herbivores (59 percent), followed by saprophages (30 percent), and predators (11 percent). A lower proportion of biomass was found for herbivores (51 percent) and saprophages (23 percent) but greater for predators (26 percent). As with other invertebrates, both numbers and biomass data are needed.

Other casual observations included

1. Biomass and numbers of nematode herbivores and predators were much higher in the grazed treatment than the ungrazed treatment in August but were about equal in December.
2. Saprophages were higher in biomass and numbers in the grazed treatment than the ungrazed in August and September.
3. Predators were much heavier per specimen than saprophages.

Also of interest are possible correlations of nematode data with CO_2 activity, microbial activity, and biomass of other invertebrates. Collecting procedures are in question, however, and further investigation is needed before making these analyses.

Temporal Stability

The stability of invertebrate populations is impossible to assess on the basis of available data. The geographical range of some leafhopper populations moves northward to a greater extent during years with high rainfall and temperatures (Blocker et al., 1972); data gathered on leafhoppers at the Osage Site support this observation. Most major species were captured every year, but other species appeared once in three years. Population numbers also varied from year to year. Abiotic factors may be most important because predation rates in a balanced ecosystem probably respond to prey population size more than they control prey populations.

Populations of other invertebrates of the True Prairie also fluctuate from year to year (e.g., the ratio of leaf-feeding beetles and weevils). Predator populations usually increase in response to an increase in prey numbers. Saprophage populations fluctuate independently of herbivore and predator numbers, perhaps in response to abiotic factors.

Energy Flow and Nutrient Relations

No information on energy flow and nutrient relations in the True Prairie was located for this disucssion. The results of Van Hook (1971), however, who studied population dynamics, energy

TABLE 7.5 Effect of Range Treatment and Depth on Number and Biomass of Nematodes at the Osage Site (Data from Dr. J. D. Smolik, South Dakota State University, Brookings, 1973).

Depth (cm)	Herbivorous		Predaceous		Saprophageous	
	No. m^{-2} (×1000)	Biomass (mg m^{-2})	No. m^{-2} (×1000)	Biomass (mg m^{-2})	No. m^{-2} (×1000)	Biomass (mg m^{-2})
28 August 1972--Grazed[a]						
0-5	845	50	94	20	1,086	80
5-10	549	30	58	10	358	30
10-20	432	30	41	10	150	10
20-30[b]	464	20	23	10	256	20
30-60[b]	577	30	45	10	341	20
Total	2,867	160	261	60	2,191	160
Total numbers (v1000)	5,319					
Total biomass (mg)	380					
28 August 1972--Ungrazed[a]						
0-5	733	60	186	40	541	40
5-10	1,242	110	204	50	277	20
10-20	1,009	110	170	40	279	10
20-30[c]	732	50	94	20	115	10
30-60[c]	683	50	69	10	180	10
Total	4,399	380	723	160	1,392	90
Total numbers (v1000)	6,511					
Total biomass (mg)	650					

TABLE 7.5. Continued.

Depth (cm)	Herbivorous		Predaceous		Saprophageous	
	No. m^{-2} (×1000)	Biomass (mg m^{-2})	No. m^{-2} (×1000)	Biomass (mg m^{-2})	No. m^{-2} (×1000)	Biomass (mg m^{-2})
			5 December 1972--Grazed[d]			
0-5	600	50	242	50	634	60
5-10	443	30	67	10	228	20
10-20	473	40	111	30	148	10
20-30	385	30	51	10	162	10
30-60	479	40	97	20	216	10
Total	2,380	190	568	120	1,388	100
Total numbers (v1000)	4,336					
Total biomass (mg)	410					
			5 December 1972--Ungrazed[b]			
0-5	537	60	221	50	313	20
5-10	428	50	91	20	177	20
10-20	276	30	86	20	168	10
20-30	304	30	67	20	69	10
30-60	284	20	49	10	108	10
Total	1,830	190	514	120	835	60
Total numbers (v1000)	3,177					
Total biomass (mg)	370					

[a] Average of four replications.

[b] Average of five replications.

[c] Estimated on basis of 5 December sample.

[d] Average of three replications.

budgets, and nutrient fluxes in arthropod components of an eastern Tennessee successional grassland ecosystem, are summarized here. The arthropod community consisted primarily of herbivores (Melanoplus sp., Conocephalus sp., and several Homoptera-Hemiptera), and omnivore (Pteronemobius sp.), and a predator (Lycosa sp.). Nutrient concentrations for these major groups were determined: whole-body concentrations of sodium (Na), calcium (Ca), and potassium (K) of one family of Hemiptera (mirid grass bugs) and of one family of Homoptera (leafhoppers). No significant differences were obvious among species or through time for sodium, calcium, and potassium for the four species shown in Figure 7.6. Concentrations of sodium and calcium in the mirids and leafhoppers did not differ from the concentrations in the other herbivore species, but potassium concentrations were higher in mirids and leafhoppers. Leafhoppers and mirids (sucking insects) ingest intracellular juices (high in potassium), but chewing herbivores dilute potassium intake with macerated plant parts.

The mean annual concentration of sodium in the predator trophic level was higher than the concentration found in the herbivore level. Sodium concentration of herbivores was much higher than the concentration of primary producers. Omnivores had a much higher sodium concentration than the detritus food base, and the mean annual whole-body content of calcium did not increase significantly from the herbivore to predator trophic levels. No difference between herbivore and predator trophic levels for potassium was determined, but both levels were much lower than of the primary producer level. Low concentrations of calcium in arthropod trophic levels and rapid elimination of this element from each trophic level suggest that calcium is not as limiting as sodium and potassium in this arthropod community (Figure 7.7).

Total net primary production (Figure 7.8) was 1,274 kcal m^{-2}, of which herbivores and omnivores consumed 9.6 percent. Annual net secondary production by the insect community was 32.05 kcal m^{-2}, and total net tertiary production by spiders was 2.26 kcal m^{-2}. Total energy flow (total assimilation) through the community was 75.6 kcal m^{-2}, of which herbivores accounted for 79.8 percent; omnivores, 12.2 percent; and predators, 8.1 percent. Consumption of vegetation was the only output considered for primary producers in constructing the energy budget.

Herbivores reached a standing crop high of 927 mg m^{-2} and were responsible for 85 percent of the sodium turnover, 76 percent of the calcium turnover, and 78 percent of the potassium turnover by arthropods (Figure 7.7). Omnivores fed equally well on fresh vegetation or litter and reached a biomass high of 194 mg m^{-2}. They utilized 10 percent of the sodium, 22 percent of the calcium, and 20 percent of the potassium. Predator biomass reached a high of 146 mg m^{-2} and was responsible for 5 percent of the sodium, 2 percent of the calcium, and 2 percent of the potassium used by arthropods. Wolf spiders consumed 21 percent of the total net secondary production.

Coleman et al. (1977) reported on energy flow in a shortgrass ecosystem: Catabolic activity was dominated by microbes in the belowground system; energy flow through biophages was insignificant compared with the flow through saprophages; grazing by nematodes, however, may have a major impact on primary production. Ninety

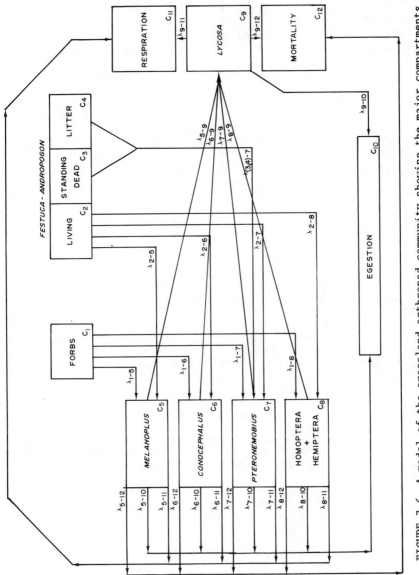

FIGURE 7.6 A model of the grassland arthropod community showing the major compartments and pathways of energy flux (redrawn from Van Hook, 1971).

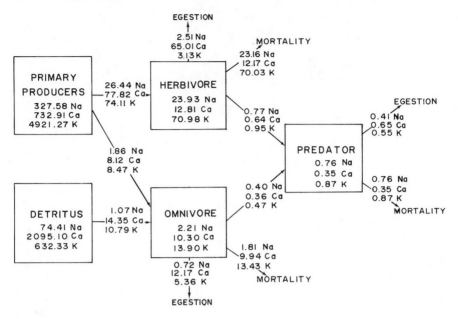

FIGURE 7.7 Annual nutrient budget (mg m^{-2}) for sodium, calcium, and potassium showing the maximum standing crops of nutrient and nutrient fluxes through the arthropod and vegetation components of an eastern Tennessee grassland ecosystem (redrawn from Van Hook, 1971).

percent of the invertebrate energy transfer was belowground, and more than 50 percent of the total energy was processed by nematodes. Saprophagic activity by invertebrates was low and dominated (90 percent) by nematodes. No comparable calculations are available from the True Prairie, but estimates of trophic levels of invertebrates coupled with biomass figures are given in Table 7.6. These estimates indicate that the ratio of aboveground energy transfer by invertebrate herbivores is similar in shortgrass and the True Prairie and that invertebrate saprophages have a more dominant role in True Prairie. Evidence also indicates that less total energy is processed by nematodes in the True Prairie, although this difference is small. In the True Prairie, saprophagic activity is not dominated by nematodes to the extent found in shortgrass prairie. The similarity of the ecosystems, however, is noteworthy, and additional data are needed to corroborate these estimates.

Trophic Relationships

Placement of invertebrates into trophic levels is difficult because information frequently is not available and the functional role of many organisms change during the life cycle. However, from data collected at the Osage Site, percentage estimates were made based on seasonal biomass averages (Table 7.7). Most data were

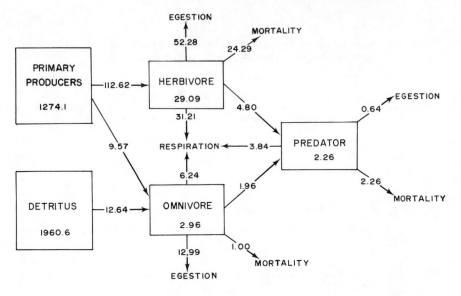

FIGURE 7.8 Annual energy budget of energy fluxes and net production (kcal m⁻²) by the arthropod and vegetation components of an eastern Tennessee grassland ecosystem. The values inside compartments represent net production (exception: for detritus, value equals standing crop of standing dead and litter), and the values on arrows represent energy fluxes (redrawn from Van Hook, 1971).

taken during the growing season, and collecting methods have not been refined.

VERTEBRATE CONSUMERS

To a degree, the distinction between vertebrate and invertebrate consumers is artificial--a taxonomic convention--and may divert attention from some basic commonalities in the roles played by various vertebrates and invertebrates in grassland ecosystems. Still, some differences are fundamental. As a group, vertebrates are usually orders of magnitude larger than invertebrate consumers and have a greater range of individual mobility and a greater degree of buffering against small-scale environmental influences (Wiens, 1976). Although exceptions do exist, invertebrates seem to be closely attuned to various chemical compounds in their environments, while vertebrates frequently respond more strongly to the physical features of their surroundings.

In True Prairie systems, only birds and mammals have received close quantitative study. The lack of systems-oriented or community-oriented investigations of reptiles in the True Prairie represents a major gap in our understanding of consumers. Studies on the Shortgrass Prairie in northeastern Colorado, however, indicate that snakes attained an estimated biomass peak of 84.3 g ha⁻¹ (Bauerle, 1972). Assimilation by this group, however, was

TABLE 7.6   Estimates of Trophic Levels of Invertebrates Coupled with Biomass Figures for Seasonal Averages.

Trophic levels	Invertebrates (mg m^{-2})		
	Nematodes	Other aboveground	Other belowground
Herbivores	230	40	210
Predators	120	10	160
Saproghages	100	20	160

TABLE 7.7   Estimates of Trophic Levels, Based on Seasonal Biomass Averages.

Trophic level	Invertebrates (percent)		
	Nematodes	Other aboveground	Other belowground
Herbivores	51	60	40
Predators	26	15	30
Saprophages	23	25	30

less than 0.05 kcal m^{-2}, compared to 1.6 kcal m^{-2} by bird populations with similar biomass in the same area (Andrews et al., 1974). Little doubt exists that reptiles are less abundant in most True Prairie locations than in this Shortgrass Prairie, so their role, at least as gauged by energy processing, may be quite minor.

AVIAN CONSUMERS

Birds are relatively scarce in True Prairies. A series of plot-based breeding censuses provides the information needed to consider these avifaunas in quantitative detail. Censuses have been conducted at ten locations in True Prairie habitats (Figure 7.9), and, collectively, twenty-four censuses are available for analysis (Table 7.8). These censuses were conducted during the breeding season on plots of 5-10 ha in which the positions of singing males were recorded repeatedly to provide an approximation of the number of territories present, or territories were mapped by repeated sequential flushings of singing males (Wiens, 1969). Counts were then converted to total numbers of individuals of each species

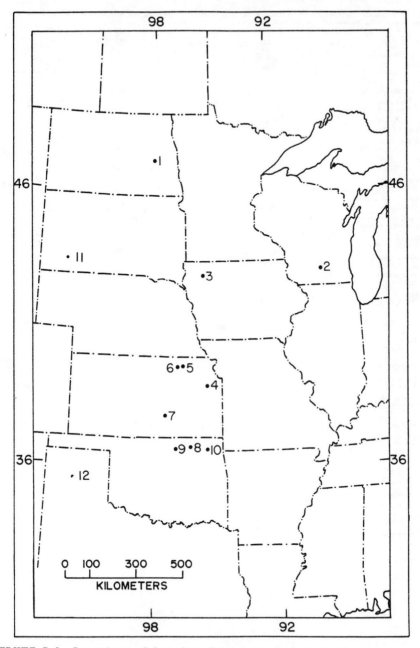

FIGURE 7.9   Locations of breeding bird census plots in True
Prairies.   Census locations: 1 = Stutsman Co., ND; 2 = Fitchburg,
WI; 3 = Iowa; 4 = Douglas Co., KS; 5 = Riley Co., KS; 6 = Donaldson
Pastures, KS; 7 = Elmdale, KS; 8 = Washington Co., OK; 9 = Osage
Site, OK; 10 = Inola, OK; 11 = Cottonwood Site, SD; 12 = Pantex
Site, TX.   (Sites 11 and 12 are not located within the True
Prairie.)

TABLE 7.8 Densities (Individuals km^{-2}) and Biomass (g ha^{-1}) (in Parentheses) of Breeding Bird Populations in True Prairie Census Locations. Numbers following locations coincide with those shown in Figure 7.10.

Species	Stutsman Co., ND[a] (1)			Fitchburg, WI[b] (2)			Iowa[c] (3)	Douglas Co.,[d] KS (4)	Riley Co.,[e] KS (5)		
	1972	1973	1974	1964	1965	1966		1974	1965	1966	1967
Ring-necked pheasant (Phasianus colchicus)	--	--	--	--	--	--	69.2	--	--	--	--
Bobwhite quail (Colinus virginianus)	--	--	--	--	--	--	--	--	--	13.0	--
Marbled godwit (Limosa fedoa)	10.3 (15.4)	41.2 (61.8)	41.2 (61.8)	--	--	--	--	--	--	--	--
Upland plover (Bartramia longicauda)	--	--	--	--	--	--	--	--	--	--	--
Common nighthawk (Chordeiles minor)	--	--	--	--	--	--	--	--	--	--	--
Mourning dove (Zenaida macroura)	--	--	--	--	--	--	--	--	64.2 (80.2)	13.0 (16.2)	39.5 (49.0)
Short-billed marsh wren (Cistothorus platensis)	--	--	41.2 (2.5)	--	--	--	--	--	--	--	--
Common yellowthroat (Geothlypis trichas)	--	--	--	--	--	--	--	79.2 (8.7)	--	--	--
Eastern meadowlark (Sturnella magna)	--	--	--	57.9 (60.8)	50.2 (52.7)	92.7 (97.3)	--	79.2 (83.1)	49.4 (51.9)	74.1 (77.8)	80.0 (84.0)
Western meadowlark (Sturnella neglecta)	41.2 (41.2)	41.2 (41.2)	41.2 (41.2)	18.5 (18.5)	24.7 (24.7)	6.2 (6.2)	29.7 (29.7)	--	24.3 (24.3)	80.3 (80.3)	67.5 (67.9)
Red-winged blackbird (Agelaius phoeniceus)	20.6 (13.4)	10.3 (6.7)	102.9 (66.7)	--	--	--	--	136.6 (88.8)	118.6 (77.1)	148.3 (96.4)	108.0 (70.0)
Bobolink (Dolichonyx oryzivorus)	61.7 (18.5)	--	102.9 (30.9)	30.9 (9.3)	55.6 (16.7)	49.4 (14.8)	79.1 (23.7)	--	--	--	--
Dickcissel (Spiza americana)	--	--	--	--	--	--	--	198.0 (53.4)	108.7 (29.4)	78.1 (21.4)	158.1 (42.7)
Savannah sparrow (Passerculus sandwichensis)	102.9 (19.6)	20.6 (3.9)	--	160.0 (30.5)	173.0 (32.9)	231.7 (44.0)	--	--	--	--	--
Grasshopper sparrow (Ammodramus savannarum)	10.3 (1.8)	--	10.3 (1.8)	160.6 (27.3)	105.0 (17.8)	185.3 (31.5)	79.1 (13.4)	--	64.2 (10.9)	39.5 (6.7)	39.5 (6.7)
Henslow's sparrow (A. henslowii)	--	--	--	24.7 (3.7)	--	24.7 (3.7)	--	--	--	--	--
Vesper sparrow (Pooecetes gramineus)	--	--	--	--	--	30.9 (7.4)	--	--	--	--	--
Field sparrow (Spizella pusilla)	--	--	--	--	--	--	--	--	--	--	--
Chestnut-collared longspur (Calcarius ornatus)	--	10.3 (2.0)	--	--	--	--	--	--	--	--	--

[a] Van Velzen (1972, 1973, 1974)

[b] Wiens (1969)

[c] Kendeigh (1941)

[d] Van Velzen (1974)

[e] Zimmerman (1965, 1966, 1967)

[f] Wiens (1974a)

[g] Wiens (1971, 1973, 1974a, unpublished); Wiens et al. (1974)

[h] Wiens (unpublished)

| Donaldson[f] Pastures, KS (6) | Elmdale,[f] KS (7) | | Washington Co.,[d] OK (8) | Osage Site, OK[g] (9) | | | | | | | Inola, OK[h] 1974 (10) | |
| | | | | Plot 1 | | | | Plot 2 | | | | |
1968	1967	1968	1974	1970	1971	1972	1974	1971	1972	1974	Ungrazed	Grazed
--	--	--	--	--	--	--	--	--	--	--	--	--
--	--	--	--	--	--	--	--	--	--	--	--	--
--	--	--	--	--	--	--	--	--	--	--	--	--
--	--	7.0 (9.1)	--	9.5 (12.4)	6.9 (9.0)	7.1 (9.2)	9.8 (12.7)	29.1 (37.8)	18.6 (29.2)	--	--	--
--	--	--	--	--	--	5.9 (4.4)	7.3 (5.5)	--	--	--	--	--
--	--	--	--	--	--	--	--	--	--	--	--	--
--	--	--	--	--	--	--	--	--	--	--	--	--
--	--	--	--	--	--	--	--	--	--	--	--	--
93.0 (92.0)	74.0 (73.0)	80.0 (79.0)	95.9 (100.7)	88.3 (86.6)	80.3 (78.8)	74.4 (77.4)	74.1 (77.0)	124.0 (121.6)	75.1 (78.2)	103.7 (107.9)	68.0 (70.7)	96.0 (99.8)
--	--	--	--	--	--	--	--	--	--	--	--	--
--	--	--	23.3 (15.2)	--	--	--	--	--	--	--	--	--
--	--	--	--	--	--	--	--	--	--	--	--	--
46.0 (13.0)	66.0 (18.0)	91.0 (25.0)	226.7 (61.2)	80.9 (22.2)	94.0 (24.9)	8.6 (2.2)	427.0 (10.9)	282.3 (74.3)	73.5 (18.7)	187.4 (47.9)	17.0 (4.3)	398.0 (101.6)
--	--	--	--	--	--	--	--	--	--	--	--	--
67.0 (12.0)	144.0 (25.0)	98.0 (17.0)	244.2 (41.5)	71.4 (12.0)	58.3 (10.0)	36.9 (6.3)	115.4 (19.7)	83.2 (14.2)	130.6 (22.3)	150.3 (25.7)	--	57.0 (9.8)
--	--	--	2.9 (0.4)	--	--	--	--	--	--	--	--	--
--	--	--	--	--	--	--	--	--	--	--	--	--
--	--	--	2.9 (0.5)	--	--	--	--	--	--	--	--	--
--	--	--	--	--	--	--	--	--	--	--	--	--

present in the plot, from which overall density and biomass values were calculated. Because few quantitative studies have been done in grassland habitats during nonbreeding periods, our analyses must be confined to consideration of breeding populations and communities. The neglect of wintering populations is unfortunate, as grassland habitats may support large wintering concentrations of some species, especially in southern sections of the prairie, and overwinter survival may be critically important to the population dynamics of many bird species (Fretwell, 1972). Bird populations of the True Prairie are emphasized here, and broader analyses of North American grassland bird communities have been presented by Cody (1968), Wiens (1973, 1974a), and Wiens and Dyer (1975).

Although the census locations in Table 7.8 do not represent a random or complete sampling of True Prairie avifaunas, they do reveal patterns of species distributions and co-occurrence that are probably typical of the variation in the True Prairie. A more sensitive image of the avifaunal relationships of these sample plots can be obtained by examining the overall similarities based on relative densities of bird species present in the plots averaged over all years sampled. Using similarity index values of Horn's (1966) R_o (which ranges from 0 in the case of total exclusiveness of compared plots to 1 in the case of complete identity), we conducted an unweighted pair-group cluster analysis (Sokal and Sneath, 1963). The clustering of census plots (Figure 7.10) indicates close similarity among the sites in northeastern Oklahoma and south-central Kansas. The two census plots at Inola (in a somewhat isolated and agriculturally dissected grassland in extreme northeastern Oklahoma) were not similar to nearby plots; the ungrazed plot at Inola showed greater affinity with the northeastern Kansas plots than with the geographically nearer plots, including a grazed plot 1 km away. However, this small ungrazed plot contained only two breeding species (Table 7.8) and is probably not typical of the True Prairie in this area. However, so-called typical True Prairie plots are now rare. The Donaldson plot clustered with the northeastern Oklahoma plots, probably because it was located on an exposed slope with shallow soil and was thus more xeric than the other grasslands censused in this region. The Mixed-grass (Cottonwood) and Shortgrass Prairie (Pantex) locations emerge as a separate cluster in this analysis.

Species Abundance Patterns

Census estimates of total breeding densities (individuals km^{-2}) of bird species present in these sample plots are listed in Table 7.8. Nineteen species were recorded in the twenty-four censuses, but only eight were recorded in more than 15 percent of the counts. Eastern meadowlarks were present in substantial numbers at all but two sites, in which western meadowlarks were plentiful. The two meadowlark species overlapped at several northern sites, but only eastern meadowlarks were present on the census plots in southern Kansas and Oklahoma. These distributions accord with the general distribution of sympatry of these species, which has been documented in detail by Lanyon (1956). The co-occurrence of species populations over the range of census plots may be expressed in an

SIMILARITY INDEX (R_o)

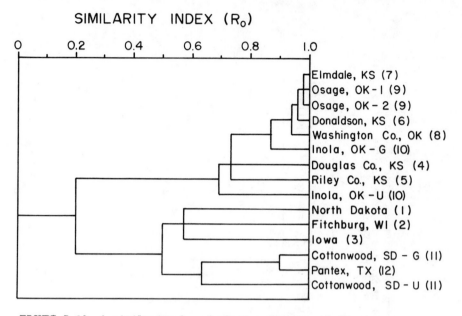

FIGURE 7.10 A similarity-based cluster diagram of the census
locations of Figure 7.9, based on average avian density values
calculated from Table 7.16. Numbers in parentheses correspond to
the locations of Figure 7.9.

index of "niche overlap" (Colwell and Futuyma, 1971) in a manner
parallel to the method used to consider distributional "niche
breadths" of species disucssed in Chapter 3. Values indicate the
relatively low degree of co-occupancy of local plots by the two
meadowlark species (Table 7.9). On the other hand, eastern
meadowlarks exhibited a relatively high degree of overlap with
dickcissels and grasshopper sparrows, again pointing to the overall
importance of these three species in True Prairie habitats.

Dickcissels were present as breeding species on plots in Kansas
and Oklahoma but were absent from the more northern plots. The
species does occur in these areas (Emlen and Wiens, 1965; Hurley and
Franks, 1976; Sealy, 1971; Wiens and Emlen, 1966) but apparently
occupies the more marginal or ecotonal habitats and is not found on
large prairie grassland plots. Dickcissel abundance varied
erratically over the plots in which it was recorded, from less than
10 to nearly 400 individuals km^{-2}. Bobolink distribution was the
mirror image of dickcissel occurrence in the plots, as indicated by
the zero overlap value (Table 7.9). Both species are characteristic
of semidisturbed, early stages of prairie succession (Martin, 1974;
Wiens, 1974b; Zimmerman, 1965) and have fluctuating densities and
distributions through the True Prairie. Some evidence shows that
these distributional variations may be inversely linked (Wiens and
Emlen, 1966), but the relationships between these species have not
been studied closely. Of the sparrows, only the grasshopper sparrow
occurred over most of the True Prairie plots.

The most obvious feature of Figure 7.10 is the separate
clustering of the southern sites and the more northern locations

TABLE 7.9 Distributional "Niche Overlap" Index Values, Representing the Degree of Overlap in Distributions and Abundances of the Major Bird Species in True Prairie. Calculations Were Made from the Data of Table 7.8, with the Procedures of Colwell and Futuyma (1971).

	Eastern meadowlark	Western meadowlark	Bobolink	Dickcissel	Savannah sparrow	Grasshopper sparrow
Upland plover	0.25	0.00	0.00	0.23	0.00	0.17
Eastern meadowlark		0.28	0.25	0.62	0.25	0.68
Western meadowlark			0.55	0.17	0.27	0.32
Bobolink				0.00	0.44	0.39
Dickcissel					0.00	0.42
Savannah sparrow						0.41

(North Dakota, Wisconsin, and Iowa). The general patterns of plot similarities were such that these plots clustered earlier in the analyses with studies of Mixed -grass Prairies in South Dakota (Cottonwood) and Texas Shortgrass Prairies (Pantex) than with the True Prairie sites farther south. To some degree, all clustering procedures exhibit this pattern; plots are paired according to similarity whether or not the clustering is natural. Table 7.8 confirms that northern sites do differ substantially from the more southern locations. Three species (marbled godwit [Limosa fedoa], bobolink [Dolichonyx oryzivorous], and Henslow's sparrow [Passerherbulus henslowii]) are limited in breeding distribution to northcentral areas. Nowhere, however, are these species dominant components of the avifaunas, and the remaining species are all broadly distributed. A broader range of census locations is needed to determine the biological validity of the clusterings of Figure 7.10.

Community Structure

     Breeding plot censuses provide information that can be used to measure several features of the avian communities occupying these sites. The definition of community used here is purely operational: an assemblage of breeding individuals of all species present within a sample plot (which is somewhat akin to MacArthur's [1971] view of a community as "any set of organisms currently living near each other and about which it is interesting to talk"). Mean values of various community attributes for the sample locations in the True Prairie are given in Table 7.10. The average number of species breeding on sample plots varied from two to more than six. No trend to the values among the sites is apparent, but detailed measurements of vegetation structure or productivity that might permit closer definition of patterns of species numbers (e.g., Cody, 1968) are lacking. At the Osage Site, where studies have been most intensive, species numbers ranged from three to five. Here, as elsewhere, most variation in species numbers was attributable to the presence or absence of forms that rarely accounted for more than 10 percent of the total density, so breeding assemblages seemed to be quite stable locally, at least in species composition.

     Total densities (individuals km^{-2}) ranged from under 100 to nearly 600 among the plot locations, but here, also, no clear patterns emerged. Relatively high densities were recorded in Wisconsin, northern Kansas, and northeastern Oklahoma; low densities, in Iowa, northcentral and central Kansas, and northeastern Oklahoma. Thus, local variations in densities were just as great as regional variation within True Prairie habitats. At the Osage Site, for example, replicate census plots were situated in different parts of the lightly grazed grassland, perhaps 75 m apart. Total densities in plot 2, however, averaged nearly twice as great as in plot 1, despite similar species compositions. Plot 2 plot was situated on an area having a somewhat deeper soil and supporting a more luxuriant growth of emergent forbs, especially ironweed (Vernonia baldwini). These forbs provided singing perches for territorial males and may have contributed to the greater densities of some species, especially dickcissels (Table 7.8).

TABLE 7.10 Attributes of Breeding Bird Communities in the True Prairie Locations of Figure 7.9. Values are Means, with Standard Deviations in Parentheses.

Site	Years sampled	Number of species	Density (individual km^{-2})	Biomass (g wet wt ha^{-1})	H'[a]	J'[a]	Dominance[b]
North Dakota	3	5.7 (0.58)	236.7 (108.4)	143.3 (53.3)	1.51 (0.08)	0.87 (0.04)	23.2
Fitchburg, WI	3	6.0 (1.00)	494.2 (112.0)	166.6 (33.3)	1.47 (0.07)	0.83 (0.05)	38.1
Iowa	1	4.0	257.1	not known	1.33	0.96	30.8
Douglas Co., KS	1	4.0	493.0	234.0	1.31	0.95	40.1
Riley Co., KS	3	6.3 (0.58)	456.9 (33.4)	305.7 (27.6)	1.68 (0.02)	0.91 (0.04)	27.4
Donaldson, KS	1	3.0	206.0	116.0	1.06	0.96	45.1
Elmdale, KS	2	3.5 (0.71)	279.5 (6.36)	122.5 (9.2)	1.11 (0.11)	0.90 (0.06)	43.2
Washington Co., OK	1	6.0	595.9	219.5	1.21	0.67	41.0
Osage, OK, plot 1[c]	4	4.5 (0.58)	217.9 (56.9)	120.2 (14.4)	1.20 (0.04)	0.81 (0.07)	36.4
Osage, OK, plot 2[c]	3	3.7 (0.58)	419.3 (112.1)	191.3 (53.2)	1.14 (0.08)	0.90 (0.08)	43.2
Inola, OK, Ungrazed	1	2.0	85.0	75.0	0.50	0.72	80.0
Inola, OK, Grazed	1	3.0	551.0	211.2	0.77	0.71	72.4

[a] See text, page 222.

[b] Proportion of total individuals recorded that belong to the most common species, averaged over all samples.

[c] Replicate census plots. See text, page 214.

Body sizes differ substantially among typical True Prairie species (see Table 3.13), and measures of standing-crop biomass in avian communities thus may exhibit different variation patterns than density measures. The True Prairie plot averages in Table 7.10 suggest a positive, but coarse, relationship between biomass and density. Overall, standing-crop biomass varied from 75 to over 300 g wet wt ha^{-1}. The lack of a close association between biomass and density indicates that plots differ in the frequency of occurrence of birds of different sizes.

Species numbers and total densities apparently vary locally within the same general range of values in most North American grassland types (Wiens, 1974a; Wiens and Dyer, 1975), although densities seem to be reduced in grassland-like agricultural habitats to the east of the Great Plains. Within grassland types, however, standing-crop biomass apparently decreases with generally decreasing annual precipitation and primary production. True Prairie plots consistently support significantly greater biomass than Shortgrass Prairie, Palouse Prairie, or Western Shrub Steppe plots (Wiens, 1974b). Agricultural habitats generally support a relatively low community biomass. By comparison, North American coniferous forests generally contain two to four times as many species as prairie plots, while densities average nearly 3.5 times the grassland densities and biomass roughly twice the biomass in the grasslands (Wiens, 1975).

General community attributes of prairie avifaunas are relatively uniform, at least compared to other vegetation types. Cody (1966a) expressly noted the consistency of species diversity values over a wide range of grassland types and inferred that these avifaunas were saturated; that is, they contained the maximum number of possible coexisting species and had reached maximum diversity. These views of ecological saturation were overly simplified and have been challenged (Wiens, 1974b). But the fact remains that the range of expression in community features in grassland bird communities is decidedly limited.

Measures of species diversity and equitability for the True Prairie census plots are averaged in Table 7.10 for each site. In these analyses, diversity (H') was measured by

$$H' = -\Sigma \ p_i \ \ln \ p_i$$

where p_i = proportion of all individuals present that are represented in the ith species. Equitability (J') is then given by

$$J' = \frac{H'}{H'max}$$

where H'max = ln S (S = number of species present). Diversity values are dimensionless and have meaning only in comparisons. In this regard, avian communities in the more northern locations in the True Prairie (North Dakota, Wisconsin, Iowa, and northeastern Kansas) are apparently more diverse than communities in the more southern locations (central Kansas and northeastern Oklahoma). The H' diversity index is influenced by species richness and by the evenness of distribution of individuals among species. Because the more northern locations contain no more species, on the average, than the southern areas, it seems likely that their greater

diversity is primarily a consequence of greater species equitability. Equitability indices for the northern locations are, in fact, relatively high, but so are some indices for southern sites with lower diversity. Equitability is an inverse measure of <u>species dominance</u>, the degree to which most individuals present belong to the single most abundant species. The greatest degree of single-species dominance occurred in the grasslands at Inola, which has a simple avifauna (Table 7.10); otherwise, dominance values ranged between 23 and 45 percent. In general, low dominance values were associated with relatively high equitability and diversity, as expected, but exceptions were noticeable (e.g., Donaldson Pasture, Kansas).

One can easily become deeply involved in the possible meaning of diversity and equitability measures and values and read into them biological meanings they do not possess. In this case, diversity and equitability values for the True Prairie plot census locations provide quantitative verification of our earlier qualitative observations that breeding bird communities in the True Prairie are comprised of relatively few species and that most of those species are locally quite abundant. Dominance is shared by several of these species, and subordinate (numerically rare or uncommon) species are few. Furthermore, because the relatively few dominant species are present over a broad area of the True Prairie, species replacements are not ordinarily evident and local or regional variations in species composition are slight. Relative densities of the most common species do vary in space, often on a quite local scale, but these variations are rather subtle compared with the general theme of avifaunal uniformity.

Temporal Stability

Bird populations and the communities they form fluctuate in abundance and composition through time, and any characterization of avifaunas of the True Prairie and their possible roles in ecosystem functioning must consider the degree of temporal stability of these populations. Stability or instability may be considered in several time frames, each of which has distinct ecological implications. On one scale, flux in the density or presence of bird populations during the breeding season may be critical; on another scale, interseasonal variations may be substantial; and a third scale may consider annual or year-to-year changes.

Flux in Density

One assumption underlying plot censuses like the censuses just discussed is that the breeding densities sampled are stable and represent true breeding season densities. In fact, however, densities ebb and flow within breeding populations because of territorial shifts, mortality, early abandonment of nestlings, or late renestings. Studies at Fitchburg, Wisconsin (Wiens, 1969) provide data on changes in territory size and numbers of complete territories in a 32-ha area from the arrival of birds in mid-April to the establishment of stable breeding populations in early June

(Figure 7.11). Because plot density estimates are derived from counts of territorial males, these values should at least approximate total density changes. Obviously, patterns of change in territory sizes and numbers occurred in all dominant species, although the flux was more conspicuous in bobolinks and grasshopper sparrows. Censuses taken at other times during these seasonal changes would produce markedly different results. Perhaps the most reliable indication of true breeding densities would be the census taken at the peak of breeding activity, when most nests are active and territories are most stable. The censuses conducted at other True Prairie locations (Table 7.8) were taken at about the same stage in the local breeding phenology. Studies by R. A. Ryder and

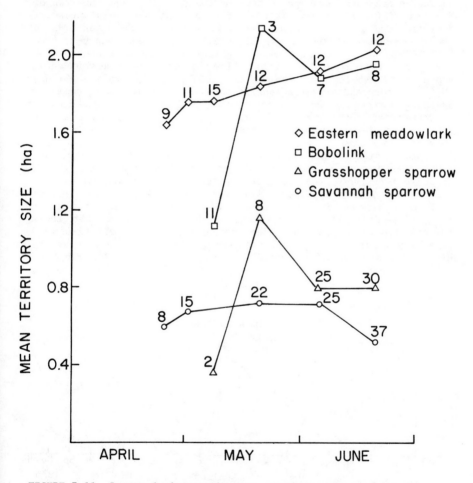

FIGURE 7.11 Seasonal changes in mean territory sizes of breeding birds at Fitchburg, Wisconsin. Numbers are the number of territories mapped at a given time (data from Wiens, 1969, Table 7, page 46).

his students in a Shortgrass Prairie in northeastern Colorado (Wiens and Dyer, 1975) also reveal large within-breeding season density variations. Unfortunately, because no other censuses have been taken from True Prairie locations, the degree of intraseasonal variation cannot be assessed.

Interseasonal Variations

Interseasonal variations in species abundances and distributions are largely products of the degree of residency of the species populations. Because roughly two-thirds of the species breeding in the True Prairie are migratory (see Table 3.14), extensive seasonal turnover in species composition seems inevitable. The best information available to evaluate the degree of seasonal flux in True Prairie bird populations comes from standardized roadside censuses taken at the Osage Site four times each year from 1971-1975. These censuses were taken along a 22-km route through grasslands typical of the Osage Site by standard roadside count procedures (Robbins and Van Velzen, 1967; Rotenberry and Wiens, 1976). Interseasonal variations in abundances of all three dominant breeding species at Osage were extreme (Figure 7.12). Dickcissels exhibited the most extreme seasonal variations, being present only for a short period (generally May to July) and thus were recorded only in the summer roadside counts. Grasshopper sparrows were also seasonal residents but always were recorded in both spring and summer censuses. In 1972 and 1973, spring densities were highest, while in 1971, 1974, and 1975, summer densities were highest. These differences likely depend on whether the spring censuses coincided with the passage of large numbers of migrants through the area. Eastern meadowlarks were resident at the Osage Site, but their numbers fluctuated substantially during the year. Densities were generally greatest in spring, then decreased to a winter low as the birds became both more widely dispersed over a broader range of habitat types and thus less conspicuous (and so less readily recorded by roadside count techniques).

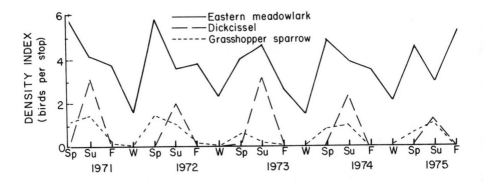

FIGURE 7.12 Interseasonal variations of dominant bird species abundance near the Osage Site.

Roadside censuses provide only an approximation or index of actual densities rather than absolute counts. Thus, although the censuses may provide valuable insights into seasonal and annual changes, they do not reveal the magnitude of these changes. Only intensive study of local, carefully defined areas, such as the plots used in breeding censuses, would provide reliable density estimates. Such information is available only for the two plots at Inola, Oklahoma, for only part of one year (Figure 7.13). Nonbreeding censuses at these locations used the strip census (transect) technique of Emlen (1971). In general, estimates made during the breeding season with this technique correlated closely with estimates obtained from the more careful (and time-consuming) plot censuses. Census estimates clearly show the influx of migrant grasshopper sparrows then and the reduction to lower breeding densities as well as the rapid increase in dickcissel numbers to high breeding densities. Grasshopper sparrows were present only in the grazed plot at Inola (Table 7.8), but other species were present also in the ungrazed plot. Most notably, eastern meadowlarks heavily used the grazed plot as wintering habitat. The ungrazed plot supported only sparse winter densities, and meadowlark numbers increased in this habitat into the spring, perhaps as birds established territories and dispersed from prime wintering habitat. These censuses also revealed the presence of great numbers of wintering Savannah sparrows on both plots. Densities increased between February and late March, perhaps as migrants from farther south began moving northward. Densities were still high in the grazed plot in early May, but a month later this species was absent from study plots at Inola.

Roadside counts also provide information useful for assessing seasonal changes in avian community features. The roadside census route near the Osage Site covers a broader area and a wider range of habitats than are sampled by the plot censuses, so a broader array of species is potentially available for recording. Total number of species recorded in roadside counts at the Osage Site varied from thirteen to thirty-three over the five years (Figure 7.14), but for any given year, lowest species numbers were consistently recorded in winter and highest numbers in summer. These values, however, do not indicate the actual interseasonal changes in species occurrences, only species numbers. Perhaps a more sensitive measure of avifaunal stability can be obtained by measuring interseasonal species turnover (T), the net change in species composition from one season to the next. This measure is given by

$$T = \frac{S_i + S_{i+1}}{S_c + S_i + S_{i+1}}$$

where S_i = number of species unique to count i, S_{i+1} = number of species unique to the next successive count (next season), and S_c = number of species common to both counts. Values of T for the Osage Site counts are plotted in Figure 7.14 and indicate substantial turnover (>50 percent) in species composition between all seasons. Spring-summer turnovers were generally lowest, probably because of the relatively early arrival of many breeding species. Fall-winter

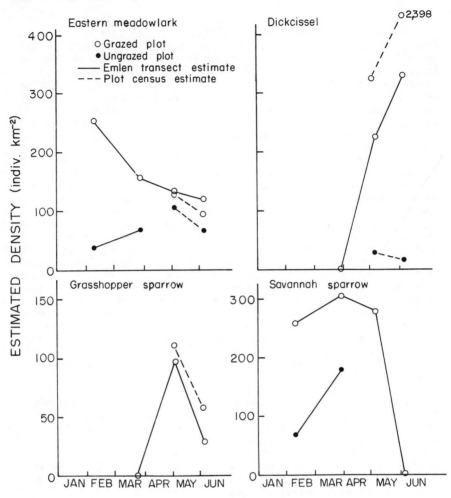

FIGURE 7.13 Temporal changes in estimated density of major bird species at the Inola, Oklahoma site in 1974.

and winter-spring turnovers were generally high, and patterns were relatively consistent from year to year.

Some of the species recorded in roadside censuses are peripheral to the prairie habitat because they are more closely associated with thickets, hedgerows, or other road-type habitates. These species may be defined by their absence from prairie census plots or by intuitive evaluation of their habitat affinities (Wiens et al., 1974). The roadside counts therefore may be divided into "native" and "peripheral" species components. The analysis in Figure 7.14 reveals that the native species component in the True Prairie is at least moderately stable, varying from four to fourteen species. The peripheral species component is much less consistent, varying from four to thirty-two species. Hence, much of the

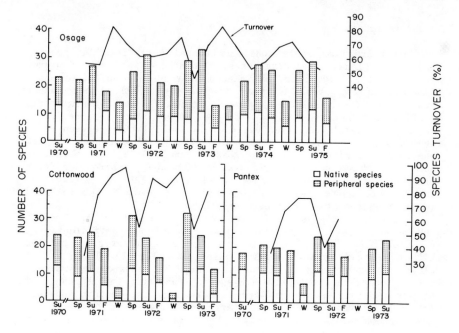

FIGURE 7.14 Number of species recorded in successive seasonal roadside censuses in True Prairie (Osage), Mixed-grass (Cottonwood), and Shortgrass (Pantex) locations. The line plots the index of intercount seasonal turnover in species composition.

interseasonal variation in species numbers and species turnover at the Osage Site is attributable to the peripheral species, many of which occur in low frequency and abundance in the counts.

These variation patterns may be placed in a broader perspective by comparing them with data similarly collected at IBP sites in the South Dakota Mixed-grass (Cottonwood) and Texas Shortgrass (Pantex) Prairies by volunteer observers (Figure 7.14). Because these comparisons include effects of both latitude and habitat, they are not as easily interpreted as one might like. The Mixed-grass Prairie had a larger seasonal range in species numbers, including the native component, largely because of the general scarcity of birds present during the winter on the South Dakota plains. Seasonal species-turnover values, especially the winter-spring turnover, were accordingly higher than the values at the Osage Site. On the Shortgrass Prairie, on the other hand, seasonal variation in total species recorded was less, although few species were encountered in the one winter census. Peripheral species accounted for a smaller proportion of the avifauna than at the other prairie locations, and interseasonal species turnover, although relatively high in fall-winter and winter-spring, was somewhat lower than elsewhere.

Annual Variation

The third time scale considers annual variations in breeding bird populations. Breeding season abundances of the dominant species at the Osage Site, as indexed by roadside counts, were far from stable from year to year (see Figure 7.12). Such annual variations are reflected more precisely in values from breeding-season census plots at the Osage Site (Figure 7.15). Because meadowlark densities varied relatively little between years (contra Figure 7.12), they are not depicted. Grasshopper sparrow densities on replicate plot 1 decreased from 1970-72, then more than doubled by 1974; on plot 2, less than 100 m away, sparrow densities were generally higher and increased from 1971-72. Dickcissel numbers were quite variable, especially on plot 2, but followed similar trends on the adjacent plots.

A broader analysis of the relative stability of species populations over the True Prairie considers coefficients of variation (CV) of breeding density estimates from several sites for which more than a single year's value is available (Table 7.11). Values represent only a sparse sampling, but apparently indicate that whereever the two meadowlark species co-occur, the abundance of the western species varies more than that of the eastern species. Dickcissels appear to be highly variable in abundance in the Kansas and Oklahoma plots, even though these areas represent apparently

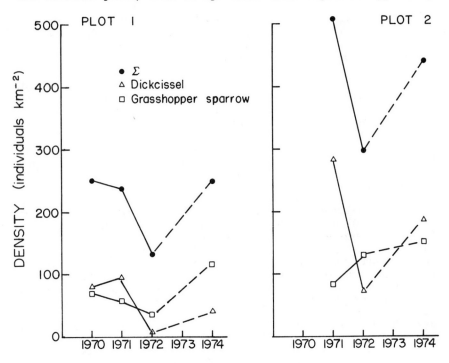

FIGURE 7.15 Yearly density values from breeding census plots on the Osage Site. No censuses were conducted in 1973.

TABLE 7.11 Coefficients of Annual Variation in Density (CV = 100s/$\bar{x}$) of Several Dominant True Prairie Breeding Species, from Plot Counts Conducted in Successive Years at Some of the Sites Listed in Table 7.8.

Sites	Eastern meadowlark	Western meadowlark	Red-winged blackbird	Dickcissel	Grasshopper sparrow
North Dakota	--	0.0	113.8	--	86.6
Fitchburg, WI	33.8	57.2	--	--	27.4
Riley Co., KS	24.1	51.2	16.5	34.6	29.9
Elmdale, KS	5.5	--	--	22.5	26.9
Osage, OK - Plot 1	8.4	--	--	68.4	47.0
Osage, OK - Plot 2	24.3	--	--	57.7	28.4
Average where present[a]	19.2 (11.9)	36.1 (31.4)	65.2 (68.8)	45.8 (21.0)	41.0 (23.6)

[a] Standard deviation in parentheses.

optimal breeding habitat within the main range of this species. Surprisingly, the local abundance of grasshopper sparrows seems to be nearly as variable as that of dickcissels, although less so at the Osage Site than elsewhere. Red-winged blackbirds (Agelaius phoeniceus), one of the peripheral grassland species, exhibited substantial annual variation in abundance, at least in the North Dakota site.

The same approach may be used to evaluate relative annual stability of various attributes of entire avian breeding communities in these True Prairie locations (Table 7.12). Species number and diversity appeared to vary most at the Elmdale site in the central Kansas Flint Hills and in the second census plot at the Osage Site and least in the northeastern Kansas location and at the first Osage Site plot. Relative stability of avian densities varied widely among these sites, and these variations were generally paralleled by differing degrees of stability in total standing crop biomass of birds. Annual density and biomass variations were greatest at the North Dakota Mixed-grass Site and least in the Kansas locations. At the Osage Site, annual density variations in the two plots were similar, but biomass was more stable in plot 1 because the species contributing most heavily to total biomass (upland plovers and eastern meadowlarks) had more stable population densities on plot 1 than on plot 2 (see Table 7.9).

Relatively few measurements are available, but on the average, total breeding densities in the True Prairie appear to be somewhat more variable than densities in the northwestern Palouse Prairie or montane grasslands and less variable than densities in western arid shrub steppe areas (Wiens and Dyer, 1975). Interestingly, bird communities in coniferous forests in many areas of North America appear to be no more stable on an annual basis than the low-diversity communities of the True Prairie, despite the apparently greater climatic moderation in forests (Wiens, 1975).

TABLE 7.12   Average Coefficients of Annual Variation of Avian Community Features in True Prairie Plots Censused in Successive Years.

Site	N	Species no.	Density	Biomass	Diversity
North Dakota	3	10.1	45.8	37.2	5.26
Fitchburg, WI	3	16.7	22.7	20.0	4.43
Riley Co., KS	3	9.2	7.3	9.0	0.91
Elmdale, KS	2	20.2	2.3	7.5	10.19
Osage, OK; plot 1	4	12.8	26.1	12.0	3.66
Osage, OK; plot 2	3	15.6	26.7	27.8	7.09

Thus, annual variations in total breeding densities and in the population levels of the constituent species of True Prairie avifaunas appear to be relatively large. However, the species composition of these communities remains relatively unchanged from year to year over large regions of grassland, especially if only the dominant or characteristic species are considered. Interpreting these variations in breeding bird communities is not a straightforward task. Conventional wisdom suggests that avifaunal variations should be closely related to contemporary climatic variations, but because of the differences in plots subjected to various grazing intensities or separated by short distances in a single area, close tracking of climatic variables by bird populations or communities is difficult to discern. Furthermore, because many species are migrants, their abundance at a location in any given year may result from past and present conditions elsewhere in the species' range (a wintering area) as well as the conditions of the breeding area. In fact, Fretwell (1973) suggested that the large annual variations in dickcissel breeding densities in North American breeding areas may reflect conditions on wintering grounds in South America.

Another possibility is that variations recorded in plot censuses from year to year are caused not by real changes in population sizes but by differences in local distribution and dispersion (Wiens and Dyer, 1975). Thus, the number of individuals in an area of several square kilometers may change little from year to year, but the samples from 0.1-km^2 plots may change greatly because of yearly changes in dispersion patterns within the population, which may be largely unrelated to local plot conditions. Consider the analogy of individual territories as checkers on a checkerboard with walled edges. The board is shaken every year to redistribute the checkers, and while the total number of checkers on the board may remain constant, the number encountered on any sample plot of, say, eight squares may change. Moreover, the direction of change on adjacent eight-square plots might well differ. Obviously, the extent of such change would be related to the packing of checkers (territories) on the board (habitat). On a densely packed board, little room would be available for yearly redispersal, and plot counts would indicate relative stability. This analogy suggests that variations in population densities in plot censuses are closely linked to real or causal environmental variations only if the available habitat is fully occupied or saturated. If the habitat is not saturated, the variations recorded in plot censuses may or may not reflect true population changes, and the search for factors either statistically or causally correlated with the measured density variations may well be in vain.

How common are conditions of nonsaturation or incomplete packing of available habitat in the True Prairie? Cody's analyses (1966a, 1968, 1974) assumed saturation, documented by relative constancy in species numbers and diversity over widely distributed grasslands. But this is evidence of saturation only at the general level of species packing, not at the finer level of individual densities or population occupancy of suitable habitat. North American grasslands are characterized by climatic regimes that are both severe and unpredictable (Wiens, 1974b). Recurrent but unpredictable large magnitude variations in climate (and thus

production) may impose "bottlenecks" on bird populations in prairies, reducing population sizes and filtering out marginally adapted species. During periods between these occasional extremes, resources may be plentiful during the breeding season and population sizes may be too small to occupy all the habitat available to them. Under such conditions, much irrelevant variation may occur in avian community features (Wiens, 1977a). Based on climatic (i.e., precipitation) predictability and variation, these extreme conditions may be expected to hold more frequently in the more arid Shortgrass Prairies than in the True Prairie. Nevertheless, the climatic irregularity of the True Prairie probably forestalls complete spatial packing, or saturation, of habitats by breeding bird populations.

Ecological Structure

The foregoing comments and analyses have treated True Prairie bird communities as groups of taxonomic units (species), but additional insights may emerge if species designations are discarded in favor of ecological attributes such as seasonal mobility, general dietary habits, or body size.

Migratory Tendencies. Information from plot censuses (Figure 7.16a) indicates that over the region, slightly over 60 percent of the species are seasonal migrants that occur at a given site only during the breeding season. This value agrees closely with values derived from more general faunal analyses (see Table 3.14). Intuitively, we assume that migratory tendencies should be expressed most strongly in the more northerly locations and that full-year residency should be greatest in the southern areas. Figure 7.16 indicates otherwise, however. The degree of sedentariness of the avifauna does vary among locations, but the greatest proportions of permanent residents occur in northeastern Kansas and Iowa, while a greater proportion of the breeding species are migrants at the Osage Site than elsewhere. A more general analysis of grassland bird communities (Wiens, 1974b) indicated that the True Prairie and Mixed-grass Prairies contained a significantly greater proportion of migratory species than did shortgrass prairies, and suggested that these patterns might reflect the greater food supply (chiefly seeds) in shortgrass prairies during the winter, which promotes full-year residency.

Dietary Habits. If such relationships do exist, they may be expressed in the general dietary habits of the birds during the breeding season. Categorizing species roughly as omnivores or carnivores (Wiens, 1974b) revealed that omnivory--that is, some dependence on seeds during the breeding season, which is presumably related to a nearly complete seed diet during winter--is more strongly expressed in bird populations inhabiting Shortgrass Prairies than in other grassland populations. Within the True Prairie the expression of omnivory in breeding avifaunas varies locally (Figure 7.16b), although in all cases except one (the atypical Inola, Oklahoma ungrazed plot), omnivory is substantially greater than carnivory. No overall patterns in dietary habits emerge, however, and even at the Osage Site, the degree of omnivory of bird communities occupying the two census plots differed.

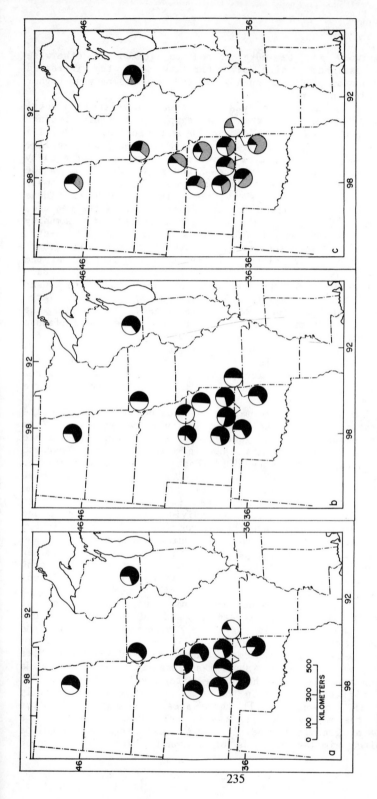

FIGURE 7.16 Proportions of migrant (solid) and resident (clear) bird species in True Prairie census locations (a). Proportions of omnivore (solid) and carnivore (clear) bird species in True Prairie census locations (b). Proportions of small- (≦ 25 g; black), medium- (26–80 g; stippled), and large- (> 80 g; clear) sized species in True Prairie census locations (c).

235

However, omnivory may be more common on grazed or otherwise "stressed" grasslands, which may correlate with the increased abundance of annual plants on such areas and the accompanying higher levels of seed production.

Body Size. Total density was only coarsely related to total standing crop biomass in the plot samples, probably because differences in the body sizes of constituent species (Table 7.10) dilute this relationship. Over a broad range of grasslands, small-, medium-, and large-sized species seem to be more or less equally abundant in the True Prairie and Mixed-grass Prairies, while medium-sized forms predominate in Shortgrass Prairies, and small species dominate Shrub Steppe avifaunas (Wiens, 1974b). These general findings are confirmed for the True Prairie by the censuses (Figure 7.16c), but as with the other measures of community structuring, distribution of individuals varies among body-size classes as well. At many locations, the distribution of individuals among the three body-size classes was relatively equal, but at the Fitchburg, Wisconsin site nearly 75 percent of all individuals present belonged to small species, while at the northeastern Kansas locations and in the grazed plot at Inola, Oklahoma, large proportions of the avifauna were medium-sized. Except for the Wisconsin area, body-size composition of the breeding avifaunas varied substantially (Figure 7.17), although dominance by any single size class was unusual. Although plausible explanations may be offered for regional variations in body-size ratios throughout the North American grasslands (Wiens, 1974b) or broader habitat zones (Wiens, 1975), the determinants of these within-True Prairie variations remain elusive.

Seasonal Mobility. Information on seasonal changes in the ecological structuring of True Prairie avifaunas may be gleaned from the roadside census results (Figure 7.18). Because more species are encountered in roadside counts than in intensive plot censuses, broader ecological groupings (guilds) (Root, 1967) seem appropriate. At the Osage Site, small ground-feeding species (e.g., most sparrows, horned larks) accounted for 10 to 30 percent of the individuals, peaking in importance in summer. Meadowlarks contributed the greatest share of individuals during spring and decreased in relative abundance through summer and fall into winter. Other large ground-dwelling species (chiefly upland plover and greater prairie chicken [Tympanuchus cupido]) followed similar trends but were never so abundant. Air-foraging forms were absent in winter (all are insectivores) and contributed a rather consistent 7 to 10 percent of the avifauna in other seasons. Species associated with brushy roadside areas comprised 70 percent of the individuals in winter and between 20 and 30 percent in other seasons. All species in this guild are peripheral species in True Prairie ecosystems.

Similar data from roadside counts at a Mixed-grass Site (Cottonwood) and the Texas Shortgrass Prairie (Pantex) provide additional perspective on the Osage Site results (Figure 7.18). Contributions of meadowlarks to the total avifaunas were strikingly similar at all seasons in all three areas, suggesting that the dynamics of the meadowlark role may be relatively uniform over a broad range of grassland conditions. Air and large ground guilds also exhibited similar patterns of seasonal change in the three

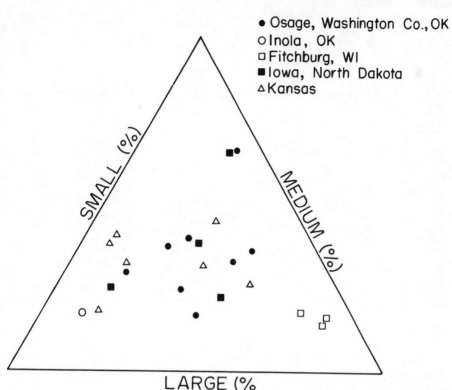

FIGURE 7.17 Distributions of body sizes of breeding birds in True Prairie avifaunas, according to the size categories of Figure 7.16. The perpendicular distance of a point from each side is a measure of the proportion of the total breeding avifauna in that size class.

areas, but small ground-feeding forms were proportionally more plentiful at the Shortgrass Site than at the Osage Site in spring and during the summer-fall plateau, and declined sharply in relative abundance in winter. Trends in relative contribution of this guild were opposite in the Mixed-grass Prairie, where small ground-feeding birds were relatively scarce in spring and summer but composed nearly half of the fall-winter avifauna. Roadside brush species in the Shortgrass Prairie followed generally similar seasonal trends as at the Osage Site, but in Mixed-grass Prairies such birds were proportionally more abundant in spring and summer and declined to low wintering levels.

Energy Flow

Information on avian species' presence, abundance, and community structure may be obtained from direct field observations and censuses, and biomass values may be calculated from samples of species' weights. These measures provide information on the status

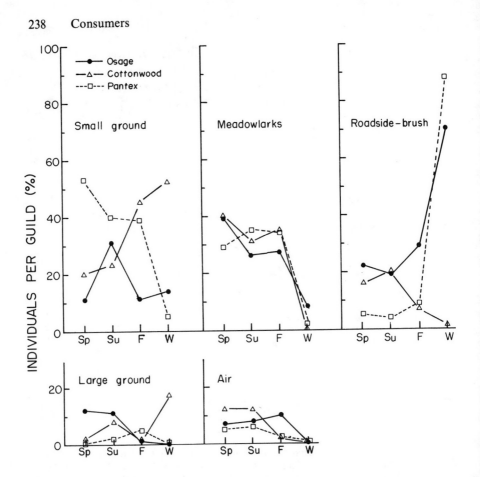

FIGURE 7.18 Average percent of individuals recorded in roadside censuses representing several major foraging guilds in spring, summer, fall, and winter periods.

of bird populations as components of the True Prairie ecosystem but do not link bird populations to other system components in any meaningful fashion. Functionally, birds are integrated into the fabric of ecosystem dynamics through their roles in energy flow and trophic dynamics. At present, however, obtaining direct measures of energy dynamics of free-living bird populations is virtually impossible (e.g., Gessaman, 1973). As an alternative, an indirect approach based on consideration of individual metabolism and projections of population dynamics may be followed, relying heavily on computer modeling. In addition to computational ease, such an approach has the advantage of permitting detailed dissections of the energy flow patterns in simulated bird populations and communities. Various components of the ELM model (see Chapter 9) treat consumers only in general, primarily with reference to mammalian consumers. The more realistic and detailed analyses of energy flow through bird populations presented here used a separate simulation model (BIRD) specifically developed to model the population dynamics and

bioenergetics of grassland birds (Wiens and Innis, 1973, 1974). The details of the structure and the assumptions of the model are given in the Wiens and Innis papers and will be summarized only briefly here.

Two basic submodels compose the BIRD model (Figure 7.19). In one--the population submodel--information on various life history attributes of a population (e.g., clutch size, hatching success, immigration and emigration timings and patterns, reproductive phenology) is coupled with field estimates of breeding population density to generate model simulations of the density of each age class of each species at any given time during the breeding season. The second--the energy submodel--estimates energy demands placed upon the ecosystem by individuals of these populations through time. Individual energy demands are determined from calculations of existence energy metabolism (EMR), following Kendeigh (1970):

$$EMR_0 = 4.337 \ W^{0.53} \ \text{at } 0°C \text{ for all species}$$
$$EMR_{30} = 1.572 \ W^{0.62} \ \text{at } 30°C \text{ for passerines}$$
$$EMR_{30} = 0.540 \ W^{0.75} \ \text{at } 30°C \text{ for nonpasserines}$$

where W = body weight (g). These functions have been revised recently to include effects of photoperiod and the results of additional studies (Kendeigh et al., 1977; Wiens and Dyer, 1977), but the changes do not drastically affect model output. Through these functions, the existence energy demands (i.e., the energy expended in standard metabolism, specific dynamic action, and limited locomotor activity, in kcal bird^{-1} day^{-1}) of individuals of different age classes of different species may be calculated for any given date and ambient temperature, with field values of temperature and body weights. The existence energy demand values are then adjusted to reflect the additional energy costs of growth (for young), activity, molt, and egg production. The adjusted individual metabolic energy demand estimates are then multiplied by 1.43, which reflects a 70 percent assimilation efficiency (Kale, 1965), to produce an estimate of energy demands placed on the ecosystem by individuals. Integrating the energy estimates with the estimates of the population dynamics submodel permits calculation of the energy demand of the entire population of a species on any given day. The BIRD model thus generates daily estimates of population density in each age class, total biomass in each age class, and the bioenergetic demands placed on the food resources by each age class for all of the species in an area.

Model analyses were conducted of the bird populations breeding at the Osage Site only, where field studies provided a broad framework of ecosystem dynamics within which the results of BIRD simulations could be considered. Only energy flux patterns during the breeding season (1 April to 31 August) were considered, because quantitative measures of bird populations at the Osage Site during the winter were scarce. These analyses provide unidirectional estimates of energy flows only, in that the amount of food consumed by birds, as projected by the model, does not feed back to influence the amount of food available to them.

The results of model analyses of energy flow through breeding bird populations at the two replicate Osage Site study plots are summarized in Table 7.13. Total energy flow through breeding

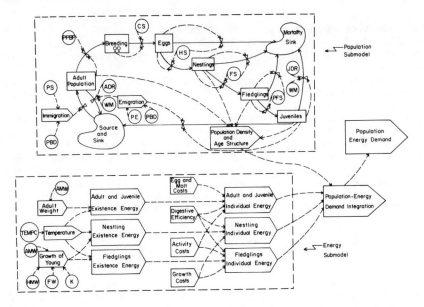

FIGURE 7.19 Generalized compartmental diagram of the BIRD model
structure. Rectangular boxes indicate state variables; five-sided
boxes, computational controls; circles, input variables. Solid
arrows indicate flows of materials or energy or changes of state;
dashed arrows indicate controls or computational transfers. (ADR =
adult daily mortality rate in breeding season; AMW = adult mean body
weight; CS = clutch size; FS = fledging success; FW = fledgling
weight; HMW = weight at hatching; HS = hatching success; JDR =
juvenile daily mortality rate; K = growth rate; PBD = population
breeding density; PE = population at end of run; PFS = post-fledging
survival; PPBF = proportion of population that is breeding females
(first brood); PS = population at start of run; TEMPC = temperature;
and WM = adult mortality rate in winter season) (from Wiens and
Innis, 1974; copyright 1974 by the Ecological Society of America).

populations of all species present on these plots during the 150-day
breeding period ranged from about 1 to 3 kcal m^{-2} season^{-1}. As
expected from the greater density and biomass of birds present (see
Table 7.10), total energy flow in plot 2 was greater than in plot 1.
Over a broader range of grassland and Shrub Steppe conditions,
breeding season energy flow averaged between 1.1 and 2.1 kcal m^{-2}
season^{-1} (Wiens and Dyer, 1975; Wiens, 1977b), so the average energy
flow at the Osage Site (1.75 kcal m^{-2} season^{-1}) is toward the higher
end of that range. But variations in the magnitudes of energy flow
through grassland bird populations show little apparent pattern.
The bird community in the mesic True Prairie at the Osage Site
processed more energy, on the average, than the breeding bird
community in the Mixed-grass Prairie plots in South Dakota, but the
birds breeding in dry Shortgrass Prairies in the Texas Panhandle
required just as much energy over the breeding season as did the
Osage Site avifauna (Table 7.13). Breeding season energy flow at
the Osage Site varied somewhat more over sample plots and years than

TABLE 7.13 Energy Flow Through Breeding Bird Populations at Two Osage Site Study Plots.

Site	Year	Total energy*	Percent of total production	Excretion	Respiration	Percent to				Percent to seasonal residents
						Egg	Nestling	Fledgling	Σ Reproduction	
Osage 1	1970	1.56	0.024 (1.5)	0.46	1.08	0.39	4.62	13.34	18.35	22.9
	1971	1.49	0.021 (1.4)	0.44	1.03	0.38	4.32	12.90	17.60	26.5
	1972	1.08	0.017 (1.5)	0.32	0.74	0.41	4.41	14.36	19.18	10.9
Osage 2	1971	2.92	0.041 (1.4)	0.87	2.02	0.38	4.33	11.82	16.53	48.4
	1972	1.70	0.025 (1.4)	0.50	1.17	0.38	4.42	12.41	17.21	24.8
Osage										
X̄		1.75	0.026	0.52	1.21	0.39	4.42	12.97	17.77	26.7
SD		0.70	0.009	0.21	0.48	0.01	0.12	0.96	1.03	13.6
CV		39.8	35.2	39.9	39.8	3.34	2.73	7.43	5.77	50.9
Cottonwood (N=8)										
X̄		1.33	0.019	0.39	0.92	0.40	4.02	12.57	16.99	54.2
SD		0.35	0.006	0.10	0.24	0.12	0.58	1.82	2.41	40.5
CV		26.1	35.5	25.9	25.9	30.47	14.4	14.49	14.17	74.7
Pantex (N=8)										
X̄		1.77	0.018	0.53	1.23	0.25	3.32	9.33	12.90	0.1
SD		0.58	0.006	0.17	0.40	0.07	0.31	1.42	1.75	0.3
CV		32.5	34.6	32.5	32.5	26.51	9.31	15.20	13.60	265.2

* (kcal m^{-2} season^{-1}).

in other grasslands, and variation in energy flow through bird populations appears to decrease as aridity increases (Wiens, unpublished). In general, however, similarities in energy flow magnitudes between plots and years at the Osage Site or between a broad range of grassland sites are more impressive than the differences. Of course, one might expect these similarities if local avifaunas are not closely tracking the carrying capacity of their environment and if population densities and biomass are determined by factors unrelated to proximate site conditions.

The absence of well defined patterns in the magnitude of energy flux through bird populations at the Osage Site, or in grasslands as a whole, does not imply that broader patterns do not exist. In an analysis of breeding season energy flow through bird communities in northwestern coniferous forests, Wiens and Nussbaum (1975), using the BIRD model, reported values of 10.5 to 20.8 kcal m^{-2} season^{-1} (for a 190-day breeding season). Thus, energy flow through bird populations in coniferous forests is roughly ten times the energy flow in grasslands, although standing crop biomass is only two to three times greater. The magnitude of difference reflects the dominance by smaller-sized species and the generally cooler climate in coniferous forests.

The BIRD model simulates daily energy demands, so daily as well as seasonal patterns can be considered. Daily energy flux in plot 1 at Osage varied from about 0.005 to 0.015 kcal m^{-2} day^{-1} (Figure 7.20). Although daily energy demand varied throughout the breeding season because of fluxes in species and age class composition on the site, these variations were relatively slight. On the other hand, in plot 2 at the Osage Site daily energy flow varied substantially through the breeding season, ranging from 0.005 to nearly 0.030 kcal m^{-2} day^{-1}.

Because model simulations of energy fluxes are partitioned by age classes, partitioning of energy flow within breeding bird

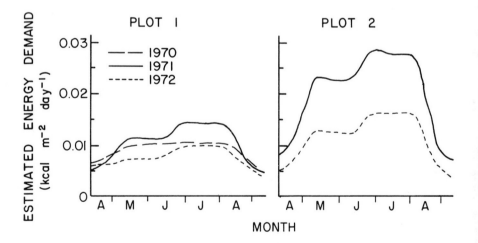

FIGURE 7.20 Estimated daily energy demands of the total assemblage of birds breeding on the two grazed plots at Osage, 1970-72.

communities can be estimated in general. Percentages allocated to respiration and excretion (i.e., costs of digestive processing and assimilation) are dictated by the assimilation efficiency factor (70 percent) built into model analysis, but allocation of energy to production is less apparent. This allocation includes energy costs of production and growth (but not maintenance) of offspring as well as of changes in body weight. Because adult weight was considered to be constant during the breeding season, "production" refers strictly to the production of eggs and growth of young. At the Osage Site (Table 7.13), production rather consistently accounted for 1.4 to 1.5 percent of the total energy flow during the breeding season.

The energy allocation to production during the breeding season is only one element of an overall energy commitment to reproduction (as distinct from self-maintenance). The magnitude of energy allocation to reproduction is a central element of the reproductive strategy of a population (Cody, 1966b; Gadgil and Solbrig, 1972; Pianka and Parker, 1975), but few calculations of the energy associated with reproduction are available for bird populations. We can obtain a conservative estimate of the energy channeled into reproduction related processes by combining the costs of egg production with the total energy demands (growth plus maintenance) of offspring. However, such a calculation ignores the energy costs of reproductive behavior of adults. At the Osage Site, egg production costs amounted to about 0.33 percent of the total breeding season energy flow through the bird communities, while nestlings accounted for nearly 4.5 percent and fledglings for almost 13 percent (Table 7.13). Overall, reproduction accounted for nearly 18 percent of the total breeding season energy flow, a proportion slightly higher than that characterizing the bird community at the Mixed-grass Site and substantially greater than that in the Shortgrass Prairie. This allocation to reproduction was relatively constant among samples at the Osage Site, unlike the variability of this measure in the Mixed-grass and Shortgrass Prairie locations.

Even though roughly 75 percent of the species breeding at the Osage Site were seasonal migrants (Figure 7.16a), an average of only 25 percent of the breeding season energy flux passes through seasonal resident populations (Table 7.13). This low percentage is apparently because, although a relatively small proportion of the species inhabit the Osage Site as permanent residents, the birds that do are large-sized and abundant (chiefly meadowlarks). At the more northerly Mixed-grass Site in South Dakota, over 50 percent of the breeding season energy flow was to seasonal residents, while at the Texas Shortgrass Site virtually all the energy flow was to permanent residents (Table 7.13).

Results of BIRD model analyses also can be used to consider aspects of energy dynamics of individual species at the Osage Site. In all samples (plots and years), eastern meadowlarks contributed most to the total energy flux through the Osage Site bird populations, while upland plovers contributed least (Table 7.14). Energy flow through dickcissel and upland plover populations varied greatly; the flow through meadowlark populations was fairly constant. Of the four major species breeding at the Osage Site, only meadowlarks were full-year residents, which is reflected in the pattern of daily energy demands (Figure 7.21). Changes in daily

TABLE 7.14 Total Energy Flux (kcal m^{-2} season^{-1}) Through the Osage Site Bird Populations.

Species	Plot 1			Plot 2		$\bar{X}$	SD	CV
	1970	1971	1972	1971	1972			
Upland plover								
Σ	0.079	0.059	0.061	0.250	0.160	0.122	0.083	67.8
R (percent)[a]	17.3	15.9	15.8	15.9	16.0	16.1	0.63	3.91
Eastern meadowlark								
Σ	0.976	0.900	0.834	1.300	0.842	0.970	0.193	19.9
R (percent)	20.8	20.6	20.6	22.0	20.6	20.9	0.61	2.91
Dickcissel								
Σ	0.280	0.355	0.031	1.004	0.262	0.382	0.366	95.9
R (percent)	12.2	11.6	11.6	11.6	11.6	11.7	0.27	2.29
Grasshopper sparrow								
Σ	0.229	0.195	0.124	0.208	0.437	0.239	0.111	49.2
R (percent)	20.4	29.9	14.6	19.5	14.4	19.8	6.28	31.7

[a] Percent of reproduction.

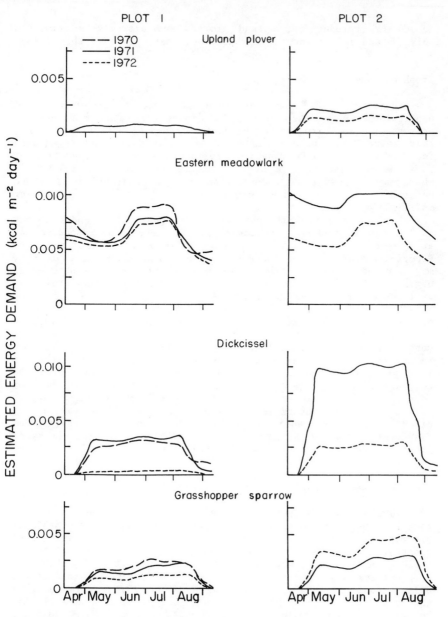

FIGURE 7.21   Estimated daily energy demand of breeding bird
populations on two grazed plots at Osage, 1970-72.

energy demands were most spectacular for dickcissel populations,
which built to high densities shortly after arriving in the area.
The species also differed in the proportion of total breeding season
energy flow allocated to reproduction.   Meadowlarks and grasshopper
sparrows, on the average, used 20 percent of the seasonal energy

demand in reproduction; upland plovers and dickcissels allocated only 16 and 12 percent to reproduction, respectively (Table 7.14).

Trophic Dynamics

These patterns and magnitudes of energy flow tie bird populations into the functional integrity of True Prairie ecosystems. Yet birds do not directly consume kilocalories; rather, they search for, pursue, and capture individual prey organisms. Food webs that typify rangeland avifaunas form the foundation of their involvement in ecosystem structure and the framework within which most ecosystem processes occur, and disruption of these natural food webs may precipitate ecosystem imbalance or modification.

Food is an essential resource to all populations and one that is frequently assumed to be limiting. As such, food resources may at least theoretically represent niche dimensions that are especially sensitive to competitive adjustments and that, therefore, may play an important role in determining community structure and organization. Thus, documentation of the trophic relationships of at least the major species is essential.

Composition of the diet of a population results from a complex process of diet selection that is influenced by various characteristics of the consumer, the prey, and the environmental context (Ellis et al., 1976). Certain kinds and quantities of food resources are present in an area, and the way this resource spectrum is exploited by the consumers can be estimated by examining the contents of their digestive tracts. But the relationship between food presence and dietary composition is far from direct; a variety of factors may intercede to bias diet selection in various directions. Thus, when we analyze the dietary composition of consumers, we are examining the end result of a complex series of processes about which we know relatively little.

Dietary Composition of True Prairie Birds

The general analyses presented in Chapter 3 suggested that roughly two-thirds of the birds breeding in True Prairie habitats are omnivorous during the breeding season and consumed a mixture of plant (primarily seed) and animal prey (see Table 3.14). Although these food habits contrast with the predominantly insectivorous propensities of birds breeding in woodland or forest habitats (e.g., Holmes and Sturges, 1975; Wiens and Nussbaum, 1975), this level of analysis lacks the detail needed to consider the trophic relationships of birds in the True Prairie. More complete information comes from analysis of stomach contents of individuals collected at the Osage Site from 1970 to 1972. Before reviewing results of these analyses, we should note two sources of bias. First, the specimens were collected during a one- to two-week period at the peak of the breeding season in early June. Because avian diets are known to fluctuate substantially during a season in response to changing patterns of prey availability and changing nutritional demands of birds, estimates generated from a short

interval during this season may be incomplete.   Second, these
analyses were based on the identification and enumeration of prey
items (actually fragments of prey items) found in the digestive
tract.   Contents of the digestive tract of individuals were
preserved immediately after collection in the field, but the
digestibility of prey types differs.   Thus, prey items that are
digested slowly by birds will remain in the stomach longer than
items that are digested rapidly and will appear in the collections
in disproportionately high frequencies.   Although these biases are
not insurmountable, the nature of the studies did not permit
sufficiently intense sampling to deal with them.   The results
discussed in the next paragraphs must thus be considered as only
general indications of trophic patterns.

The compositions of the diets of the major bird species
breeding at the Osage Site are given in Table 7.15 and depicted more
generally in Figure 7.22.   Upland plovers were not abundant on the
site and were collected in only two years.   Beetles, especially
weevils and carabids, dominated the plovers' diets, and crickets and
(in 1970) isopods (Crustacea) contributed significant fractions.
During the sampling period, plovers were almost exclusively
carnivorous.   Eastern meadowlarks also were carnivorous but consumed
a greater variety of prey taxa.   Lepidopterous larvae comprised
nearly half the meadowlark diet in the 1970 sample but were much
less important in 1971 and 1972.   In 1971, carabid beetles and
orthopterans (both grasshoppers and crickets) were important diet
constituents; in 1972, crickets were absent from the samples,
carabids accounted for over half the total prey consumed, and
scarabaeid beetles were more important than in previous years.
Dickcissels were consistently the most omnivorous species breeding
at the Osage Site, with 25 to 33 percent of the diet consisting of
seeds (almost entirely of grasses).   Lepidopterous larvae were a
substantial part of the diet in 1970 and 1971 but, as in the
meadowlark diet, were a much smaller part in 1972.   Grasshoppers

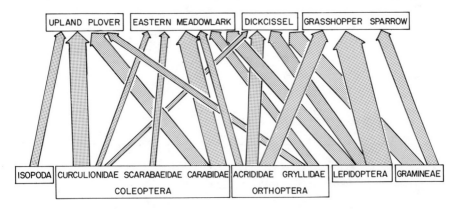

FIGURE 7.22   Diet relations of the major bird species at the Osage
Site, 1970-72 average.   Width of the bar is proportional to the
contribution of a prey group to the diet of a species.   Taxa
contributing less than 5 percent to the diet or present in only one
year are not shown.

TABLE 7.15 Composition (Percent Dry Weight) of the Diet of the Major Bird Species Breeding at the Osage Site (T = <1 Percent).

Prey taxa	Upland plover 1970	1971	Eastern meadowlark 1970	1971	1972	Dickcissel 1970	1971	1972	Grasshopper sparrow 1970	1971	1972
INSECTA											
Hymenoptera		1	3	T	4	1		1		T	T
Formicidae		1	3	T	4			T		T	T
Others			T		T	1		1		T	
Coleoptera	10	76	29	39	77	6	24	3	7	9	5
Curculionidae	18	44	9	7	5	4	17	1	5	6	3
Cerambycidae			1		0	1	5				
Scarabaeidae		2	10	1	18						T
Carabidae	21	30	7	29	53		1	T	2	2	2
Tenebrionidae											
Chrysomelidae	1		T	T							T
Others			2	2		T	1	1	T	1	
Orthoptera	17	9	20	41	15	29	7	31	29	14	19
Acrididae			4	13	15	26	6	31	29	14	19
Tettigoniidae											
Gryllidae	17	9	16	28		3	1				
Others											
Hemiptera	3	3	T	T	T	1	T	T	3	T	2
Miridae											
Pentatomidae	1		T						3		
Others	2	3	T	T	T	1	T	T		T	2
Lepidoptera		4	46	16	4	34	34	4	53	58	26
Diptera				T		T					
Neuroptera						T					1
Odonata								33			
Unidentified insects				T							T
ARACHNIDA											
Araneida	6			2	T		1	T		1	2
Others			T			T			1		
CRUSTACEA	32		T	T							
ANGIOSPERMAE											
Gramineae	2	7		T	T	25	33	26	8	17	39
Others					T	4			T	T	T

248

accounted for a significant portion of the diet in 1970 and 1972, but in 1971 beetles (chiefly weevils) were most significant. The dietary opportunism of the dickcissel may be revealed by the yearly variations in diet. In 1972, for example, dragonflies comprised a third of the diet but were absent from the samples in other years. The Osage Site included a small pond, so the occurrence of such prey in the diets in 1972 may be related to an emergence of these aquatic insects in that year that did not occur during the 1970 and 1971 sampling periods. Grasshopper sparrows fed mostly on lepidopterous larvae, although, again, the contribution of such forms to the diet decreased in 1972. Grasshoppers comprised a significant share of the sparrows' diets, and seeds (especially grass seeds) were important, especially in 1972. Beetles were not a major dietary component in any year.

Although diets of the bird populations breeding at Osage differ from year to year, there is still an overall consistency to their composition. These relationships may be quantified by calculating overall dietary similarity between years for each species, using Horn's (1966) R_o similarity index and the information in Table 7.15. Index values (Table 7.16) are not easily interpreted because deciding how similar entities must be to be considered essentially the same is a subjective judgement. Diets of upland plovers in 1970 were moderately similar to the diets in 1971. Between-year diet similarity was generally greater in other species, and in most cases, successive years were characterized by greater overall similarity than the 1970-72 comparison. Dickcissel diets in 1972, however, bore less than 60 percent resemblance to diets of previous years. In general, the analyses suggest that the dietary composition of these species does not vary greatly between years. Although much discussion could be made of yearly as well as individual variations, only general dietary relationships will be considered in the following analyses, and yearly variations will be ignored for the most part.

Although all species breeding at the Osage Site exploit a common resource pool, species differences were evident in feeding patterns. Using Horn's similarity index again, we calculated values of overall dietary similarity between species for each year (Table 7.17). Overall similarity between the small sample of upland plovers and the dickcissels was low, and similarity of plovers with eastern meadowlarks was moderate. Meadowlark diets showed moderate similarity to dickcissel diets, and similarity between the diets of the two finch species was relatively high. In general, the degree of dietary similarity was associated with similarity in body size.

The foregoing analyses of dietary composition were based on taxonomic categories of prey, but it is unlikely that the birds will respond to the same differences between prey that taxonomists have used to classify taxa. Grouping prey taxa into functionally defined categories may be more meaningful. Table 7.18 represents one way of grouping the frey taxa into categories. As noted, grass seeds were the major component of the plant prey for all species; forb seeds made only minor contributions. For some birds, beetles were important in the diet. Placing beetles in three functional groups reveals the relative importance of phytophagous forms among the rather small total quantities of beetles consumed by grasshopper sparrows and dickcissels and the relatively equal contribution of

TABLE 7.16 Similarities (Horn's R_o) Between Yearly Samples of the Diets of Breeding Species at the Osage Site, from the Dietary Information of Table 7.15.

Upland plover		1971	
	1970	0.70	

Eastern meadowlark		1971	1972
	1970	0.81	0.57
	1971		0.71

Dickcissel		1971	1972
	1970	0.87	0.68
	1971		0.58

Grasshopper sparrow		1971	1972
	1970	0.93	0.80
	1971		0.87

all three groups of beetles to upland plover diets. Meadowlarks consumed a larger proportion of ground-dwelling beetles than other bird species. Phytophagous forms dominated the beetle portion of their diet during 1970, and predaceous beetles were more important in 1971 and 1972. Chewing phytophagous insects are the most important animal prey group for all species in all years. Other animal prey categories were relatively unimportant in the diets of dickcissels and grasshopper sparrows, but predaceous and omnivorous invertebrates comprised a significant portion of upland plover and eastern meadowlark diets. Sucking phytophagous forms and flower-feeding or scavenging insects were generally not important elements in the diets of any of the breeding species.

Analyses of the dietary information of Tables 7.15 and 7.18 using niche-breadth metrics in the same fashion as before reveal a general intermediacy of diet niche breadths of the breeding species at the Osage Site (Table 7.19), especially for diet taxa. By this measure, grasshopper sparrows had the most restricted diet and

TABLE 7.17   Dietary Similarity (Horn's R_o) Among Breeding Bird
Species at the Osage Site, Based Upon the Dietary
Information of Table 7.15.   Values are Averages of
All Years Sampled.

Species	Eastern meadowlark	Dickcissel	Grasshopper sparrow
Upland plover	0.58	0.38	0.29
Eastern meadowlark		0.49	0.55
Dickcissel			0.82

dickcissels and upland plovers occupied broader trophic niches of
prey functional groups than the other species.

The foregoing analyses consider averages of all individuals in
a given collection period and ignore interindividual variation.
Dietary collections of course represent point samples of individual
feeding habits.   Nevertheless, patterns of variation among
individuals in a collection may indicate the apparent flexibility or
opportunism of individual diets.   Calculated percentages of animal
prey items in each of three major functional groupings are plotted
for individual specimens of each species shown in Figure 7.23.   For
eastern meadowlarks, the scattering of points (individuals) is
impressive.   This pattern contrasts markedly with the dickcissel and
grasshopper sparrow patterns, in which most individuals collected
contained essentially similar proportions (at least of the three
major categories) of prey forms.   Perhaps meadowlarks are
individually flexible and generalist consumers of animal prey items,
while dickcissels and grasshopper sparrows are more restricted in
both overall dietary composition and individual dietary variation.

Trophic Patterns and Niche Relationships

Cody (1968, 1974) inferred from his studies of breeding avian
communities in North American grasslands and elsewhere that
communities are rather tightly structured, with species differing
from one another within a relatively small set of niche dimensions.
These differences (in habitat vegetational height and density and
features of bill morphology) are presumed to relate closely to the
resources critical to the species, and suffice to circumvent intense
competition.   Studies at the Osage Site were not designed to test
Cody's intuitions, but data on trophic relationships do permit some
analysis of species relationships on this niche dimension, which is
often presumed to be critical.

According to ecological theory, similar species (i.e., members
of the same guild) that co-occur should differ in types of prey
consumed, sizes of prey consumed, locations and/or behaviors for
procuring prey, or some combination of these elements.   Theory,
unfortunately, does not say how much any one of these features must

TABLE 7.18 Bird Diet Composition in Terms of Functionally Defined Ecological Categories, Osage Site, 1970-1972 (Values are Percent Dry Weight).

Prey category	Upland plover		Eastern meadowlark			Dickcissel			Grasshopper sparrow		
	1970	1971	1970	1971	1972	1970	1971	1972	1970	1971	1972
Seeds											
Grass	--	7.4	--	0.2	0.2	37.6	38.3	54.6	6.3	18.1	45.5
Forb	--	--	--	--	0.1	--	--	--	--	0.4	--
Animal											
Predaceous	38.6	31.4	8.3	34.6	41.2	1.1	2.2	9.9	3.4	4.1	8.1
Omnivorous	31.3	8.0	21.4	18.6	6.4	3.3	2.1	1.3	0.5	2.2	0.2
Phytophagous chewing	24.1	48.7	63.5	44.7	44.2	47.0	56.9	34.1	85.0	74.7	42.3
Phytophagous sucking	6.0	3.6	0.8	0.3	0.2	1.7	0.5	0.2	4.7	0.3	2.8
Flower-feeding	--	--	--	0.1	--	0.4	--	--	--	--	--
Scavenging	--	0.8	6.0	1.2	7.7	0.1	--	--	--	--	--
Beetles only[a]											
Ground-dwelling	28.3	33.0	20.2	35.2	61.8	1.2	7.1	1.1	2.2	1.7	2.4
Predaceous	28.3	31.4	7.8	33.5	40.9	--	0.9	1.1	2.4	4.1	2.4
Phytophagous	24.1	44.3	20.2	11.4	37.9	7.0	22.0	1.6	9.7	8.3	4.1

[a] Categories are a subset of those listed under "Animal."

TABLE 7.19   Index Values of Dietary Niche Breadth of the
             Breeding Bird Species at the Osage Site,
             According to Taxonomic and Functional
             Categorizations of Prey, 1970-72 Averages.

Species	Prey categories	
	Taxonomic	Functional
Upland plover	0.38	0.52
Eastern meadowlark	0.36	0.39
Dickcissel	0.36	0.55
Grasshopper sparrow	0.29	0.38

differ to reduce food-related competition between the species to
unimportance. This problem sharply reduces the testability of
competition-related hypotheses, as differences between species
inevitably will be found if they are studied closely enough. Tables
7.15 and 7.18 or Figures 7.20 to 7.22 indicate that overlap in the
types of prey consumed by breeding bird species at the Osage Site
was considerable. This impression can be sharpened by expressing
diet relationships among species as niche overlap indices, following
the procedures used earlier. Although the matrices of values
(Tables 7.20 and 7.21) are somewhat bewildering, several points do
emerge. First, food niche overlap between upland plovers and
eastern meadowlarks and between dickcissels and grasshopper sparrows
was generally great, but overlap between the two groups was
generally lower. Annual variation can be shown in these patterns,
however. For example, dickcissels generally exhibited greater
distinctiveness from the other species in 1972 than in other years.
Frequently, niche overlap in the diets of several species in a given
year was greater than between diets of a single species in different
years. This difference reinforced a previous conclusion: Breeding
species at the Osage Site apparently consume a rather broad variety
of prey types, overlap extensively in food preferences at any one
time, but differ substantially in diets through time. Clear-cut
patterns of species differences in diet composition do not emerge.
    Even if diet composition is similar among coexisting species,
however, sufficient trophic niche difference may remain if the
species differ in the size of prey consumed. Few studies have
measured size distributions of prey actually consumed by birds, and
much of the current theory of avian trophic niche relationships is
based on the assumption that bill size and shape are closely related
to prey size and type and that, therefore, niche relationships can
be inferred from bill measurements (Wiens and Rotenberry, in
preparation). Studies at the Osage Site recorded dimensions of prey
items in dietary samples, and the frequency distributions of animal
prey types for breeding species over the three years are depicted in

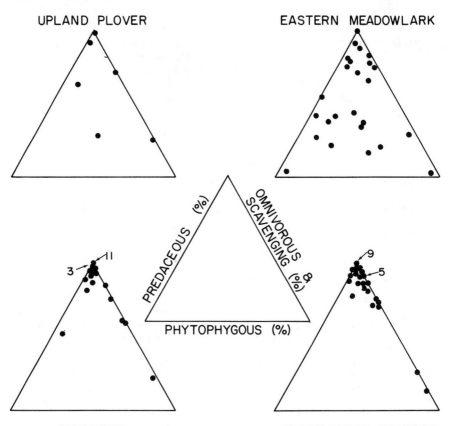

FIGURE 7.23 Proportions of three major animal prey groups in
individual samples of the breeding bird species at the Osage Site.
Points represent individual specimens; the perpendicular distance of
a point from a side of the triangle is proportional to the
percentage of that prey group in the diet of an individual.

Figure 7.24. The bird species differ substantially in body size (17
to 128 g), and bill sizes and shapes are correspondingly diverse
among the four breeding species. Nonetheless, the frequencies of
different-sized prey in the diets are remarkably similar, and the
similarities persist throughout the study. Prey size is apparently
not an important mode of niche differentiation among these species,
despite differences in bill morphology. These data strengthen the
inference that the bird species may exploit the prey resource pool
more or less opportunistically, consuming what they encounter within
a wide variety of prey types. Unfortunately, data are lacking on
the actual size distributions of prey available in the Osage Site
prairie, so whether all bird species are selecting prey from some
portion of the available size spectrum is impossible to determine.
     Another dimension of trophic niche differentiation--foraging
location and/or behavior--remains uninvestigated at the Osage Site.

TABLE 7.20 Matrix of Niche Overlap Values for Taxonomic Composition of Diets of Breeding Bird Species at the Osage Site. Values in the Inset Are Means for Years (All Species) or Species (Years Combined).

Species	Year	Upland plover 1971	Eastern meadowlark 1970	1971	1972	Dickcissel 1970	1971	1972	Grasshopper sparrow 1970	1971	1972
Upland plover	1970	.58	.31	.55	.37	.08	.17	.04	.08	.08	.08
	1971		.32	.60	.55	.19	.28	.16	.19	.20	.19
Eastern meadowlark	1970			.54	.27	.55	.56	.11	.69	.69	.43
	1971				.59	.35	.34	.17	.35	.36	.35
	1972					.18	.15	.17	.20	.20	.20
Dickcissel	1970						.82	.55	.76	.75	.81
	1971							.43	.62	.70	.80
	1972								.35	.33	.53
Grasshopper sparrow	1970									.89	.62
	1971										.65

Inset:

1970 = .41 Upland plover = .58
1971 = .41 Eastern meadowlark = .46
1972 = .30 Dickcissel = .60
Grasshopper sparrow = .72

255

TABLE 7.21 Matrix of Niche Overlap Values for Functional Prey Groups. Values in the Inset Are Means for Years (All Species) or Species (Years Combined).

Species	Year	Upland plover 1971	Eastern meadowlark 1970	Eastern meadowlark 1971	Eastern meadowlark 1972	Dickcissel 1970	Dickcissel 1971	Dickcissel 1972	Grasshopper sparrow 1970	Grasshopper sparrow 1971	Grasshopper sparrow 1972
Upland plover	1970	.67	.61	.86	.74	.17	.17	.20	.22	.22	.21
	1971		.52	.80	.79	.43	.46	.39	.54	.54	.42
Eastern meadowlark	1970			.64	.53	.25	.28	.20	.57	.52	.24
	1971				.87	.25	.28	.21	.34	.35	.24
	1972					.25	.28	.21	.34	.35	.24
Dickcissel	1970						.96	.88	.46	.72	.92
	1971	1970 = .38	Upland plover = .47					.87	.50	.76	.91
	1972	1971 = .53	Eastern meadowlark = .44						.42	.66	.95
Grasshopper sparrow	1970	1972 = .47	Dickcissel = .48							.73	.46
	1971		Grasshopper sparrow = .50								.70

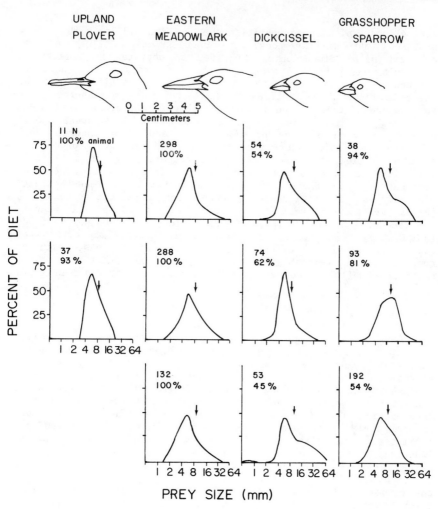

FIGURE 7.24 Size-frequency distributions of animal prey of the four
major bird species at the Osage Site. Arrow indicates mean prey
size.

Cody's studies (1968) indicated foraging differences among some
species, and studies by Wiens (1969) in a True Prairie in southern
Wisconsin also indicated species differences in the vegetational
features of areas used for foraging and foraging activities. But in
both studies, similarities among species were perhaps more
impressive than differences, and incidental observations during
studies at the Osage Site also suggested that the four species were
fairly similar in overall foraging characteristics.

     The information on trophic relationships of the species
breeding at the Osage Site does not fit well into conventional
ecological theory, but simple logic should prompt us to question

some elements of niche/competition theory (Wiens 1977a). The theory proposes that similar coexisting species must differ in some critical niche dimension(s) in order to obviate the effects of continuing competition over limited resources. If, however, resources are not continuously limiting, competition must be intermittent. Hence, times must occur when resources are superabundant, competition relaxed or absent, and niche overlap not detrimental. Determining how often such conditions occur should have priority in studies of competition and niche theory. Such conditions may hold rather frequently in prairies, where production is closely tied to climate and where climatic regimes are highly variable from year to year (Chapter 5). This may act to produce frequent superabundances of prey, as higher-level, low-fecundity consumers such as birds cannot produce strong numerical responses to yearly variations in resource conditions (Wiens, 1974b). If studies are conducted at such times, extensive niche overlaps may be found (as at the Osage Site). The remaining differences may or may not relate to proximate effects of species interactions, such as competition, but certainly the patterns of species relationships are neither as well defined nor as neat as some theoreticians suggest.

Prey Consumption Rates

Defining the composition of the diets of major True Prairie bird species at one time during the breeding season is one thing, but quantifying the rates of consumption of the prey types represented in the diet is quite another. However, the actual consumption of prey defines the role of bird populations within the context of the prairie ecosystem. Unfortunately, direct measures of consumption rates of birds are virtually impossible to obtain under natural field conditions. Thus, we must again resort to indirect estimates. Calculations of the metabolic requirements of individuals and the energy demands these requirements place on the system, derived from a simulation model such as BIRD, may be combined with periodic samples of stomach contents, which indicate the proportional composition of the diet. If we know the normal caloric values of unit weights of the various prey types (Cummins and Wuycheck 1971), we can convert energy demands to prey biomass consumed. This procedure was used for the Osage Site studies, although, given the constraints on diet sampling during the breeding season, the results must be considered as only general estimates. Because of these constraints, it seems best to consider consumption of general prey types rather than specific prey taxa (Table 7.22). Total estimated consumption of various prey groups during the 150-day breeding season varied among species, reflecting differences in both diets and energy demands. Dickcissels and grasshopper sparrows consumed the most grass seed, meadowlarks consumed the greatest quantities of both predaceous insects and omnivorous invertebrates, and eastern meadowlarks and dickcissels consumed the greatest quantities of phytophagous chewing insects.

Although the compositions of the diets of species breeding at the Osage Site differed somewhat, the overall effect of avian consumers at the Osage Site is perhaps best measured by estimating the entire breeding avifauna's consumption of various prey groups

TABLE 7.22   Estimated Consumption of Prey Categories by the Four
Major Bird Species at the Osage Site During the
Breeding Season. Values Are Averages for Both
Plots, 1970-72.

Prey category	Upland plover	Eastern meadowlark	Dickcissel	Grasshopper sparrow
	(km dry wt km^{-2} season^{-1})			
Plant				
Grass seed	1.6	0.4	38.7	17.2
Forb seed	0.0	0.1	0.0	0.1
Total	1.6	0.5	38.7	17.3
Animal				
Predaceous	8.4	59.6	2.7	2.9
Omnivorous	3.7	26.9	1.6	0.4
Phytophageous				
Chewing	10.7	89.1	41.8	28.5
Sucking	1.1	0.6	0.4	1.1
Flower-feeding	0.0	0.1	0.1	0.0
Scavenging	0.1	7.9	0.0	0.0
Total	24.0	184.2	46.6	32.9

(Figure 7.25).   Total biomass consumption was greater in plot 2 than
in plot 1, especially in 1971, when 535 kg km^{-2} season^{-1} of prey
were taken.   Chewing phytophagous invertebrates accounted for the
greatest share of the total prey consumed in all years on both
plots, with predaceous invertebrates of secondary importance.   Seed
consumption was greater in plot 2 than in plot 1, and more
omnivorous invertebrates were eaten on both plots in 1971 than in
other years.

The overall effects of these consumption rates on ecosystem
properties are difficult to determine.   Ideally, consumption should
be related to the availability of each prey group in order to assess
the potential impact of avian prey consumption on prey populations.
Because both the estimates of prey consumption and the measures of
prey abundance at the Osage Site are only general, however, our
comparisons of these measures can be only general.   Estimated weekly
consumption rates of major prey groups by the total breeding bird
community at the Osage Site are compared with measures of standing
crops of these prey groups present in June and July samplings
(Table 7.23).   The two measures obviously are not directly

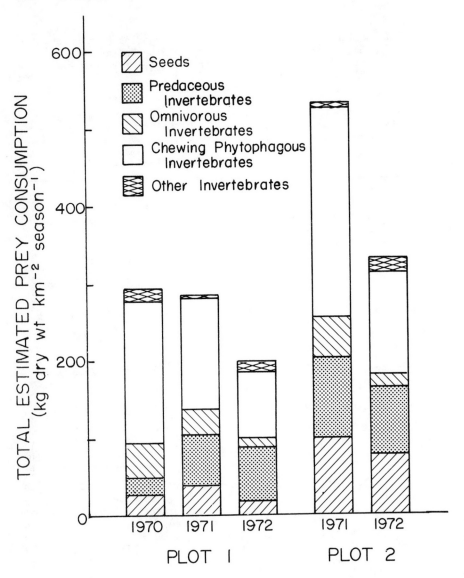

FIGURE 7.25 Total estimated prey consumption by birds at the Osage Site, 1970-72.

comparable, largely because the invertebrate measures are of standing crops at one time, while the consumption measures cover a seven-day period. If the turnover rates of the various prey groups were known, the comparisons could be sharpened. The measures suggest, however, that the total weekly prey consumption of the birds may represent as much as 25 percent of the measured standing crop of the prey groups. Phytophagous chewing insects

TABLE 7.23 Estimated Weekly Consumption Rates of Major Prey Groups
by the Community at Osage Site Plots 1 and 2 Compared
With Measures of Standing Crops of These Prey Groups
Present in Samples Taken on 3 July 1970, 19 June 1971,
and 7 June 1972 (T = Trace).

Prey group	Standing crop			Estimated weekly consumption				
				Plot 1			Plot 2	
	1970	1971	1972	1970	1971	1972	1971	1972
Phytophageous								
Chewing	129.7	34.6	28.4	8.5	6.8	4.0	12.6	6.2
Sucking	13.1	17.9	3.8	0.2	T	0.1	0.2	0.2
Flower-feeding	2.1	14.2	3.9	T	T	0.0	T	0.0
Scavenging	0.4	2.6	0.4	0.5	0.1	0.6	0.2	0.6
Predaceous	5.2	13.6	2.7	1.1	3.0	3.3	4.8	4.0
Omnivorous	163.1	110.8	47.5	12.4	11.5	8.6	20.3	11.8

accounted for the greatest proportion of standing crop (31 to
80 percent) as well as of total prey consumed (47 to 69 percent).
In 1972, when early June prey standing crop was lower than in 1970
and 1971, weekly estimated consumption of scavenging and predaceous
insects exceeded the standing crops, and avian consumption of
scavenging forms generally exceeded supply. These discrepancies
more likely represent inadequate sampling of the forms by the
collection procedures than a nearly complete depletion of standing
crops by the birds. A similar analysis using information on grass
seed standing biomass (see Table 6.11) suggests that the birds may
consume only about 0.04 percent of the seed biomass per week at the
peak of breeding season.

Energy Flow and Food Consumption in Agricultural Habitats

Most of the eastern part of the original True Prairies has been
replaced in the last century by cultivated croplands or other
managed surrogate grasslands, such as pastures. Mention has been
made of the apparent changes in bird populations and communities
accompanying these vegetational modifications, but one recent
analysis (Wiens and Dyer, 1977) permits consideration of some
aspects of energy and trophic dynamics of such bird populations.

Shortly after the turn of the century, Forbes and Gross made a systematic avifaunal survey of Illinois. Fifty years later, in 1957-58, Graber and Graber (1963) repeated the surveys, using the same methods in much the same areas. A wide variety of habitats were considered and many species encountered, but Wiens and Dyer paid attention only to the guild of predominantly granivorous birds occupying agricultural habitats most closely resembling native grasslands: pastures, fallow fields undergoing secondary succession, mixed hayfields, corn croplands, and small grain (wheat-oat) fields. Ten to nineteen granivorous species occurred in these habitats during the year, with the greatest number of species generally present in pastures and the lowest number in wheat-oat fields. The set of granivorous species generally accounted for well over half of all species present in the areas during the year (Table 7.24), and proportional numbers of granivores were roughly constant over the fifty-year period. Average annual densities of these granivorous species were, in most cases, near the high end of the range of breeding season densities of all species in native True Prairie locations to the west (see Table 7.10). In some cases, however, densities of granivores in the agricultural habitats were fairly high, especially in hay fields, in which density levels more than doubled in most areas over the fifty-year period (Table 7.24). This increase seems to be largely associated with tremendous increases in the abundance of red-winged blackbirds (one of the peripheral prairie species) in this habitat.

Wiens and Dyer (1977) used the BIRD simulation model to estimate magnitudes of energy demand and food consumption by the granivorous bird groups in these habitats for various sections of the state at the two survey times. Because detailed information on reproductive levels was lacking, simulations considered only adults throughout the year. The estimates are thus conservative and not directly comparable with estimates made for the Osage Site breeding avifaunas. General information about the proportions of plant and animal prey in the diets of the species at various times of the year was obtained from Martin et al. (1961).

Estimates of the collective energy demand of all granivores present in these habitats and regions ranged from 0.96 (northern Illinois corn, 1909) to 15.21 kcal m^{-2} yr^{-1} (central hay fields, 1958). Total annual energy demands were greatest in hay fields and fallow fields and lowest in corn fields in the 1958 surveys. Changes over the fifty-year period between surveys were especially great in hay fields in all regions of the state and in fallow fields in southern areas (early surveys from the central and northern sections were not available). In these areas blackbird densities increased dramatically, which certainly contributed a large share of the increase in energy flow.

Although these values cannot be directly compared with Osage Site estimates (see Table 7.13), a crude picture of the patterns can be derived by assuming that total annual energy demands of the birds at the Osage Site are roughly two times the estimated breeding season demands. These values range from 2.15 to 5.84 kcal m^{-2} yr^{-1}. If 70 percent of the avifauna in the Illinois habitats is assumed to be granivorous, the estimated energy demands of granivores can be adjusted to the estimated energy demands of the entire bird community. These adjusted values, which still do not include the

TABLE 7.24   Proportions and Densities of Granivorous Birds in
the Avifaunas of Several Grassland-like Agricultural
Habitats in Northern, Central, and Southern Illinois,
1909 and 1958.

Agricultural type	Section of Illinois	Granivores in avifaunas (percent)		Density of granivorous species (individuals km^{-2})	
		1909	1958	1909	1958
Corn	North	77	77	131	214
	Central	65	89	143	263
	South	37	26	357	434
Wheat-oats	North	91	96	398	420
	Central	79	66	238	393
	South	73	71	300	503
Fallow	North	--	76	--	712
	Central	--	80	--	845
	South	48	65	348	967
Hay fields	North	84	79	509	1340
	Central	78	93	509	2453
	South	77	78	482	941
Pasture	North	69	78	510	566
	Central	68	78	553	555
	South	48	50	334	611

energy associated with reproduction, range from 2.28 to 6.60 kcal
m^{-2} yr^{-1} over all agricultural types and regions except hay fields
and fallow fields in 1958.

This procedure is fraught with assumptions, and the results
should be considered only as coarse approximations. Still, except
for hay and fallow fields, the magnitudes of energy flow through
bird populations occupying habitat types that have replaced native
prairies do not appear to differ appreciably from their counterparts

in the more natural settings to the west. Until recently, the
avifaunas of hay and fallow fields may also have processed roughly
the same quantities of energy annually. Since the turn of the
century, however, bird populations in these habitats have increased
tremendously as blackbirds and several other species have expanded
their habitat ranges into these types, with consequent effects on
energy demands. We know virtually nothing of the circumstances
responsible for this habitat expansion; but if overall productivity
in these habitat types has not increased appreciably (as seems
likely), this expansion provides one more thread of evidence
supporting the contention that the avifaunas of prairies (and
perhaps of their agricultural counterparts as well) are not
saturated or do not exist at levels approaching resource-defined
carrying capacities.

The BIRD model analyses of the Illinois granivorous birds
included estimated consumption rates of plant (seed) and animal
prey. To attempt to compare these values with those derived for the
breeding populations at the Osage Site (Table 7.22) makes little
sense, since the Illinois analyses were predisposed to emphasize
seed consumption. But one extension of the Illinois treatment is
worth mentioning, for it suggests something of the magnitude of prey
consumption rates over a large region. Graber and Graber (1963)
obtained general approximations of the total area of each
agricultural type over the entire state during 1907-09 and 1957-58.
Although these values are not precise, they do provide a basis for
extending the model results to estimate total annual consumption of
insects and seeds by granivorous species in each habitat type over
the entire state (Table 7.25). The area in agricultural use for the
five types decreased by over 40 percent between 1909 and 1958,
largely as a result of decreases in area of pasture, wheat-oats, and
corn. During the same period, total granivore abundance increased
markedly in hay fields, slightly in corn and fallow fields, and
decreased substantially in pastures. In 1909, projected seed
consumption was high and relatively similar ($\sim$10,000 to 12,500
metric tons yr^{-1}) in pasture, wheat-oat fields, and corn and
relatively low in hay and fallow fields. Fifty years later, overall
seed consumption was relatively unchanged in corn fields (despite
the significant reduction in area under cultivation), lower in
wheat-oat fields and pastures, somewhat higher in fallow fields, and
over 250 percent greater in hay fields (despite the lack of any real
change in the area of hay fields in Illinois). In all five
agricultural types combined, total statewide insect and seed
consumption increased slightly over the fifty-year period, in spite
of the substantial decrease in area cultivated.

MAMMALS

Site-Specific Quantitative Patterns and Temporal Stability

Quantitative data reflecting spatial and temporal patterns of
small mammal communities are available from a northern and southern
site within the True Prairie. The southern Osage Site is detailed

TABLE 7.25 Estimated Food Consumption Rates by Granivorous Birds in Several Agricultural Habitats in Illinois, 1909 and 1958.

Agricultural type	Year	Total area (km^2)	Animal prey Average (kg km^{-2} yr^{-1})	Animal prey Total over state (metric tons yr^{-1})	Seeds Average (kg km^{-2} yr^{-2})	Seeds Total over state (metric tons yr^{-1})
Corn	1909	42,087	60	2,525	246	10,353
	1958	33,994	169	5,744	368	12,510
Wheat-oats	1909	26,304	152	3,998	448	11,784
	1958	17,401	215	3,741	596	10,371
Fallow	1909	6,070	277	1,681	284	1,723
	1958	4,047	345	1,396	1,268	5,132
Hay fields	1909	10,100[a]	399	4,030	382	3,858
	1958	10,100[a]	929	9,383	1,445	14,595
Pasture	1909	24,686	262	6,467	504	12,441
	1958	8,094	334	2,703	415	3,359
Total	1909	109,265	--	18,701	--	40,159
	1958	62,726	--	25,967	--	45,967

[a] Estimated

265

here, and the mammalian study techniques used there were given by
Birney (1974a). The northern site is the Cedar Creek Natural
History Area in Isanti County, Minnesota, which was farmed from
about 1900 until 1966 (Birney et al., 1976). During the period of
study, grasses dominated; Kentucky bluegrass (Poa pratensis),
timothy (Phleum pratense), quackgrass (Agropyron repens), and
panicum (Panicum spp.) were especially common in the more densely
covered areas. Threeawn (Aristida spp.), little bluestem, and in
some places bristlegrass (Setaria spp.) were common on the drier
sections of the grid. The meadow vole (Microtus pennsylvanicus) was
the most abundant small mammal. Other species captured on or near
the grid included masked shrew (Sorex cinereus), short-tailed shrew
(Blarina brevicauda), eastern chipmunk (Tamias striatus),
thirteen-lined ground squirrel (Spermophilus tridecemlineatus),
plains pocket gopher (Geomys bursarius), deer mouse (Peromyscus
maniculatus), white-footed mouse (P. leucopus), southern red-backed
vole (Clethrionomys gapperi), house mouse (Mus musculus), meadow
jumping mouse (Zapus hudsonius), and ermine (Mustela erminea).

The 2.72-ha grid was studied with techniques similar to the
methods used at the Osage Site by Birney et al. in 1976. Density at
both sites was estimated by dividing the number of individuals
trapped one or more times during a five- or ten-day sampling period
by the area of the grid. This technique may underestimate density
if individuals are living on the grids but avoiding capture, or it
may overestimate density if the area of trap-grid influence is
greater than the area of the grid. Biomass density was estimated by
multiplying mean body weight by density and converting to dry weight
by assuming 70 percent water in live animals (Golley, 1960).

Total dry weight biomass values at the Osage Site for the eight
sampling periods from 1970 to 1972 are presented in Figure 7.26.
Two species, prairie vole (Microtus ochrogaster) and hispid cotton
rat (Sigmodon hispidus), contributed most significantly to the
total. The density changes of both species tend to be somewhat
cyclic (Haines, 1971; Krebs and Myers, 1974), and the species may
compete with each other (Terman, 1974) for some important resources.
Prairie voles apparently had two-year cycles at Osage, but the
pattern of fluctuation for cotton rats and any pattern of
competition between the two during the study are unclear.

Prairie voles at the Osage Site were at a relatively high
density (502 g ha^{-1}) in May 1970, decreased slightly during that
summer, were at low density (9 g ha^{-1}) by May 1971, then increased
to about 49 g ha^{-1} by October. Apparently as a result of highly
successful winter breeding, voles were at a high density of 1,160 g
ha^{-1} by May 1972. During the 1972 growing season, the vole
population again decreased, but the cotton rat population increased
(from 210 g ha^{-1} in May to 684 g ha^{-1} in October) to become the
major component of small mammal biomass by autumn.

Trends in small mammal biomass fluctuation at the northern True
Prairie site are shown in Figure 7.27. Mean biomass at the northern
site was consistently lower than at the southern site and never
exceeded 150 g ha^{-1}. The pattern at Cedar Creek tended to be
seasonal and basically similar for three years. Meadow voles and
deer mice contributed most to total biomass at Cedar Creek.

Whereas mean biomass density at the southern sites for the
first spring sample was estimated at 679 g ha^{-1} and was highly

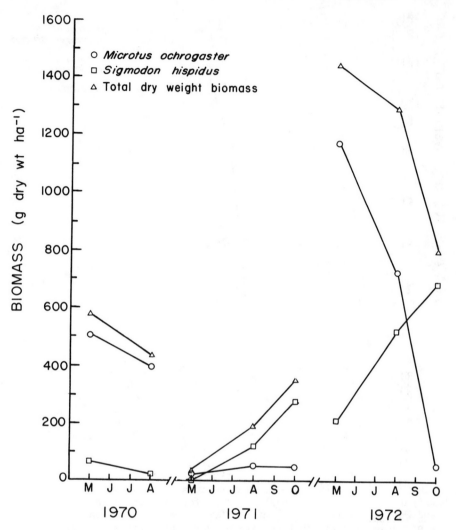

FIGURE 7.26 Yearly seasonal fluctuations in total small mammal dry weight biomass and dry weight biomass for <u>Microtus ochrogaster</u> and <u>Sigmodon hispidus</u> at the Osage Site. The gaps between years represent periods when small mammal populations were not censused.

variable between years (CV = 106), density for the northern grid was only 28 g ha^{-1} and was much less variable between years (CV = 51). Although invariably low in June at the northern site, biomass density increased by July (average = 111 g ha^{-1}; CV = 3), continued to increase until August (average = 119 g ha^{-1}; CV = 37), then decreased in 1971 and in 1972, when September samples were available. No mammal species at Cedar Creek appeared to be undergoing multi-year population cycles; instead, fluctuations in small mammal biomass seemed to be attributable to annual phenomena.

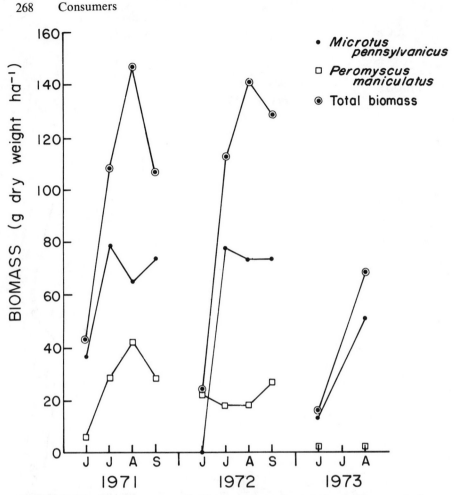

FIGURE 7.27 Yearly seasonal fluctuations in total small mammal dry
weight biomass and dry weight biomass for Microtus pennsylvanicus
and Peromyscus maniculatus at the Cedar Creek Site in east-central
Minnesota. The gaps between years represent periods when small
mammal populations were not censused.

Populations were low in spring, increased by midsummer, then leveled
off before autumn. Birney (unpublished) has three years (1974-76)
of additional data from the same grid, and this annual pattern is
prevalent for all six years.

Why does small mammal biomass fluctuate on an annual rhythm at
a northern location but not at a southern location within the True
Prairie?   The prairie vole apparently was undergoing multi-year
population fluctuations at the Osage Site, whereas at Cedar Creek,
the meadow vole, also considered a cyclic species (Krebs et al.
1969), was not. Birney et al. (1976) attributed the two patterns to
differences in vegetative cover at the two sites, not to any
fundamental phenomenon resulting from difference in latitude.
Vegetation at the Osage Site was homogeneously dense, whereas

vegetation at Cedar Creek was heterogeneous, with some parts sparsely vegetated and others densely vegetated. Birney et al. (1976) showed that at Cedar Creek the vast majority of vole captures occurred at stations having dense cover. We hypothesize that most successful breeding voles at Cedar Creek live in areas of dense cover and that subadults and young animals disperse from local areas of prime habitat to areas of sparse vegetation. Mortality is probably high and reproductive success low in such secondary habitats. Because areas of sparse vegetation would seem to be especially unsuitable for winter survival, winter mortality would be high and spring populations small. All areas on the Osage grid, however, apparently provided adequate summer and winter habitat. Winter breeding at the Osage Site resulted in high spring populations of voles in 1970 and 1972 and may be attributable at least in part to the effect of latitude. However, winter breeding by microtine rodents has been reported in the subnivean (beneath the snow) space much farther north than the Cedar Creek site (Krebs and Myers, 1974). Birney et al. (1976) also studied one _Microtus_ population in Minnesota where vegetative cover was dense and the vole population became extremely high, but the population fluctuation again seemed to be annual. Data for two additional years (Baird, unpublished) show the same trend. These observations argue against habitat heterogeneity as the basic cause for annual fluctuations (rather than the more commonly described multi-year cycles) at these two northern sites.

Cotton rats contributed significantly to total small mammal biomass at the Osage Site during some sampling periods. The weight used here to calculate young adult cotton rat biomass was about 100 g. At even a moderately high density, these relatively large rodents have a major impact on biomass dynamics within the community. Cotton rats do not live as far north as Cedar Creek, and no comparable species exists there. Thus, when the _Microtus_ population was low at Cedar Creek, biomass invariably was low. In addition to voles, only the much smaller deer mouse regularly contributed to total biomass at Cedar Creek. Although deer mice were present at every sampling period, their biomass density never exceeded 45 g ha^{-1}. This species also occurred regularly on the grid at the Osage Site but was never observed to reach high densities at either site.

Figure 7.28 presents a comparison of means for dry weight biomass, number of species, diversity (H'), and equitability (J'). Three-year averages of biomass densities at the Osage Site fail to show the drastic seasonal fluctuation observed for individual years. The long-term picture there is of mean seasonal equality with drastic annual fluctuation. The hint that average biomass density is higher in spring than in summer and fall may have resulted from the extremely high _Microtus_ densities in May 1970 and 1972 that were not matched in autumn during the three years of study. Whether that pattern is real--and _Microtus_ populations always reach highs in springs that follow winters of successful breeding--or is an artifact of the limited data base is unclear.

Because the pattern of seasonal change in biomass density at Cedar Creek was repeated in all three years of data collection, it is not surprising that seasonal means follow the previously described sequence: small mammal biomass density is low in June,

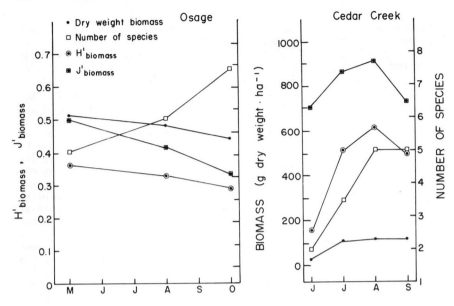

FIGURE 7.28 Mean seasonal changes in total small mammal dry weight biomass, number of species trapped, biomass H', and biomass J' at the Osage Site and at the Cedar Creek Site in east-central Minnesota.

increases by July, then changes little until high winter mortality rates in the absence of population recruitment causes it to decrease.

A summary of biomass comparisons shows that small mammal biomass averaged about seven times higher at an ungrazed True Prairie site in Oklahoma than at an ungrazed site in Minnesota. Total variation in biomass between years, calculated either for each season separately or for all samples, was much greater at the Oklahoma site, but a coefficient of variation (CV) calculated from seasonal means was much lower for the Oklahoma site (8.4) than for the Minnesota site (47.0). These data apparently reflect the seasonally variable but predictable changes at the northern site compared with multi-year fluctuations at the Osage Site. Effects of latitude, differences in degree of habitat heterogeneity, and differences in species composition cannot be separated in the existing data to explain adequately the trends observed.

Mean number of species trapped (Table 7.26) on the grid at the Osage Site for all eight sampling periods was higher but less variable (average = 6.0; CV = 25) than the mean for the ten sampling periods at Cedar Creek (average = 3.8; CV = 40.8). The number of species at the Osage Site ranged from four to eight. At Cedar Creek, only two species were captured in June in each of the three years, and the greatest number of species taken there during a single sampling period was six, in both August and September of 1972. During three years of study, a total of only eight species was trapped on the Cedar Creek grid, whereas eleven were captured on the Osage grid (Table 7.27). The pattern of variation in species

TABLE 7.26   Diversity (H') and Equitability (J') of Small Mammals
Trapped on 2.72 ha Ungrazed Grids at the Osage Site and
at the Cedar Creek Natural History Area, Minnesota.

Date	Species (no.)	Individuals		Biomass	
		H'	J'	H'	J'
Osage Site					
1970					
May	4	0.19	0.32	0.19	0.32
August	4	0.24	0.40	0.17	0.28
1971					
May	6	0.76	0.98	0.64	0.82
August	7	0.70	0.82	0.44	0.53
October	8	0.70	0.78	0.34	0.37
1972					
May	5	0.28	0.40	0.25	0.36
August	7	0.42	0.50	0.37	0.43
October	7	0.51	0.60	0.24	0.29
Total average	6.0	0.48	0.60	0.33	0.42
Total CV	25.0	47.8	39.5	47.2	42.1
Cedar Creek Natural History Area					
1971					
June	2	0.25	0.85	0.18	0.61
July	3	0.35	0.73	0.39	0.61
August	5	0.56	0.80	0.55	0.79
September	4	0.40	0.67	0.32	0.54
1972					
June	2	0.13	0.44	0.12	0.41
July	4	0.57	0.94	0.40	0.66
August	6	0.64	0.83	0.54	0.69
September	6	0.65	0.83	0.44	0.56
1973					
June	2	0.24	0.80	0.17	0.58
August	4	0.33	0.54	0.33	0.55
Total average	3.8	0.41	0.74	0.34	0.60
Total CV	40.8	44.4	20.6	43.7	16.9

TABLE 7.27 Species of Mammals Taken at Osage and at Cedar Creek. Numbers in the Body of the Table Represent Rank Order of Each Species to the Total Biomass Density of That Species. The Number in Brackets After Each Species Name Is the Weight in Grams per Individual Used in Biomass Calculations.

Species	Osage 1970 May	Osage 1970 Aug	Osage 1971 May	Osage 1971 Aug	Osage 1971 Oct	Osage 1972 May	Osage 1972 Aug	Osage 1972 Oct	Cedar 1971 June	Cedar 1971 July	Cedar 1971 Aug	Cedar 1971 Sept	Cedar 1972 June	Cedar 1972 July	Cedar 1972 Aug	Cedar 1972 Sept	Cedar 1973 July	Cedar 1973 Aug
Masked shrew (Sorex cinereus) [5]										3				4	5	6		
Least shrew (Cryptotis parva) [5]				5	5													
Short-tailed shrew (Blarina brevicauda) [18]	3	4	4	4	3	4				3	2	3	2	3	2	3		4
Southern short-tailed shrew (B. carolinensis) [13]							3	3										
Thirteen-lined ground squirrel (Spermophilus tridecemlineatus) [125]																		2
Hispid pocket mouse (Perognathus hispidus) [40]				3														
Fulvuous harvest mouse (Reithrodontomys fulvescens) [14]	4	3				5	6	5										
Plains harvest mouse (R. montanus) [10]			2		6													
White-footed mouse (Peromyscus leucopus) [22]			3	7	4	3	4	4	2	2	2	2	1	2	3	2		3
Deer mouse (P. maniculatus) [2]	2	2																
Hispid cotton rat (Sigmodon hispidus) [100]	1	1	1	1	1	2	2	1										
Prairie vole (Microtus ochrogaster) [40]	1	1	1	1	2	1	1	2	1	1	1	1	1	1	1	1		1
Meadow vole (M. pennsylvanicus) [42]										5						5		1
House mouse (Mus musculus) [11]			5	6	7		7	6										
Meadow jumping mouse (Zapus hudsonius) [16]											5				4			6

272

richness at both sites increased during the growing season but was most pronounced at Cedar Creek, where June samples invariably consisted of only two species.

Diversity (H') takes into account both species richness and evenness. Table 7.26 presents H' values for each sampling period at both the northern and southern True Prairie sites. Diversity was compared initially using only the proportional contribution each species made to total biomass (biomass H'). Because species such as ground squirrels and cotton rats are large and others such as shrews and harvest mice are small, a second set of calculations was made based on the number of individuals of each species (individual H') regardless of biomass.

For most sampling periods and in all cases where seasonal site means are compared, biomass H' is lower than individual H'. The three species that tended to dominate, prairie voles (Microtus ochrogaster) and hispid cotton rats (Sigmodon hispidus) at Osage and meadow voles (Microtus pennsylvanicus) at Cedar Creek, are all relatively large compared to most of the less common species. Thus, capture of several Cryptotis or Reithrodontomys at the Osage Site or of several Sorex or Zapus at Cedar Creek appreciably increased individual H' but had little effect on biomass H'.

Mean seasonal changes in biomass H' are shown in Figure 7.28 for both sites. At Cedar Creek, trends in H' followed trends in number of species and biomass for June, July, and August but decreased in September. Number of species at the Osage Site increased from May to August and from August to October, when both biomass H' and total biomass decreased. Thus, the reduction in biomass H' during the summer clearly reflects the shift to preponderance of Microtus and Sigmodon in the October sample. In May 1971, only seven individuals of six different species were trapped. Thus, diversity was high at that time; and, although diversity was low in May 1970 and 1972, the mean for May was greatly influenced by this single datum. The seasonal trend of slight reduction in biomass H' at the Osage Site (Figure 7.28) reflects the sharp reduction in that value from one trapping period to the next during 1971 as the densities of Microtus and Sigmodon increased more rapidly than the densities of other species living on the grid. Variability in both biomass H' and individual H' on the two sites was consistently high, with CVs between 43.5 and 48.0 in all four cases (Table 7.26).

Equitability (J') was consistently less variable than diversity as measured by the CVs. At Cedar Creek, CVs for J' were less than half as high as CVs for H'. At the Osage Site, however, variability associated with the two parameters was similar, with J' slightly lower. CVs were relatively much higher (39.5 and 42.1 for individual J' and biomass J', respectively) at the Osage Site than at Cedar Creek (20.6 and 16.9, respectively). This difference in variability between the northern and southern True Prairie sites clearly resulted from the fact that Microtus frequently (and Sigmodon occasionally) dominated the Osage community, whereas at other times (e.g., May 1971) the fauna consisted of nearly equal proportions of several species. At Cedar Creek, Microtus was the most common species in nine of ten samples but never overwhelmingly dominated the small mammal fauna. This observation most likely reflects idiosyncrasies of these sites, such as habitat homogeneity

versus habitat heterogeneity, rather than a pervasive effect of latitude on mammalian communities of the True Prairie.

Functional Attributes and Adaptations

The True Prairie is not characterized by a particular mammalian fauna, and no species found there is limited to the True Prairie in its distribution. Therefore, the fact that neither unique mammalian adaptations nor unique mammalian functional attributes can be associated exclusively with True Prairie mammals is not surprising. Some nonunique attributes of these organisms, however, do merit attention.

Very young mammals obtain all their dietary essentials from milk. Milk composition varies widely, both intra- and interspecifically, in the relative amounts of water and total solids, in protein content and composition, and in energy content, especially energy stored as fat (Jenness, 1974). We are tempted to speculate that the milk of each species should be precisely adapted to its natural history and thus would like to ascertain whether similar grassland mammals have milk of similar composition. The milks of two families of large grassland herbivores, horses and bovids, are dissimilar. In both cases, the young are precocial, nurse on a schedule combining demand by the young and rationing by the mother, and partake of supplementary food early in life. Yet, horse milk is much more watery and less concentrated than bovid milk (Jenness, 1974). Jenness also noted that the milks of some True Prairie lagomorphs, cottontails and jackrabbits, are similar, despite fairly different diets and degree of maturation at birth and, hence, different durations of lactation.

For most adult grassland herbivores, grasses and other high cellulose plants provide most dietary requirements. The evolution of high-crowned teeth having a wide variety of grinding surfaces was discussed in great detail by Sloan (1972) and Vaughan (1972). The vast array of mammalian dental patterns and formulae and the relatively short time spans during which these extensive dental evolutions have occurred attest to the importance of selection on feeding efficiency in mammals.

Digestive efficiency, especially that associated with deriving energy from cellulose, also has been an important object of selection in grassland mammals. As discussed in Chapter 10, the rumen of artiodactyl stomachs--stomachs with three or four chambers--is an elaborate anatomical modification that has evolved in response to the need for digesting grasses. (For further details see Moir [1968] and Spedding [1971].) Peden et al. (1974) showed that a native North American grassland species, the bison, is highly selective of, and more efficient at, extracting energy from food plants than are domestic cattle, which did not evolve in the grasslands of North America.

Some grassland mammals (e.g., horses) do not have a rumen but do have a large caecum (pouch) between the small intestine and the colon that functions like a rumen in that bacteria act on cellulose and other resistant materials in it. The caecum usually has thin walls through which liberated materials may be absorbed into the blood. Many small nonruminating herbivores, such as rabbits, pocket

gophers, and some other rodents, reingest large, soft caecal feces
as they pass from the anus, after having been acted on by bacteria
(Eden, 1940; Harder, 1949; Ingeburg et al., 1951). These animals
thus derive additional energy as well as B-complex vitamins
synthesized in the caecum by the microflora (Davis and Golley,
1963).

Life-history strategies are invariably deeply interwoven with
food habits and feeding efficiency. Timing of breeding, development
of young, frequency and duration of lactation, and even less obvious
characteristics such as solitariness versus gregariousness, mating
systems and harem sizes, and migratory versus nonmigratory habits
all relate to food availability and feeding efficiency. Jarman
(1974) studied these interactions among large grassland herbivores
in Africa and proposed interesting behavioral and ecological
theories. No comparable study of North American herbivores is
available (Geist, 1974), nor is such a study possible, because North
American grassland mammal faunas no longer exist in anything
resembling a pristine state (Martin, 1975).

Grassland habitats present an essentially two-dimensional
environment to a mammal, unless the mammal is small and agile enough
to climb on the stems of grass or morphologically equipped for
burrowing. A few species, such as the harvest mice
(Reithrodontomys) are climbers. These small animals (usually <15 g)
are extremely quick and agile and feed on both seeds and insects.

Creating a third dimension by burrowing is a common habit among
grassland mammals. Some species, such as pocket gophers
(Geomyidae), are obligate burrowers that come aboveground only for
brief periods to forage or to disperse. Other mammals, such as
prairie dogs and ground squirrels (Sciuridae), forage almost
exclusively aboveground, sometimes at great distances from burrow
entrances, but depend on a burrow for shelter, escape from
predators, rearing of young, and in some cases hibernation.

Grant and Birney (1979) classified the small mammal communities
of six major North American grassland types (Tallgrass, Mixed-grass,
Shortgrass, Desert grass, Montane grass, and Bunchgrass) into guilds
according to selected functional attributes of each species: body
size, reproductive strategy, life form, seasonality of activity, and
diet. They studied mammals captured in small live traps set in
2.72-ha grids over a broad geographic area, but the Osage Site and
the Cedar Creek Site were the only True Prairie sites included.
Biomass density was calculated by an elaborate technique using
movement patterns of individual species to estimate species-specific
areas of trap grid influence (French et al., 1976). Sites were
compared with Horn's R_o similarity index, which was discussed in the
section on avifauna.

Proportional distribution of small mammal biomass within the
defined subcategories of the five functional groupings is shown in
Figure 7.29. Note that at the Osage Site, one subcategory contains
over 50 percent of the total biomass in all guilds and between 90
and 100 percent in three of the five. At Cedar Creek, on the other
hand, small mammal biomass tended to be spread into at least two
subcategories for all functional groups except seasonality of
activity. The predominance of prairie voles, with occasional high
biomass density of hispid cotton rats, resulted in the uneven
distribution of biomass within functional groups at the Osage Site.

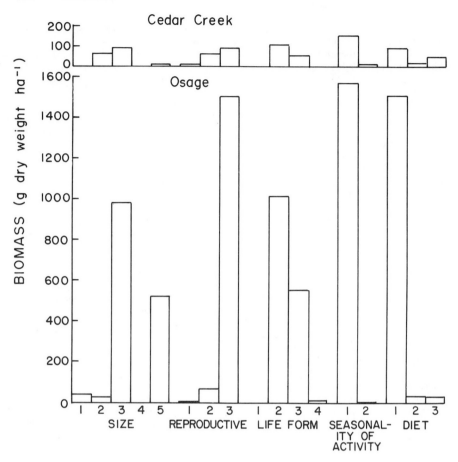

FIGURE 7.29 Histograms showing proportional distribution of small
mammal biomass in five functional groupings at a northern (Cedar
Creek Site, east-central Minnesota) and a southern (Osage Site,
northeastern Oklahoma) locality in the True Prairie. Size: 1 = <
g, 2 = 16-30 g, 3 = 31-45 g, 4 = 46-60 g, 5 = >60 g. Reproductive
strategy: 1 = 1 < 0.1, 2 = 0.1 < 1 < 0.5, 3 = 1 > 0.5. Life form:
1 = fossorial, 2 = subsurface or litter, 3 = surface, 4 = vegetation
climber. Seasonality of activity: 1 = year-round activity, 2 =
seasonal activity. Diet: 1 = herbivore, 2 = carnivore, 3 =
omnivore. (Redrawn from Grant and Birney, 1979.)

For some groups, both of these two most abundant species were in the
same subgroup. Similarly, because no species was so completely
dominant at Cedar Creek and because the two most abundant species,
meadow vole and deer mouse, generally did not fall into the same
functional subcategory, biomass tended to be more evenly distributed
at the northern site.
      When Grant and Birney (1979) compared geographic patterns in
composition of functional groups (Table 7.28), similarity

TABLE 7.28   Categories to Which the Fifteen Species of Small Mammals Were Assigned for Analyses Based on Functional Groups.   Site Codes Are N for the Northern Site (Minnesota) and S for the Southern Site (Oklahoma).   (Modified from Grant and Birney, 1979.)

Species	Site	Size[a]	Reproductive strategy[b]	Life form[c]	Seasonality of activity[d]	Diet[e]
Masked shrew (Sorex cinereus)	N	1	2	2	1	2
Short-tailed shrew (Blarina brevicauda)	N	2	2	2	1	2
Southern short-tailed shrew (B. carolinensis)	S	1	2	2	1	2
Least shrew (Cryptotis parva)	S	1	3	2	1	2
Thirteen-lined ground squirrel (Spermophilus tridecemlineatus)	N	5	1	3	2	3
Hispid pocket mouse (Perognathus hispidus)	S	3	1	3	2	1
Plains harvest mouse (Reithrodontomys montanus)	S	1	2	4	1	3
Fulvous harvest mouse (R. fulvescens)	S	1	2	4	1	3
Deer mouse (Peromyscus maniculatus)	N,S	2	2	3	1	3
White-footed mouse (P. leucopus)	N,S	2	2	3	1	3
Hispid cotton rat (Sigmodon hispidus)	S	5	3	3	1	1
Meadow vole (Microtus pennsylvanicus)	N	3	3	2	1	1
Prairie vole (M. ochrogaster)	N	3	3	2	1	1
House mouse (Mus musculus)	N,S	2	2	3	1	1
Meadow jumping mouse (Zapus hudsonius)	N	2	2	3	2	1

[a] 1 = $\leq$ 15 g; 2 = 16-30 g; 3 = 31-45 g; 4 = 46-60 g; 5, > 60 g

[b] 1 = I $\leq$ 0.1; 2 = 0.1 < I < 0.5; 3 = I $\geq$ 0.5

[c] 1 = fossorial; 2 = subsurface or litter; 3 = surface; 4 = vegetation climber

[d] 1 = year-round; 2 = seasonal

[e] 1 = herbivore; 2 = carnivore; 3 = omnivore

relationships between grassland types differed markedly, with few pervasive trends. The grassland types studied were found to differ least in composition of functional groups when the analysis was based on life form. Progressively greater differences were observed when analyses were based on diet, seasonality of activity, size, and reproductive strategy. The high mean similarity for life form (categories were fossorial, subsurface or litter, surface, and vegetation climber) indicates that the balance between life forms within grasslands persists relatively independently of whatever particular species compose a given community; that is, allocating small mammal biomass to various life forms reduces the deviation from a hypothetical situation in which all grassland types have identical life-form compositions.

On a gradient from tallgrass to mixed-grass to shortgrass to desert grass, the small mammal communities shift domination from herbivores (mostly microtine rodents) to a combination of omnivores and granivores (mostly heteromyid rodents) to granivores (Grant and Birney, 1979). The two True Prairie sites (Osage and Cedar Creek) were both dominated by herbivores, but the proportion of omnivores at the northern site was the greater, primarily because of the greater proportion of biomass contributed by Peromyscus at that site.

The small mammal fauna at both True Prairie sites consisted almost exclusively of species that are active year round. A slightly greater proportion of the biomass hibernates at the northern site than at the southern site. Overall similarity among all grassland types was only slightly less for seasonality of activity than for diet.

Overall similarity of grassland mammalian communities differs appreciably in size composition. True Prairie communities tended to be slightly more variable than the average. Hagmeier and Stults (1964) indicated that the average size of mammals in the True Prairie is larger than average for North America but not as large as at such places as northwestern coastal Alaska and parts of the Cascade and Rocky Mountain regions.

Reproductive guilds subdivided into three categories of a Reproductive Index (RI), calculated as

$$RI = \frac{(mean\ litter\ size)\ (maximum\ number\ of\ litters\ in\ 180\ days)}{(minimum\ age\ of\ initial\ breeding)},$$

showed lower similarity values than other functional groups (Grant and Birney, 1979). The array of reproductive strategies within a small mammal community in the North American grassland may be a function of the predictability of resource availability. The general trend is toward seasonal breeding, but this pattern is interrupted in desert grassland, for example, where species that are more long-lived, have relatively small litters, and breed opportunistically (e.g., when annual grasses produce a seed crop following a rain) tend to prevail.

Overall, the southern True Prairie site had the most distinctive small mammal fauna if all functional groups are considered simultaneously. That grid was characterized by much taller vegetation and denser vegetative cover than any other site and apparently provided a more nearly unique environment for a small

mammal fauna than did any other site. Although species composition at the Osage Site differed from the composition at other sites more in quantity than in quality, the quantitative differences do demonstrate the functional group uniqueness of the site. In comparisons of functional groups, the small mammal fauna at the northern Tallgrass Site was generally most like that of a Montane Grassland.

Despite the functional distinctiveness of the small mammals at the Osage Site, overall results of the study of functional groups support the conclusion that the True Prairie does not have a distinctive mammalian fauna. This conclusion was based initially on a taxonomic distributional analysis of grassland mammals but remains valid even when taxonomy of the small mammals is ignored and only characteristics reflecting general modes of life or functional roles are considered. Grant and Birney (1979) found that the two True Prairie sites were not especially similar and that the northern small mammal community resembled in a functional sense a Montane Grassland, which does not support the notion of uniqueness within the True Prairie.

Viewing ecological communities by the guild, or functional group, approach presents an interesting alternative to the species-by-species, or taxonomic, approach. The composition of small mammal communities does not evolve necessarily into predictable arrays of guilds. Current understanding of both competition and convergent evolution suggests, however, that whenever a set of resources in a given environment is not being used or is being used inefficiently, species that can take advantage of the resources will do so. This tendency would be expected to result in patterns of similar functional communities evolving wherever environments are similar and common resources are available.

Energy Flow and Trophic Relationships

Small mammal biomass at the two True Prairie sites, Osage and Cedar Creek, was divided into dietary categories of herbivore, omnivore, and carnivore (Figure 7.29). Small mammal food habits in five grassland types (including the Osage Site, labeled "Tallgrass") have been summarized (Figure 7.30).

Energetics of small mammals in grassland ecosystems were compared by French et al. (1976). They used much the same data base for small mammal population and dietary data as did Grant and Birney (1979) and included the Osage Site as the only True Prairie site studied. The following discussion of energy flow in True Prairie mammals is taken from French et al. unless otherwise indicated.

Table 7.29 provides averages for the following parameters at six grassland sites: small mammal biomass, average annual energy of respiration, production, and consumption. Respiration, production, and consumption generally vary directly with biomass.

Support efficiency (the ratio of biomass to consumption) of the small mammal community at the True Prairie site is less than the average for grasslands. Both prairie voles (<u>Microtus ochrogaster</u>) and hispid cotton rats (<u>Sigmodon hispidus</u>), the predominant species, are nonhibernating herbivores with low assimilation efficiencies. Small mammal production at the True Prairie site is high, more than

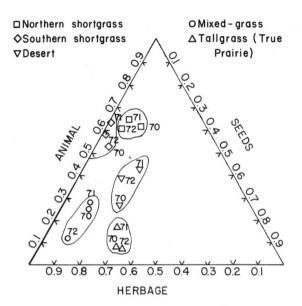

FIGURE 7.30 Diagrammatic representation of sources of energy utilized by small mammals, based on dietary analyses. Values were computed as proportional utilization of three food categories by the total population for the years 1970, 1971, and 1972. (Data from French et al., 1976.)

twice as high as at any other site except the desert grassland. Apparently, therefore, at least ungrazed True Prairie captures energy in a form palatable for small mammals at a rate capable of supporting an exceptionally high secondary productivity. Even so, French et al. calculated that an average of only 4 percent of the herbage available at the Osage Site was consumed by small mammals. Small mammals at the True Prairie site were estimated to have consumed slightly greater-than-average proportions of herbage and seeds and relatively less animal matter than the small mammals at other grassland types (Figure 7.30).

Based on the relationship between production, respiration, and biomass and the low level of small mammal biomass observed at the Cedar Creek site in east-central Minnesota, however, the small mammal community at that northern site was much less productive and probably more efficient than at Osage. Although primary productivity at Cedar Creek appears to be much less than at Osage, most of the vegetation at Cedar Creek appears to decompose without being used by consumers.

French et al. concluded that small mammals often do not consume enough of the herbage in grasslands for consumption to be a limiting factor, but they might use a much greater proportion of the available seeds. Seed drop failures would not often limit small mammals in the True Prairie but would limit them, not surprisingly, in a Desert Grassland, where small mammal biomass usually is high, primary productivity is relatively low, and a large proportion of the small mammal fauna is granivorous. French et al. also concluded

TABLE 7.29   Small Mammal Average Biomass, Annual Respiration, Production, and Consumption for 1970-72 at Grassland Sites, and the Ratio of Biomass to Consumption. (From French et al., 1976; copyright 1976 by the Ecological Society of America.)

Grassland sites	Biomass (g ha-1) live weight)	Respiration (kcal ha-1) × 103	Production (kcal ha-1)	Consumption (kcal ha-1) × 103	Biomass/consumption v 100
Bunchgrass	299	41	841	52	0.58
Mixed-grass	76	11	199	17	0.44
Desert grassland	665	81	1,822	98	0.67
Tallgrass (Osage Site)	813	127	3,225	197	0.41
Southern shortgrass	255	49	998	69	0.37
Northern shortgrass	247	24	461	32	0.76

281

that the animal matter available to small mammal populations in North American grasslands is heavily utilized, including the animal matter at the True Prairie site. However, observations at both the Osage Site and at Cedar Creek indicate that insect populations are high throughout the growing season. The relative amount of animal matter that eventually passes through the small mammal community is unknown.

French et al. (1979) studied trophic relationships of mammals at the Osage Site, where, even though standing-crop biomass was much greater than at any other grassland site studied (Figure 7.30), mammalian biomass contributed little to the total pyramid. French and his co-authors considered all mammals as being either primary or secondary consumers and thought that primary consumers were 100 percent biophagic (which probably is high, considering the food habits of these animals in winter). They classified some secondary consumers as partly biophagic and partly saprophagic but considered all carnivorous small mammals as being 100 percent biophagic.

Energetics and trophic relationships of True Prairie mammals can be summarized in the following ways. Movement of energy through small mammals in a True Prairie may be substantial (French et al. [1976] estimated that annual consumption at Osage is 19.7 kcal m^{-2}); and, because of the relatively high metabolic rate of homeotherms and the energy expended in movement and thermoregulation, the amount of energy flow through small mammals is high relative to their biomass but is still low relative to total energy flow in the ecosystem.

The shape of the trophic pyramid at the True Prairie site differs from those at other grassland sites, but the difference in small mammal biomass at the various sites had little effect on pyramid shapes.

While "making a living" in the True Prairie, small mammals do have an impact about which biologists have long conjectured. Early natural historians always included a notation on the economic importance of each species included in a faunal study. Only recently have North American ecologists attempted to classify and quantify the interactions of those small mammals that actually influence community structure and function (see Chew, 1978).

Having given special attention to grasslands, Grant (1974) concluded that the functional role of small mammals is to regulate the rate of mineral cycling. Small mammals clearly seem to affect rates of movement of nutrients and energy in grasslands, but the relatively small percentage of herbage they consume (perhaps an average of 4 percent in ungrazed True Prairie) or otherwise have an impact on (see Petryszyn and Fleharty [1972] for quantitative impact caused by small mammals clipping but not consuming vegetation) is so small that to support that hypothesis is difficult. Grant also discussed the importance of upward translocation of soil, which has a significant effect on the percent of bare surface soil and quantity of nitrogen in the top soil layer. He thought that the consumption of arthropods by small mammals has an important effect on the dynamics of arthropod biomass but noted that even at relatively high density, the direct consumption of live plants has little or no effect on primary production. However, McNaughton (1976) demonstrated that large mammalian herbivores in a natural grassland system in Africa can drastically reduce standing crop in a

short time, prevent senescence of the vegetation, enhance primary productivity, and thus directly facilitate energy flow. Simulation models of grazing by small mammals at the Osage Site (discussed in Chapter 10) indicated a similar trend, but, of course, the impact was much less than in Africa because of the difference in the amount of vegetation consumed. Grant found no evidence that small mammals significantly alter gross ecosystem structure in North American grasslands except when the mammal populations are extremely dense. The potential effects of grazing by small mammals should be least in the True Prairie because that prairie has a greater potential for primary productivity than other North American grasslands.

8

Decomposers

Two approaches generally can be used to describe the microbial decomposition of organic material in soil (Tribe, 1961). The first approach describes decomposition as the process of breaking down material by microorganisms responding to environmental conditions, such as moisture and temperature, but does not consider the taxonomy of the microbes. The second approach, the biological analysis of decomposition, regards decomposition as a result of a succession of microbes growing on an organic substrate. In an ecosystem approach, the relationship between the total rate of production and the rate of decomposition, regardless of what organisms are responsible, is of overall importance (Odum 1971). Since this book is the result of an interdisciplinary study in ecosystem analysis of the True Prairie, this chapter will emphasize the function of microorganisms in decomposition rather than the taxonomy and biological succession of specific microorganisms.

Taxonomically, the decomposers are the most diverse group in the True Prairie. Decomposers are mostly microorganisms--the bacteria, fungi, algae, actinomycetes, and viruses. Of the microflora, the importance of the bacteria and fungi in decomposition is well recognized (Dickinson and Pugh, 1974), and their structure and function in grasslands is the best documented of all microbial groups (Clark and Paul, 1970). In the earlier studies on the Osage Site, the bacteria and fungi were found to compose the greatest biomass (Harris, 1971). Thus, these two groups will receive the greatest discussion in this chapter. However, the complete process of decomposition cannot be fully appreciated without an understanding of the role played by the invertebrates. The structure and role of invertebrates in the True Prairie were discussed in Chapter 7.

Although microorganisms are found in all parts of the True Prairie, their roles in the phyllosphere (surface of plant stems and leaves) and the rhizosphere (the area immediately adjacent to the roots) are not well documented. Neither is the role microorganisms play as animal and plant parasites. Perhaps the best summary of the phyllosphere and rhizosphere in grasslands is by Clark and Paul

(1970). No attempt will be made in this chapter to review these important aspects of the True Prairie.

MICROBIAL BIOMASS

Microbial biomass in terrestrial ecosystems is far more difficult to measure than biomass for producers and consumers. The ubiquity of these organisms and the microscopic cells and filaments embedded in their nutritional sources impose a difficult problem for the microbial ecologist.

Traditionally, microbial ecologists have relied upon two methods to estimate microbial biomass in the soil. Direct microscopic observation on litter and soil particles buried in the soil has been widely used (Allen, 1957; Casida, 1971; Cholodny, 1930; Rossi and Riccardo, 1927). The microscopic size of the organisms makes this method tedious, and active and dormant forms are difficult to distinguish. Fluorescent microscopy and the use of the scanning electron microscope have greatly improved direct microscopic observation (Babiuk and Paul, 1970; Todd et al., 1973; Trolldenier, 1973). The second method that soil microbiologists have relied upon is the plate count method, in which soil samples are diluted and plated upon microbial media in the laboratory. Viable colonies are counted as an indication of the numbers of microbes per gram of soil. The number of colonies obtained is influenced by the type and concentration of the media used (Clark, 1965). Dormant cells and spores usually grow well in such artificial environments as the Petri plate and further confuse the results of these studies.

Microbial ecologists have attempted to improve their biomass estimates by relating microbial biomass to ATP (adenosine triphosphate) levels in the soil (Holm-Hansen, 1973; Sparrow and Doxtader, 1973). Although this method can be done more rapidly and measures metabolically active cells, ATP from root hairs and soil fauna will influence the results. Labelled compounds in the soil have usually been used to measure decomposition rates and to follow the movement of plant decomposition products through the soil (Jansson, 1960; Jenkinson, 1965, 1971). Shields et al. (1973) used labelled carbon and nitrogen together with direct and viable plate counts to determine active microbial biomass and turnover rates. They reported that only 20 percent of the total biomass measured was active.

ATP was extracted from soil samples at the Osage Site in an attempt to measure biomass by correlating it with plate counts and carbon dioxide evolution. Although correlation of ATP to cell plate counts and carbon dioxide evolution was too variable to interrupt (K. G. Doxtader, personal communication, 1973), certain trends were observed in the top 10 cm. ATP was consistently higher in the first 2.5 cm of soil (1195 to 441 μg g^{-1} oven-dried soil) than in the 2.5- to 5-cm level (819 to 38 μg g^{-1} oven-dried soil) or the 5 to 10 cm level (389 to 33 μg g^{-1} oven-dried soil). ATP concentration was correlated better with soil water than with temperature, since ATP was higher when soil water was higher.

The results of plate count studies made on the bacterial population at the Osage Site (Harris, 1971) showed an average of 9

million bacteria per gram of soil at a depth of 0 to 5 cm and 6 million bacteria per gram of soil at a depth of 10 to 20 cm. Although the highest number of bacteria was found in the first 5 cm of soil, viable numbers of bacteria were present to a depth of 50 cm. Direct counts of bacteria do not exist for the Osage Site; therefore, bacterial biomass has not been determined experimentally.

Fungal hyphae were measured directly during the growing season by Harris in 1971. Measurements on hyphal length were made with a camera lucida and a map-measuring device. Values of the hyphal length for both the ungrazed and grazed sites (Table 8.1) fall within the range from 1,000 to 2,000 m g^{-1} found in other recent measurements (Clark and Paul, 1970). Fungal biomass was calculated for the ungrazed treatment from these measurements (Risser, 1972). The assumed average hyphal diameter was 5 mm, with a water content of 90 percent, and specific gravity of 1.2, giving a calculated value for fungal biomass at the Osage Site of 635 g m^{-2} dry weight for the top 50 cm of soil. Most estimates of microbial biomass (Clark and Paul, 1970; Latter et al., 1967) suggest that fungal biomass is at least twice the amount of bacterial biomass. If this estimate is true for the Osage Site, bacterial biomass should approach 317 g m^{-2} for the upper 50 cm, giving a total microbial biomass of 952 g m^{-2} for the top 50 cm of soil. Clark and Paul (1970) reported a total biomass of 200 g m^{-2} for the upper 30 cm of the Shortgrass Prairie at the Matador Grassland Site in Saskatchewan. Although the True Prairie value is much higher, it represents an additional 20 cm of measurement under much warmer and wetter soil conditions.

Warcup (1957) stated that only 23 percent of the fungal hyphae studied in a wheat field were active. Examination of total and viable microbial populations in a Brown Chernozemic soil (Shields et al., 1973) indicated that only 20 percent of the bacterial biomass was active. Although active biomass was not distinguished from total biomass at the Osage Site, if the relationship there was similar to that found in the Brown Chernozemic soil of Saskatchewan, only 190 g m^{-2} for the upper 50 cm is functional.

MICROFLORA OF PRAIRIE LITTER AND SOIL

The taxonomic composition of microbial flora in the prairie has been studied primarily by the identification of colonies isolated by

TABLE 8.1   Total Hyphal Length for the Ungrazed and Grazed
Prairie, Osage Site, June 1970 (Harris, 1971).

Pasture	(m g^{-1} soil)				
	0-5 cm	5-10 cm	10-20 cm	20-30 cm	30-50 cm
Ungrazed	1,060	1,270	1,650	1,506	1,071
Grazed	1,801	1,296	1,563	1,607	1,680

the laboratory plate count method. Because the bacteria, fungi, and actinomycetes are more easily isolated, grown, and identified by this method, more is known about the taxonomy of these groups than about the protozoa and algae. Since, for the same reasons discussed in the last section, dormant cells and spores grow on laboratory media, microbial ecologists interpret with caution the frequency of the various isolated species. Obviously, one cannot relate growth on a laboratory Petri plate with ecological significance in the field, but data obtained by this method may be useful. Species identification made from plant counts is valuable when comparing seasonal trends, different treatments, succession, alleopathy, and different ecosystems.

Bacteria are the most numerous microorganisms in the prairie ecosystem, although most estimates of microbial biomass indicate that fungal biomass is greater than bacterial biomass (Clark and Paul, 1970). Soil bacteria are frequently classified into functional groups, such as decomposing, cellulytic, denitrifying, nitrogen-fixing, and nitrogen-oxidizing bacteria. Morphological groups are also used. Table 8.2 summarizes the morphological types of bacteria found at the Osage Site.

The isolates described in Table 8.2 were tested for their ability to hydrolyze starch and protein (casein) (Harris, 1971). Of 172 cultures tested, 46 percent hydrolyzed starch, 54 percent hydrolyzed protein, 25 percent hydrolyzed neither starch nor protein, and 33 percent hydrolyzed both starch and protein. Bacillus species contributed significantly to this bacterial population. Bacillus cereus was the most frequently identified Bacillus, followed by B. subtilis and B. megaterum. Bacillus species are comparable in number at all levels in the upper 20 cm of soil. Other genera representative of the prairie soil are Arthrobacter, Pseudomonas, Aerobacter, and Clostridium.

Actinomycetes composed a greater number of the total plate counts in grassland than in nongrassland soil (Alexander, 1961; Küster, 1967), but lower numbers of actinomycetes were found in undisturbed grassland soil than in cultivated soil (Robinson and MacDonald, 1964). In the prairie of Oklahoma, the actinomycetes

TABLE 8.2   Bacterial Morphological Types Found in Osage
            Soil.

Morphological types[a]	Percent
Gram-negative rods	61.5
Gram-positive rods	18.9
Gram-negative and positive rods	18.9
Gram-negative cocci	0.7

[a] Compilation of 172 isolates made from freshly isolated cultures (Harris, 1971).

accounted for only 10 percent of the total plate count, although the pH of the soil was slightly less than other soils in the Kansas-Oklahoma area (Harris, 1971). Streptomyces was the dominant genus of actinomycetes and Nocardia was the second.

Fusarium species reportedly are the most frequently isolated fungi in prairie soil (Clark and Paul, 1970; Orpurt and Curtis, 1957 a and b). Identification of fungal genera made from the isolation of mycelium from litter in a simulated laboratory ecosystem in the True Prairie is shown in Table 8.3. Fusarium species was found 17 percent of the time. Paecilomyces was found on all occasions. England and Rice (1957) found that the most frequently encountered species varied with the season in the prairie: Mucor janssenie in the spring, Monila bunnea in the summer, and Fusarium diversisporium in the fall.

Decomposition of Plant and Animal Tissue

The biodegradation of plant and animal tissue is one of the major functions of microorganisms in the prairie ecosystem. Although decomposition was known to contribute significantly to the total flow of energy and biomass within any ecosystem, an appreciation of just how much energy and biomass moves through the decomposer compartment has come only with the recent emphasis on ecosystem modeling.

Although field methodology in microbial ecology is still being developed, two traditional methods are used by most ecologists studying decomposition. Gravimetry, or the measurement of weight lost by a buried substrate, is used most frequently to determine the rate of decomposition (Clark and Paul, 1970; Jenkinson, 1965; Waksman and Gerretsen, 1931; Witkamp and Olson, 1963). Gravimetry is reliable for long-term studies of decomposition but does not

TABLE 8.3   Frequency Distribution of Fungi
Identified from Litter in Simulated
Ecosystem (May, 1974).

Fungi	Frequency
Paecilomyces	100.0
Fusarium	66.6
Tricothecium	33.3
Alternaria	33.3
Cladosporium	16.6
Penicillium	16.6
Circinella	16.6

measure the short-term responses of microbes to changes in the environment.

The measurement of chemical changes in the substrate (Dickinson and Pugh, 1974; Gray and Williams, 1971; Sorenson, 1967) or chemical responses in the soil system, such as carbon dioxide evolution or oxygen uptake (Edwards and Sollins, 1973; Humfeld, 1930; Kucera and Kirkham, 1971; Reiners, 1968; Swift and French, 1972), is the second method used to measure decomposition and microbial activity in both the laboratory and the field. The evolution of carbon dioxide was used by the investigators in the IBP Grassland Biome in an attempt to standardize methodology in the various ecosystems. The measurement of carbon dioxide from the soil in a field study may be criticized, since it includes root respiration as well as microbial and soil invertebrate respiration. However, total carbon dioxide evolution is useful in comparing soil respiration between ecosystems and in monitoring the response of soil organisms to changes in the soil environment.

Results from a laboratory ecosystem showed that in the range of soil water values from 0 to -4 bars, which is comparable to the soil water values measured during the growing season at the Osage Site, the optimal temperature for decomposition was 26°C. Under these conditions, plant litter lost 17 percent of its weight in 10 days. Decomposition continued at temperatures higher than 26°C, but the overall rate was less than the rate observed at 26°C. The maximum temperature at which the experiment was run was 35°C, since soil temperature at the Osage rarely reached higher temperatures. At 35°C the decomposition rate at optimal soil water was 10.5 percent for 10 days. Plant litter had a 4 percent decomposition rate for 10 days at 4°C under optimal soil water but did not show appreciable loss below 4°C. At 26°C the optimal soil water for decomposition was -1 bars. The range of soil water that supported appreciable decomposition at 26°C was from slightly below 0 to -4 bars. Below -4 bars soil water decomposition decreased rapidly.

Under optimal conditions--26°C and -1 bars--native plant litter from the Osage Site lost 90 percent of the soluble carbohydrates, 95 percent of the soluble amino acids, 16 percent of the starch, and 60 percent of the protein in 10 days after the litter was added to the soil.

Under field conditions the annual rate of decomposition of plant litter was 60 percent. This rate of decomposition, determined by gravimetry, approximates the rate found by Jenkinson (1971). The decay of plant material was gradual and sporadic. During the 108-day growing season, the Osage Site showed a higher decomposition rate for the first 45 days than for the remaining 63 days. Seventeen percent of buried litter disappeared in the first 45 days, compared with a total of 33 percent in the 108-day experiment. Cellulose was rapidly broken down, disappearing at a rate of 54 percent in the first 45 days. Cellulose was 96 percent decomposed after 3 months in the soil.

Cellulose decomposition in the prairie has been well documented (Lengkeek and Pengra, 1973; May and Risser, 1973; Pieper et al., 1972; Thayer, 1972). However, since the cellulose used in laboratory experiments is often pure--Whatman filter paper--the results of these studies do not indicate the actual decomposition of cellulose in the native plant litter. Comparative studies between

treatments, other ecosystems, and of the effects of the environment on decomposition can be made with cellulose, however. Table 8.4 compares the rate of cellulose decomposition among several grassland ecosystems. The True Prairie, represented by the Osage Site, has the highest decay rate for cellulose of the ecosystems shown in the table.

Carbon dioxide evolution was measured in situ at the Osage Site by the KOH absorption method. The purpose was to obtain data that would be useful in determining a carbon balance between the decomposer and the producer compartments, as well as to show relative differences in the microbial activity between ungrazed and grazed treatments in the prairie. These measurements were also useful in monitoring responses of the soil habitat to changes in the environment, particularly soil temperature and soil water.

The average rates of carbon dioxide evolution at the Osage Site were 9.15 and 12.86 g m^{-2} day^{-1} for the ungrazed and grazed prairie, respectively. The highest rates were measured in August and September when the soil temperature was the highest (20°C in the ungrazed treatment and 23°C in the grazed treatment to a depth of 5 cm). The averages for these two months were 14.51 and 21.79 g m^{-2} day^{-1} in the ungrazed and grazed prairies, respectively. The lowest carbon dioxide measured was during the winter--November through February. Values obtained during this period averaged 3.73 g m^{-2} day^{-1} in the ungrazed and 4.20 g m^{-2} day^{-1} in the grazed. The total carbon dioxide produced annually in the ungrazed prairie was 2,173 g m^{-2} and 2,684 g m^{-2} in the grazed prairie.

Table 8.5 summarizes some of the rates for carbon dioxide evolution from various ecosystems throughout the world. The Osage Site showed the highest rate of carbon dioxide evolution from the soil for any of the temperate ecosystems. This information

TABLE 8.4   Half-life (T½) and Seasonal Decay Constant (k) for Cellulose Decomposition During the Growing Season. (Calculated from Semi-logarithmic Plots of Herman and Kucera, 1975).

Site	T½ (days)	k (percent per day)
True Prairie (Osage, OK)	40	1.73
Shortgrass Prairie (Pawnee, CO)	55	1.26
True Prairie (Tucker, MO)	78	0.88
Desert Grassland (Jornada, NM)	80	0.87
Mixed-grass Prairie (Cottonwood, SD)	85	0.82
Mixed-grass Prairie (Matador, Saskatchewan, Canada)	250	0.28

TABLE 8.5   Comparison of Carbon Dioxide Evolution Rates for the Osage Site with Rates for Other Ecosystems.

Ecosystem	Method	Highest daily production $(g \ m^{-2})$	Annual total $(g \ m^{-2})$	Reference
Osage Site, True Prairie, Oklahoma	Alkali	23.81	2,684	May and Risser (1973)
True Prairie, Missouri	Infrared	10.80	1,675	Kucera and Kirkham (1971)
Deciduous Forest, Tennessee	Infrared	5.89	1,518	Witkamp (1966)
Tropical Savanna	Alkali	7.90	2,884	Schulze (1967)
Tropical Deciduous Dry Forest	Alkali	8.95	3,266	Schulze (1967)
Oak Forest, Minnesota	Infrared	--	2,912	Reiners (1968)
Swamp, Minnesota	Infrared	--	2,710	Reiners (1968)
Fen, Minnesota	Infrared	--	2,592	Reiners (1968)

indicates that the microbial activity of the True Prairie soil is one of the highest rates so far reported.

A carbon balance for 1970, 1971, and 1972 is shown here, including the total annual carbon dioxide evolution from both the grazed and ungrazed treatments and the average NPP for those treatments, as reported in Chapter VI. The carbon balance is based on the following assumptions and calculations: (1) the plant biomass is 40 percent carbon (the percentage of C in any basic carbohydrate); (2) carbon dioxide is 27 percent C; and (3) the carbon dioxide measured from soil respiration at the Osage Site is 60 percent from microbial respiration (Wildung et al., 1975).

The average net production reported for the ungrazed treatment is 887 g m^{-2} or 354 g C m^{-2}. The average annual carbon dioxide evolution from the ungrazed treatment is 2,173 g m^{-2} or 579 g C m^{-2}. When the latter figure is adjusted for microbial respiration, the value becomes 352 g C m^{-2}. That is, 579 g C m^{-2} are produced annually and 352 g C m^{-2} are respired annually by microorganisms. For the grazed treatment, 1,077 g m^{-2} biomass or 431 g C m^{-2} are added to the ecosystem by net photosynthesis, compared with 434 g C m^{-2} evolved by microbial respiration.

Coleman (1973) stated that in well-aerated soils the annual production of carbon measured as carbon dioxide evolution equals the annual input of carbon by the primary producers plus root respiration, if the carbon content of organic and mineral soil remains constant (Witkamp and Frank, 1969). The data for the Osage Site suggest that the carbon dioxide is too low to balance this equation. This can be the case when the alkali absorption method is used (Kirita and Hozumi, 1966) unless measurements are made to assure 80 percent of the potassium oxide remains unneutralized. During carbon dioxide sampling at the Osage, no values were used that were over 20 percent neutralized during titration. If the data are correct, then almost 100 percent of the NPP of the Osage was utilized by the decomposers during the three years of study. Values in the range of 56 percent (Macfadyen, 1963) and 70 percent (Golly, 1960) have been reported for the amount of NPP utilized by microorganisms, and Wiegert et al. (1970) estimated that from 80 to 90 percent of NPP is metabolized by decomposers.

Environmental Influences on Microbial Activity

Microbial activity is known to be influenced by several environmental factors. Soil water, soil temperature, soil pH, soil structure and texture, patterns of precipitation, and grazing are all important. Over a three-year period at the Osage Site soil temperature and soil water were the most important factors influencing microbial activity, and soil temperature was perhaps the most important factor in predicting microbial decomposition rates. Figure 8.1 summarizes the correlation of carbon dioxide evolution with soil water and soil temperature for the True Prairie during the 1972 growing season. The high correlation of carbon dioxide evolution with soil temperature has been noted by other investigators. Witkamp (1966) reported that in the deciduous forest, soil temperature was more significantly correlated with carbon dioxide evolution than was soil water or age of the litter.

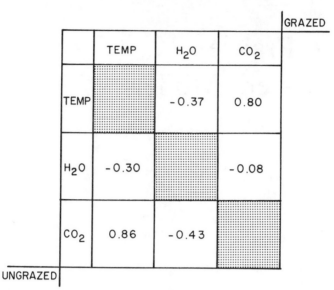

FIGURE 8.1 Simple correlation for CO_2-evolution data from 1972
growing season. Upper right shows grazed; lower left shows
ungrazed.

Kucera and Kirkham (1971) also reported that carbon dioxide
evolution from the soil of a Tallgrass Prairie responded
exponentially to increases in temperature within the range of field
conditions.

In carbon dioxide evolution investigations in arid grassland
soils, Wildung et al. (1975) found that carbon dioxide evolution
correlated better with soil temperature in the fall, winter, and
early spring and correlated better with soil water in late spring
and summer. The effect of soil water depended upon temperature and,
in the long term, the combined effect of soil temperature and water
was the most important factor influencing the flow of carbon in the
soil.

Grazing (see Chapter 10, page 346 for a description of the
treatment) increased decomposition at the Osage Site. Carbon
dioxide evolution was higher in the grazed treatment as were the
overall rates for decomposition of buried and surface litter and
cellulose. This difference is caused in part by the higher soil
temperatures measured in the grazed treatment. Soil water was
consistently lower in the grazed treatment than in the ungrazed
treatment. Soil nitrogen was not significantly higher in the grazed
treatment as might be expected. The physical state of the litter
used in the experiments was the same for both treatments, although
the effect of grazing on the overall physical composition of the
litter in the grazed treatment may have had a priming influence on
the microbial activity.

9

True Prairie
Ecosystem Models

A system is a set of interacting components and processes that include feedback and function as an integrated unit. Dale (1970), noting that systems analysis is the use of the scientific method with conscious regard for the complexity of the object of study, identified four phases of systems analysis: (1) selecting the parts that compose the system, (2) specifying the relationship between the parts, (3) specifying the mechanism by which changes in the system take place, and (4) validating the model and investigating its properties.

The most important attribute of systems analysis is that it provides a unifying structure for interrelating facts and observations. This unifying structure, generally embodied in some sort of model and defined as an abstraction of the real world, can be expressed in many different dimensions: mental, pictorial, physical, compartmental, graphical, or mathematical. Mathematical and compartmental models are most frequently used in the study of ecosystems.

Before a mathematical model of any system can be developed, the objectives of the modeling activity must be specified. The importance of the objective is emphasized by realizing that models are judged not on an absolute scale that condemns them for failing to be perfect but on a relative scale that approves them if they meet the specified objectives. Typical objectives of an ecosystem model might include improving understanding about the ecosystem, providing a tool for communicating information about the system, suggesting further research, and improving management of ecosystems. Generally, ecosystem models are considered successful if they open the road to improving the accuracy with which reality can be represented.

The first step in developing a model is to specify the components of the system (Ross, 1967). Each component of the system is referred to as a state variable, which is defined as an "easily recognized component of the system which contains, at any time, a certain amount of matter and (or) energy which is measured in

294

concentration units." The state of the system at any given time is defined by the value of the state variables.

The next step is to specify the relationship between the state variables, including the determination of the pathway by which matter or energy flows through the system (from one state variable to another) and specification of the equation for the different flows. A flow diagram shows the state variables and pathways over which either energy or matter flows through the system. For example, Figure 9.1 is a flow diagram for the True Prairie system at the Osage Site, with state variables (boxes) and flows (arrows). With mathematical models, flows of material from one state variable to another are represented by an equation that is a function of the state variables or external driving variables. A driving variable is defined as an independent or extrinsic variable that causes the system to respond but is not affected by the system itself. An open system is a system that responds to driving variables, while a closed system does not respond to driving variables. Most ecosystem models are open systems with atmospheric driving variables, such as rainfall, cloud cover, wind speed, and relative humidity.

The last step in developing a model is to validate the model and investigate its properties. The validation process depends upon type of model and the objective, and although Wiegert (1975) suggests that validation should be used in the very general sense of suggesting the degree of confidence one has in the model, Caswell (1976) distinguishes between prediction and theoretical models. Prediction models are designed solely to predict the future behavior of the state variables and are validated by establishing the degree of accuracy to which the model simulates the system's behavior and the range of conditions over which the model can be used with a reasonable degree of accuracy. Theoretical models are designed to provide insight into how the real system operates and are validated by attempts to disprove the model so that confidence in theory can be established. A single instance in which the model output does not correspond to real-world data can be considered sufficient evidence for invalidation of the theoretical model. Most ecosystem models combine both theoretical and predictive functions; thus, determination of what criteria to use to validate a model is difficult.

Sensitivity analysis can be used to investigate the properties of a model. In a sensitivity analysis the control parameters are altered and the response of the system is observed. The results indicate how sensitive the model is to particular processes and show which components of the system have the greatest influence upon the responses of the system. Cooper (1976) argues that a formal sensitivity analysis is essential for models that are used to guide environmental policy, since the greatest use for models may be at points of leverage where incremental policy changes have the greatest impact on the system. An expanded discussion of the use of simulation models and systems analysis in ecology is presented by Dale (1970) and Wiegert (1975).

In the last few years, many ecosystem-level simulation models have been developed. This array of models encompasses a range of internal complexity, number of variables, and degree of resolution as well as a variety of output variables. The agricultural-systems models have been summarized by Van Dyne and Abramsky (1975) in terms

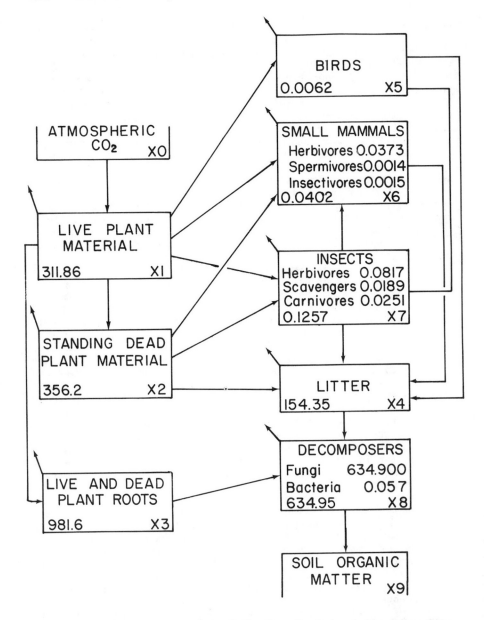

FIGURE 9.1 Compartment model of the True Prairie at the Osage Site. The compartments are numbered X0, X1, . . . , X9, and the numerical values are biomass measurements (g m^{-2}) on 1 July 1970.

of use and general model characteristics. Although some of these are grassland models, none is specifically for the True Prairie. Gutierrez and Fey (1975) used Forrester's computer-simulation method for testing and discussing hypotheses about grassland secondary

succession, but no True Prairie data were used. The two models that have used True Prairie data include the Osage Site version of the ELM (Ecosystem Level Model) grassland model and the Belowground Ecosystem model.

THE OSAGE SITE VERSION OF THE ELM GRASSLAND MODEL

ELM is a total ecosystem model of a grassland, developed by the US/IBP Grassland Biome and designed to be a generalized model applicable to different US/IBP grassland sites. The ELM model simulates the flow of water, heat, nitrogen, and phosphorus through the ecosystem and the biomass dynamics of up to five plant categories, ten consumers, and two decomposers (Innis, 1978). The model also simulates decomposition of aboveground and belowground plant and animal biomass. Driving variables for the model include daily rainfall, maximum and minimum air temperature, cloud cover, relative humidity, and wind speed. The model was developed to study the effects of the levels and types of herbivory, variation of rainfall and temperature, and the addition of nitrogen and phosphorus on grassland ecosystems.

The complete flow diagram of ELM in Figure 9.2 shows ten submodels: abiotic, water, temperature, producer, mammalian consumer, insect consumer, plant phenology, decomposer, nitrogen, and phosphorus. A complete description of the submodels is in Cole (1976), Cole et al. (1977), Hunt (1977), Innis (1978), Reuss and Innis (1977), and Rodell (1977).

The three major steps in developing the Osage Site version of the ELM model were (1) determining the model structure (e.g., which plant and consumer species to use); (2) determining values for site specific parameters required by the ELM model; and (3) comparing output from the model with observed data from the grazed and ungrazed treatments at the Osage Site. A list of the data types collected at Osage and a description of the sampling techniques were given in Swift and French (1972), and most of the data were presented by Birney (1974a), Birney et al. (1976), Blocker and Reed (1971), May and Risser (1973), Risser (1971), and Risser and Kennedy (1972a, 1975). The data included state variable information about the abiotic, producer, consumer, and decomposer components of the True Prairie at the Osage Site from 1970 through 1972.

The site specific structural components of the ELM model included the specification of the producer plant types, the consumer species, the number and depth of soil water layers, and the depth of the soil layers used in the decomposer, nitrogen, and phosphorus submodels. The five plant types simulated by the producer submodel included a warm-season perennial grass (Schizachyrium scoparius), a cool-season perennial grass (Poa pratensis), a warm-season forb (Ambrosia psilostachya), a cool-season forb (Nothoscordum bivalve), and a cool-season annual (Bromus japonicus). The mammalian consumer species simulated for the grazed site included the cow (Bos taurus), coyote (Carnis latrans), jackrabbit (Lepus californicus), cottontail (Sylvilagus floridanus), cotton rat (Sigmodon hispidus), prairie vole (Microtus ochrogaster), deer mouse (Peromyscus maniculatus), plains harvest mouse (Reithrodontomys montanus), thirteen-lined ground squirrel (Spermophilus tridecemlineatus), hispid pocket mouse

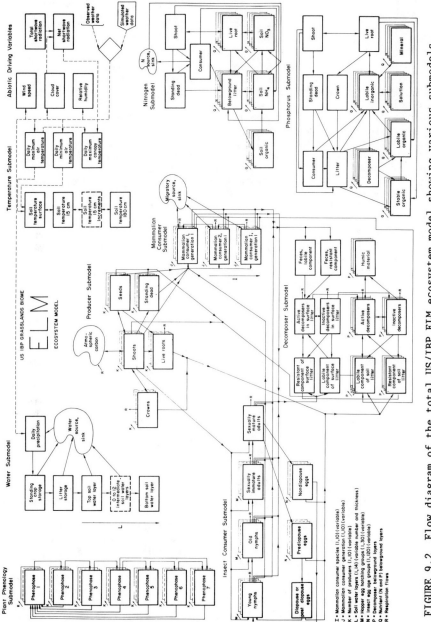

FIGURE 9.2 Flow diagram of the total US/IBP ELM ecosystem model showing various submodels, state variables, and flows.

298

(Perognathus hispidus), and shorttail shrew (Blarina brevicauda). The Osage Site version of the ELM model did not use the grasshopper submodel, because of the low grasshopper biomass at the Osage Site (Blocker and Reed, 1971), the relatively small amount of biomass consumed by grasshoppers (Scott et al., 1979), and the additional computer time required to run the grasshopper submodel. The water-flow submodel had seven soil water layers (0-5, 5-10, 10-15, 15-30, 30-60, 60-75, and 75-90 cm), the phosphorus and nitrogen submodels had four soil layers (0-5, 5-15, 15-30, and 30-90 cm), and the decomposer model had three soil layers (0-5, 5-15, and 15-90 cm). The temperature-profile submodel used the layer structure shown in the flow diagram (Figure 9.2).

Values for the site-specific parameters required by the ELM model were obtained from three sources. Where available, the appropriate information from the Osage Site was used first. If no data from the site were available, information was used from other studies conducted on the Tallgrass Prairie. Some of the values from the Osage Site were interpreted using literature information. For example, the grazed treatment was not actually grazed during the growing season at the Osage Site, so literature values were also used to arrive at some grazing parameters. Finally, where neither data from the Osage Site nor appropriate literature information was available, parameters were obtained by an iterative model-tuning process where values were determined by comparing model output with observed data, then modifying the parameter value until the difference between model output and observed data was minimized. Unfortunately, the model required many parameters, and use of the iterative model-tuning process was necessary in order to determine a fairly large number of parameters because many of the site specific parameters could not be determined directly from the existing information derived from field and laboratory studies.

In the process of developing the Osage Site version of the ELM model, the ELM model structure appeared inadequate for representing the effect of a plant canopy on light interception and water loss. Thus, the ELM model was modified slightly to include these effects.

OUTPUT FROM THE OSAGE SITE VERSION OF ELM GRASSLAND MODEL

ELM Output at the Osage Site

The Osage Site version of ELM grassland model was developed and verified with three years of abiotic and biotic data from the grazed and ungrazed treatments. The decision was made that all of the available site data would be required to develop and verify the model, thus eliminating the possibility of using Osage Site data to validate the model. That subjective decision was based upon both an analysis of the available data and the experience we gained in developing the Pawnee Site version of the ELM grassland model. Adequate abiotic and producer data from the Osage Site were available for 1970-72, plant-phenology data were available only for 1971 and 1972, and aboveground nitrogen and phosphorus standing-crop data were available only for 1972. Experience with the model also suggested that at least two to three years of abiotic and

plant-phenology data were required to determine adequately the
parameter value in the plant-phenology submodel.

Comparisons of observed and simulated data from the grazed
treatment in 1970 and the ungrazed treatment for 1970 and 1971 are
presented to show how closely the model results correspond to
observed data and to demonstrate how the abiotic and biotic state
variables in the tallgrass prairie change throughout the year.
These comparisons include a discussion of the possible errors in the
model structure that contribute to differences between observed data
and simulated model results. The biological mechanisms that are
responsive to temporal changes in state variables are also
discussed. The only true validation of the model is presented in
Chapter 10, where the response of the model to various management
practices is compared to published information about the response of
tallgrass prairies to management practices.

Producer Submodel Output

In evaluating the correspondence of observed and simulated
live-shoot and standing-dead biomass for the grazed and ungrazed
treatments (Figure 9.3), one must note that the grazed treatment on
the Osage Site was grazed only during the winter months. The
results for both treatments showed that peak live-shoot biomass
occurred in July and that most live shoots senesced by the middle of
November. In general, the observed data and simulated model output
compared very well. The ungrazed standing-dead data showed that the
simulated standing-dead biomass decreased from the beginning of the
season until the time of the peak live-shoot biomass, then increased
until the end of the year. Data from the grazed site showed that
the minimum standing-dead biomass occurred in May, while the maximum
value occurred in November. The increase in standing-dead biomass
at the end of May resulted from the death of cool-season annual
grasses (Figure 9.4). Seasonal dynamics of the simulated data for
standing dead was paralleled by the observations. The only
significant discrepancy was that standing dead for the grazed site
tended to be overestimated by the model; the reason for this
discrepancy was unclear.

The observed and simulated warm-season perennial and
cool-season annual live-shoot and standing-dead biomass for 1970 are
presented for the grazed and ungrazed treatments in Figures 9.4 and
9.5. The simulated warm-season perennial output for both treatments
had the peak live biomass occurring in the early part of July, with
most of the shoot biomass senescent by November. The seasonal
dynamics of the observed data and model output were similar for both
treatments. Simulated cool-season annual output illustrated that
the peak shoot biomass occurred in May, the shoot biomass died
rapidly in June because of senescence, and the cool season annuals
germinated and started to grow in the fall. These seasonal dynamics
compared favorably with the seasonal dynamics indicated by the
observed data, but the model tended to overestimate shoot growth in
the early part of the spring.

The warm-season perennial standing biomass (Figure 9.5) for
both treatments was lowest in early June and highest in November,
while the cool-season annual standing dead biomass was lowest in

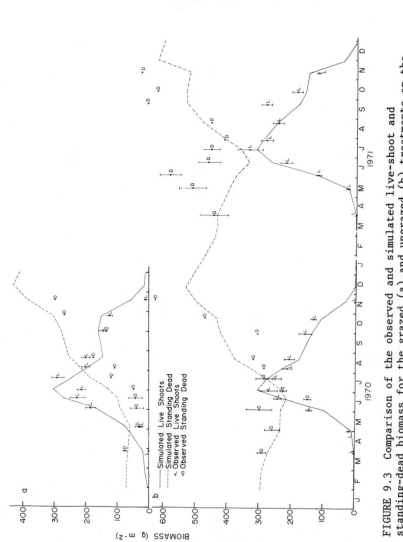

FIGURE 9.3 Comparison of the observed and simulated live-shoot and standing-dead biomass for the grazed (a) and ungrazed (b) treatments on the Osage Site for 1970 and 1971. The observed data ±1 SE (standard error) is shown for both live and standing-dead material (redrawn from Parton and Risser, 1976).

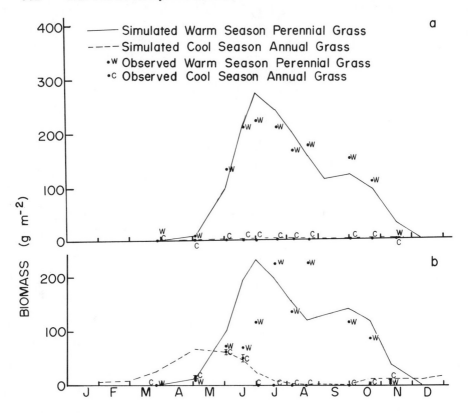

FIGURE 9.4 Comparison of the observed and simulated live-shoot
biomass for the warm-season perennial grass and cool-season annual
grass in the ungrazed (a) and grazed (b) treatments during 1970.
The standard error of observed data is shown for the cool-season
annual grass (redrawn from Parton and Risser, 1976).

April and highest in July. Simulated seasonal dynamics of the
cool-season annual and warm-season perennial standing dead agreed
with the observed data, although the cool-season annual standing
dead biomass was overestimated by the model.

Both the model output and observed data indicated that the
live-shoot biomass was greater on the grazed treatment in the spring
and the standing-dead biomass was larger on the ungrazed treatment
at all times. Model output and observed data also showed that the
warm-season perennial grasses were dominant in both treatments. In
the grazed treatment the live-shoot biomass of cool-season annuals
and warm-season perennials was very similar in the spring.

The simulated litter data for both treatments (Figure 9.6)
indicated that litter biomass decreased from the beginning of the
growing season until October, then increased. Litter biomass
decreased during the growing season because of favorable conditions
for decomposition of litter (high temperature and adequate water)
and increased during the winter months because of slow litter
decomposition rates (microbial activity is generally reduced by low

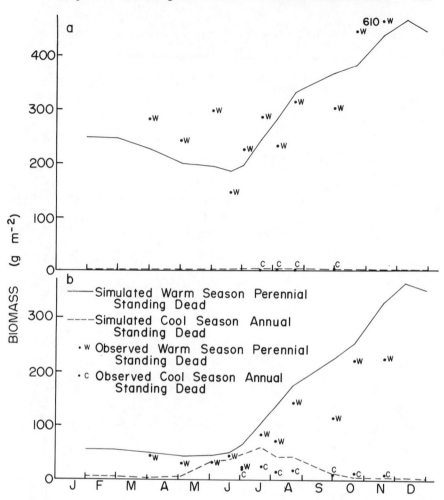

FIGURE 9.5 Comparison of the observed and simulated warm-season perennial and cool-season annual standing dead in the ungrazed (a) and grazed (b) treatments in 1970 (redrawn from Parton and Risser, 1976).

temperatures). An analysis of observed litter data revealed considerable variation between sample dates. The litter biomass in 1971 increased 300 g m^{-2} from July to August, then decreased 300 g m^{-2} in September. The high standard error (±42 g m^{-2}) of the observed data and the large variance between sample dates made determination of any clear seasonal patterns difficult. A comparison of the mean simulated litter biomass for the ungrazed and grazed treatments and observed data showed that the mean simulated litter biomass was 20 g m^{-2} less than the observed biomass for the ungrazed treatment and 30 g m^{-2} greater than the observed biomass for the grazed treatment. The mean values of the observed grazed and ungrazed litter data were 400 and 241 g m^{-2}, respectively.

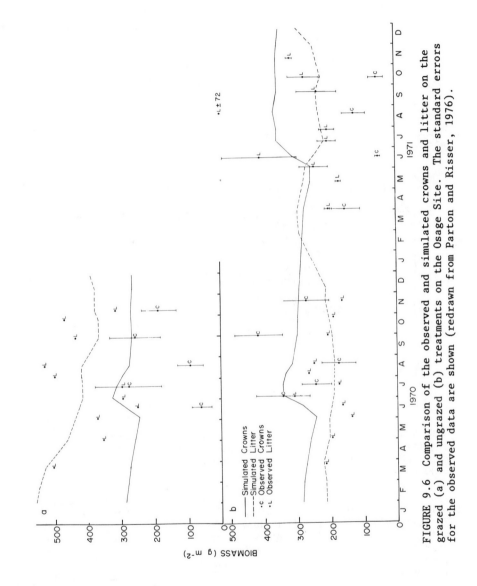

FIGURE 9.6 Comparison of the observed and simulated crowns and litter on the grazed (a) and ungrazed (b) treatments on the Osage Site. The standard errors for the observed data are shown (redrawn from Parton and Risser, 1976).

304

Improvement in the ability of the model to simulate litter dynamics depended upon developing a better technique to sample litter biomass and improving our understanding of critical processes that control litter dynamics (e.g., the transfer from standing dead to litter).

The simulated crown data (Figure 9.6) reached a minimum in early June and a maximum in July or August. Crown biomass increased during the growing season because of translocation of photosynthate or products of photosynthesis from the live shoots to the crowns, while the decrease in crown biomass during the nongrowing season was caused by crown death and respiration during the time period when the plant was not actually producing new plant material. The extreme variability of observed crown biomass between sampling dates (differences as great as 239 g biomass m^{-2} within a month) and the high mean standard error (± 57 g biomass m^{-2}) again made it difficult to determine any seasonal patterns. The simulated means for crown biomass were greater by over 100 g m^{-2} than the observed means for both treatments; however, the significance of this difference was not clear, because of the extreme variability of observed crown data between sampling dates and years. A change was made in the technique used to sample crowns in 1972 because all techniques suffer from the difficulty of routinely separating crowns from roots. The ability of the model to simulate crown dynamics will be improved only if a better technique to sample crowns is developed and if our understanding of the processes that control crown dynamic processes (e.g., crown death and translocation to the crowns) is improved.

A comparison of the simulated crown and litter results for the grazed and ungrazed treatments showed crown biomass to be greater on the ungrazed site and litter biomass greater on the grazed treatment during 1970. These results were verified by a comparison of observed data for the two treatments.

Simulated total root data (Figure 9.7) showed a minimum value in early June and a maximum in July or August. In the model, root biomass increased during the growing season because the translocation of photosynthate to the roots exceeded the loss of dead roots due to decomposition, while the general decrease in root biomass in the nongrowing season was caused by loss of root biomass due to respiration by live roots and decomposition of dead roots during the time period when translocation to the roots was lowest. Observed data for both treatments compare favorably, considering that the mean standard error for the observed data was ± 116 g biomass m^{-2} and that the root biomass observed on 31 March 1971 on the ungrazed treatment was probably an overestimate, since root biomass would not likely increase during the nongrowing season. Both observed data and simulated model output demonstrated that the root biomass on the grazed treatment was less than on the ungrazed site. The simulated mean seasonal root biomass for the grazed and ungrazed treatments was 30 and 102 g m^{-2}, respectively, which was much less than the observed means (1,050 g m^{-2}, grazed; 1,212 g m^{-2}, ungrazed).

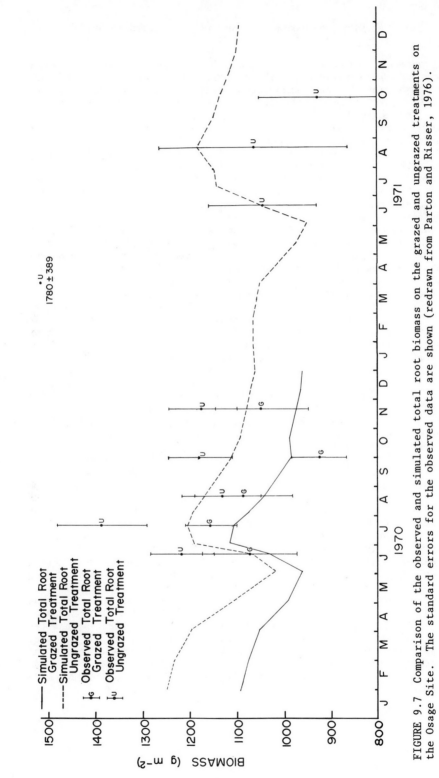

FIGURE 9.7 Comparison of the observed and simulated total root biomass on the grazed and ungrazed treatments on the Osage Site. The standard errors for the observed data are shown (redrawn from Parton and Risser, 1976).

Decomposer Submodel Output

Simulated and observed values of CO_2 evolution from soil are compared for the ungrazed treatment during 1972 in Table 9.1. A considerable discrepancy is evident between the observed and simulated values of CO_2 evolution for several dates, especially during the summer months (June-August), when the model underestimated CO_2 evolution by 50 percent. A comparison of observed and simulated values of the two abiotic variables that influence the CO_2-evolution processes in the model--soil water and soil temperature--showed close agreement between the observed and simulated values of soil temperature and soil-water content. The discrepancy between the observed and simulated soil temperature and soil water did not explain the difference between observed and simulated CO_2 evolution. At present, we feel that this difference is caused by the inability of the model to correctly predict the time when substrate (litter biomass and dead roots) becomes available for decomposition by microbes, since the total CO_2 evolution output for 1972 appears to be reasonable.

Carbon dioxide evolution originated from respiration by live roots and respiration by soil and litter microbes. Separation of the CO_2 that came from these different components in the field was difficult; however, an analysis of model results for root respiration indicated that 30 percent of the annual CO_2 evolution came from root respiration and 70 percent from microbial respiration. The model assumed that respiration was primarily a function of soil temperature (increasing with increasing temperature), with highest values of root respiration occurring during the summer. Microbial respiration in the model was largely a function of soil water, rainfall, and temperature (microbial activity was greatest with high temperature and adequate soil water), and the process had very large temporal variations. In the model, respiration of microbes increased very rapidly after a rain event, and rainfall was primarily responsible for the day-to-day variation in total CO_2 evolution. Although CO_2 resulting directly from a rain event may be influenced by the bicarbonate absorbed in the precipitation, neither field data nor model results considers this. This rapid increase after rain events was caused by favorable moisture conditions and an increase in the substrate available for microbial activity (the model assumed that transfer from standing dead to litter was directly proportional to rainfall amounts). The respiration process of soil microbes in the model was greatest in the summer, was reduced by low soil-water conditions, and was less variable than litter respiration because the moisture conditions were less variable within the soil. These model results cannot be verified by field data because of a lack of an adequate data base.

Consumer Submodel Output

A simulated time series of small mammal biomass for the ungrazed treatment in 1972 is compared with observed data (Birney, 1974a; Birney et al., 1976) in Figure 9.8. In the simulation, minimum biomass for small mammals occurred in April and the maximum in November. The same pattern was simulated by the model for

TABLE 9.1 Simulated and observed CO_2 evolution soil water at 0 to -10 cm and average daily soil temperature (0 to -15 cm) for the ungrazed treatment at the Osage Site during 1972.

Date	CO_2 evolution (g CO_2 m^{-2} day^{-1})		Soil water (cm H_2O)		Average daily soil temperature (°C)	
	Simulated	Observed	Simulated	Observed	Simulated (0 to -15 cm)	Observed (-5 cm)
22 April	13.66	5.76	3.25	3.75	16.2	14.0
16 May	4.97	4.44	2.00	3.01	19.1	15.0
6 June	5.86	11.89	2.16	1.91	21.8	17.5
7 July	5.83	9.23	2.45	2.49	18.4	18.5
8 August	4.49	13.81	1.34	1.94	21.7	19.8
2 September	16.40	15.21	3.25	2.93	19.8	19.3
4 November	6.30	3.73	3.23	2.49	9.2	9.0

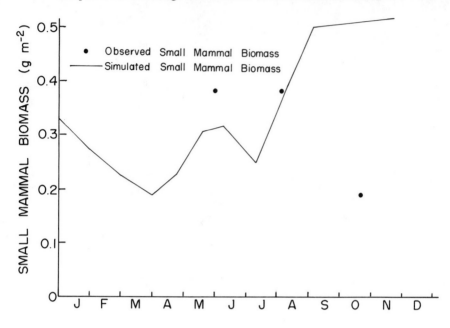

FIGURE 9.8 Simulated and observed small mammal biomass on the ungrazed treatment at the Osage Site during 1972.

1970-72 for both the grazed and ungrazed treatments. The observed data showed that in 1970 and 1972 small mammal biomass was highest in spring and least in the fall, while in 1971 small mammal biomass increased throughout the growing season. These results suggested that the small mammal submodel was not capable of accurately simulating the seasonal dynamics of the small mammal biomass. The model assumed that food limitation and a density-dependent mortality rate were the primary factors for controlling the population dynamics of small mammals. Our analysis of field data indicated, however, that climatic conditions were important controls where food supply was always adequate during the growing season. As discussed by Birney et al. (1976), the causes for seasonal changes in small mammal biomass are poorly understood at this time. The annual mean simulated small mammal biomass compared favorably with observed mean (0.32 versus 0.36 g wet wt m^{-2}).

Abiotic Submodel Output

The simulated and observed data of soil water content from 0-45 cm in grazed and ungrazed sites indicated highest soil water content in the spring, a decrease to minimum in the summer, then an increase in the fall (Figure 9.9). The major difference between the observed and simulated data was that the model underestimated soil water in April and May. The measured soil water content was greater in the ungrazed treatment than in the grazed treatment, and the greatest difference was observed in the early part of the growing season. These seasonal dynamics were reflected in the model that

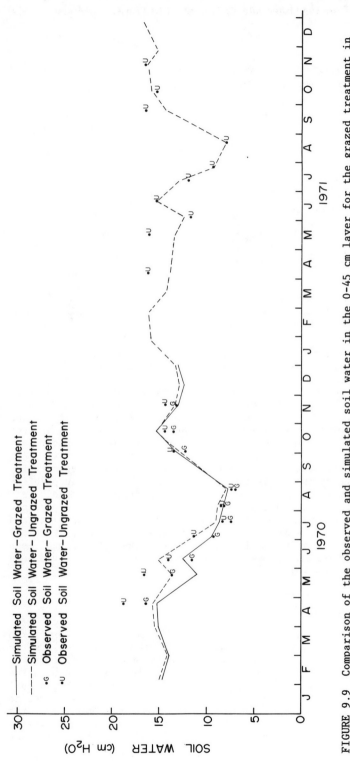

FIGURE 9.9 Comparison of the observed and simulated soil water in the 0-45 cm layer for the grazed treatment in 1970 and for the ungrazed treatment in 1970 and 1971 (redrawn from Parton and Risser, 1976).

represented lower evaporation and transpiration water loss during the spring in the ungrazed treatment (Figure 9.9). Decreased evaporation in the ungrazed treatment for the model was caused by decreased solar radiation and wind speeds near the ground (see pages 142-150, Chapter 5), while the decreased transpiration was produced by a later start of growing season and lower shoot biomass in the spring.

The simulated soil temperature for the grazed treatment at 0 and -30 cm may be compared to observed soil temperature at -2.5 and -25 cm, respectively (Figure 9.10). In simulated output and observed data the soil temperature increased in the spring to a maximum in August, then decreased in the fall. The soil temperature at the upper layer (0 or -2.5 cm) was from 2 to 3°C warmer in the spring and summer, and 1 to 2°C cooler in the fall. The major discrepancy between the observed and simulated data was that soil temperature was overestimated (1 to 3°C) at both soil depths in early spring and was underestimated (1 to 4°C) in the fall. A comparison of soil temperatures simulated for the grazed and ungrazed treatments in 1970 showed that the soil temperature at 0 and -30 cm was 0.76 and 0.61°C, respectively, warmer in the grazed treatment. May's (1973) soil temperature data for the grazed and ungrazed treatments in 1972 indicated that the soil temperature was warmer on the grazed treatment. Warmer temperatures in the model on the grazed treatment were a result of greater penetration of solar radiation to ground surface in the grazed treatment (lower standing-crop biomass).

The computer model results were used to calculate cumulation of monthly water flow for 1970 (Figure 9.11). Evapotranspiration water loss peaked in June; a secondary maximum occurred in September with minimum values during the winter months. The evapotranspiration water loss included evaporation (bare-soil evaporation plus evaporation of intercepted water) and transpiration. Transpiration was low during the winter months, increased to a peak in June, then decreased until the end of the growing season. Evaporation water loss was generally greater than transpiration except during late spring and early summer. The subsurface water flow included interflow (horizontal flow of water) and drainage (downward flow of water) and was lowest (less than 0.6 cm month^{-1}) during all months except April. A detailed analysis of the data shows that drainage was the major component (98 percent) of subsurface water flow for all months except April, when most (85 percent) of the subsurface water flow was interflow. In the model, interflow occurred only when the total soil-water content exceeded the water-holding capacity of the soil. Changes in soil water storage for the grazed and ungrazed treatment (Figure 9.11) demonstrated that the spring and fall were the two major periods when soil water was recharged, with most of the soil water recharge occurring in September and October. Soil water loss occurred primarily during the summer. Model outputs for 1970-72 indicated that in the spring of each of these years the soil profile was completely saturated during April or May and in two of these three years excess rainfall during the spring was lost from the system as interflow. These results suggested that the Osage Site generally started the growing season with a completely saturated soil profile and that combined fall, winter, and spring rainfall was generally more than enough to completely saturate the soil profile in the spring.

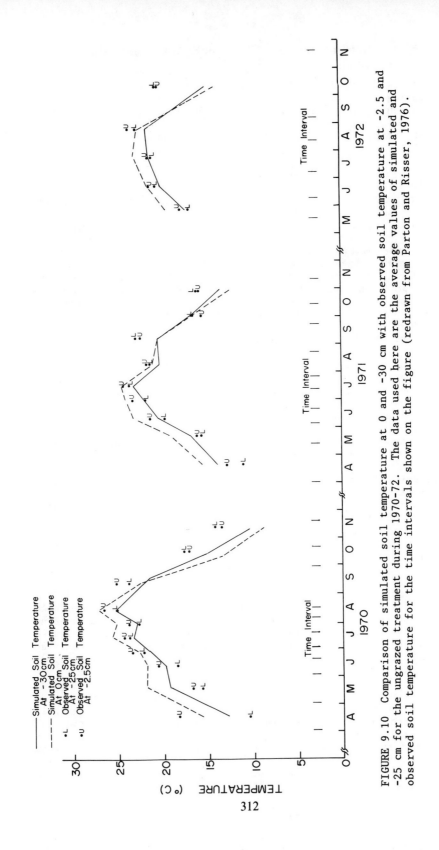

FIGURE 9.10 Comparison of simulated soil temperature at 0 and -30 cm with observed soil temperature at -2.5 and -25 cm for the ungrazed treatment during 1970-72. The data used here are the average values of simulated and observed soil temperature for the time intervals shown on the figure (redrawn from Parton and Risser, 1976).

312

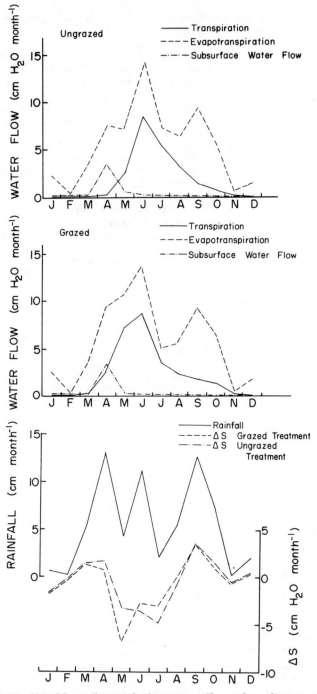

FIGURE 9.11 Monthly values of the water flows for the grazed and ungrazed treatments during 1970 at the Osage Site.

Evaporation and transpiration water loss was greater in spring in the grazed treatment and greater in summer in the ungrazed treatment, which then resulted in greater soil water loss in grazed treatments during the spring and a larger loss of soil water during the summer in ungrazed treatments. The soil water recharge during the spring and fall was similar for both treatments. In the model this lag in water loss for the ungrazed treatment was caused by the higher standing-crop biomass and the later start in the growing season. (See discussion in the last section and on pages 142-150, Chapter 5).

The simulated cumulative yearly flows of water (1970-72) on the grazed treatment showed the greatest water loss occurring as transpiration, followed by evaporation of intercepted water, bare-soil evaporation, and subsurface water flow (Table 9.2). In the ungrazed treatment the largest water loss occurred as evaporation of intercepted water, followed by transpiration, bare-soil water loss, and subsurface water flow. The three years of data for the ungrazed treatment also suggested that the buildup of standing-dead vegetation associated with the absence of grazing would cause the intercepted water loss to increase and the bare-soil water loss to decrease. The ungrazed treatment data also showed a larger loss of water storage in 1970, a significant increase in 1971, and a slight decrease in 1972. These data were consistent with the below-normal rainfall in 1970, the above-normal rainfall in 1971, and the near-normal rainfall of 1972.

The heat flows for the Osage Site in 1970 were calculated for the grazed and ungrazed treatments using model output from the

TABLE 9.2   Simulated cumulative yearly flow of water in the grazed and ungrazed treatments at the Osage Site.

Water flow	Grazed	Ungrazed		
(cm yr^{-1})	1970	1970	1971	1972
Rainfall	65.25	65.25	94.60	82.65
Interception	23.68	28.33	44.66	39.86
Bare-soil water loss	16.75	15.40	9.86	8.51
Transpiration	28.23	22.46	27.99	28.10
Subsurface water flow[a]	5.14	6.27	3.69	6.70
Change in water storage	-8.55	-7.21	8.40	-0.52

[a] Subsurface water flow includes both interflow (horizontal flow of water) and drainage (downward flow of water).

soil-temperature and water-flow models. Equation (9.1) was used to calculate heat-flow terms as follows:

$$Q_n = Q_g + Q_h + Q_e \qquad (9.1)$$

where Q_n = net all-wave radiation (ly day^{-1})

Q_g = transfer of heat through the ground (downward flow of heat is positive [ly day^{-1}])

Q_h = turbulent transfer of sensible heat to the atmosphere (upward flow is positive [ly day^{-1}]);

Q_e = contribution of latent heat of evapotranspiration (upward flow of water is positive [ly day^{-1}]).

The latent-heat flux (Q_e) was determined directly from model output variables, while soil-heat flux (Q_g) was calculated using the monthly average temperature data as input into equation 9.2 and below:

$$Q_g = \rho c \int_0^{90 \text{ cm}} \frac{\partial T}{\partial t} \, dz - k \frac{(T_{90} - T_{115})}{15} \Delta t \qquad (9.2)$$

where ρ = density of soil (g m-3);

c = specific heat (cal g^{-1} $^{\circ}C^{-1}$);

k = thermal conductivity (cal cm^{-1} sec^{-1} $^{\circ}C^{-1}$);

T_{90} = average monthly soil temperature at -90 cm ($^{\circ}$C);

T_{115} = average monthly soil temperature at -115 cm ($^{\circ}$C);

Δt = time step (86,400 sec).

The net all-wave solar radiation was calculated with the following equations:

$$Q_n = Q_s - t_n \qquad (9.3)$$

and

$$t_n = (1.0 - 0.8C) (0.2447 - 0.195 \sigma T^4) \qquad (9.4)$$

where

Q_s = net shortwave solar radiation (downward flow of radiation energy is positive [ly day^{-1}]);

t_n = net longwave radiation (upward flow of radiation energy is positive [ly day^{-1}]);

C = cloud cover (fraction);

σ = Stefan-Boltzman constant (0.813×10^{-10} cal m^{-2} $^{\circ}K^{-4}$ min^{-1});

T = air temperature at 2 m ($^{\circ}$K).

Net shortwave solar radiation was a model output variable, while the equation for net longwave solar radiation was determined by using Swinbank's equation (1963) for longwave radiation from clear skies, Horwitz's equation (1941) for the impact of cloud cover on net longwave radiation, and the Stefan-Boltzman law (Hess, 1959).

The sensible-heat flux was calculated by substituting values of Q_n, Q_e, and Q_g into equation (9.1). Equations (9.1) and (9.2) were derived from equations presented by Munn (1966).

The monthly-heat flux terms are shown in Figure 9.12 for the grazed and ungrazed treatments in 1970. Latent-heat flux was low during the winter, reached a maximum during June, decreased to a minimum in August, then increased to a secondary maximum in September. The sensible-heat flux was fairly uniform throughout the year with maximum values occurring in early spring and late summer. The soil-heat flow had its greatest positive value in May and its greatest negative value in November, indicating rapid heating in May and cooling during November. Soil-heat flux was fairly low compared with the sensible- and latent-heat fluxes; sensible-heat flux was the dominate term during winter months, and during the rest of the year the latent-heat flux was the largest. The minimum value of net all-wave radiation occurred in December and the maximum in July. During the spring the sensible-heat flux was greater on the ungrazed treatment, while during the summer months it was greater on the grazed treatment. These results were consistent with the results of the water-flow analysis, showing that the evapotranspiration water loss was greater on grazed treatments during the spring and larger on ungrazed treatments during the summer. The heat fluxes were also calculated for the ungrazed treatments for 1971 and 1972 and showed very similar patterns to patterns observed in 1970.

Nutrient Submodel Output

The simulated net flow of nitrogen (N) through the grassland for the ungrazed site in 1972 (Figure 9.13) illustrates that most of the root uptake is transferred to live shoots. The live-shoot nitrogen was cycled through the standing dead to the litter, then returned to either the soil NH_4^+ or soil organic pools as a result of decomposition of litter.

Simulated input of atmosphere nitrogen to the system (0.98 g m^{-2} yr^{-1}) compared favorably with the data for a Tallgrass Prairie in Missouri, where from 0.5 to 1.0 g N m^{-2} yr^{-1} was added from rainfall (Dahlman et al., 1969). Root uptake, shoot uptake, nitrogen transfer from shoots to standing dead, nitrogen transfer from standing dead to litter, and the release of nitrogen from litter (dead roots and aboveground litter) were compared with observed data for the ungrazed treatment in 1972 (Table 9.3). Nitrogen root uptake, nitrogen transfer from standing dead to litter, and nitrogen released from dead litter compared favorably. However, the simulated shoot uptake and nitrogen transfer from shoots to standing dead were two to three times greater than the observed data. R. G. Woodmansee (personal communication, 1977) indicates that the error may be caused by model underestimation of the amount of nitrogen that is recycled in the live root/shoot system of the plant, and McKendrick et al. (1975) showed that in big bluestem and Indiangrass a significant portion of the nitrogen used in early growth came from the rhizome and crown storage areas. The flow of soil organic nitrogen N to soil NH_4^+, the flow of dead litter nitrogen to soil organis nitrogen, the flow of NH_4^+ to NO_3^-, and the root uptake of nitrogen all seem to be overestimated by the model.

Observed and simulated nitrogen content of the live shoots, standing dead, and litter for the ungrazed treatment in 1972 are compared in Table 9.4. Simulated shoot nitrogen increased very

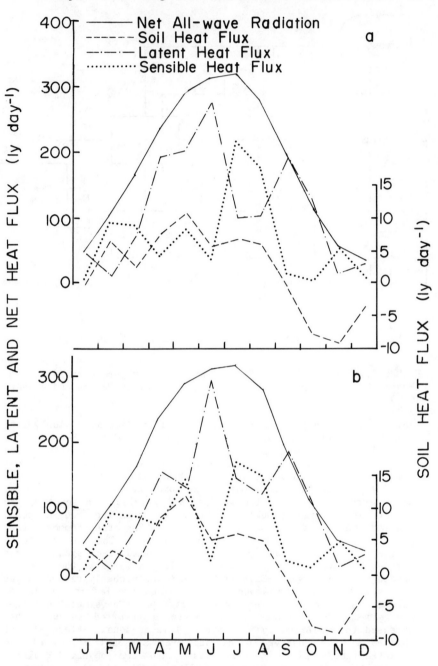

FIGURE 9.12 Monthly values of heat fluxes for the ungrazed (a) and grazed (b) treatments during 1970 at the Osage Site.

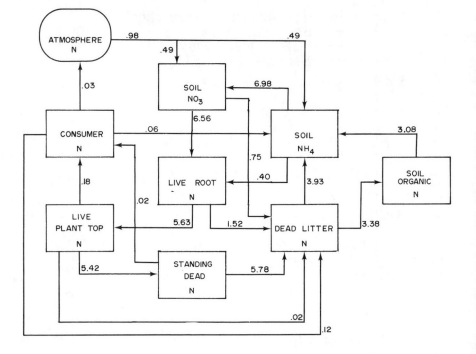

FIGURE 9.13 Simulated net flow of nitrogen for the ungrazed treatment at the Osage Site in 1972 (redrawn from Parton and Risser, 1976).

rapidly in spring to a maximum in June, then decreased slowly into September. Simulated data compared favorably with observed data, except for the overestimate of live-shoot nitrogen in November. The large overestimation in November was probably because the shoot biomass was not senescing fast enough at this time of year, the amount translocated from shoot to rhizomes and crown was too low, and an error occurred in the way phenology influenced shoot uptake of nitrogen. High simulated values of standing-dead nitrogen occurred in early spring, decreased during the summer, then increased at the end of the growing season. The observed data verified this seasonal pattern. The simulated standing-dead nitrogen, however, was approximately twice as large as the observed standing-dead nitrogen. This discrepancy was probably caused by the underestimation of the amount of nitrogen recycled in root/shoot systems, as noted in the preceding paragraph. The simulated litter data showed that maximum litter nitrogen occurred in May, then decreased to a minimum in September. This seasonal pattern was not verified by the observed data, and mean seasonal values of the simulated-litter nitrogen were less than the observed values. Uncertainty about the quality of litter nitrogen data makes it difficult to make inferences about the difference between the observed and simulated data. Simulated root nitrogen data (data not presented) had a mean value of 12.04 g m^{-2} for the growing season. A comparison with observed data was not possible, because of the

TABLE 9.3   Comparison of simulated and calculated[a] values of
the net annual root and shoot uptake of nitrogen,
the net annual flow of live shoot nitrogen to
standing-dead nitrogen and standing-dead nitrogen to
litter nitrogen, and the annual release of nitrogen
from the litter, roots, and crowns for 1972.

Nitrogen	Simulated	Observed[a]
	$(g\ N\ m^{-2}\ yr^{-1})$	
Total root uptake of nitrogen	6.96	5.35
Shoot uptake of nitrogen	5.63	2.01
Nitrogen transfer from shoots to standing-dead nitrogen	5.42	1.76
Nitrogen transfer from standing-dead to litter nitrogen	5.78	5.12
Nitrogen release from roots, litter, and crowns	7.31	6.72

[a] The values were calculated by Bokhari and Singh (1975) using
1972 harvest data from the Osage Site.

poor quality of the root nitrogen data; and information from R. G.
Woodmansee (personal communication, 1977) indicated that much
uncertainty existed about the amount of nitrogen contained in live
and dead roots.

Most of the apparent problems with the nitrogen submodel have
been corrected in the most recent version of the model. Further
improvements to the nitrogen model are being implemented at the
present time, along with a thorough analysis of the nitrogen data
from Osage Site. A complete description of the revised nitrogen
submodel and an analysis of the nitrogen data will be presented by
Parton and Risser (in press).

Phosphorus Submodel Output

Output from the phosphorus submodel was presented for the
ungrazed treatment during 1972. The net annual cumulative
phosphorus flows (Figure 9.14) showed that no external flow of
phosphorus (P) went into the system and that inflow and outflow of
phosphorus were approximately the same for most of the state
variables. The sole source of phosphorus for the microbes and live
plants was the labile inorganic pool; of the total uptake of
phosphorus from the labile inorganic pool, 18 percent went into the

TABLE 9.4 Comparison of simulated and observed values of nitrogen contained in the live shoots, standing dead, litter, and roots for the ungrazed treatment in 1972.

Date	Live-shoot nitrogen (g m^{-2})		Standing-dead nitrogen (g m^{-2})		Litter nitrogen (g m^{-2})	
	Simulated	Observed	Simulated	Observed	Simulated	Observed
22 April	0.06	0.13	6.87	4.14	2.30	--
15 May	0.59	0.31	6.42	2.78	2.37	2.97
6 June	2.90	1.32	6.28	2.62	2.36	5.14
5 July	2.71	2.74	6.42	3.46	2.26	4.63
8 August	2.07	1.86	6.74	2.72	2.08	7.39
2 September	1.60	1.58	7.08	3.37	1.98	--
14 November	1.81	0.01	6.94	3.78	2.69	--

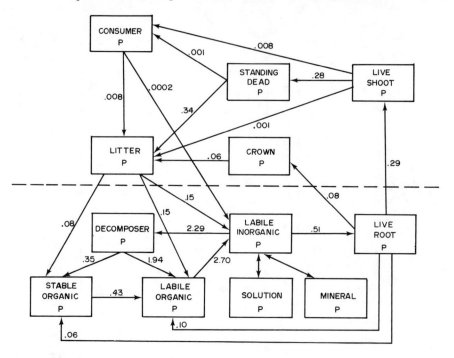

FIGURE 9.14 Simulated net flow of phosphorus for the ungrazed treatment at the Osage Site in 1972 (redrawn from Parton and Risser, 1976).

live roots and 82 percent went into the microbial phosphorus pool. Approximately 66 percent of phosphorus uptake by roots went into live shoots, and the remaining 33 percent went into the labile organic and stable organic pools.   Unfortunately, observed net phosphorus-flow data for the Osage Site were not available to compare with the simulated data.   C. V. Cole (personal communication, 1976) suggests that net root uptake of phosphorus should be approximately 10 percent of the net root uptake of nitrogen.   With this approximation, the observed and simulated values of net root uptake of phosphorus equaled 0.54 and 0.51 g m^{-2} yr^{-1}, respectively.

Simulated live-shoot phosphorus increased in the spring to a maximum value in June, then decreased until September (Table 9.5). The observed data show an increase through August.   The data for November showed that the simulated shoot phosphorus was generally overestimated. This overestimation was caused by an underestimation of the death rate of live shoots during this time period and by an error in the way that phenology influenced plant uptake of phosphorus.   Both   the   phosphorus   and   nitrogen   submodels overestimated live-shoot phosphorus and nitrogen at the end of the growing season.   The simulated model output showed a decrease of phosphorus in standing dead from the beginning to the end of the growing season. This pattern was not verified by observations, and the model tended to overestimate the amount of phosphorus contained

TABLE 9.5 Observed and simulated values of the phosphorus contained in live shoots, standing dead, litter, and crown vegetation for the ungrazed treatment in 1972 at the Osage Site.

Date	Live shoot P (g P m^{-2})		Standing dead P (g P m^{-2})		Litter P (g P m^{-2})		Crown P (g P m^{-2})	
	Simulated	Observed	Simulated	Observed	Simulated	Observed	Simulated	Observed
22 April	0.001	0.015	0.44	0.30	0.043	--	0.35	--
15 May	0.018	0.035	0.42	0.20	0.050	0.089	0.35	0.063
6 June	0.122	0.124	0.40	0.201	0.045	0.126	0.35	0.049
5 July	0.114	0.121	0.40	0.266	0.049	0.138	0.35	0.002
8 August	0.080	0.167	0.40	0.537	0.046	0.181	0.34	0.092
2 September	0.053	0.143	0.41	0.200	0.049	0.140	0.34	0.167
14 November	0.118	0.0005	0.36	0.230	0.070	--	0.35	--

in the standing dead. Model output indicated very little seasonal variation in the phosphorus contained in litter, and the model tended to underestimate the phosphorus in the litter. Simulated phosphorus in the crowns was constant, but observed phosphorus showed a decrease in the late spring and an increase in the fall, possibly because of translocation from and to the crowns, respectively, in the spring and fall. Reasons for these discrepancies between observed data and model output from the standing dead, litter, and crown are unclear. The total amount of phosphorus contained in roots is not presented here because the phosphorus contained in the dead roots is considered part of the labile and stable organic pools and cannot be separated within the present model structure. A thorough analysis of the phosphorus data for the Osage Site and modification of the phosphorus submodel is proceeding at the present time.

The results of comparison of model output with observed data, summarized in Table 9.6, give the mean difference between observed data and ELM model output (mean error), the mean of the absolute value of the differences between observed data and ELM model output (absolute mean error), the mean and standard error for the observed data, and the ratio of absolute mean error to observed mean value for Osage Site data on the ungrazed treatment. The mean error gives an indication of model bias (tendency to over- or underestimate), the absolute mean error gives an estimate of the mean deviation between observed data and simulated output, and the ratio of the absolute mean error to the mean of the observed data shows the magnitude of error relative to the observed mean.

The live-shoot, standing-dead, and root biomass data showed that the model underestimated these variables by 4, 40, and 96 g m^{-2}, respectively, and that the absolute mean error was less than ±18 percent of the observed mean values. The litter biomass was underestimated by 15 g m^{-2}, while the absolute mean error was ±26 percent of the mean litter biomass. The model overestimated the crown biomass by 115 g m^{-2}, while the absolute mean error was very large (±154 g m^{-2}). The fairly large difference between observed data and model output for the crown and litter was attributed partly to problems associated with sampling litter and crowns.

The aboveground-nitrogen state variable data showed that the nitrogen content was overestimated for live shoots and standing dead and underestimated for the litter and that absolute mean error ranged from ±50 percent (shoot nitrogen) to ±104 percent (standing crop nitrogen) of the observed mean. The overestimate of nitrogen content and the large absolute mean error was caused primarily by an underestimate of the amount of nitrogen recycled in the root/shoot system. The phosphorus data indicated that the shoot and litter phosphorus were underestimated, while the standing-crop phosphorus was overestimated. The absolute mean error ranged from ±56 percent (live shoot) to ±79 percent (standing-crop phosphorus) of the observed mean. The reason for the discrepancy between the observed phosphorus data and model results is unknown.

Abiotic model results illustrated that the model slightly underestimated soil water and soil temperature and that absolute mean error was less than 10 percent of the observed mean for these variables. CO_2 evolution was underestimated by the model, and observed mean error was 48 percent of the observed mean. The

TABLE 9.6 Comparison of the mean difference between observed data and ELM Model output (mean error), the mean of the absolute value of the difference between observed data and ELM Model output (absolute mean error), and ratio of absolute mean error to the mean of the observed data with mean values and standard error for observed standing-crop biomass (1970-72), nitrogen (1972), and phosphorus and observed soil water (1970-71), CO_2 evolution (1972), and soil temperature data (1970-71) from the ungrazed treatment.

State variable	Mean of observed data (MOD)	Standard error of observed data	Mean error[a]	Absolute mean error[b] (AME)	$\frac{AME}{MOD}$
Live-shoot biomass (g m^{-2})	165	14	+4.0	+23.0	.14
Standing-crop biomass (g m^{-2})	419	28	+40.0	+72.0	.17
Litter biomass (g m^{-2})	241	42	+15.0	+62.0	.26
Crown biomass (g m^{-2})	203	41	-115.0	+154.0	.76
Root biomass (0-70 cm) (g m^{-2})	1,212	129	+96.0	+77.0	.15
Live-shoot N (g m^{-2})	1.14	--	-0.54	0.55	.50
Standing-dead N (g m^{-2})	3.27	--	-3.41	+3.41	1.04
Litter N (g m^{-2})	5.02	--	+2.75	+2.75	.55
Live-shoot P (g m^{-2})	0.09	--	+0.01	+0.05	.56
Standing-crop P (g m^{-2})	0.28	--	-0.07	+0.22	.79
Litter P (g m^{-2})	0.13	--	+0.09	+0.09	.69
Soil water at 0-45 cm (cm H$_2$O)	13.16	--	+0.26	1.01	.08
Soil temperature at 0 cm (°C)	20.22	--	+0.03	1.88	.09
Soil temperature at 30 cm (°C)	19.14	--	+0.30	1.41	.07
CO_2 evolution (g m^{-2} day^{-1})	9.15	--	+0.94	4.42	.48

[a] Mean error: $\dfrac{\sum\limits_{i=1}^{N} (X_i - S_i)}{N}$, where X_i = observed data; S_i = simulation results; and N = number of observations.

[b] Absolute mean error: $\dfrac{\sum\limits_{i=1}^{N} ABS(X_i - S_i)}{N}$, where ABS = absolute value and the other definitions are as given above.

underestimation of CO_2 evolution was primarily caused by an underestimation during the summer months. The reason for the significant discrepancy between the observed data and model output is poorly understood at this time. However, since the simulated total CO_2-evolution output for 1972 and the abiotic driving variables for decomposition appeared to be reasonably well simulated, the cause of the underestimation of CO_2 evolution was believed to be an inability of the model to correctly predict the time when substrate (litter and dead-root biomass) became available for decomposition by microbes.

Simulated soil water, soil temperature, and live-shoot, standing-crop, and root biomass compared favorably with the observed data, while simulated CO_2 evolution, phosphorus and nitrogen variables, and crown biomass deviated substantially from the observed data. Further discussion of the differences between observed field data and simulated model results was presented in Parton and Risser (1976).

THE BELOWGROUND ECOSYSTEM MODEL

The Belowground Ecosystem Model represents the intra- and interseasonal dynamics of various herbage compartments, both above- and belowground, in a Shortgrass and Tallgrass Prairie. The model was developed to combine recent field information about the translocation of photosynthates to crowns and roots, amounts of metabolically functional root biomass, and patterns of root growth and mortality (Ares, 1976; Singh and Coleman, 1973; 1974). The model was also designed to be used as a tool for guiding future research into the dynamics of the belowground ecosystem.

The structure of the model was very similar to the structure of the Ares and Singh (1974) model, which was driven by an unlimited carbon pool available for translocation to the belowground plant parts, and the various flows in the model were nonmechanistic. The Belowground Ecosystem Model formulated the flows in a mechanistic way and replaced the aboveground translocation carbon pool with the simulation of photosynthesis and aboveground-biomass dynamics. Emphasis, however, is still on the belowground plant compartments and associated processes.

A compartmental representation of the herbage system (Figure 9.15) shows the flows and state variables considered in the model. The aboveground state variables included live-shoot, new and old standing-dead, and litter biomass. The live and dead root compartments were further subdivided into juvenile, nonsuberized, and suberized roots, following Ares (1976) and Ares and Singh (1974). Juvenile roots are young, light in color, capable of rapid elongation, and the primary location of root hairs. As the juvenile roots age and increase in diameter, they are referred to as nonsuberized roots. The characteristics of nonsuberized roots are a light-brown or yellowish color, an average diameter of 0.126 mm, and high activity with respect to water absorption. The older roots, referred to as suberized roots, are characterized by a high degree of suberization, dark-brown color, a mean diameter of 0.148 mm, a lower capacity for water uptake than nonsuberized roots, and an ability to act as important storage organs. Each belowground state

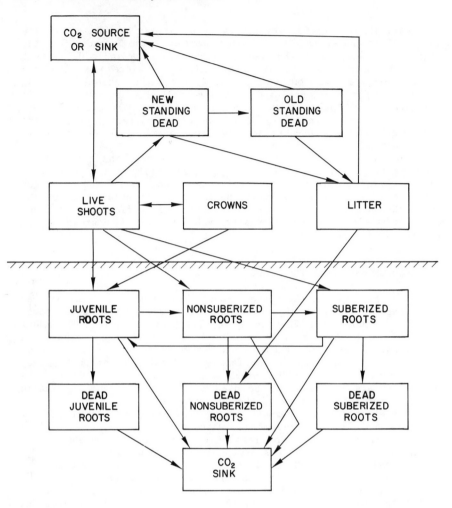

FIGURE 9.15   Flow diagram for the Belowground Ecosystem Model.

variable is considered for six soil layers (0-5, 5-15, 15-30, 30-45, 45-60, and 60-90 cm).   All state variables and rate processes are given in terms of dry weight (g biomass m^{-2}).

The abiotic section of the model consisted of a water flow submodel and a temperature profile submodel.   The water flow submodel simulated the flow of water through the plant canopy and the soil layers, and the allocation of water in the soil layers and the evapotranspiration of water were the important processes.   The temperature profile submodel simulated daily solar radiation, maximum canopy air temperature, average daytime canopy air temperature, and average daily soil temperature at thirteen points in the soil profile (0-180 cm).   A detailed description and verification of the abiotic submodel is given by Parton (1976) and Parton and Risser (1976).

The atmospheric driving variables required by the abiotic model included daily rainfall, maximum and minimum air temperatures (2 m), and long-term monthly average values of cloud cover, relative humidity, and wind speed. Daily rainfall recorded at Foraker, Oklahoma (1970-72), daily maximum and minimum air temperatures at Pawhuska, Oklahoma (1970-72), and long-term average monthly values of cloud cover, relative humidity, and wind speed at Tulsa, Oklahoma (1966-1972) were used for the Osage version of the model.

The model was first constructed for the Shortgrass Prairie at the Pawnee Site in Colorado because more information and experience were available there. The model was then adapted to the Osage Tallgrass Prairie site. A complete description of both the Osage and Pawnee Site versions of the Belowground Ecosystem Model and output is presented by Parton and Singh (1976) and Parton et al. (1978).

Belowground Model Output

The Osage Site version of the Belowground Model was used to generate a three-year simulation (1970-72) of the ungrazed pasture at the Osage Site. The simulation used the same atmospheric driving variables as in ELM computer runs. The Belowground Model assumes that the different plant species of the ungrazed site can be represented by a single warm-season perennial grass, since over 80 percent of the peak live-shoot biomass consists of warm-season perennial grasses. The Belowground Model was specifically designed to study the dynamics of roots in grassland systems, and the results of the model are presented to demonstrate the dynamics of root systems in a Tallgrass Prairie. The model output was compared with observed data whenever possible, but most of the simulated dynamics of root systems cannot be verified by observed data because of the lack of a sufficient data base.

Figure 9.16 shows the three-year simulated time series of live-shoot, crown, total standing-dead, and litter biomass for the Osage Site. Peak live-shoot biomass ranged from 315 to 360 g m^{-2} and occurred in early July. The simulations also showed that shoot growth started in April and that aboveground plant growth ended in November. The seasonal dynamics and the occurrence of the peak live biomass were represented well by the model when compared with observed data; however, the model tended to overestimate the peak live-shoot biomass (+45 g m^{-2}).

Minimum values of standing dead occurred in July, the maximum values occurred in November, and standing-dead biomass tended to increase over the three-year simulation. Observed data verified both the seasonal dynamics and increasing biomass for the three-year simulation. The maximum litter biomass occurred in April and the minimum value generally occurred in late August but increased throughout the simulation. The seasonal dynamics indicated by the model were not verified by the observed data, although an increase in litter from 1970 to 1972 was shown. Litter biomass was quite variable between observation dates and determination of an organized seasonal pattern was very difficult.

Crown biomass was lowest in May and highest in July. This trend was not derived from observed data. The variability of

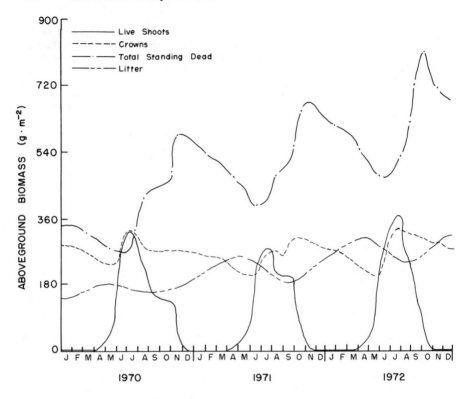

FIGURE 9.16 Three-year simulated time series of live-shoot, crown, standing-dead, and litter biomass for the ungrazed prairie at the Osage Site (Redrawn from Parton and Singh, 1976).

litter-biomass data was very high each year (in 1970 the crown biomass increased from 175 to 414 g m^{-2} within a month) and between years. The simulated mean of crown biomass for the three-year study was 40 g m^{-2} greater than the observed biomass.

Simulation of total root biomass, the total live-root biomass, and the total dead-root biomass showed that minimum live-root biomass occurred in May and the peak generally occurred in July or August, while the intraseasonal fluctuations of biomass were around 300 g m^{-2} in all three years (Figure 9.17). The increase of live roots from May to August was caused by translocation of photosynthate to the roots, while the decrease in biomass during the remainder of the year was caused by root death and root respiration during the time that translocation was minimal. The time series for the dead-root biomass indicated that the variation in dead roots was smaller (intraseasonal fluctuations less than 100 g m^{-2}) and that the maximum values generally occurred during the winter months, with the minimum in June and July. During the time when live-root biomass was increasing most rapidly, dead-root biomass was decreasing. The increase of dead roots in winter was caused by larger root-death and respiration rates compared to the rate of root decomposition (unfavorable soil temperature for decomposition),

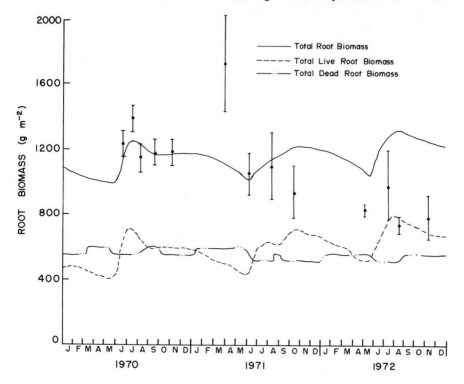

FIGURE 9.17 Simulated live, dead, and total root biomass at the Osage Site for 1970-72. Observed total root biomass ±1 SE is also indicated.

while the reverse condition was responsible for the decline of dead roots in the summer (adequate soil water and soil temperature patterns increase the decomposition rate). The total root biomass was at a minimum in May and reached a maximum in August, while intraseasonal fluctuations averaged around 250 g m^{-2}. Most of the fluctuations in total root biomass were produced by live roots, while the dead roots reduced the seasonal dynamics of the total root because dead roots had the opposite seasonal dynamics of the live roots. The live fraction of total root biomass ranged from a minimum of 40 percent in May to a maximum of 60 percent in August. The seasonal dynamics of total root data for the model tended to be verified by the observed data, but the model overestimated total root biomass in 1972. This discrepancy appeared to be related to an underestimation of root decomposition rate. The observed and simulated mean values of the total root biomass for the three-year simulation were within 30 g m^{-2} of each other. A comparison of live- and dead-root biomass with observed data was not possible since the live and dead roots were not separated by the sampling techniques.

The juvenile roots were very dynamic, reaching a peak value of around 40 g m^{-2} in early July and decreasing to very small values during the winter (Figure 9.18). The simulated rapid seasonal

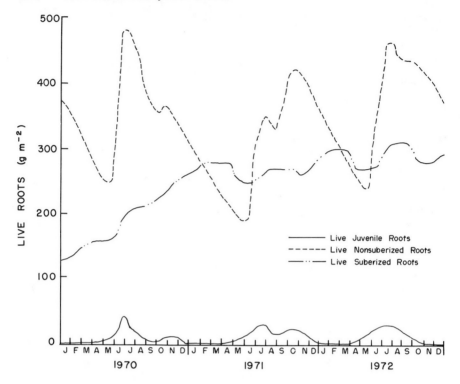

FIGURE 9.18   Three-year time series of live juvenile, nonsuberized, and suberized roots at the Osage Site (Redrawn from Parton and Singh, 1976).

dynamics of the juvenile roots were a result of the fast response of growth and mortality of young roots to fluctuations in soil water potential and the rapid aging of juvenile roots (juvenile roots matured to nonsuberized roots in twenty days). The nonsuberized roots increased from a minimum of around 200 g m^{-2} in May to a maximum of more than 400 g m^{-2} in July. The suberized roots tended to reach maximum values in February and August and minimum values in May and June. The increase in suberized roots from 1970 to 1972 was probably caused by underestimates of the initial suberized root biomass. Intraseasonal fluctuations of the nonsuberized roots were much greater (200 versus 50 g m^{-2}) than the fluctuations of the suberized roots.

      Values of the net root, crown, and aboveground production were also calculated for the three-year simulation. A comparison of estimated values based upon harvest technique (Sims et al., 1978) and simulated values is shown in Table 9.7. The simulated total net production ranged from 799 to 863 g m^{-2} yr^{-1}, comparing favorably to estimated values (847 to 933 g m^{-2} yr^{-1}). Thirty-six to forty-three percent of the total annual net production went to the roots and crowns. The simulated net root and crown production did not compare well with the estimated values from the harvest technique. The root and crown data used as input to the harvest techniques were highly

TABLE 9.7 Estimated[a] and simulated values of the net
production variable of the ungrazed pasture
at the Osage Site.

Production	Observed data (g m^{-2} yr^{-1})			Simulated data (g m^{-2} yr^{-1})		
	1970	1971	1972	1970	1971	1972
Net root production	222	361	185	287	256	228
Net crown production	380	70	407	55	82	82
Net aboveground production	331	416	290	457	525	542
Total net production	933	847	882	799	863	852

[a] The estimated values are based upon the results of the
harvest technique presented by Sims et al. (1978).

variable, and the dynamics indicated by the data could easily result
from sampling error. The simulated net aboveground production
ranged from 457 to 542 g m^{-2} yr^{-1}, generally higher than the values
estimated from the harvest technique. If the net crown and
aboveground production are combined, the estimates from simulation
and harvest technique become reasonably close.

The results of the Belowground Model showed that the root
system was quite dynamic, with the live juvenile and nonsuberized
roots changing very rapidly through the growing season. The results
also showed that variation in total root biomass was caused by
changes in live-root biomass, that dead-root biomass was fairly
homogeneous throughout the year, and that 40 percent of net annual
primary production went to the roots and crowns. The difficulty of
determining anything about the dynamics of the belowground root
system by simply observing the total root biomass is illustrated by
the model results. Researchers must determine the proportion of
live and dead roots and further subdivide the roots into other
functional groups (juvenile, nonsuberized, and suberized roots) in
order to make significant advances in our understanding of the
belowground root system.

10

Ecosystem Responses
to Stresses

In previous chapters we have been discussing both the abiotic and biotic components of the True Prairie ecosystem. This chapter describes the system responses to stresses--grazing, nutrients, water, fire, and pesticides. All categories but the last contain field data summaries, ELM simulation results, and comparisons between simulated and observed results.

The Osage version of the ELM model was used to simulate the impact of different management practices on the Tallgrass Prairie. The ELM grassland model was specifically designed to simulate the impact of range management practices on grassland systems and considers different management practices as driving variables for the model.

The range management manipulations simulated by the model included:

1. Altering the grazing intensity, the grazing regimes, and the grazing time periods;
2. Changing the species composition;
3. Adding nitrogen and phosphorus;
4. Adding water during the growing season; and
5. Spring burning of the prairie.

The management practices were simulated by imposing different management practices on three-, five-, and six-year computer simulations. Three-year runs were used for all of the management manipulations except the fire runs, which used five- and six-year runs.

Each of the computer runs started with identical initial conditions. The abiotic driving variables included daily air temperature data from Pawhuska, Oklahoma (1970-1973), daily rainfall data from Foraker, Oklahoma (1970-1973), and long-term monthly average values of relative humidity, cloud cover, and wind speed from Tulsa, Oklahoma. The effects of the various management practices on the True Prairie as represented by the Osage Site are summarized by comparing average values for selected output variables. The presented output variables were selected from over

two-hundred variables simulated by the model because they summarized the state of the system. Included in the group are abiotic, primary, and secondary production, standing crop, and nitrogen and phosphorus variables. The results are summarized in four tables that consider grazing manipulation (Table 10.15), species manipulation (Table 10.16), impacts of fertilizer and irrigation (Table 10.26), and spring burning (Table 10.45).

Clearly the model is not a perfect representation of the True Prairie. Chapter 9 carefully analyzed the simulated values and compared model results and observed data. Significant discrepancies occurred in some parameters, but many other areas were well simulated. Data taken at the Osage Site were appropriate for some of these comparisons, but a large and complex model like the ELM requires many more measurements than could ever be made in one set of laboratory and field studies. Furthermore, as we have discussed earlier, the Osage Site cannot represent the entire True Prairie grassland, nor can three to six years of data represent the prevailing climate over decades. Finally, no fertilizer or irrigation experiments were initiated at the Osage Site, and simulated grazing regimes were different from the operational regimes ones used on the ranch where the Osage Site is located.

These comments are not meant to deprecate the model or its results. They are intended to clarify the role and purpose of the model described in this section. Construction of the model required an understanding of the structure and function of the True Prairie ecosystem. This understanding was obtained from field and laboratory experiments and experience, mostly from current studies at the Osage Site. As much of the data as possible from the Osage Site were used to construct the model, but literature information certainly entered judgmental decisions about parameter values as well as mechanisms and processes that were incorporated into the model structure.

Once a model is operational, it needs to be used and tested to determine how well it simulates the real-world situation. With a complex model like the ELM, simply constructing integrated and comprehensive experiments to obtain answers to compare with model output is impossible. Therefore, only partial validation is possible by comparing simulated results with what would be expected under various environmental conditions or management manipulations. These expectations can be obtained only by independent studies conducted elsewhere. The advantage of these literature studies is that they were not used directly to construct the model and therefore are largely independent standards against which the model results can be compared. The disadvantage of using the results from these literature studies is that they were obtained at sites different from the Osage and the experiments were not performed to answer the specific questions asked by the model.

We should remember that the models are simply ideas or hypotheses about the structure and function of the True Prairie. A comprehensive model like the ELM only provides a framework for these ideas. Even though model results cannot be rigorously tested by comparing them with completely independent, comparable field or laboratory observations, the simulations can be compared with our best understanding of the True Prairie. Both congruities and discrepancies between observed and simulated results are invaluable

in organizing information, framing ideas, and suggesting future
experiments and management strategies.

GRAZING

The grasslands of the world have evolved with the process of
grazing or consumption by herbivores as an integral feature. As a
result, not only are the herbivores adapted for the growing habits
of the plants, but the morphology and life history strategies of
most plants are adapted for continuing survival under grazing
pressure (Allis and Kuhlman, 1962; Bell, 1973). Too often,
"grazing" summons a picture of beef or dairy cattle on a pasture or
native grassland. However, as seen previously, "grazers" are not
limited to large domestic herbivores. In fact, a wide variety of
herbivores including such diverse forms as jackrabbits, subterranean
nematodes, and scale insects inhabit the native grasslands. Thus,
the grassland is really a highly integrated system, composed of the
producer or green plants that support--energetically--a host of
heterotrophic organisms.

Early Grazing and Settlement Patterns

The introduction of livestock by early settlers did not
constitute a new biological process, since large herbivores had
grazed these areas for hundreds of years. Though the total number
may have been less than the number established by present-day
stocking rates (Halloran and Shrader, 1960), locally the large
native herbivores were found in high numbers. Some herds apparently
grazed almost continually in one place, although many large herds
fed in one location and when the forage was exhausted they moved
elsewhere.

Cortez supposedly introduced the modern horse to the North
American continent when he landed in Mexico in 1515 and in 1521.
Gregario Villalobos brought cattle to eastern Mexico, but these
cattle did not reach the United States territory. The first record
of cattle importation was with Coronado in 1540, when he traveled
from western Mexico northeastward through Arizona, New Mexico, and
Colorado to Kansas. His expedition included 1,000 horses, 500
cattle, and 5,000 sheep; strays from these herds began the stocking
of the ranges. Although during the next century introductions
continued, the livestock industry was not really established until
the first half of the nineteenth century. We have little
information about these early introductions, but Stoddard and Smith
(1955) suggested that they must have "represented a motley array of
kinds, shapes, and breeds."

The livestock industry began in about 1830, when the settlers
from the East came to the Mississippi Valley and joined the
livestock movers who were traveling northward from Texas. Although
large areas of grassland were available, the industry was limited by
the lack of transportation and market facilities as well as by the
resistance from North American Indians. These limitations
diminished as forts, railroads, and settlements were constructed
(Stoddard and Smith, 1955).

Periodic stimuli affected the initial stages of the beef cattle industry. The first stimulus occurred during the Civil War, when the Confederate Army's need for supplies provided a ready market for cattle that could be brought from Texas. The Union blockade soon curtailed this movement, but at the end of the war, with monetary inflation and few beef cattle in the East, reasons for endeavoring to move animals to the eastern markets were abundant. This need for beef cattle led to the Texas trail drives, perhaps the most colorful (and movieland-commercialized) era of the livestock business. The first of the northward drives began in 1846, when Edward Piper took a herd from Texas to Ohio (Wellman, 1939). By 1866, this well-established activity was further encouraged by soldiers returning from the Civil War, who saw an opportunity for instantaneous wealth.

The actual drives consisted not of single trails but numerous small trails that coalesced at river fords and mountain passes. The trails were of various distances (Figure 10.1), and the cattle movements, in the form of drives, frequently lasted several years. Two- to four-year-old stock might leave Texas and arrive at the

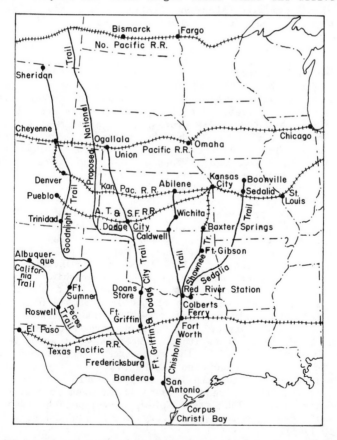

FIGURE 10.1 Major cattle drive trails (——) and railroads (++++) across the Central Plains (Redrawn from Stoddard and Smith, 1955).

railroad exchanges as 4- to 6-year-olds or older (Stoddard and Smith, 1955).

These trips, contrary to popularized romantic portrayals, were difficult and arduous.   The cattle were wild and unmanageable, especially while being moved through unfamiliar country.   Cattle were frequently lost or drowned at river crossings.   In addition, frequent altercations with native Indians and white settlers interrupted the drives because these people reacted adversely to the introduction of Texas livestock to their lands.   Despite the conflict over the land the early ranchers, who needed animals for breeding purposes, frequently bought the cattle.

After the Civil War the cattle industry flourished as cattle herders and ranchers settled in the Midwest.   Cattle sold for high prices in the East, so money was easily available not only from the East but also from abroad.   This upward trend came to an abrupt halt with the unprecedented severity of the winter of 1885-86, when in many areas 85 percent of the cattle died.   The northern plains suffered a severe drought the next summer, followed by a severe blizzard the following January.   Many herds were nearly eliminated and most herds suffered extensive losses.   Much of the earlier money was withdrawn, but subsequently the cattle industry began rebuilding.

With westward migration, conflicts ensued between cattle herders and settlers.   Open range was vital to the cattle herders, who needed to move their cattle unhindered over lands to which they had no legal rights.   The settlers wanted to plow the land, fence and protect property from stock other than their own, and build towns.   The most desirable sites, especially sites near water, were settled first, but these were precisely the areas most vital to the cattle herders.   During the 1870s, when barbed wire became cheap and readily available, free movement of the cattle was severely restricted.   The farmers also fenced lands that were not theirs, thus leading Congress to enact a law in 1885 that permitted the cutting of illegally constructed fences (Stoddard and Smith, 1955).

The Homestead Act, passed in 1862, allowed the acquisition of 160 acre (64 ha) tracts after five years of residence on the land and certain improvements.   Because the Act contained a commutation clause that enabled the settler to purchase the land for $1.25 per acre after a six-month residence, many of the tracts were subsequently unoccupied.   By 1870, most of the highly productive lands had been settled and were in private ownership.   The Enlarged Homestead Act of 1909 increased the homestead size to 320 acres (128 ha) and provided that one-fourth of the land should be cultivated.

Because most early settlers came from Europe and the eastern United States where trees were common, the settlers found the grassland to be quite inhospitable.   Water was not adequate and the sod was difficult to till.   The cast iron plows brought from the east would not scour when used in the prairie.   Large plows, with wooden moldboards plated with iron strips, would turn furrows, but plowing now required as many as six yoke of oxen to pull the plows through the sod.   The steel plow, developed by John Deere (1837-1840), proved more satisfactory in tilling the prairie soils (Edwards, 1948).

By 1830, most of the eastern True Prairie had been settled and plowed.   Maps of Ohio vegetation in 1827 show that only a few wet

prairies remained, virtually all the tillable land in Michigan had been plowed, and only a few grasslands remained in west and northwest Indiana. At about this time a considerable amount of prairie remained in the northern two-thirds of Illinois, especially in the northwest portion. Wisconsin was settled by about 1830, the southwestern part characterized by sizable tracts of prairie with large bur oaks (Quercus macrocarpa). By 1850, Iowa was settled, though most of the state was still prairie; and Missouri, though first settled much earlier, was also largely settled and had relatively few large tracts of natural prairie remaining. Minnesota was unfenced and unbroken until about 1870. Kansas and Oklahoma were largely settled by about 1880, and these two states now contain most of the unplowed True Prairie.

Characteristics of Mammalian Herbivores

Mammalian herbivores can be conveniently divided into ruminants and nonruminants. Ruminants, which include cattle and sheep, are the most important agriculturally. Nonruminants include horses, pigs, and many nonagricultural animals such as rodents and rabbits.

Mammalian herbivores do not have digestive enzymes to break down cellulose cell walls. This process is performed in their stomachs and intestines by bacteria and protozoa that do possess celluloytic enzymes (Hungate, 1975). However, the breakdown is a slow process that is further retarded by the presence of cellulose and lignin. Therefore, the plant cell wall is an obstacle to the extraction of the cell contents, and this impediment is quantitatively enhanced by thickened, lignified cell walls.

The two primary nutritional ingredients in plant cells are protein and carbohydrate. The relatively small amounts of protein occur primarily in the cytoplasm, so protein is more abundant in younger cells. Energy can be obtained from cytoplasm and broken-down cell walls. However, herbivores must obtain sufficient protein from a food supply containing a superabundance of carbohydrates, primarily in the form of cellulose, which is difficult to break down and effectively compartmentalizes protein when cells are intact (Bell, 1971).

The digestive tract of ruminants includes a stomach that is divided into four compartments: the rumen, the reticulum, the omasum, and the abomasum. The first two compartments form one large organ (95 percent of the total digestive capacity of the cow) in which bacterial fermentation of the food takes place. This process is assisted by a continuous flow of saliva, rhythmic contractions of the rumen wall, regurgitation and cud-chewing, and vast numbers of bacteria (10^9-10^{10} ml^{-1} of rumen contents) and protozoa (10^6 ml^{-1}). About 70 percent of the digestible dry matter entering the rumen is converted by these microorganisms into soluble and gaseous compounds; some of this dry matter is absorbed and some lost by eructation (Spedding, 1971).

When food is ingested, it is masticated and exposed to the salivary enzymes. Even when swallowed, the particles may be fairly large with many cells still intact. In the rumen these large particles float on the surface of the rumen liquor, are passed into the diverticulum, and then are regurgitated. The cycle is repeated

until the cells are sufficiently disintegrated to sink into the rumen liquor and pass to the intestine, where protein absorption occurs (Bell, 1971).

The rate of passage through the gut is limited by the rate of breakdown of material in the rumen; this rate is a function of the quantity and quality of cell wall in the food (Byerly, 1977). When food contains much cell wall, particularly lignified cell wall, the rate of passage is quite slow. Under these conditions the degree of protein extraction is high because of the relatively long residence time in the gut, but the overall rate of assimilation of protein is low because relatively small amounts of food move through the gut. When cell wall constituents reach a high concentration in the gut, the ruminant fails to assimilate enough protein to meet its maintenance requirements. Then the animal selects herbage components with thin cell walls and high amounts of protein such as young leaves. The strategy of the ruminant is based on high efficiency and utilization of protein but at the expense of a high rate of intake and processing of food, which results in the need to select high-protein plant components (Bell, 1971).

Much less information is available about the nutrition of nonruminants. For example, the horse has only a simple stomach, but the caecum and colon are both enlarged and are inhabited by microorganisms that perform the same kind of functions as the microorganisms in the rumen (Spedding, 1971). The protein is extracted and assimilated in the relatively small and simple stomach. Because the repeated mechanical maceration and celluloytic attack characteristic of the ruminant does not occur, the extraction of protein in the nonruminant is decreased even more by an increase in the quantity of cell wall material. Furthermore, once the protein has been extracted in the horse's stomach, it is immediately assimilated as amino acids and is not subjected to the recycling and reconstitution that increase efficiency of protein utilization in the ruminant (Bell, 1971).

The horse is less efficient than the ruminant in utilizing protein in its food; but unlike the ruminant's system, the system in the horse has no mechanism that restricts the rate of passage of material through the gut. Material passes about twice as fast through the gut of a horse (30 to 45 h) as the gut of a cow (70 to 100 h). Because the nonruminant has a high rate of food passage, the nonruminant can survive on a diet that is too low in protein to support a ruminant, when all other factors are equal. Thus, the nonruminant is much more tolerant to cell wall material in the diet but must maintain a high rate of intake to survive on this low-quality forage. Under these conditions, food selectivity is much less intense for the nonruminant, and the work by Bell (1971) in the Serengeti (Tanzania) on zebra and wildebeest indicate that these patterns are widespread.

Chemical Composition of Herbage

The chemical composition of herbage is largely a function of phenological stage, botanical composition, nutrient status of the soil, and prevailing climatic conditions (Daniel, 1935; Daniel and Harper, 1934; 1935; Gallup and Briggs, 1948; Harper et al., 1933;

Maynard and Loosli, 1956; Murphy, 1933; Murphy and Daniel, 1936).
The range in chemical composition as a percent of the dry matter has
been measured in a number of these studies and is summarized in
Table 10.1 (Spedding, 1971).

A small difference in forage quality can make a large
difference in animal production. For example, forage of 70 and 60
percent IVDMD (in vitro dry matter digestibility) might produce
steer gains of 1 and 0.5 kg day^{-1}, respectively, but forage of 50
percent IVDMD might be only a maintenance ration (Maynard and
Loosli, 1956).

In addition to the characteristics of the digestive tract,
certain plant attributes determine diet selection. Attributes such
as external spines and hairs determine the palatability of a
species. Palatability can be determined by the use of an esophageal
fistula with which samples of actual ingestion can be removed
periodically and analyzed microscopically to determine plant species
composition. Species fragments can be identified by examination for
such characteristics as surface texture, color, epidermal hairs,
leaf venation and margins, and floral parts. The percentage
composition of each plant species in the diet does not provide an
assessment of its relative palatability, unless species availability
is also considered (Van Dyne and Heady, 1965). A relative
preference index (Buchanan et al., 1972) can thus be calculated as:

$$SP = \frac{D}{C}$$

where SP = preference of a species, D = average percent in diet
(fistula samples), and C = average percent composition in the
pasture (green weight).

Ruminants require calcium, phosphorus, potassium, sodium,
chlorine, sulfur, magnesium, iron, zinc, copper, manganese, iodine,
cobalt, molybdenum, and selenium; they may also need fluorine,
bromine, barium, and strontium. Though toxicity is rarely a
problem, some elements are toxic when taken in excess, particularly
copper, selenium, molybdenum, and fluorine. Most required minerals

TABLE 10.1   Summary of chemical composition of herbage as a percent
of dry matter (Spedding, 1971).

Chemical composition	Dry matter
Crude protein (nitrogen × 6.25)	3-30
Crude fiber	20-40
Total carbohydrate	4-30
Cellulose	20-30
Hemicelluloses	10-30
Water content	65-85

are present in herbage, and when deficiencies occur, excesses of other nutrients (such as molybdenum and sulfate in the case of copper deficiency) or nonavailability of minerals (magnesium deficiency) are frequently the cause. Other common deficiencies are of iodine and cobalt and imbalances of calcium and phosphorus (Spedding, 1971).

The urine is rich in nutrients since it contains some 70 to 80 percent of the excreted nitrogen, sulfur, and potassium. Fecal material is usually rich in phosphorus--much of it transformed into the inorganic form--as well as calcium and magnesium. Beef cattle may excrete up to 95 percent of the consumed nitrogen (Maynard and Loosli, 1956).

Fecal samples can also be used to determine the amount of material that is not assimilated. Of the energy ingested by a ruminant as food, losses occur as methane and ammonia from the rumen and as energy in fecal material and urine. The remainder of the energy is lost as excess body heat, is utilized as maintenance energy, or forms new tissue or fat reserves (Spedding, 1971).

Approximately 9 kg of air-dry material are required each day to maintain a 450 kg mature animal depending on climatic conditions, quality of forage, and age of the animal. When an animal receives adequate amounts of carbohydrates and fats, the amount of required protein is minimized. Young animals, however, require more protein than mature animals. High-quality forage containing 6 percent crude protein, which is about 16 percent nitrogen, is adequate for mature cattle, but calves need higher amounts. A crude protein content of greater than 16 to 18 percent confers no additional advantage. Retarded growth will result if the calcium content of the forage is below 0.25 percent or if the phosphorus content of the forage is below 0.12 percent (pregnant or lactating cows require 0.20 percent). The minimum requirement for carotene is approximately 1.65 ppm of the total feed intake (Stoddard and Smith, 1955), but because carotene is stored in the fat, excess amounts may be available when cattle are consuming low carotene forage. Other nutrients may also be stored and then utilized when the diet is deficient.

A 450 kg steer has a gross composition of 54 percent water, 15 percent protein, 26 percent fat, and 4.6 percent mineral (of which 0.74 percent is phosphorus and 0.19 percent is potassium). So, removal of 450 kg of beef from the system represents a removal of 10.9 kg nitrogen, 3.4 kg phosphorus, and 0.9 kg potassium (Maynard and Loosli, 1956).

During the 1930s the calcium, nitrogen, phosphorus, and magnesium constituents of tallgrasses were well characterized (Harper et al., 1933), but the most comprehensive analysis of nutritional qualities of the tallgrasses has been done at the Southern Plains Experimental Station, which is on a sandy soil just west of the True Prairie in Oklahoma (Savage and Heller, 1947). Species that were studied are listed in Table 10.2.

Almost all range plants contain more calcium than is required by beef cattle (0.23 percent of dry weight), but many warm-season grasses are deficient in phosphorus (required amount is 0.13 percent dry matter) during the dormant season. The average protein content of cool-season grasses exceeded that of warm-season species in every month except July, although cool-season species are usually dry and

TABLE 10.2   List of species used in the study of the True Prairie
(Savage and Heller, 1947).

Common name	Scientific name

Perennial grasses

Warm-season
Blue grama	Bouteloua gracilis
Buffalograss	Buchloe dactyloides
Sand lovegrass	Eragrostis trichodes
Sand paspalum	Paspalum stramineum
Sand bluestem	Andropogon halli
Sideoats grama	Bouteloua curtipendula
Switchgrass	Panicum virgatum
Sand dropseed	Sporobolus cryptandrus
Little bluestem	Schizachyrium scoparius
Big sandreed	Calamovilfa gigantea
Blowoutgrass	Redfieldia flexuosa
Fall witchgrass	Leptoloma cognatum
Hairy grama	Bouteloua hirsuta
Caucasian bluestem	Bothriuchloa intermedius var. caucasiens
King Ranch bluestem	Bothriochloa ischaemum
Weeping lovegrass	Eragrostis curvula

Cool-season
Texas bluegrass	Poa arachnifera
Western wheatgrass	Agropyron smithii
Canada wildrye	Elymus canadensis

Annual grasses

Purple sandgrass	Triplasis purpurea
Little barley	Hordeum pusillum
Sixweeks fescue	Festuca octoflora

Forbs

Bush morning glory	Ipomoea leptophylla
Virginia tephrosia	Tephrosia virginiana
Horseweed fleabane	Erigeron canadensis
Showy partridgepea	Cassia fasciculata
Wild-alfalfa	Psoralea tenuiflora

Shrubs

Skunkbush	Rhus trilobata
Sand sagebrush	Artemisia filifolia

relatively unpalatable during most of the summer. Fatty constituents of grass are an excellent source of energy, but fat content is usually low and not very significant (Figure 10.2). However, the energy value is 2.25 times the value of protein and carbohydrate (nitrogen-free extract).

Nitrogen-free extract is a principal source of heat and energy in plants. That carbohydrate fraction is composed chiefly of starches, sugars, and pentosans, which are easily digestible. (See Figure 10.2.) Crude fiber represents about 35 percent of the composition of dry range grasses and constitutes much of the structural part of the plant (Figures 10.2 and 10.3). In general, the greater the quantity of fiber, the lower its nutritional quality.

In the case of most range forage species, ash content, which contains essential minerals and inert material, varies much the same way that protein, fat, and phosphorus do (compare Figures 10.2 and

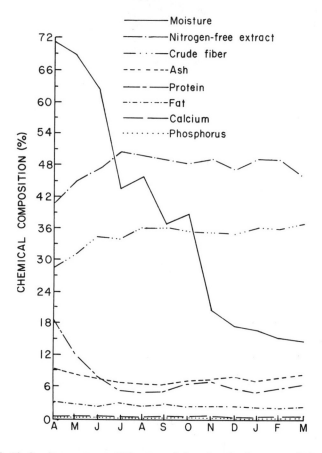

FIGURE 10.2 Average monthly trend in chemical composition of fourteen perennial grasses on an oven-dry basis for a five-year period, Spring 1940 to Spring 1945.

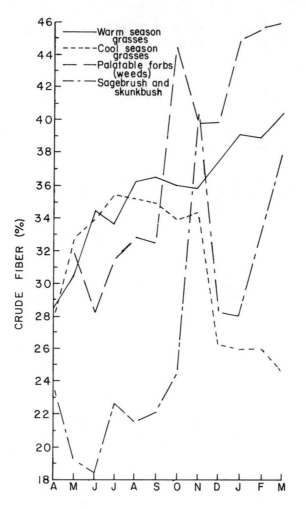

FIGURE 10.3 Average monthly trend in fiber content (oven-dry basis) of four classes of range forage for a five-year period, Spring 1940 to Spring 1945 (Redrawn from Savage and Heller, 1947).

10.4). This trend is generally the inverse of the trend for crude fiber.   The percent moisture of warm-season grasses decreases throughout the growing season and the percent moisture of cool-season grasses parallels periods of active growth (Figure 10.5).

Grazing Systems

An animal alters the habitat it grazes by reducing forage quantity. The quality of the forage may be decreased if a species compositional shift to a less palatable species occurs.   However,

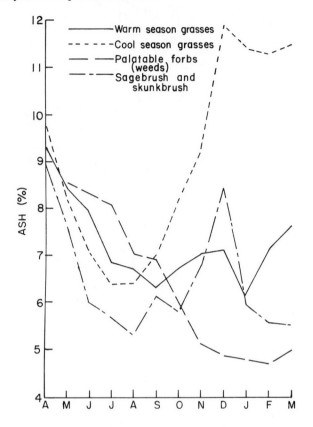

FIGURE 10.4 Average monthly trend in ash content (oven-dry basis) of four classes of range forage during a five-year period, Spring 1940 to Spring 1945 (Redrawn from Savage and Heller, 1947).

forage quality on a properly stocked range increases as a result of grazing, because the regrowth is of high quality. Thus, grazing systems are designed by recognizing the regenerative capacity of the grasslands and the grazing characteristics of the herbivores in order to ensure the long-term maintenance of the range while simultaneously maintaining optimum animal production per head and per hectare.

Grazing capacity has been defined in various ways, but essentially it is the number of animals that can produce the greatest return without damage to physical resources (Heady, 1975) and is largely determined by the quantity and seasonal availability of food. In order to reach grazing capacity during a year, range livestock operators may combine feeds from native forages, seeded ranges, seeded pastures, crop aftermath, planted and native hay, and concentrates.

Grazing capacity refers to the number of animals over long periods of time, and stocking rate is the number of animals in a pasture area for a stated time period--usually for the entire grazing season. Stocking density is the number of animals per unit

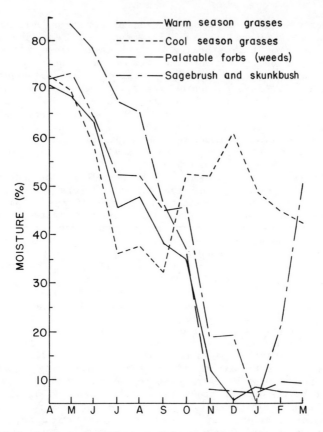

FIGURE 10.5 Average monthly trend in moisture content (oven-dry basis) of the field-collected forage from four classes of range plants for a five-year period, Spring 1940 to Spring 1945 (Redrawn from Savage and Heller, 1947).

area at a given time. Animal unit (AU) refers to a mature cow with calf or their equivalent. Adult bulls are 1.25 AUs, and young animals between weaning and maturity vary from 0.6 to 0.9 AU. Horses, sheep, and goats are commonly converted, respectively, to 1.25, 0.2, and 0.17 AUs. AUs for bison, white-tailed deer, pronghorn antelope, and elk are 1.0, 0.13, 0.10, and 0.53, respectively (Seton, 1929). An animal unit month (AUM) is the amount of forage required by an animal unit for one month of grazing.

At low stocking rates, per-animal performance shows little response to changes in animal numbers per area because food supply exceeds demand and animal selection for forage quality is not limited. Under these conditions, realized animal production is a function of animal performance rather than forage production. However, especially in the southern Tallgrass Prairie, low stocking rates may permit the accumulation of coarse plant parts and consequently less homogeneous utilization of the available forage.

As numbers of animals are increased per area, closer, more homogeneous utilization and smaller amounts of remaining herbage result. Over time high stocking densities may cause a change in species composition, resulting in forage of poorer quality. Animals show lower gain per head, and weight gains of individual animals are more varied. At higher stocking rates, the gains per animal are reduced rapidly with small increases in the number of animals per area (Heady, 1975). Weight gain per head is a function of stocking rate. Pastures in excellent range condition but heavily stocked for long periods of time produce low gains per head, and pastures in poorer condition but lightly stocked may produce high gains per head.

The amount of plant material consumed by the herbivores, expressed as a percentage of the current herbage crop, has been referred to as: range utilization, degree of use, actual use, herbage use, and range use. Time frames are important distinctions (Heady, 1975). Stocking is a daily phenomenon, range forage utilization is seasonal, and grazing has a longer time reference. So overstocking can be remedied in a day and overutilization (or overuse) can be corrected in a growing season, but the results of overgrazing may require many years to remedy. If the major dominant grasses have not been eliminated, the True Prairie recovers relatively quickly. However, more xeric grasslands require a longer recovery time, so pastures on dry sites or at the southern and western boundaries of the True Prairie recover more slowly than pastures on mesic sites in the center of the tallgrass region.

Stocking rates on bluestem pastures have steadily decreased since first used extensively for grazing. Old grazing records show that before 1900 most bluestem pastures were stocked at a rate of 1 ha for one mature cow or steer for a grazing season of six months beginning 1 May. By 1933, or just before the drought years, the best pastures carried only one mature animal on 2 ha, and the average carrying capacity was 2.2 ha per animal (Anderson, 1940).

Herbel and Anderson (1959) used yearling steers for stocking, each considered to equal approximately 0.67 AU. Two hectares per AU for six months beginning 1 May is considered moderate stocking for the Flint Hills. The beef production on this range, with no supplements, is shown in Table 10.3.

On the Osage Site, yearling steers gained 7.81 g m^{-2} when grazed at 2.4 ha AU^{-1}. Similar gains were realized from year-round

TABLE 10.3 Beef production on stocked range.

Stocking rate	Beef production (g m^{-2})
1.5 ha/AU	4.82
2.0 ha/AU	5.66
3.0 ha/AU	3.02
2.0 ha/AU (deferred rotation)	3.79

operations if protein supplements were used and the stocking rate was 3.2 ha per cow.

Deferred-rotational plans are used on relatively few hectares in the tallgrass rangelands; for cow-calf operations, continual grazing is most common. Under these conditions, the diet is supplemented during the winter with alfalfa or prairie hay and/or high-protein concentrates. Most steers on grass in the Flint Hills are not supplemented during the winter because supplementation is not economically profitable (C. E. Owensby, personal communication, 1977).

Basically, four methods (Smith 1953) of beef production are practiced in the Flint Hills region of the True Prairie: (1) summer grazing, (2) wintering and grazing, (3) deferred full-feeding, and (4) continual grazing of cow herds for feeder calves and creepfed calves. Cow-calf operations are designed to maintain the cow herd and produce calves, whereas growth systems are designed to produce efficient, rapid growth of steers and heifers. With a protein supplement, livestock can be satisfactorily wintered on bluestem pastures. Steer calves wintered on dry bluestem pasture supplemented with protein gain an average of 50 kg. In the deferred full-feeding practice, steer calves can be wintered to gain about 110 kg when grazing from 1 May to before 1 August and 30 September without supplemental feed and finished in a dry lot. Late spring calving is practiced by many operators, wintering the cows on grass and protein supplement. In the southern parts of the True Prairie, where complementary fall and spring green forage such as fescue is available, fall and winter calving are not unusual but spring calving is preferable; July to September is the least desirable period because of excessive heat and danger from insects.

The ratio of bulls to breeding cows on the range where the Osage Site is located is about 1:25. The average length of service for bulls is four years, and here the annual breeding season extends over two months. The bulls are placed with the cows around 15 May and calves are dropped from 15 February to 15 April. Heifers are bred to drop their first calves at two years of age. Calves are weaned in October at 7 to 8 months of age and about 250 kg, turned on the range, and fed protein pellets until grass begins its spring growth (Gardner, 1958).

Stocking rates, forage production, and expected livestock production in the True Prairie are summarized in Table 10.4 (Harlan 1960a). Range conditions will be discussed subsequently, and "historical grazing level" describes grazing conditions over a long period of time so that heavy grazing, for example, has led to "fair" range conditions.

Effects of Grazing on Abiotic Variables

Infiltration. The physical conditions and properties of the top few centimeters of the soil mass control the rate of water infiltration into a prairie soil. The amount of infiltration is inversely proportional to the rate of runoff. Conditions that decrease the rate of evaporation and the rate of overland flow enhance the capacity of the soil to absorb and retain water.

TABLE 10.4 Historical grazing levels, range condition, stocking rates to maintain condition, forage production, and expected livestock production from southern True Prairie native grasslands (Harlan, 1960a).

Historical grazing level	Range condition	Stocking rates	Forage production	Expected livestock production	
				Gain per head	Gain per acre
Light	Excellent	16 acre AU cow^{-1} (6.5 ha AU cow^{-1}) 10 acre AU yearling^{-1} (4.0 ha AU yearling^{-1})	2,000–3,000 lb acre^{-1} (224–336 g m^{-2})	450 lb calf (204 kg calf) 200 lb yearling (91 kg yearling)	28 lb (12.7 kg) 20 lb (9.0 kg)
Moderate	Good	10 acre AU cow^{-1} (4 ha AU cow^{-1}) 6 acre AU yearling^{-1} (2.4 ha AU yearling^{-1})	1,500–2,500 lb acre^{-1} (168–280 g m^{-2})	450 lb calf (204 kg calf) 200 lb yearling (91 kg yearling)	45 lb (20.4 kg) 33 lb (15.0 kg)
Heavy	Fair	8 acre AU cow^{-1} (3.2 ha AU cow^{-1}) 4.7 acre AU yearling^{-1} (1.9 ha AU yearling^{-1})	1,000–2,000 lb acre^{-1} (112–224 g m^{-2})	400 lb calf (181 kg calf) 175 lb yearling (79 kg yearling)	50 lb (22.7 kg) 37 lb (16.8 kg)

Large grazing animals directly affect infiltration since physical properties are changed by compaction, and the rate of infiltration decreases as bulk density increases. Kelting (1954) measured respective bulk density values of 0.9 and 1.2 g cm^{-3} in an ungrazed and grazed prairie in Oklahoma. On coarse or sandy soils this effect may be minimal. On fine-textured soils, however, the rate of infiltration may be reduced by half under heavy stocking rates (Rauzi and Smith, 1973). The infiltration rate changes relatively slowly--that is, more slowly than changes in vegetation in response to grazing. Compaction often disappears or decreases after seasonal wetting and drying or freezing and thawing.

Both plant standing crop and litter on the soil surface, which are functions of grazing, have a marked effect on infiltration. This relationship is complex and the actual effect depends not only on characteristics of the plant and soils but also on precipitation events. Plant cover increases infiltration by limiting soil surface sealing (Hawks 1965). Also, soils with high organic material permit greater infiltration of water, just as coarse-textured soils have higher infiltration rates than fine-textured soils. Similarly, percolation through the soil profile is greater in coarse-textured soils. Once moisture reaches the soil surface, litter and plant canopy retard water loss by evaporation, but plant cover also intercepts rainfall, especially light rainfall (Clark, 1940), and thus effectively decreases the availability of water to the soil. Living plants transpire large quantities of water from the soil. In fact, the ratio between the amount of water transpired and the amount of herbage produced is frequently between 1,000 and 700:1 (Stoddard and Smith, 1955).

In general, the amount of water entering the soil is positively correlated with the amount of standing crop and litter on the soil. In Mandan, North Dakota Rauzi (1963) found that 73 percent of the variance in the infiltration rate was accounted for by the weight of live plants and litter. Near Hastings, Nebraska Dragonn and Kuhlman (1968) showed that runoff rates under heavy grazing were 1.7 to 2.8 times greater than runoff rates under light grazing.

Rauzi and Smith (1973) examined infiltration rates on three sites with three grazing levels on the Shortgrass Prairie in northeastern Colorado. On fine-textured soils, infiltration rates were higher on lightly and moderately grazed than on heavily grazed grassland. On coarse-textured soil, infiltration rates were no different at light and moderate grazing levels. In the same study, infiltration rates over short time intervals (first ten minutes) showed no differences, and only after at least fifteen minutes were grazing differences detectable. On shortgrass ranges ground cover is usually relatively sparse compared to tallgrass ranges, and this sparse canopy magnifies the differences that result from grazing intensity.

Beebe and Hoffman (1968) measured soil water under ungrazed and grazed treatments in a Tallgrass Prairie in southeastern South Dakota. Their measurement was taken on one date in midsummer; the soils were shallow silt loams, but two different soil types were represented in both the grazed and ungrazed stands. Soil water in the top 15 cm was about 1 percent higher on the grazed treatment, but water content was considerably greater in the ungrazed treatment at all of the deeper levels (Table 10.5).

TABLE 10.5 Soil water content on ungrazed and grazed treatments at different soil depths (Beebe and Hoffman 1968).

Soil depth (cm)	Soil water (percent)	
	Ungrazed	Grazed
0-15	17.6	18.7
15-30	27.3	17.3
30-45	24.8	15.1
45-60	21.1	14.3

On the Osage Site, soil water was almost always greater on the ungrazed treatment, but if the grazed treatment had been grazed during the summer with the transpiration consequently reduced, at least the top soil layers on the grazed treatment would have greater amounts of soil water.

Organic Matter and Nitrogen. Other abiotic factors across the True Prairie are not so consistent, although organic matter and nitrogen in the soil are usually higher in the ungrazed treatment. Beebe and Hoffman (1968) measured organic matter and Kjeldahl total nitrogen and averaged these values over two ungrazed sites and three grazed pastures (Table 10.6). Both percentages of organic matter and total nitrogen were lower in the grazed condition at all soil depths. The decrease with depth followed, in both parameters, the decrease in root biomass with depth.

Kelting (1954), working in the Oklahoma True Prairie, found that soil water was higher in the ungrazed site, but that organic matter was higher in the grazed pasture (Table 10.7).

TABLE 10.6 Organic matter and total nitrogen at various soil depths in a South Dakota tallgrass prairie (Beebe and Hoffman 1968).

Soil depth (cm)	Organic matter (percent)		Total nitrogen (percent)	
	Ungrazed	Grazed	Ungrazed	Grazed
0-15	6.23	3.48	0.34	0.21
15-30	4.98	1.87	0.23	0.11
30-45	2.78	0.86	0.15	0.07
45-60	2.36	0.61	0.14	0.05

TABLE 10.7   Organic matter on ungrazed and
grazed treatments at different
soil depths (Kelting 1954).

Soil depth (cm)	Organic matter (percent)	
	Ungrazed	Grazed
0-15	2.28	2.96
30-45	1.14	1.38

Garnert (1936) reported higher organic matter percentages in pastures subjected to frequent clipping (clippings removed) than in the ungrazed Tallgrass Prairie in north central Oklahoma. However, greater root biomass was found in the ungrazed treatment, and he concluded that organic matter was not associated with root biomass. Although the judgment is made from relatively few studies with the latter case notwithstanding, soil nitrogen and organic matter both appear to parallel root biomass.

Effects of Clipping

Because of the difficulties of experimentally imposing the actual grazing process, investigators have sought to evaluate the effect of herbage removal by simply clipping forage at different intervals and plant heights. The responses have been measured in terms of food storage materials in roots and rhizomes, belowground biomass, and production by aerial plant parts. However, the effect of grazing on plants is a complicated process compounded by effects other than the simple removal of plant parts (Dyer and Bokhari, 1976; Harris, 1974; Reardon et al., 1972; Weinmann, 1952).

The principal carbohydrate reserves in grasses are sugars, fructosans, and starch. Grasses containing fructosans are usually cool season grasses; grasses accumulating sugars and starch are warm season grasses. Unlike cellulose and pentosan, hemicellulose shows some seasonal fluctuations and may play a role in food storage (May, 1960). These reserves are utilized during early spring growth then accumulated sometime during the active period of growth (Aldous, 1930a; 1933; Anderson, 1960). The behavior of these reserves may ultimately be particularly important in explaining the differential response of grasses to clipping and grazing (Smith, 1975; Weaver and Hougen, 1939).

In Wisconsin, Neiland and Curtis (1956) studied the phenology of root production, stem elongation, and anthesis (Figure 10.6) of seven Tallgrass Prairie species in response to different clipping frequencies. More frequent and more severe clippings reduced carbohydrate reserves (total available carbohydrates) more than less frequent and less severe clippings. However, some differences were apparent between the species, and neither sideoats grama nor big

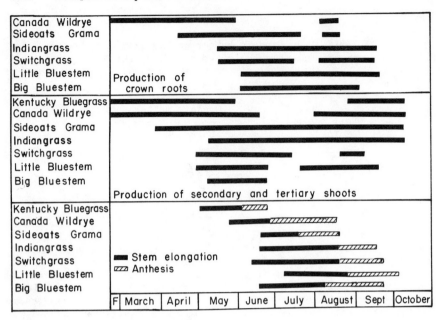

FIGURE 10.6 Seasonal development of roots, shoots, and anthesis in seven Tallgrass Prairie species (From Neiland and Curtis, 1956; copyright 1956 by The Ecological Society of America).

bluestem showed a very close relationship between clipping and carbohydrate reserves. In fact, after being clipped seven times with a 35 percent removal each time, absolute amounts of carbohydrate reserves in big bluestem suffered a smaller relative reduction than the reserves of other species.

A subsequent and similar study conducted in Oklahoma with little bluestem (Schizachyrium scoparius), big bluestem, Indiangrass (Sorghastrum mutans), switchgrass (Panicum virgatum), sideoats grama, and King Ranch bluestem (Andropogon bothriochloa) involved four clipping regimes: (1) annual clipping at the end of January; (2) annual clipping at the middle of July; (3) two clippings annually at the first of July and September; and (4) four clippings annually at the first of June, July, August, and September. Annual clipping in January had the least effect, but frequent clipping, especially at thirty-day intervals, generally reduced original planted stands. Stand density and plant vigor decreased and broadleaf weeds and annual bromes (Bromus spp.) increased under increased frequency of clipping. The greatest forage yield for all species occurred under the annual July clipping; Indiangrass showed the greatest forage yield of all species. Switchgrass was by far the most susceptible to injury from clipping and even demonstrated considerable damage from clipping during the dormant season. King Ranch bluestem was the only species for which the stand was not reduced. The percentage of original grass stand of big bluestem remaining after six years of clipping treatments was somewhat less than for Indiangrass, but for resistance to clipping and forage production, the two grasses were very similar. Of all species

studied, these two showed the greatest potential for both forage production and ability to withstand mowing (Dwyer et al., 1963).

Although many studies have indicated that early-season defoliation reduced food storage in grasses less than late-season defoliation (see Owensby and Smith, 1977), clipping big bluestem and little bluestem to 4 cm after 1 September reduced aboveground production less than clipping earlier in the season (Aldous, 1930b). Herferd (1951) reported that decreased production resulted from clipping little bluestem at heights of 5 and 15 cm before seedstalks were formed. Clipping after seedstalk formation, however, resulted in increased vegetative production the first year. In a continuation of the Herferd study Jameson and Huss (1959) showed the importance of distinguishing between leaf and stem removal in clipping studies. In their study, carbohydrate percentages of crowns were decreased by removing leaves, but a lesser decrease occurred when stems were clipped at the same time. The major influence of clipping treatments on the roots was apparently to stop further growth rather than to utilize carbohydrates already in roots. Weaver (1954) mowed a big bluestem prairie at a 4 cm height four times a year to simulate grazing. Height decreased 33 to 50 percent and the total yield was reduced 39 percent the first year and 59 percent the second. Gay and Dwyer (1965) in the Osage Hills of Oklahoma showed that two seasonal clippings increased foliage significantly ($1,464$ g m^{-2}) over one seasonal clipping (339 g m^{-2}). However, two or more clippings per season reduced all these growth characteristics. Many studies have found that moderate clipping, especially under moist conditions, results in an increase in aboveground production (Aldous, 1930b; Biswell and Weaver, 1933; Duvall, 1962; Kelting, 1954; Peterson and Hagan, 1953; Tomanek, 1948).

Other investigators (Bulksey and Weaver, 1939; Garnert, 1936; Harrison, 1939; Troughton, 1957; Weaver, 1930; 1950a) have shown that the lower the cutting heights and the shorter the cutting interval, the greater the reduction in root weight and the greater the depletion of food reserves. Three different strains of Kentucky bluegrass (Poa pratensis) were shown to have different rates of tiller emergence after clipping (Youngner et al., 1976). Crider (1955) removed the top growth of smooth brome (Bromus inermis), reed fescue (Festuca arundinacea), orchardgrass (Dactylis glomerata), Florida paspalum (Paspalum floridanum), King Ranch bluestem, switchgrass, blue grama, and Bermudagrass (Cynodon dactylon) in 10 percent increments from 0 to 90 percent. Removal of more than 40 percent of the top in a single clipping stopped root growth of all species. The larger the percentage removed, the longer the period of root growth stoppage. As the severity of clipping increased, the number of roots that did not resume growing increased. Tillers behaved as individuals in their reaction to clipping. Root growth of unclipped tillers was not affected, even though the remainder of the plant was severely clipped.

In Dwyer et al. (1963), big bluestem showed the greatest root production under the July clipping, while sideoats grama showed the least. Generally, less root production occurred as the clipping became more severe, but an exception was King Ranch bluestem, whose root production increased under all clipping frequencies.

Reduction of the root system of grasses by defoliation is usually attributed to a shortage of carbohydrates for root growth. The shoot is thought to have priority over other plant parts for the utilization of carbohydrates from both reserves and current photosynthate. According to this theory, with frequent defoliation reserves are translocated from roots, rhizomes, and leaf bases to the shoot for the development of new top growth, thus filling the deficiency of current photosynthate created by defoliation. Although many studies have shown loss of carbohydrates from the root system following defoliation, the relative amounts used for regrowth and respiration have not been determined (Youngner, 1972). May (1960) has suggested that plants do not benefit from reserves greater than needed to support early regrowth of tops and to meet the respiratory requirement of underground organs during defoliation.

In recent years the theory of plant growth stimulation by the actual grazing process has been suggested. For example, cattle saliva has been shown to enhance regrowth, and thiamine has been implicated as the active ingredient (Reardon et al., 1972). Harris (1974), noting an increase in the number of tillers being produced by the crown after insect grazing, hypothesized that the mechanism was probably hormonal from the affected meristematic tissue. Dyer and Bokhari (1976) showed that grasshopper consumption was greater on younger blue grama plants; food intake was greater, but digestive efficiency was lower with younger plants. Injection of compounds in saliva appeared to act as a physiological stimulus to plant growth. The pH of the growth medium decreased, and this decrease was attributed to increased root respiration and excretion of organic acids brought about by grazing. These last three studies are cited as a reminder that simply clipping vegetation does not simulate all conditions associated with the actual grazing process.

Vegetation Composition and Production Responses

Vegetation Composition

Not all plant species respond the same way to grazing pressure (Gardner et al., 1957; Kucera, 1956; Tomanek et al., 1958; Weaver and Darland, 1948; Weaver and Hansen, 1941; Weaver and Tomanek, 1951). Plant characteristics that permit persistence under defoliation by grazing (Neiland and Curtis, 1956) are: (1) possession of rhizomes; (2) capacity for production of lateral shoots; (3) small height and erectness of growth habit, (4) lateness of seed germination and spring growth; (5) slow growth rate; (6) high fertile:sterile stem ratios; and (7) lateness of elevation of the stem apex above the minimum point of grazing. Total plant production of many grasses is greatest when the initiation of reproductive shoots is retarded (Hyder, 1972). Modification to horizontal stem growth from auxillary buds, in order to produce rhizomes or stolons, generally occurs about midseason. Defoliation after initiation of growth by terminal rhizomes of switchgrass and sideoats grama causes rhizomes to turn upward and produce aerial shoots that otherwise would not arise until autumn or even the next

year.   Investigators have argued that little bluestem is somewhat more resistant to grazing pressure than big bluestem, partly because little bluestem is less preferred early in the season and because shoots elongate somewhat later (Gardner et al., 1957; Kucera, 1956).

Most dominants and certain minor species of the undisturbed True Prairie decrease under close grazing, but other members of the community become more conspicuous.   If grazing pressure continues even these other species begin to decrease, and invasion by weeds takes place   (Herbel and Anderson, 1959).   Several early investigators recognized these compositional shifts (Dyksterhuis, 1949; Smith, 1940; Tomanek et al., 1958; Weaver and Hansen, 1941) and classified plants as decreasers, increasers, and invaders. Decreasers decrease under grazing, increasers increase under grazing, and invaders invade an area when the otherwise closed community has been disturbed by heavy grazing.   The concept of increaser must be regarded with some caution since in some cases, especially in the center of the True Prairie, these species do not increase either in terms of absolute number or relative contribution to the species present (C. E. Owensby, personal communication, 1977).   Whether a species is regarded as an increaser or a decreaser depends to some extent on where the prairie is located.   For example, on the eastern side of the prairie, sideoats grama is an increaser, but farther west in the mid-grass region, it is considered a decreaser.   Weaver (1954) produced a generalized list of species that decrease, increase, or invade in response to grazing (Table 10.8).

In a degenerating range near Lincoln, Nebraska, Weaver and Tomanek (1951) noted that in excellent range condition, big and little bluestem alone composed 72 percent of the vegetation but contributed only 45 percent in good condition classes and 7 percent in fair range.   Grazing preference was highest for big bluestem in all range condition classes, and preference was equally high for little bluestem in fair range condition but decreased in good pasture and even more in excellent range.   Preference for bluegrass increased from excellent to fair range.

Herbel and Anderson (1959) have listed the major decreasers, increasers, and invaders in the Flint Hills portion of the True Prairie (Table 10.9).

Field data illustrating these responses are given in Tables 10.10 (data from central Oklahoma) and 10.11 (data from just south of the Kansas border in the Osage Hills).   The Osage Hills data (Hazel, 1967) were taken very near the actual sample locations of the Osage Site.

Changes in species composition under grazing are partly caused by the palatability and growth form of the species.   Also, grazing behavior of cattle (Dwyer, 1961; Weaver and Tomanek, 1951) and cattle distribution in the pastures are important in determining responses to grazing.   Even light grazing may cause a species change compared to an ungrazed pasture, especially in the drier parts of the True Prairie, and drought conditions further enhance the shift in species composition (Voight and Weaver, 1951; Weaver and Darland, 1948).

A recent concept indicates that antiherbivore mechanisms are not limited to attributes of any particular prey plant species but depend on chemistry, morphology, distribution, and abundance of

TABLE 10.8 Species that decrease, increase, or invade in response to grazing (Weaver, 1954).

DECREASERS

Perennial grasses

Big bluestem (Andropogon gerardi)
Little bluestem (Schizachyrium scoparius)
Indiangrass (Sorghastrum nutans)
Switchgrass (Panicum virgatum)
Tall dropseed (Sporobolus asper)

Canada wildrye (Elymus canadensis)
Prairie cordgrass (Spartina pectinata)
Porcupine needlegrass (Stipa spartea)
Prairie Junegrass (Koeleria cristata)
Prairie dropseed (Sporobolus heterolepis)

Legumes

Purple-flowered leadplant (Amorpha canescens)
Canada milkvetch (Astragalus canadensis)
Groundplum milkvetch (Astragalus crassicarpus)
American licorice (Glycyrrhiza lepidota)
Roundhead lespedeza (Lespedeza capitata)

Illinois tickclover (Desmodium illinoense)
White prairieclover (Petalostem candidum)
Purple prairieclover (Petalostem purpureum)
Silverleaf scurfpea (Psoralea argophylla)
Indian breadroot (Psoralea esculenta)
Prairie scurfpea (Psoralea florbunda)
Canada tickclover (Desmodium canadense)

Other forbs

Onion (Allium spp.)
Long-fruited anemone (Anemone cylindrica)
Smooth aster (Aster laevis)
Willow aster (Aster praealtus)
Inland ceanothus (Ceanothus ovatus)
Tickseed (Coreopsis palmata)
Pale echinacea (Echinacea pallida)
Floweringspurge (Euphorbia corollata)
Sunflower (Helianthus spp.)

Sunflower heliopsis (Heliopsis helianthoides)
Falseboneset (Kuhnia eupatorioides)
Dotted gayfeather (Liatris punctata)
Tall gayfeather (Liatris aspera)
Halfshrub sundrops (Oenothera serrulata)
Poisonivy (Toxicodendron radicans)
Prairie rose (Rosa suffulta)
Tall goldenrod (Solidago altissima)
Noble goldenrod (Solidago speciosa)

INCREASERS

Grasses

Kentucky bluegrass (Poa pratensis)
Western wheatgrass (Agropyron smithii)

Sideoats grama (Bouteloua curtipendula)
Blue grama (Bouteloua gracilis)
Hairy grama (Bouteloua hirsuta)

Buffalograss (Buchloe dactyloides)
Scribner panicum (Panicum scribnerianum)

Wilcox panicum (Panicum wilcoxianum)
Penn sedge (Carex pennsylvanica)
Purple lovegrass (Eragrostis spectabilis)

TABLE 10.8 Continued.

Forbs

Common yarrow (*Achillea millefolium*)
Field pussytoes (*Antennaria neglecta*)
Louisiana sagewort (*Artemisia ludoviciana*)
Whorled milkweed (*Asclepias verticillata*)
Heath aster (*Aster ericoides*)
Plains wildindigo (*Baptisia leucophaea*)
Plains poppymallow (*Callirhoe alcaeoides*)
Annual fleabane (*Erigeron annuus*)
Prairie fleabane (*Erigeron strigosus*)

Rush skeletonplant (*Lygodesmia juncea*)
Common oxalis (*Oxalis stricta*)
Wholeleaf rosinweed (*Silphium integrifolium*)
Compassplant (*Silphium laciniatum*)
Missouri goldenrod (*Solidago missouriensis*)
Ashy goldenrod (*Solidago mollis*)
Stiff goldenrod (*Solidago rigida*)
Baldwin ironweed (*Vernonia baldwini*)
Western ironweed (*Vernonia fasciculata*)

INVADERS

Grasses

Oldfield threeawn (*Aristida oligantha*)
Chess brome (*Bromus secalinus*)
Downy brome (*Bromus tectorum*)
Hairy brome (*Bromus commutatus*)
Tumble windmillgrass (*Chloris verticillata*)
Hairy crabgrass (*Digitaria sanguinalis*)
Stinkgrass (*Eragrostis cilianensis*)
Carolina lovegrass (*Eragrostis pectinacea*)

Foxtail barley (*Hordeum jubatum*)
Little barley (*Hordeum pusillum*)
Common witchgrass (*Panicum capillare*)
Canada bluegrass (*Poa compressa*)
Sand paspalum (*Paspalum stramineum*)
Tumblegrass (*Schedonnardus paniculatus*)
Sand dropseed (*Sporobolus cryptandrus*)
Poverty dropseed (*Sporobolus vaginiflorus*)

Forbs

Ragweed (*Ambrosia* spp.)
Gumweed (*Grindelia* spp.)
Verbena (*Verbena* spp.)
Russianthistle (*Salsola kali*)
Belevedere summercypress (*Kochia scoparia*)

Thistle (*Cirsium* spp.)
Plantain (*Plantago* spp.)
Euphorbia (*Euphorbia* spp.)
Stickseed (*Lappula* spp.)

TABLE 10.9 Major decreasers, increasers, and invaders in the Flint Hills portion of the True Prairie (Herbel and Anderson, 1959).

Decreasers

Indiangrass (Sorghastrum nutans)

Switchgrass (Panicum virgatum)

Increasers

Kentucky bluegrass (Poa pratensis)
Sideoats grama (Bouteloua curtipendula)
Blue grama (Bouteloua gracilis)
Hairy grama (Bouteloua hirsuta)

Buffalograss (Buchloe dactyloides)
Tall dropseed (Sporobolus asper)
Penn sedge (Carex pennsylvanica)
Baldwin ironweed (Vernonia baldwini)

Invaders

Annual bromes (Bromus spp.)
Little barley (Hordeum pusillum)
Sixweeks fescue (Festuca octoflora)
Western ragweed (Ambrosia psilostachya)

Wavyleaf thistle (Cirsium undulatum)
Woolly verbena (Verbena stricta)
Curlycup gumweed (Grindelia squarrosa)

TABLE 10.10   Relative foliage cover in ungrazed and grazed prairie
in central Oklahoma (Penfound, 1964).

Species	Ungrazed	Grazed
	(Percent)	
Coralberry (Symphoricarpos orbiculatus)	15.0	0.8
Switchgrass (Panicum virgatum)	22.0	2.7
Big bluestem (Andropogon gerardi)	11.0	4.6
Indiangrass (Sorghastrum nutans)	8.7	6.9
Little bluestem (Schizachium scoparius)	20.4	30.5
Heath aster (Aster ericoides)	0.1	2.4
Sideoats grama (Bouteloua curtipendula)	11.4	18.5
Fall witchgrass (Leptoloma cognatum)	2.6	5.8
Scribner panicum (Panicum scribnerianum)	1.0	3.6
Hurrahgrass (Paspalum pubescens)	0.1	1.4
Western ragweed (Ambrosia psilostachya)	--	17.2
Other species	7.7	5.6
Number of species	19.0	29.0
Total percent of cover	42.2	47.3

alternative prey and nonprey plants as well.   Atsatt and O'Dowd
(1976) define a guild on the basis of herbivore feeding behavior and
suggest that guild members function as antiherbivore resources in
three major ways: (1) as insensitive plants that aid in maintaining
herbivore predators and parasites, (2) as repellent plants that
either directly or indirectly cause the herbivore to fail to locate
or reject its normal prey, and (3) as attractant-decoy plants that
cause the herbivore to feed on alternative prey.   Examples of these
three mechanisms exist, but the idea has not been studied in a
uniform manner in the True Prairie or elsewhere.

TABLE 10.11 Grass composition on lightly and heavily grazed areas in the Osage Hills (Hazel, 1967).

Species	Lightly grazed	Heavily grazed
	(Percent)	
Little bluestem (Schizachium scoparius)	74.1	6.1
Big bluestem (Andropogon gerardi)	13.5	6.1
Indiangrass (Sorghastrum nutans)	4.8	--
Switchgrass (Panicum virgatum)	1.6	--
Buffalograss (Buchloe dactyloides)	1.6	21.6
Tall dropseed (Sporobolus asper)	0.4	23.4
Silver bluestem (Andropogon saccharoides)	--	12.9
Sideoats grama (Bouteloua curtipendula)	1.6	4.0
Sand dropseed (Sporobolus cryptandrus)	0	
Scribner panicum (Panicum scribnerianum)	0.4	2.6
Blue grama (Bouteloua gracilis)	0.4	2.9
Purple lovegrass (Eragrostis spectabilis)	0.4	--
Fringeleaf paspalum (Paspalum ciliatifolium)	0.4	2.0
Tumble windmillgrass (Chloris verticillata)	--	6.1
Tumblegrass (Schedonnardus paniculatus)	--	0.9
Sedges (Carex spp.)	0.4	3.0

Production

Forage production is difficult to discuss in relation to grazing because grazing levels are not always described in full, past history and stocking rates affect measurements taken in any one year, and the amount of herbage consumed by the animals themselves is not easily calculated. The changes in total production with grazing are a function of actual trampling losses (Fryrear and McGully, 1972; Quinn and Hervey, 1970), regrowth, plant vigor, and changes in both abiotic factors and species composition (Vickery, 1972). Comparisons of ungrazed and grazed conditions from several studies throughout the True Prairie are summarized in Table 10.12. In most cases, where production is simply measured as peak standing crop, grazed sites produce less herbage, which is certainly not always the case. Calculations of NPP (see Chapter 6) on the Osage Site show that net aboveground production values under the grazing treatment were greater by 31, 26, and 28 percent during the 1970-72 period. In terms of total net production, the two treatments were equivalent in 1970, but an increase showed of 11 and 53 percent of the grazed over the ungrazed in 1971 and 1972, respectively. Peak standing-crop values for live biomass (Table 10.12) show that the grazed treatment was greater two of three years. Similarly, peak root biomass was virtually the same under both treatments (Table 10.13). Root production was 22 percent lower in the grazed treatment in 1971 but 51 and 220 percent greater in the grazed treatment in 1970 and 1972, respectively.

Weaver (1950a) found that root weights from pastures in so-called poor condition in Nebraska were only 42 percent of the weights taken from pastures in so-called good range condition. Although Herbel and Anderson (1959) did not measure root growth, they did find that six years of overgrazing decreased the amount of forage produced because of the reduced vigor of range plants and a shift to less productive increaser and invader species (Table 10.12). They estimated that at heavy stocking rates, between 35 and 75 g m^{-2} were removed by grazing.

As in other instances, actual total production on the Osage Site was greater on the grazed site. The range condition of the grazed treatment was good to excellent, and grazing did not occur during the growing season. Other studies, which have shown increased above- and belowground production with grazing, have been in situations where moisture conditions were favorable and grazing intensities were not severe. In conditions where stocking rates were high, prolonged overgrazing leads to decreased above- and belowground production.

Plants Introduced for Grazing

In general, True Prairie rangeland is managed to encourage succession and maintain the palatable warm season perennial grasses. The introduction of nonnative grasses is clearly counter to natural successional patterns. Therefore, the system is altered and subsidized to maintain the introductions, which frequently have high growth rates and palatability. If ungrazed, however, some

TABLE 10.12 Summary of ungrazed and grazed tallgrass prairie production values.
Most values are peak standing crop of aboveground biomass.

Site	Treatment	Production ($g\ m^{-2}$)	Reference
Osage County, Oklahoma	Ungrazed		
	1970	270	
	1971	335	
	1972	254	
	Grazed		
	1970	286	
	1971	314	
	1972	311	
Osage County, Oklahoma	Lightly grazed	414	Hazel (1967)
	Heavily grazed	348	
McClain County, Oklahoma	Ungrazed	316	Kelting (1954)
	1.6 ha/AU	405	
Oklahoma County, Oklahoma	Ungrazed	592	R. C. Anderson (1976: personal communication)
Donaldson Pasture, Kansas (1955 season)	Year-round		Herbel and Anderson (1959)
	1.5 ha/AU	148	
	2.0 ha/AU	196	
	3.0 ha/AU	233	
	Deferred rotation		
	2.0 ha/AU	239	
	2.0 ha/AU	206	
	2.0 ha/AU	216	
Lincoln, Nebraska	Lightly grazed	344	Weaver and Tomanek (1951)
	Moderately grazed	432	
	Heavily grazed	168	
Union County, South Dakota	Ungrazed	500	Beebe and Hoffman (1968)
	Ungrazed	566	
	1.3 ha/AU	286	
	1.8 ha/AU	244	
	0.4 ha/AU	169	

TABLE 10.13   Peak root biomass at 0-90 cm
under grazed and ungrazed
treatments at the Osage Site.

| Year | Root biomass (g m^{-2}) | |
	Ungrazed	Grazed
1970	1166	1055
1971	1022	1018
1972	673	771
Mean	954	948

introduced species (e.g., weeping lovegrass [Eragrostis curvula])
produce unpalatable, poor quality, dormant-season forage.

The major introduced species in the True Prairie are
Bermudagrass, red fescue, and weeping lovegrass, but several other
perennials are also introduced grasses: Bahiagrass (Paspalum
notatum), perennial ryegrass (Lolium perenne), smooth bromegrass,
tall wheatgrass (Agropyron elongatum), orchardgrass, and reed
canarygrass (Phalaris arundinacea) (McMurphy, 1976). Pastures are
sometimes overseeded with legumes. The most common legumes are
hairy vetch (Vicia villosa), arrowleaf clover (Trifolium
vesiculosum), and white clover (T. repens). Legumes have been
reported to fix between 8 and 35 kg ha^{-1} of nitrogen, and pastures
overseeded with vetch produced as much forage as pastures that
received 14 kg ha^{-1} of nitrogen (Erdman, 1959; McMurphy, 1976).

Bermudagrass is not high quality forage, but nitrogen
fertilizer greatly affects carrying capacity and beef production.
When 18, 36, and 54 kg ha^{-1} nitrogen were added to Bermudagrass
pastures in eastern Oklahoma, annual beef production was 57, 98, and
112 kg ha^{-1}, respectively. Without nutrient amendments in most
soils, Bermudagrass would not yield as well as native grasses
(Stansel et al., 1939). As a rule, with the addition of fertilizer
and water as necessary, cattle gain per hectare on these improved
pastures would be three to five times greater than on the native
range. If Bermudagrass was given longer than three weeks deferrment
during the grazing season, unavailable leaf material accumulated.

Red fescue is not high-quality forage, but it can be fertilized
with 10 kg ha^{-1} of nitrogen in August, deferred until 1 December,
then grazed from 1 December to 1 June with about 2.5 cows ha^{-1} if an
additional 7 kg ha^{-1} of nitrogen are topdressed in February. This
fertilization program provided forage through the winter months with
adequate protein for older cows in a fall calving program.
First-year calving heifers should receive additional energy
supplements to maintain their rebreeding performance. Steer gains
reflected the low quality forage and average about 0.5 kg of daily

gain over the six-month growing season. Growth of reed fescue was slow during the hot weather in June, but the grass became dominant later in the summer (McMurphy, 1976). This species was used after burning in the oak-dominated areas of the eastern True Prairie because it grew earlier in the year than the oak sprouts. Red fescue caused a lameness in the form of a crippling disease called fescue foot, which was related to the alkaloids in the plant, but a rumen fungal infection might also be involved (Farnell et al., 1975).

Weeping lovegrass can yield comparable cattle production, but it produces best only on sandy, well-drained soils and is restricted to the southern areas of the True Prairie. Weeping lovegrass also requires intensive management including the application of nitrogen. Dead material must be removed each spring by mowing or controlled burning. A proper grazing program involves the deferment of lovegrass to achieve optimum growth but does not allow leaves to become so mature that they are not palatable.

Response of Invertebrates

Knutson and Campbell (1976) reviewed the effects of grazing on grasshopper populations in the True Prairie near Manhattan, Kansas and noted that grazing resulted in direct forage loss, loss of mulch, reduced plant cover, water loss by runoff, increased evaporation, accelerated erosion by wind, gullying, and loss of topsoil. These processes influence the plant population, which in turn influences food and cover for nymphs, adult grasshoppers, and their eggs.

Six of the eight predominant species were most numerous in heavily grazed range, less numerous in moderately grazed, and least numerous in lightly grazed. Phoetaliotes nebrascensis were most numerous in lightly grazed range; Hypochlora alba were most numerous in moderately grazed range.

The number of adult and nymph grasshoppers increase with grazing intensity. Eggs hatch earlier and populations increase partly because overgrazing and the accompanying decrease of plant cover result in increased soil temperatures and exposed soil areas conducive to increased oviposition.

Smith (1940) discussed changes caused by overgrazing in the Oklahoma Mixed-grass Prairie, where the total number of insects collected with a sweep net increased in overgrazed areas, although the total number of species showed a decline. Orthoptera (primarily acridid grasshoppers) showed an increase in total number of species and total number of specimens in overgrazed areas. Homoptera (including leafhoppers) increased in abundance as the degree of overgrazing increased, but Coleoptera decreased in number of species and specimens in overgrazed areas, although they were well represented in undisturbed prairie.

Bertwell and Blocker (1975), using a unit-area trapping method, reported that burning and grazing significantly reduced total Coleoptera numbers and biomass in the True Prairie near Manhattan, Kansas. Conversely, low plant species diversity, high average plant height, and greater plant biomass in the unburned-ungrazed treatment resulted in greater Coleoptera numbers and biomass. These results

correspond with Smith's (1940), even though the unit-area trapping method shows a family composition (e.g., a greater percentage of Carabidae) much different from the sweeping method used by Smith.

Mason (1973) found fewer species but more abundant numbers of leafhoppers in the unburned-ungrazed treatment in the same grasslands near Manhattan, Kansas compared to grazed areas, which indicates that areas with high foliage density contained high numbers of leafhoppers. These results contradicted Smith's (1940) results; however, Mason used a unit-area trapping method instead of the sweeping method.

Only a few recent comparative studies of the composition and species interactions have been completed (Hewitt and Burleson, 1976; Rathcke, 1976; Smolik and Rogers, 1976).

In summary, grasshopper abundance increases and Coleoptera abundance decreases under overgrazing, but the results differ concerning the effect of overgrazing on leafhoppers.

Response of Bird Populations

No direct studies of grazing effects on bird populations were made during investigations at the Osage Site, so likely responses of birds to grazing must be inferred from studies elsewhere (Tester and Marshall, 1961; Wiens, 1973). As we have seen, grazing leads to a reduced stature of vegetation, less litter, increased spatial heterogeneity of vegetation, and, at least in True Prairies, an increase in the coverage of such emergent forbs as ironweed (Vernonia spp.). Continued heavy grazing brings further vegetation changes to the point of nearly complete removal of litter and emergent forbs and a more even, almost lawn-like, vegetation. As birds apparently respond primarily to physiognomic features of vegetation in their selection of breeding habitat (Wiens, 1969), these changes should produce shifts in species composition and perhaps other attributes of breeding avifaunas. Over a variety of other grassland types in North America, bird communities tend to contain somewhat more species, have greater diversity, and in at least some cases support greater densities and biomass in ungrazed or lightly grazed conditions compared with areas receiving heavy grazing. In the True Prairie these effects are less clear. Censuses at the Inola, Oklahoma area (see Table 7.8) indicated that densities of both eastern meadowlarks (Sturnella magna) and dickcissels (Spiza americana) were less on an ungrazed field than in a moderately grazed area, and grasshopper sparrows (Ammodramus savannarum) were present only in the grazed pasture. Increases in species richness, density, and biomass in the grazed area here can readily be associated with features of vegetation structure. In the ungrazed field, vegetation cover was dense and fairly homogeneous with a deep litter layer, and few emergent forbs were present. Reduction in coverage and litter depth with grazing, increase in vegetation heterogeneity, and greater density of emergent forbs (which are used as singing perches by dickcissels and grasshopper sparrows) increased the suitability of this habitat to these species. A likely assumption seems to be that under heavy grazing pressure these patterns would be reversed, leading to a reduction in meadowlark densities, the disappearance of dickcissels and

grasshopper sparrows, and the occupancy of the area by horned larks (Eremophila alpestris), which favor heavily grazed fields with low vegetation stature (Graber and Graber, 1963; Wiens, 1973). In a Minnesota prairie, Tester and Marshall (1961) noted no consistent effects of moderate grazing on breeding densities of bobolinks (Dolichonyx oryzivorus) and LeConte's sparrows (Ammospiza leconteii), although savannah sparrow (Passerculus sandwichensis) densities were slightly lower on grazed than on ungrazed plots.

On the basis of rather scanty evidence, then, apparently grazing at moderate intensities may increase the species diversity, density, and biomass of breeding bird communities compared to ungrazed conditions and may lead to larger populations of some of the more representative True Prairie species. Heavy grazing pressure, on the other hand, leads to a reduction in diversity and probably density and biomass and promotes species replacements.

Response of Small Mammals

Some effects of grazing in the True Prairie can be seen from results of small mammal studies on three grids at the Osage Site in 1972 (Figure 10.7). One grid was ungrazed throughout 1970-72 at the Osage Site and had not been grazed for several years previously (Hoffmann et al., 1971). The second grid was in a moderately grazed pasture 200 m away; it had been grazed during previous years and was grazed throughout the summer of the study. The third grid was adjacent to the second in the grazed pasture and had been treated identically until May, at which time a fence was erected to create a grazing exclosure. This grid was not grazed during the summer until after the October census of small mammals was complete.

Seasonal changes in aboveground plant biomass on the three grids also are shown in Figure 10.7. Plant biomass increased sharply between May and August where grazing was prohibited. Although a heavy litter layer was absent, superficially the exclosure looked more like an ungrazed than a grazed pasture by August.

Species composition and density of each species of mammal on each grid at each of three censusing dates are provided in Table 10.14. The number of species was highest on the ungrazed grid throughout and slightly lower in the grazing exclosure than in the grazed pasture. Species such as the short-tailed shrew (Blarina carolinensis) and the fulvous harvest mouse (Reithrodontomys fulvescens), both present at all three trapping periods on the ungrazed grid, apparently were not living in the grazed pasture. Individuals did not invade the exclosure during the summer, nor did they likely encounter it. Two species found in the grazed pasture but not in the exclosure, the thirteen-lined ground squirrel (Spermophilus tridecemlineatus) and the hispid pocket mouse (Perognathus hispidus), both tend to restrict their activities to relatively more open habitats within areas of shorter grass and less litter.

Densities of individual species were highly variable, especially on the ungrazed grid (mean density in individuals per ha for all species at each censusing period was 13.0; CV = 190.6). The grazing exclosure was intermediate in variability and density

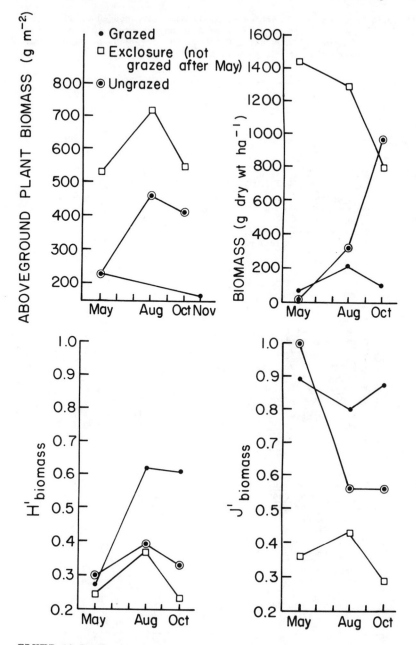

FIGURE 10.7 Seasonal change in aboveground plant biomass, small mammal biomass density, diversity (biomass H'), and equitability (biomass J') of a grazed, ungrazed, and a grazing exclosure within a grazed pasture at the Osage Site in 1972.

TABLE 10.14   Species and densities of small mammals captured on a 2.72 ha ungrazed grid, a 0.99 ha grid in an exclosure that prevented grazing from May until October, and a 1.73 ha grid in a grazed pasture at the Osage Site in 1972.

Date	Species	Density (individuals ha^{-1})		
		Ungrazed	Exclosure	Grazed
May	Blarina carolinensis	5.9		
	Reithrodontomys fulvescens	0.4		
	R. montanus		4.0	
	Peromyscus maniculatus	6.3	2.0	7.5
	Sigmodon hispidus	7.0		
	Microtus ochrogaster	97.4		
	Perognathus hispidus			1.7
	Mean density	23.4	3.0	4.6
	CV	177.1	47.1	89.2
August	Blarina carolinensis	6.3		
	Reithrodontomys fulvescens	0.4		
	R. montanus	0.7	1.0	
	Peromyscus maniculatus	3.3	2.0	6.4
	Sigmodon hispidus	17.3	7.1	0.6
	Microtus ochrogaster	60.3	9.1	2.9
	Mus musculus	0.4		
	Cryptotis parva		4.0	0.6
	Perognathus hispidus			1.7
	Spermophilus tridecemlineatus			4.0
	Mean density	12.7	4.6	2.7
	CV	172.4	73.5	83.2
October	Blarina carolinensis	11.0		
	Reithrodontomys fulvescens	0.4		
	R. montanus	0.4		
	Peromyscus maniculatus	1.5	3.0	6.4
	Sigmodon hispidus	22.8	20.2	0.6
	Microtus ochrogaster	4.8	28.3	2.9
	Mus musculus	0.4		
	Cryptotis parva		4.0	2.3
	Spermophilus tridecemlineatus			1.2
	Perognathus hispidus			1.7
	Mean density	5.9	13.9	2.7
	CV	142.3	89.6	115.4
	Inclusive mean density	13.0	7.7	3.0
	Inclusive CV	190.6	112.9	89.9

(average 7.7; CV = 112.9), whereas the grazed grid was least variable and had the lowest mean density (average 3.0; CV = 89.9). Thus, both mean density and variability associated with density of small mammal species increased as grazing decreased. One or two species tended to increase markedly in the absence of grazing and increased both mean density and variability, which are associated with species density.

Total small mammal biomass on the three grids for 1972 is shown also in Figure 10.7. Biomass densities on the exclosure and on the grazed pasture were similar at the time the exclosure was constructed in May. By August, biomass density in the exclosure was roughly 60 percent higher than on the grazed grid. This increase was true despite a density of four ground squirrels per hectare in the grazed pasture and no ground squirrels in the exclosure. Both hispid cotton rat (Sigmodon hispidus) and prairie vole (Microtus ochrogaster) had begun to increase markedly in the absence of grazing, with the tall, dense grasses growing on the grid. Biomass density on the ungrazed grid decreased throughout the summer of 1972 (see Chapter 7), but clearly this decrease was not related to grazing.

By October, the biomass of small mammals was higher in the grazing exclosure than on the ungrazed pasture. Cotton rats and prairie voles had reached high densities in the exclosure. Although the cotton rat population was still increasing in the ungrazed grid, the prairie vole population had reached a very low density (4.8 individuals ha^{-1}). Fewer ground squirrels were trapped in the grazed pasture, perhaps because part of the population was already in hibernation, and total small mammal biomass detected by trapping was lower than in August.

With the cessation of grazing, diversity (biomass H') and equitability (biomass J') of a small mammal community may change in a single season from diversity and equitability typical of a grazed pasture to characteristics more in keeping with an ungrazed pasture (Figure 10.7). Biomass H' was low on all grids in May but increased sharply for the grazed grid by August. Populations of several species apparently had recovered somewhat from low spring densities. Because one or two species tended to dominate both the exclosure and ungrazed grids, biomass H' remained low on both. Both conditions showed a slight increase from May to August and a decrease by October.

The pattern of seasonal change in equitability differed sharply from the pattern of diversity. Biomass J' was low on the ungrazed grid because of the predominance of prairie voles early in the season and of cotton rats later in the season. A slight increase in J' occurred in August, when the contribution of both species was nearly equal as one declined and the other increased. No species dominated the grazed pasture, and J' was relatively high there throughout the summer of study. That picture prevailed on the exclosure in May before the vegetation had responded to the cessation of grazing. By August, two species had begun to dominate, and J' had decreased sharply. However, because the number of species occurring in the exclosure was lower than on the ungrazed grid, J' remained somewhat intermediate between J' of a grazed pasture and of an ungrazed pasture.

In summary, mammalian species composition in the True Prairie differs with amount of vegetative cover, both standing and in the litter layer. Grazing reduces vegetative cover, which results in fewer total species and a shift away from species of the litter layer toward species of more open habitats that occur primarily in the mid- and shortgrass prairies. The latter frequently reach their distributional margins within the True Prairie or at its eastern edge. Although densities of most species remain low, the one or several that reach periodic high densities in ungrazed True Prairie result in higher mean density and increased variability associated with individual species density whenever the True Prairie remains ungrazed.

The total small mammal biomass on a True Prairie grazing exclosure can increase within a single growing season to become equivalent to the biomass in an ungrazed True Prairie. The brevity of the lag time between cessation of grazing and increase in small mammal biomass density results from the high potential for primary productivity within the True Prairie. Both diversity and equitability of small mammal communities tend to be lower in ungrazed than in moderately grazed True Prairie habitat. This tendency is true despite the higher number of species that occur in the ungrazed pastures and results from one or two herbivorous mammals that are dependent on the vegetation (especially as cover and perhaps also as food) being able to increase rapidly in density when grazing is suspended. No species showed such an ability in the grazed pasture at the Osage Site.

As noted earlier, the small mammal fauna of the True Prairie undoubtedly evolved in association with large, grazing herbivores. Thirteen-lined ground squirrels, hispid pocket mice, perhaps harvest mice, and other species are most successful in the grazed areas and probably would not survive long in uninterrupted expanses of ungrazed tallgrass. Other species such as the prairie vole and cotton rat may survive and even breed in the moderately grazed areas but build up high populations only where ample vegetation persists. Such areas are spatially heterogeneous today because of agricultural practices but, most likely, were temporally variable in pre-Columbian times, when, for example, large herds of grazers might have overgrazed severely one year, missed an area completely the next, and moderately grazed it during a third. Other True Prairie species such as the deer mouse (Peromyscus maniculatus) are essentially ubiquitous within the True Prairie and survive and breed both in the grazed and ungrazed habitats.

Thus, grazing apparently has a profound effect on the small mammals of the True Prairie, both on individual species and on the total small mammal community. Hypothetically, grazing has a greater effect on small mammal communities in the True Prairie than in other grasslands (Birney et al., 1976). Some species are most successful under one grazing regime and the associated level of vegetative cover--others, under a different grazing regime--but all species respond to grazing, and lag time appears to be relatively short in the True Prairie, probably because of the high cover provided by tallgrasses, even before a litter layer has developed.

Response of Microorganisms

Based on studies at the Osage Site, microbial activity of the soil increases under moderate grazing. Carbon dioxide evolution in ungrazed and grazed treatments is compared in Figure 10.8. Figure 10.9 summarizes the weight loss by buried and surface substrate for the same treatments.

Several possible reasons exist for this increased microbial activity in the grazed prairie. Certainly the mechanical breakdown of the grass caused by feeding pattern and animal movement over the grass may be important. Investigators have frequently observed that the addition of new plant material to the stabilized microbial system has a priming effect on microbial activity. Also, the addition of extra nitrogen by animal feces may prime the decomposition process. However, the exact effect of added nitrogen from feces has not been measured. Another observation on the Osage Site and elsewhere has been that the soil temperature is consistently higher in the grazed prairie than in the ungrazed prairie during the growing season. Because the fact has been established that microbial activity correlates more significantly with soil temperature than with other environmental parameters measured, the higher soil temperature in the grazed prairie might account for its increased microbial activity.

ELM Grazing Simulations

The grazing runs of the ELM simulation model were designed to evaluate the impact of grazing with different grazing schemes, changing the grazing intensity, and grazing during different time periods. The grazing systems imposed on the model include year-round cow-calf grazing at light (6.5 ha/AU), moderate (4.0 ha/AU), and heavy (3.2 ha/AU) stocking rates and season-long steer grazing at light (4.0 ha/AU), moderate (2.4 ha/AU), and heavy (1.9 ha/AU) stocking rates. In the cow-calf grazing system the calves were born in February and were removed from the range on 1 October, and the cows were allowed on the range year-round. In the season-long steer grazing system yearling steers (250 kg initial weight) were grazed from 1 May through 30 September. Stocking rates represent characteristic rates for the southern True Prairie (Harlan, 1960a).

Unless stocking rates are high or various management techniques are used, cattle do not graze pastures uniformly but tend to graze in selected areas, leaving a large percentage of the pasture essentially ungrazed (Herbel and Anderson, 1959; Klipple and Costello, 1960). This distributional problem is particularly true for light and moderate grazing intensities, but utilization increases greatly for heavy grazing intensities. For the grazing runs, cattle were assumed to graze on 50 percent of the pasture, while the remainder of the pasture would stay ungrazed. This assumption tends to overestimate the lightly grazed area and to underestimate the heavily grazed area. Steer-grazing simulations were also run with and without the assumption that grazing would be uniform over whole pastures to show how nonuniform grazing patterns influence grassland characteristics.

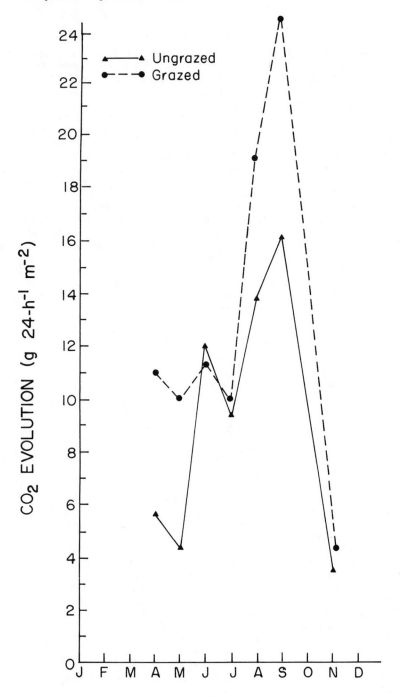

FIGURE 10.8 Seasonal carbon dioxide evolution rates from the
ungrazed and grazed treatments on the Osage Site, 1972.

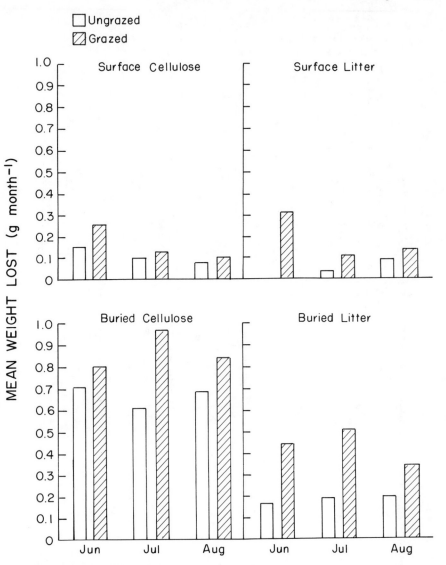

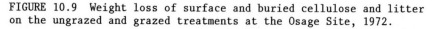

FIGURE 10.9   Weight loss of surface and buried cellulose and litter
on the ungrazed and grazed treatments at the Osage Site, 1972.

Season-Long Steer Grazing

Results for the steer-grazing runs in Table 10.15 showed that
increasing the grazing intensity from light to heavy caused
aboveground and belowground primary production and cattle weight
gain per head to decrease.   The decrease in aboveground primary
production was consistent with the results of Herbel and Anderson
(1959) in Kansas, where they found that increasing grazing level

TABLE 10.15 Three-year average values for output parameters from the grazing simulations.

Grazing runs	Primary production (g m^{-2} yr^{-1})		Animal weight gain			Peak live biomass (g m^{-2})		
	Net aboveground	Net belowground	Weight gain per head (kg head^{-1})	Forage consumption (g m^{-2} yr^{-1})	Average nitrogen content of forage consumed (percent nitrogen)	Warm-season grass	Warm-season forbs	Cool-season annuals
No grazing	542	291	-	-	-	237.5	22.5	47.5
Steer grazing								
Light	544	279	64.5	54	1.01	225.0	27.5	42.5
Moderate	527	255	60.9	90	1.00	192.5	37.5	37.5
Heavy	471	216	44.0	114	1.04	142.5	35.0	37.5
Cow-calf grazing								
Light	559	298	233.3	88	0.86	237.5	27.5	40.0
Moderate	559	288	233.3	141	0.86	207.5	37.5	37.5
Heavy	548	293	183.7	153	0.86	181.5	40.0	35.0
Uniform steer grazing								
Light	545	288	64.4	54	0.99	230.0	35.0	42.5
Moderate	544	284	64.4	90	0.99	222.5	37.5	40.0
Heavy	542	277	64.5	115	1.01	207.5	42.5	37.5

TABLE 10.15 continued

Grazing runs	Biomass (g biomass m^{-2})				Abiotic (cm H$_2$O yr^{-1})			Nutrients (g m^{-2} yr^{-1})			Biomass loss by microbial activity (g m^{-2} yr^{-1})
	Total aboveground	Standing dead	Litter	Total roots	Transpiration	Bare soil water loss	Standing-crop interception	Phosphorus root uptake	Nitrogen root uptake	Net nitrogen balance	
No grazing	276	483	405	924	28	11	37	0.58	6.15	0.95	836
Steer grazing											
Light	268	434	385	899	27	12	35	0.59	6.31	0.79	828
Moderate	242	386	392	856	27	15	32	0.60	6.25	0.54	803
Heavy	193	306	350	756	25	21	25	0.60	5.85	0.35	760
Cow-calf grazing											
Light	278	419	384	923	28	15	34	0.60	6.40	0.55	828
Moderate	257	435	384	896	29	15	30	0.62	6.53	0.40	793
Heavy	241	349	374	863	28	17	28	0.64	6.54	0.36	764
Uniform steer grazing											
Light	274	435	381	906	28	13	35	0.59	6.36	0.73	832
Moderate	268	402	364	886	27	14	34	0.59	6.46	0.58	826
Heavy	255	377	367	865	27	14	33	0.60	6.47	0.46	816

from light to moderate and moderate to heavy caused a decrease in aboveground herbage production of -5 percent and -28 percent, respectively. These results compare with simulated decrease in aboveground production of -3 percent and -13 percent.

The simulated decrease in cattle weight gain per head with heavy grazing (-32 percent) was consistent with the data presented by Harlan (1960a); however, the model predicted a 6 percent decline in weight gain with moderate grazing, while Harlan showed that weight gains were the same for light and moderate grazing intensities. The model underestimated weight gain per head at all of the grazing levels compared to Harlan's averages for Oklahoma.

Simulated peak total aboveground biomass, warm-season perennial grass, and cool-season annual grass decreased, but warm-season forbs increased as grazing levels increased from light to heavy. Cool-season annual grass would decrease because of grazing pressure early in the season before growth of warm-season species. Warm-season forbs, many of which are not very palatable, will increase with increasing grazing pressure, especially in the southern portion of the True Prairie, where broomweed (Gutierrezia dracunculoides) characteristically increases significantly. This increase in warm-season forbs may not be characteristic of the northern Flint Hills region, which is beyond the geographical range of some of the most responsive forbs.

The standing crop parameters show that standing dead biomass and total root biomass decreased with increased grazing pressure, and litter biomass also decreased with heavy grazing. Virtually all previous studies from the Osage Site and the True Prairie show a decrease in standing dead under grazing conditions, and the decrease in root biomass with heavy grazing was supported by Weaver (1950a), who found that root weights taken from pastures in poor condition in Nebraska were only 42 percent of root weights from pastures in good range condition. The Osage Site data showed lower root production under grazed conditions in one of three years, but because the site was not grazed during the summer, decreased root production would not be expected. The model results show that decreased total root biomass with grazing was a result of decreases in both the live- and dead-root biomass.

The abiotic and nutrient data show that increasing grazing intensity produces a decrease in transpiration, interception, nitrogen uptake by the roots, and the net nitrogen balance but produces an increase in bare soil evaporation. Increasing grazing levels produced a slight increase in soil water content during the growing season, a result supported by Owensby et al. (1970), who showed that clipping of the live forage during the growing season caused an increase in soil water content. The data of Owensby et al. and the model suggest that the increased soil water results from a decrease in live leaf area and the consequent decrease in transpiration water loss.

Field data that verify output results for the nutrient parameter do not exist. The implication of the model results, however, is very important. The simulation output shows that the net accumulation of nitrogen in the grassland system is 0.95 g m^{-2} in ungrazed conditions; nitrogen decreases to 0.35 g m^{-2} under heavy grazing. The decrease in the net nitrogen balance with higher grazing levels is a result of larger amounts of forage being

consumed by cattle and resulting increases in volatilized nitrogen from cattle urine and feces. Forage consumption by the cattle increased the turnover rate of nutrients in aboveground forage (cf., Dahlman et al., 1969; Wecklow, 1975).

The microbial activity was greater in the ungrazed pasture and decreased with increased grazing intensity. This decrease in microbial activity with grazing was caused by the increase in cattle biomass and a reduction in aboveground biomass available for the microbes to decompose. Although aboveground microbial activity decreased with grazing, microbial activity in the soil increased. Data from the Osage Site showed increased microbial activity of the soil with grazing. Both the model results and the Osage Site data suggest that the higher soil temperature in the grazed prairie is responsible for the increased microbial activity in the soil.

Cow-Calf Grazing System

The results for the cow-calf grazing runs in Table 10.15 show that increasing grazing intensity from light to heavy has a similar impact on the output variables as changing grazing levels had with the steer grazing system. The major difference is that increasing the grazing intensity from light to moderate with cow-calf grazing has little impact on cattle weight gain and aboveground primary production, while belowground production was only slightly decreased (<2 percent) with moderate and heavy grazing levels. Increasing steer grazing from light to heavy resulted in a steady decrease in aboveground and belowground production and cattle weight gains. Heavy grazing intensities caused a significant decrease in cattle weight gains and aboveground production for both grazing regimes.

Cow-calf grazing was more beneficial than steer grazing from the standpoint of the vegetation, since above- and belowground production and warm-season perennial production and nutrient uptake by plants were all lower with steer grazing. The model response arose because steer grazing represents a higher stocking rate during the growing season than the cow-calf grazing operation. Total forage consumption at a given grazing intensity was greater for the cow-calf operation, but remember that cow-calf grazing was spread over the whole year, while steers utilized forage only from May through September.

Another interesting difference between the two grazing regimes was that increased stocking rates reduced plant nitrogen uptake with steer grazing but increased nitrogen uptake with cow-calf grazing. Decreased root nitrogen uptake for steer grazing runs resulted from lower live root biomass with increasing grazing intensity. Increased nitrogen uptake in the cow-calf grazing system resulted from live root biomass being only slightly decreased with grazing, while consumption of forage by cattle increased the rate at which nutrients in the aboveground plant material were returned to the soil. Nutrients contained in the urine and feces were returned to soil nutrient pools much more rapidly than nutrients in standing dead and litter biomass (Dahlman et al., 1969; Wicklow, 1975).

Observed data to verify the model results are not available, but McIlvain and Shoop (1969) concluded that five different grazing

systems in western Oklahoma were not superior to continuous year-long grazing at the same stocking rates.

One of the important assumptions of the grazing runs is that the cattle do not graze a pasture uniformly but tend to graze in selected areas, thus leaving a large portion of the pasture ungrazed (50 percent of the area is assumed to be grazed in this study). The results of uniform steer grazing runs (Table 10.15) show that increasing grazing intensity from light to heavy does not have the negative impact of reducing above- and belowground production and cattle weight gains per head and actually increases the root nitrogen uptake. These results indicate that were cattle managed to graze pastures more uniformly, cattle weight gains and primary production could increase significantly. Some proposed methods for stimulating uniform grazing of a pasture include moving salt blocks and water supply, spring burning of the pasture, and cross-fencing, which ensures short-term heavy stocking rates.

The importance of including the spatial characteristics of grazing in the model is emphasized when the present model's results are compared with Parton and Risser's (1978), which did not consider the spatial effects. Parton and Risser reported that increasing grazing intensity from light to heavy caused aboveground production to increase, while the results of this model show that light grazing results in maximum aboveground production. The fact that the present model's results match the observed field data suggests that grazing system models should indeed consider the spatial grazing patterns.

A summary of the results for the grazing runs indicate:

1. A year-round cow-calf grazing system at a low average stocking rate is more beneficial to the grassland than seasonal steer grazing (April-October) because cow-calf grazing increases the above- and belowground primary production and plant nutrient uptake.
2. Cattle weight gain per head, above- and belowground plant production, transpiration water loss, standing dead, and the net nitrogen balance all decrease with increasing grazing intensity, while soil water content and bare soil water loss increase.
3. Uniform grazing of the pasture would not decrease above- and belowground plant production and cattle weight gain per head as much as the random grazing of the steer and cow-calf methods.

Species Manipulations

At the Osage Site cool season annual and warm season forbs are more common in overgrazed pastures. The impact of removing these species from moderate cow-calf and steer grazing was evaluated with a series of computer runs shown in Table 10.16. The steers were grazed at an intensity of 2.4 ha/AU from 1 May to 30 September and had an initial weight of 250 kg. The cow-calf grazing system had a grazing intensity of 4.0 ha/AU. The cows grazed in the pasture year-round, while the calves that were born in February were removed from the range on 1 October.

Removing cool season annual grass from the cow-calf grazing regime increased belowground production and production of warm season perennial grass and forbs, while net aboveground production and weight gain per head decreased slightly. Results indicate that over the entire season, aboveground cool season annual production was replaced by warm season grass production. Reduction in weight gain per head occurs because cool-season annual grasses provide a high quality forage early in the growing season, before warm season grass growth is inititated.

The removal of warm-season perennial forbs slightly reduces aboveground production and root uptake of nitrogen and phosphorus, while production of warm-season perennial grasses was increased. Removal of forbs has little effect upon the system, since the warm season grass production replaces forb production. The impact might be greater with heavy and extra-heavy grazing, since production of forbs increases at these grazing levels.

Removing cool-season annuals from the steer grazing run increased belowground production, nitrogen content of consumed forage, and forb production, but cattle weight gain per head and aboveground production were decreased. Removal of cool season annual has a similar effect for the cow-calf grazing system except that the reduction in cool season annual production is replaced by warm season forb production instead of warm-season grass production. Eliminating forbs from the steer and the cow-calf grazing systems has the same effect: aboveground production and root uptake of nitrogen and phosphorus all decreased, and warm season grass production was increased.

A summary of the model results indicated that removal of cool-season annuals reduced cattle weight gain because of the removal of a high-quality forage early in the growing season before growth of warm-season grasses had started. This information is supported by Duvall and Whitaker's (1964) results for slender bluestem ranges in Louisiana that suggest that the cool-season grass is important in providing high-quality forage for cattle during the early part of the growing season. Elimination of forbs had little impact on either grazing system. Also, removal of forbs or cool season annuals generally increased the growth of the warm-season perennial grasses.

NUTRIENTS

The role of nutrients in grassland ecosystems has received considerable attention (Henzell and Ross, 1973; Mott, 1974; Pomeroy, 1970; Wight, 1976) primarily from the standpoint of increasing forage production (Whitehead, 1970). Descriptions of nitrogen and phosphorus budgets and responses of the True Prairie to nutrient additions follow in the next sections.

Response of Plants

Soil nutrient levels have been related to plant species composition and vegetation changes. Little bluestem was invading the ridges of an abandoned field of Kentucky bluegrass in a drained

TABLE 10.16 Three-year average values of the output variables for the species manipulation simulation.

Species manipulation	Primary production (g m^{-2} yr^{-1})		Animal weight gain			Peak live biomass (g m^{-2})			
	Net aboveground	Net belowground	Weight gain per head (kg head^{-1})	Forage consumption (g m^{-2} yr^{-1})	Average nitrogen content of forage consumed (percent)	Warm-season grass	Warm-season forbs	Cool-season annuals	Total aboveground
									Moderate steer grazing
Control	527	255	60.9	90	1.00	192.5	37.5	37.5	242
Cool-season annuals	506	271	59.9	92	1.06	185.0	52.5	0.0	241
Warm-season forbs	519	246	60.7	90	1.01	212.5	0.0	40.0	225
									Cow-calf grazing
Control	559	288	233.3	141	0.86	207.5	37.5	37.5	257
Cool-season annuals	548	310	233.0	141	0.86	220.0	42.5	0.0	276
Warm-season forbs	551	288	233.2	141	0.87	220.0	0.0	37.5	245

TABLE 10.16 continued

Species manipulation	Standing crop (g biomass m^{-2})			Abiotic (cm H$_2$O yr^{-1})			Nutrients (g m^{-2} yr^{-1})		
	Standing dead	Litter	Total roots	Transpiration	Bare soil water loss	Standing crop interception	Phosphorus root uptake	Nitrogen root uptake	Net nitrogen balance
Control	386	392	856	27	15	32	0.60	6.25	0.54
Cool-season annuals	411	346	849	25	17	30	0.66	6.39	0.53
Warm-season forbs	375	378	811	27	16	31	0.58	6.14	0.55
Control	435	384	396	29	15	30	0.62	6.53	0.40
Cool-season annuals	409	361	934	27	16	31	0.66	6.63	0.40
Warm-season forbs	362	376	870	28	16	30	0.61	6.47	0.39

marsh in the sand plains of Wisconsin (Wuenscher and Gerloff, 1971). The depressions between the ridges contained more available phosphorus than the ridges, which contained only between 0.36 and 0.90 g m^{-2} of available phosphorus. The yield and phosphorus absorption of these two species were compared when they were grown in low phosphorus soil and in nutrient sand culture at various levels of added phosphorus. Little bluestem consistently had a higher yield than Kentucky bluegrass at all phosphorus levels of both soil and sand cultures. The latter absorbed only about 5 percent as much phosphorus from deficient soil as little bluestem. Little bluestem also utilized more absorbed phosphorus under stress conditions in the sand culture. At the 2.5 mg l^{-1} phosphorus level sand cultures, little bluestem produced 2.2 g and Kentucky bluegrass produced 1.4 g. The lowest tissue concentration adequate for maximum growth for little bluestem was 0.14 percent, which was considerably below the critical concentration of 0.25 percent phosphorus for Kentucky bluegrass.

Several studies have dealt with nutrients and little bluestem (Dickinson, 1964; Kilgore and Boren, 1972; Nixon and McMillan, 1964; Rao et al., 1973; Reardon and Huss, 1965; Senter, 1975; Van Amburg and Dodd, 1970; Waller et al., 1975; White, 1961). Although little bluestem has a wide tolerance to variation in soil texture (Nixon and McMillan, 1964), in phosphorus-deficient areas it responds to phosphorus fertilizer by increasing yields by 50 percent (Reardon and Huss, 1965). Soil fertility may be the most important factor in little bluestem's distribution on clay-textured soils.

In a series of experiments in Texas with a low fertility sandy loam and a moderately high fertility clay loam, Van Amburg and Dodd (1970) found that little bluestem clones collected from a clay soil were generally larger in circumference and contained more tillers than the clones from the sandy loam. However, the number of roots produced per tiller was greater in clones from the sandy loam. Pot experiments showed that net aerial production, height of tallest flowering culm, and number of flowering culms were greater in clay soil (Waller et al., 1975). In these two-year experiments moisture was added to the soil surface and dead leaves were removed from the pots. In this way, litter decomposition and throughfall, two major avenues of nutrient return to the soil, were made inoperative. Second-year growth was greater in the clay soils. Previously it was shown with ^{134}Cs, that 60 percent of the isotope was maintained in the upper 10 cm of an intact soil-plant system. In the same experiment, the isotope was slowly lost from the upper layer of the soil. So, the hypothesis is that clay soil has higher fertility and plants are not so dependent on recycling. Sandy soils have a rapid cycle, and when both litter decomposition and throughfall were removed, inadequate nutrients were available and the yield decreased the second year.

Although the addition of phosphorus at 7 g m^{-2} to a Texas little bluestem community increased forage production by 50 percent even under low rainfall, potassium reduced production, and calcium had no effect (Reardon and Huss, 1965). Cook and Rector (1964) obtained only slight increases in King Ranch bluestem with phosphorus fertilization. When nitrogen was added at the rate of 4.5 g m^{-2} to Kansas True Prairie, dry matter yields increased by 133 to 180 g m^{-2}, but phosphorus had no effect (Moser and Anderson,

1965). This response to nitrogen occurred only when precipitation
was normal or above normal. Nitrogen fertilizer also increased
crude protein of the forage, and water-use efficiency was highest on
plots receiving supplemental nitrogen (Owensby et al., 1970).

In subsequent studies, burned and unburned Tallgrass Prairies
near Manhattan, Kansas have been subjected to differential nitrogen
treatments since 1972 (Owensby and Launchbaugh, 1977). Nitrogen
treatments of 4.5 and 9.0 g m^{-2} and an untreated control were
included, and each pasture was burned in late spring to reduce cool
season grass and forb growth. Each pasture was grazed to its
potential based on the level of nitrogen added. The intent was to
leave the same amount of primary production in each pasture at
season's end. Grazing rates were 1.34 and 0.97 ha steer^{-1} for the
lower and higher nitrogen applications, respectively, and 0.76 ha
steer^{-1} for the control. Nevertheless, the effects of nitrogen and
grazing are confounded in this experiment.

Control and 4.5 g m^{-2} nitrogen treatments remained
approximately the same vegetatively, although nitrogen did promote
broadleaf forb growth. The 9.0 g m^{-2} nitrogen treatment showed a
significant change; bluestem grasses were reduced by 33 percent,
probably because of the increased stocking rate, and broadleaf forbs
increased threefold. In southwestern Kansas, Kilgore and Boren
(1972) increased yields 40 to 50 percent with the application of
nitrogen and phosphorus. In north-central Oklahoma, increases in
forage yields were attributed to nitrogen and further increases were
attributed to nitrogen plus potassium (McMurphy, 1970). However, as
Harner and Harper (1973) showed, changes in species composition
might cause greater changes in the nutrient content of forage than
the nutrient status of the soil. McMurphy indicated that low levels
of nitrogen might produce satisfactory forage yields if weeds and
other undesirables were controlled.

In Oklahoma, Harper (1957) showed that the highest average
yield of 360 g m^{-2} air-dry forage was obtained from a treatment of
3.8 g m^{-2} nitrogen, 1.8 g m^{-2} P_2O_5, and 1.1 g m^{-2} K_2O applied
annually in the spring.

Nitrogen fertilization increased water use, but plants made
more efficient use of available water. Table 10.17 shows the
increased water-use efficiency for native prairie plots in Kansas

TABLE 10.17   Fertilizer effect on water-use efficiency (Owensby
et al., 1970, Hyde and Owensby 1973).

Treatment	Water use efficiency (kg dry matter per 2.54 cm water)			
	1965	1966[a]	1967	1968
Fertilized	71	59	83	74
Control	46	52	83	51

[a] Precipitation considerably below normal.

fertilized with 4.5 g m^{-2} nitrogen annually for 4 years (Hyde and Owensby, 1973; Owensby et al., 1970).

In a heavily grazed pasture in North Dakota, 8.0 g m^{-2} nitrogen applied annually produced an average of 203 g m^{-2} dry forage compared to 118 and 67 g m^{-2}, respectively, for treatments with 2.7 g m^{-2} nitrogen and no nitrogen. On a moderately grazed pasture, 8.0 g m^{-2}, 2.7 g m^{-2}, and no nitrogen produced 179, 117, and 59 g m^{-2} forage, respectively. The increase in yield was caused primarily by an increase in the cool-season western wheatgrass (Agropyron smithii). Two years of fertilization of the heavily grazed pasture at the 8 g m^{-2} rate of nitrogen did more to improve range conditions and production than six years of complete isolation from grazing. Root weight and crude protein level in the forage was also higher with added nitrogen (Rogler and Lorenz, 1957). Depending on a number of conditions, the recovery rate for nitrogen fertilizer applied to grassland systems is seldom greater than 60 percent, perhaps because of denitrifying bacteria in the rhizosphere of plants (Dilz and Woldendorp, 1960).

The forage quality response of True Prairie grass species to nitrogen addition has not been clarified. In east-central Oklahoma, addition of nitrogen to a grassland composed primarily of little bluestem and broomsedge (Andropogon virginicus) had no effect on the crude protein of forage, nor did it raise protein or phosphorus above the winter minimum livestock requirement (Senter, 1975). Also in Oklahoma, Dickinson (1964) found that the crude protein content of big bluestem and Indiangrass was increased by nitrogen fertilization, although little bluestem did not demonstrate this protein increase. Rehm et al. (1972) did not find differences in nutrient content or dry matter digestibility with nitrogen fertilization of seeded warm season grasses in Nebraska. In Kansas Owensby et al. (1970) found crude protein higher in fertilized plots, while Dee and Box (1967) in Texas showed that nitrogen fertilizer increased the crude protein of silver bluestem (Andropogon saccharoides) at rates as high as 34 g m^{-2}, and crude protein remained higher in the forage from fertilized plots throughout the winter season.

Woolfolk et al. (1973) examined responses of the Flint Hills True Prairie to burning and nitrogen and found that the application of nitrogen fertilizer increased neutral detergent fiber, hemicellulose, crude fiber content, and lignin values, while it decreased dry matter digestibility, nitrogen-free extract, and diluted protein. Higher livestock gains per hectare were possible because of heavier stocking rates, not increased individual animal performance. Allen (1973), also working in the Flint Hills, found increased ash and lignin but decreased nitrogen-free extract, cell wall, and cellulose with nitrogen fertilizer.

The increase in cool season grasses with additional nitrogen has been found elsewhere in the True Prairie. In central Oklahoma, Huffine and Elder (1960) found that the addition of phosphorus and nitrogen caused the production of two to five times more weeds but very little increase in grass production. In the Flint Hills of Kansas, Owensby et al. (1970) also found a shift toward cool-season species, particularly Kentucky bluegrass (Poa pratensis).

In earlier work conducted very near the sample areas at the Osage Site, Gay and Dwyer (1965) examined the effect of burning and

clipping on previously grazed plots that had received 4.5 and 9.0 g m^{-2} nitrogen as ammonium nitrate. When treated with 4.5 g m^{-2} nitrogen and clipped once during the year, the yield was 219 g m^{-2}, but the yield increased to 346 g m^{-2} when clipped twice. The yields were 285 and 393 g m^{-2}, respectively, for plots treated with 9.0 g m^{-2} nitrogen. These yields imply that the grass recovered more quickly from clipping when fertilized with nitrogen, because the once-clipped and twice-clipped controls were not significantly different. When burned and fertilized with nitrogen, however, yields were greater from plots clipped once at the end of the season compared to plots clipped twice. Plots burned and fertilized with 9.0 g m^{-2} nitrogen and clipped once at the end of the season produced significantly more forage than any other treatment--439 g m^{-2} compared to 276 g m^{-2} for the control. Big bluestem and Indiangrass used nitrogen most efficiently, especially on burned treatments.

Bermudagrass is frequently used to improve pasture in the southern portion of the True Prairie. Table 10.18 depicts the response of Bermudagrass to nitrogen applications in areas of adequate rainfall in Oklahoma (Chiles, 1968). Higher application rates are usually used with irrigation. A typical fertilizer program might apply 2.7 to 5.4 g m^{-2} P_2O_5 and K_2O in the winter or early spring and 9.0 to 36.0 g m^{-2} nitrogen in the early summer.

Nitrogen Cycle

Dahlman et al. (1969) analyzed the nitrogen cycle of a True Prairie site in central Missouri. This site had never been plowed and probably represented the presettlement status with high organic matter (4 percent in the O1 horizon) and nitrogen (0.21 percent). In this system, summarized in Figure 10.10, the nitrogen content of peak standing crop of shoots is 3.6 g m^{-2} and of roots is 6-12 g m^{-2}. At the Osage Site the nitrogen content of litter is 1.88 percent and of roots is 0.66 percent. Of the nitrogen distributed

TABLE 10.18   The response of Bermudagrass to nitrogen in areas of adequate rainfall in Oklahoma (Chiles 1968).

Nitrogen (g m^{-2})	Yield (g m^{-2})
0	215
18	925
36	1328
54	1444
72	1514

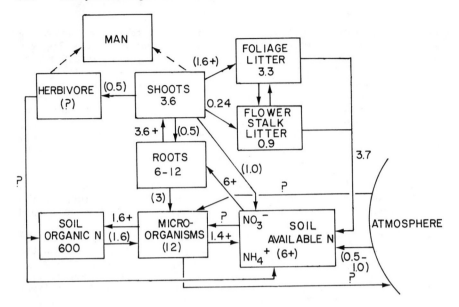

FIGURE 10.10 Preliminary estimates of nitrogen inventory and transfer relationships for different compartments of a native prairie ecosystem. Contents and rates are expressed in g m^{-2} yr^{-1}; numbers in parentheses are first-order approximations (Redrawn from Dahlman et al., 1969).

among plant biomass, litter, and soil, at least 95 percent is in the soil. Only a small part of the total ecosystem budget is associated with recent organic production--that is, plant parts and microorganisms. Apparently, new production is supported by the available nitrogen pool, much of which arises from rapid turnover of biological materials. The turnover time is weekly for microorganisms (Jansson, 1958), four years for roots (Dahlman and Kucera, 1965), and about two years for litter (Koelling and Kucera, 1965).

The authors measured the amount of nitrogen released as a result of litter breakdown by summing the amount released from different age fractions over a four-year period. This relatively large amount (3.7 g m^{-2}) probably goes to the available nutrient pool, not to the soil organic nitrogen. The evidence for the movement to the available nutrient pool is that distinctive litter layers are not apparent, which suggests a nearly complete decomposition of organic materials aboveground. The dynamics of the so-called dead-plant biomass category indicated a close relationship between carbon and nitrogen content (Dahlman et al., 1969). Assuming that a C:N ratio of about 20:1 is the highest ratio that permits significant mineralization, even after two years, only about 25 percent of the nitrogen in the forage litter had been released, and the percentage of nitrogen in the flower stalks had increased, presumably because of transfer from the foliage litter (Table 10.19).

TABLE 10.19 Carbon-nitrogen relationships in grass foliage and flower stalk litter[a] in Missouri prairie[b] (Dahlman et al., 1969).

Decomposition time (years)	Organic content of biomass (g)[c]		Nitrogen content (percent)	Carbon: Nitrogen	Nitrogen balance for litter component (g m^{-2})[e]	Net change in nitrogen (g m^{-2})
	Organic matter	Carbon[d]				
			Grass foliage			
Initial	45.6	23.1	0.41	110:1	1.4	
+1	33.0	16.5	0.51	70:1	1.3	-0.1
+2	16.6	8.3	0.80	50:1	1.1	-0.3
			Grass flower stalks			
Initial	48.9	24.5	0.15	300:1	0.24	
+1	40.9	20.5	0.26	185:1	0.33	+0.1
+2	28.7	14.4	0.47	100:1	0.42	+0.2

[a] Litter in mesh containers to permit recovery and measurement.

[b] Koelling and Kucera (1965).

[c] Based on initial quantity of 50 g.

[d] Based on 50 percent carbon in the plant organic material.

[e] Assuming that foliage constitutes 70 percent and flower stalks 30 percent of total production.

Dahlman and Kucera (1967) analyzed the carbon transfers between roots and soil and found that in the first year following ^{14}C assimilation by plants, at least 45 percent of the ^{14}C lost from the roots was measured in the soil organic matter. Four months later about 10 percent of the root loss was present in the soil (Table 10.20). So during ten months, over 90 percent of the new organic input to the soil was oxidized. Dahlman and Kucera (1965) indicated that the annual organic input of roots was at least 430 g m^{-2}, and assuming the 40 percent carbon and 0.7 percent nitrogen contents, the input of root nitrogen to the soil was 3 g m^{-2}. Considering the rapid reduction of the C:N ratio to be from 55:1 to 6:1, they estimated that about half the 3 g m^{-2} input (1.4 g m^{-2}) was released to the available pool and the rest was added to the organic nitrogen pool.

Although the quantity of available nitrogen in the soil fluctuates during the season, the total is not likely to exceed 1 to 2 percent of organic nitrogen, or 6 to 12 g m^{-2}. Dahlman et al. (1969) calculated that 3.7, 1.4, and 0.5 to 1.0 g m^{-2} for litter decomposition, root turnover, and rainfall, respectively, amounted to about 6 g m^{-2}. Additional inputs would be biological fixation (Tjepkema and Burris, 1976) and mineralization (Smith and Young, 1975). Utilization of available nitrogen by higher plants was at least 6 g m^{-2}. The demand of microorganisms was not known but thought to be at least as great as the demand for higher plants.

Koelling and Kucera showed that the nitrogen content of the current season's litter was significantly lower than the nitrogen in the mature standing crop (0.14 percent versus 0.73 percent). The difference was attributed to retranslocation within the intact plant and to leaching from the standing material. Dahlman et al. estimated retranslocation at 0.5 g m^{-2} and leaching from the standing material at 1.0 g m^{-2}.

The nitrogen cycle was examined by using data from the Osage Site for aboveground net production, crown production, and root production. The nitrogen content of plant components and soils to a depth of 20 cm were determined for each sample date. Table 10.21 gives the percentage of nitrogen in the dry weight of the various components, and the nitrogen content of these components are summarized in Table 10.22. These values are a function of biomass; values in dead aboveground material are high. The distribution of nitrogen in plant and soil components as a percentage of the total nitrogen is shown in Table 10.23. More than 95 percent of the total nitrogen is in the soil, and the amount in the litter is relatively small.

Little bluestem contributed approximately 80 percent of the biomass at the Osage Site, with a similar proportion of nitrogen (Table 10.24). This information is not surprising, since the species makes up such a high percentage of the biomass and about 95% of the biomass is contributed by warm season perennial grasses.

Total uptake of nitrogen by vegetation varied from 5.35 to 6.80 g m^{-2} (Table 10.25). In 1970 and 1971 about 53 percent of this uptake went to the belowground portion, while in 1972 belowground parts shared 62 percent of the total uptake. In 1972, only 48 percent of the nitrogen in the belowground components was released, while 118 and 217 percent were released in 1970 and 1971, respectively. Over the three years 107 percent of the total

TABLE 10.20 Turnover of root and soil carbon following a single organic input of 450 g m^{-2} to the root-soil matrix[a] (Dahlman et al., 1969).

Decomposition time (months)	Single organic input from roots (g m^{-2})	Fractional carbon transfer to soil organic matter (g m^{-2})	Carbon remaining in soil organic matter[b]		Implied turnover (g m^{-2})
			(percent)	g m^{-2}	
14	430	170	9	15	155
26	--	--	3	9	330

[a] Nitrogen content of roots was 0.7 percent.

[b] Based on ^{14}C analysis (Dahlman and Kucera, 1967).

TABLE 10.21 Percentage of nitrogen in the dry
weight of the various components
at the Osage Site.

Components	Nitrogen
Aboveground live + dead	1.35
Crown	0.72
Roots	0.66
Litter	1.88

TABLE 10.22 Nitrogen content in the above- and belowground
biomass and in the top 20 cm soil on the Osage
Site.

Components	1970	1971	1972
	(in g N m^{-2})		
Live	0.62	1.75	1.68
Dead	2.43	4.31	3.35
Live + dead	3.05	6.06	5.03
Litter	1.11	1.23	1.60
Crown	1.77	0.79	1.21
Root	3.62	4.31	2.34
Crown + root	5.39	5.10	3.50
Soil 0-20 cm	250.25	265.85	235.30

nitrogen was released to the soil. These annual transfer values
were organized into a compartment model (Figure 10.11). The total
nitrogen released through decomposition was 6.10 g m^{-2} yr^{-1},
somewhat higher than the 5.10 g m^{-2} yr^{-1} suggested by Dahlman et al.
(1969) in Missouri. The latter study also suggested a faster rate
of nitrogen uptake.

TABLE 10.23   Distribution pattern of nitrogen in the plant and soil components at the Osage Site.

Components	Percent of total nitrogen		
	1970	1971	1972
Aboveground live	0.24	0.62	0.68
Standing dead	0.94	1.54	1.36
Litter	0.16	0.43	0.64
Crown	0.67	0.27	0.48
Root	1.38	1.54	1.15
Soil	96.65	95.54	95.10

TABLE 10.24   Contribution by the two dominant species at the Osage Site to the aboveground biomass and to the nitrogen content in the aboveground biomass (Bokhari and Singh, 1975).

Year	Species	Contribution to live + dead standing crop	Contribution to nitrogen in live + dead component
		(Percent of total nitrogen)	
1970	Little bluestem	82	82
	Tall dropseed	10	2
1971	Little bluestem	73	72
	Big bluestem	16	16
1972	Little bluestem	82	79
	Big bluestem	4	4

Response of Invertebrates

Sweep net collections were taken periodically during the growing seasons of 1973 and 1974 in the True Prairie near Manhattan, Kansas.   Samples were separated into total insects, acridid grasshoppers, and leafhoppers.   Specimens were counted and weighed

Table 10.25.   Uptake, transfers, and releases of nitrogen
at the Osage Site (Bokhari and Singh, 1975).

Nitrogen component	1970	1971	1972
	(in g N m^{-2} yr^{-1})		
Uptake in ANP[a]	3.16	2.62	2.01
Uptake in BNP[b]	3.64	2.90	3.34
Total uptake	6.78	5.52	5.35
Transfer from ANP to SD[c]	1.80	2.80	1.76
Transfer from SD to litter	0.00	2.02	5.12
Release from litter	0.30	0.68	5.13
Release from root + crown	4.30	6.31	1.59
Total release	4.60	6.99	6.72

[a] ANP = Aboveground net production.

[b] BNP = Crown + root net production.

[c] SD = Standing dead vegetation.

and the total numbers and biomass for each year and for each
individual collection were analyzed. Nitrogen was applied in 1973
and 1974 at rates of 4.5 and 9.0 g m^{-2}; stocking rates were 1.34 and
0.97 ha steer^{-1}, respectively. The control, which received no
nitrogen, had a stocking rate of 0.76 ha steer^{-1} (see Torrence,
1975).

Data on total insects showed significantly fewer insects in the
control and 4.5 g m^{-2} treatments in 1973. Fewer numbers were also
found in these treatments in 1974 but the difference was not
significant at the 5 percent level. Biomass differences were not
significant for either year. In 1973, the greatest insect biomass
was found in the control and in the 9.0 g m^{-2} treatment in 1974.
Also, many significant differences were evident in numbers in
individual collections, but in 1974 differences were found in
biomass and in numbers in some individual collections.

Data on leafhoppers showed significantly higher numbers in the
9.0 g m^{-2} treatment than in the control during 1973. The same was
true in 1974, but the difference was significant only at the 25
percent level. Biomass was smaller in the control and highest in

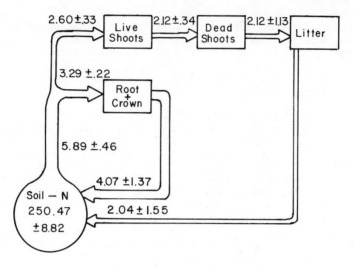

FIGURE 10.11   Compartment diagram of nitrogen cycle at the Osage Site.   The values on the arrows are transfers of nitrogen in g m^{-1} yr^{-1} ±1 standard error, averaged over 1971, 1972, and 1975 (Redrawn from Bokhari and Singh, 1975).

the 9.0 g m^{-2} treatment during both years, but the differences were not significant.   Very few instances of significant differences occurred in numbers or biomass in individual collections.   As yet, leafhoppers have not been determined to species.   Significant differences could exist at the species level just as was found with grasshoppers.

Torrence (1975) studied the response of acridid grasshoppers in the same fertilized Tallgrass Prairie and found no significant differences in numbers or biomass during either 1973 or 1974.   Very few instances of significant differences of numbers or biomass appeared in individual collections.

The major grass-feeding species were Eritettix simplex, Orphulella speciosa, Phoetaliotes nebrascensis, and Syrbula admirabilis.   E. simplex numbers were significantly higher in the nitrogen treatments, possibly because of the more vigorous growth that appeared in the pastures soon after burning.   This species overwinters as a nymph, which is uncommon among grasshoppers.   All other species decreased with the addition of nitrogen, possibly preferring the higher quality grass in the control.

The major forb feeders, Campylacantha olivacea olivacea, Hesperotettix speciosus, Hypochlora alba, and Melanoplus keeleri, all increased with the addition of nitrogen, but only H. speciosus increased significantly.   This increase may be caused by increased forb growth in the fertilized treatments.

Melanoplus femurrubrum femurrubrum is a mixed feeder and was collected in significantly higher numbers in the control, which might indicate a preference for the higher quality forage there.

For the two years of this study, we can conclude that nitrogen has little effect on acridid grasshoppers as a group.   Some individual species showed preference for one or more of the

treatments. Nothing indicated, however, that the application of nitrogen had a significant effect on numbers or biomass of leafhoppers or on the total population of insects present.

Response of Small Mammals

No studies have attempted to determine specifically the effect of nutrient enrichment or irrigation on a small mammal community within the True Prairie. Grant and French (1978) conducted such a study in the Shortgrass Prairie. They found that the addition of nutrients in the absence of irrigation did not increase primary productivity and had little or no effect on the small mammal community. Irrigation without additional nutrients increased plant biomass slowly (over four summers of study), with some concomitant increase in biomass density of small mammal density, especially prairie vole (Microtus ochrogaster) density. However, addition of both water and nutrients had a sharp, positive effect on primary productivity, followed by an increase in density of prairie voles.

True Prairie grasses require more water than do shortgrasses and certainly can suffer from drought conditions. The response of small mammals to drought in the True Prairie has not been well studied, but French et al. (1976) speculated that the low populations at Osage in 1971 resulted from the relatively dry summer of 1970, but no supporting data were available.

Addition of nutrients or water may enhance primary productivity. Once a threshold level of vegetative cover is present, small mammal species probably respond little to additional increase in vegetation (Birney et al., 1976). If that assumption is true, neither nutrients nor irrigation would be likely to affect small mammal communities in a major way. When not grazed or otherwise removed, the vegetation of the True Prairie can and does support relatively high densities of certain small mammal species.

ELM Nutrient Simulations

The effect of adding nitrogen and phosphorus to the grasslands was simulated for ungrazed pastures and pastures seasonally grazed with steers. The grazed pastures were used from 1 May to 30 September at a moderate grazing intensity (2.4 ha/AU). The pastures were fertilized in the spring (31 March) of the first year by adding 20 g m^{-2} of nitrogen and 3 g m^{-2} of phosphorus. Adding fertilizer to an ungrazed pasture (Table 10.26) caused aboveground production, cool-season annual grass production, shoot uptake of nutrients, and nitrogen content of the consumed forage to increase, while belowground production and warm-season perennial grass production decreased.

In the model, adding fertilizers to a seasonally grazed pasture (Table 10.26) increased aboveground and belowground production, nutrient uptake by the consumer and plants, nitrogen content of the forage, cattle weight gain per head, and production of cool-season annual grasses. Simulated results for both the grazed and ungrazed runs predicted that adding fertilizer increased total aboveground production (+10 percent) and cool-season annual production (+50

TABLE 10.26 Three-year average values for output parameters for the fertilization and irrigation simulations.

Water and fertilizer	Primary production (g m⁻² yr⁻¹)		Animal weight gain			Peak live biomass (g m⁻²)				Standing crop (g biomass m⁻²)			Abiotic (cm H₂O yr⁻¹)			Nutrients (g m⁻² yr⁻¹)	
	Net aboveground	Net belowground	Weight gain per head (kg head⁻¹)	Forage consumption (g m⁻² yr⁻¹)	Average nitrogen content of forage consumed (percent)	Warm-season grass	Warm-season forbs	Cool-season annuals	Total aboveground	Standing dead	Litter	Total roots	Transpiration	Bare soil water loss	Interception	Phosphorus root uptake	Nitrogen root uptake
No grazing																	
Control	542	291	--	--	--	237.5	22.5	47.5	276	483	405	924	28	11	37	0.58	6.15
+ fertilizer	580	283	--	--	--	215.0	22.5	72.5	270	478	437	894	28	11	37	1.36	9.00
+ water	877	445	--	--	--	270.0	30.0	107.5	312	640	374	1137	41	7	53	0.80	9.41
+ water and fertilizer	1030	520	--	--	--	307.5	37.5	122.5	371	722	416	1286	42	6	55	2.06	14.24
Moderate seasonal steer grazing																	
Control	527	255	60.9	90	1.00	192.5	37.5	37.5	242	386	392	856	27	15	32	0.60	6.25
+ fertilizer	584	279	72.3	84	1.57	200.0	32.5	62.5	263	400	412	874	28	13	34	1.40	9.67
+ water	900	433	65.7	89	1.05	237.5	42.5	40.0	294	597	356	1101	43	8	52	0.81	8.79
+ water and fertilizer	1054	535	72.8	84	1.61	292.5	52.5	95.0	357	710	396	1276	41	7	55	2.10	15.10

percent), while warm-season perennial production was decreased with ungrazed runs (-9 percent) and slightly increased with the grazed run (+4 percent). This common response is demonstrated by Huffine and Elder (1960) in Oklahoma and by Owensby et al. (1970) in Kansas. The predicted increase in nutrient content of the forage with fertilization was also verified by the field results of Owensby et al. (1970) and by the field and laboratory experiments of Wuenscher and Gerloff (1971). Studies to verify the predicted increase in cattle weight gain per head have not been performed on the Tallgrass Prairie region; however, studies from a Mixed-grass Prairie in North Dakota (Rogler and Lorenz, 1970) show that fertilization generally increased cattle weight gain per head.

Fertilization had little impact on the water loss variables. The model results show, however, that the average water content of the 0 to 45 cm soil layer during the growing season was decreased by fertilization. These figures are consistent with the results of Owensby et al. (1970), who showed that nitrogen fertilization in eastern Kansas reduced soil water content. The data of Owensby et al. and the model results suggest that the decreased soil water content resulted from the increased aboveground production (increased leaf area increases transpiration water loss) associated with fertilization.

These results indicate that adding fertilizer to grazed and ungrazed pastures generally has the positive effect of increasing cattle weight gain (grazed treatment), nutrient content of the forage, and aboveground production. However, the addition of fertilizer also has the negative impact of slightly decreasing warm-season perennial grass production.

IRRIGATION

Water is normally considered the primary driving variable of grassland ecosystems, and numerous early reports describe how the prairie responds to drought (Robertson, 1939; Weaver, 1950b; Weaver and Albertson, 1936). Previous discussion in Chapter 5 has already shown that the presence of grasslands in areas climatically supporting forests is largely attributable to edaphic conditions, which contribute to lower water availability in grassland situations. Toward the west, the transition to Mixed-grass Prairie is largely a function of decreasing precipitation. Within the True Prairie compositional differences at different topographic positions are mainly in response to the moisture regime. In addition, production of above- and belowground biomass, amount of flowers and seed production, and forage quality are characteristics mediated by soil water.

In Chapter 5 the climatic conditions across the True Prairie were analyzed and the microclimatic conditions in some example prairies were described. Rangeland hydrology has been discussed by Branson et al. (1972) and Taylor (1960), but most of the data are from the more western grasslands.

Response of Plants

Interception, stem flow, and throughfall have been measured in many forests, but these measurements are more difficult to make in grasslands. However, Clark (1937; 1940) placed long pans under selected vegetation and measured percent interception (Table 10.27). Clark (1940), Corbett and Crouse (1968), and Haynes (1940) calculated some storage values--amount of water held in the canopy--for several species (Table 10.28). These values provide only rough approximations, since the amount of retention is a function of a number of conditions such as amount of biomass, stage of growth, and intensity and duration of storm.

Little work has been done on the amount of interception by litter, but we know that interception is a function of the duration and intensity of the storm, the amount of litter, the water-holding capacity of the litter, and the evaporation potential during and after the storm. Interception by annual grass litter ranged from 5 to 14 percent. Algae, moss, and lichen mats also intercept moisture (Branson et al., 1972), although Booth (1941a) concluded that algae crusts did not decrease the rate of infiltration and minimized erosion.

The rate of water infiltration is a function of many factors (Branson et al., 1972). In general, infiltration rates are enhanced by low intensity storms, level topography, high soil organic material, coarse soil texture, less surface crusting, lower bulk density, well-differentiated soil structure, and low soil water. The relationship between infiltration rates and fires on the True Prairie will be discussed on page 409. Similarly, the reduction of infiltration with grazing (e.g., Rauzi and Hanson, 1966; Rhoades et al., 1964) has been discussed.

As a rule, conditions that increase infiltration decrease the amount of runoff. In Nebraska, Dragonn (1969) examined the effect of prairie species on runoff and erosion by plantings on small watersheds. Plant cover on the watersheds planted with perennial

TABLE 10.27   Percentage of interception for plant species for heavy and light showers (Clark 1940).

Species	Interception	
	Heavy showers	Light showers
Big bluestem (Andropogon gerardi)	66	97
Prairie cordgrass (Spartina pectinata)	67	80
Porcupine needlegrass (Stipa spartea)	50	
Prairie dropseed (Sporobolus heterolepis)	50	
Lowland forbs	50	66

TABLE 10.28 Storage values for selected species.

Species	Storage (cm)	References
Big bluestem (Andropogon gerardi)	0.23	Clark (1940)
Prairie cordgrass (Spartina pectinata)	0.08	Clark (1940)
Kentucky bluegrass (Poa pratensis)	0.10	Haynes (1940)
Tall wheatgrass (Agropyron elongatum)	0.03	Corbett and Crouse (1968)
Buffalograss (Buchloe dactyloides)	0.17	Corbett and Crouse (1968)

native grasses resulted in approximately 90 percent reduction in surface runoff the second year after planting; after the third year little surface runoff and no measurable soil erosion resulted (Branson et al., 1972).

Conard and Youngman (1965) studied soil water conditions in pastures planted with cool-season and warm-season grasses at Lincoln, Nebraska. The 2 ha pastures were planted with single species and mixtures and all were grazed with yearling Hereford steers each season for six years. The six warm-season pastures included one each of big bluestem, sideoats grama, and switchgrass in pure stands and one each of these grasses mixed with sand lovegrass (Eragrostis trichodes). Six cool-season pastures included two each of smooth brome, intermediate wheatgrass (Agropyron intermedium), and tall wheatgrass (Agropyron elongatum).

Less soil water accumulated under the cool-season than under the warm-season grasses in midspring each year. The amounts of water under warm-season grasses ranged from 25 to 240 percent more than under the cool-season grasses, and, as a consequence, cool season pastures suffered from midsummer drought more often than warm season pastures.

The cool-season pastures averaged only 27 steer days ha^{-1} of grazing and 51 kg ha^{-1} gain in a 90-day grazing season, 21 May to 19 August 1956. Warm-season pastures, with a reserve of about 7.6 cm of available water in the soil at the beginning of the growing season (1 May), produced 60 steer days ha^{-1} of grazing and 125 kg ha^{-1} gain in the 106-day season, 18 June to 2 October. In 1961, cool-season pastures were studied to obtain maximum allowable use of the forage during spring and early summer when forage quality was high. Steers were moved to warm-season pastures for the remainder of the summer, then back to cool-season pastures for a month in the fall. Under this system, yearling steers gained 115 kg per head during the 164-day grazing season. Cool-season pastures produced 71 steer days ha^{-1} of grazing and 240 kg ha^{-1} gain. This system has 12 percent fewer days of grazing but 54 percent more gain per m^2 than were obtained from these pastures in 1958 and 1959, when they were stocked for season-long grazing.

Weaver (1954) studied grasses and the response of each species to the six-year drought that began in 1934. At the end of the drought, he ordered the species according to their abundance, with the least abundant being the species most susceptible to drought (Table 10.29).

In a study in the southern Flint Hills, Owensby et al. (1970) examined the effect of supplemental nitrogen and water. Nitrogen was added yearly at the rate of 4.5 g m^{-2} on 1 May and water, approximately 50 cm yr^{-1} maximum, was added at irregular intervals sufficiently often to maintain adequate soil water during the four-year period. In 1965, herbage yields were increased on water-added plots above yields of control plots, but in 1966 and 1967, no increases were obtained. In 1968, the increase in herbage production on water-added plots (Table 10.30), together with similar nitrogen levels in the plants on water-added and control plots, indicates that in normal years water additions would increase herbage production, but not as much as would be produced by additional nitrogen, assuming an adequate level of water. Failure

TABLE 10.29 Susceptibility of range grasses to drought
conditions (Weaver, 1954). Least abundance
implies most susceptible.

Order of abundance	Species
Great	Western wheatgrass (Agropyron smithii) Sideoats grama (Bouteloua curtipendula) Big bluestem (Andropogon gerardi) Porcupine needlegrass (Stipa spartea)
Considerable	Blue grama (Bouteloua gracilis) Prairie dropseed (Sporobolus heterolepis)
Lesser	Little bluestem (Schizachyrium scoparius) Prairie Junegrass (Koeleria cristata) Tall dropseed (Sporobolus asper) Buffalograss (Buchloe dactyloides) Kentucky bluegrass (Poa pratensis)
Least	Plains muhly (Muhlenbergia cuspidata) Scribner panicum (Panicum scribnerianum) Indiangrass (Sorghastrum nutans)

of water and nitrogen plots to produce much greater herbage yields
than nitrogen plots in years with normal or above-normal
precipitation indicated that the nitrogen fertilization rate may not
have been adequate on plots with supplemental water. This
indication was further corroborated by water-use efficiency, which
was lower on the water plus nitrogen plots than on the nitrogen-only
plots (Table 10.31).

In a greenhouse experiment, Majerus (1975) used clones of blue
grama, western wheatgrass, and little bluestem to determine the soil
water level at which shoots and roots stopped growing (Table 10.32).
When growth terminated in both leaves and roots, more water was in
the lower soil layers than in the surface soil layer. Unlike the
other two species, little bluestem stopped growing when soil water
was higher. In all three species, roots stopped growing at higher
soil water levels and shoots continued to grow under lower soil
water availability.

Responses of Invertebrates

Apparently no information is available on the response of
arthropods to irrigation of the True Prairie. A succession of dry
years is generally followed by increased grasshopper numbers, but
the causative factors are not well understood (Smith, 1954). White
(1976) noted that drought increased total available carbohydrate by

TABLE 10.30   Approximate herbage yields in 1965-1968
from plots with different water and
nitrogen treatments (Owensby et al., 1970).

Treatment	1965	1966	1967	1968
	$(g\ m^{-2})$			
Nitrogen	678	254	605	518
Water	534	254	420	411
Water + nitrogen	751	605	630	641
Control	436	218	420	354

TABLE 10.31   Water-use efficiency for different nitrogen
and water treatments during 1965-1968
(Owensby et al., 1970).

Treatment[a]	1965	1966	1967	1968
	(in g dry matter $m^{-1}\ cm^{-1}$ water)			
Nitrogen	2.8	2.3	3.3	2.9
Water	1.9	1.1	1.8	1.8
Water + nitrogen	2.7	1.7	2.7	2.3
Control	1.8	2.1	3.3	2.0

[a] Nitrogen added at rate of 4.5 $g\ m^{-2}$ on 1 July of each
year; water added at approximately 50 cm yr^{-1} at
irregular intervals.

53 percent and leaf nitrogen by 88 percent in some grasses and
suggested that these increases might enhance insect outbreaks during
drought.

In the Colorado Shortgrass, Pfadt and Dodd (1974) reported that
the three most abundant species in a nonirrigated area did not occur
in an irrigated treatment. Four other species either tolerated or
were favored by irrigation.   A considerable change in species
composition occurred in the fertilized-irrigated plots. The "other

TABLE 10.32   Soil water potential when leaf and root growth terminated
(Majerus, 1975).

Soil depth (cm)	Blue grama (bars)		Western wheatgrass (bars)		Little bluestem (bars)	
	Plant part leaves					
5	<-80.0 b,c[a]	2,3,4[b]	-30.0 a,c	2,3,4	-24.3 a,b	2,3,4
15	-26.0 c	1,3,4	-23.6 c	1,3,4	-9.8 a,b	1,3,4
25	-11.0 b,c	1,2	-16.8 a,c	1,2	-3.4 a,b	1,2
35	-8.4 b,c	1,2	-15.3 a,c	1,2	-3.0 a,b	1,2
	Roots					
5	-16.6 b,c		-7.8 a	2,3,4	-9.2 a	3,4
15	-14.7		-10.6	1	-11.1	3,4
25	-10.7 c		-9.6 c	1	-5.0 a,b	1,2
35	-14.5 c		-13.8 c	1	-5.0 a,b	1,2

[a] Letters indicate the species that have significantly different water
potential at the 5 percent level using a t-test of the replication
means.

[b] Numbers indicate the soil levels within each species that have
significantly different water potential at the 5 percent level
using a t-test of the replication means.

species" category in Table 10.33 represented the greatest density of grasshoppers.

ELM Irrigation Simulations

        The effect of irrigating the True Prairie was simulated by adding water to pastures that are ungrazed and seasonally grazed (1 May to 30 September) with steers at a moderate stocking rate (2.4 ha/AU). Water was added from May through September if the soil water tension in the 10 to 15 cm soil layer was less than -5 bars. At each irrigation event 1.25 cm of water was added by a sprinkler-type irrigation system. The total amount of water added to the grassland averaged 27.5 cm per year for the three-year simulation. Model results (see Table 10.26) for both grazed and ungrazed pastures show that adding water caused the above- and belowground production, cattle weight gain per head (grazed), production of warm-season perennial grasses and cool-season annual grasses, standing-dead and root biomass, transpiration, standing-crop interception, and shoot uptake of nutrients to increase, while bare soil water loss and litter decreased.
        Owensby et al. (1970), who examined the response of the Flint Hills to supplemental water, found that total herbage production increased only slightly with irrigation (<6 percent) and observed that most of the increased production with irrigation occurred during the latter half of the growing season (July-October). The model results suggest that irrigation would produce a much greater increase in aboveground production (+70 percent). A factor possibly contributing to this discrepancy is that the plants at the Osage Site experience more water stress than the plants at the Flint Hills site, since the potential evapotranspiration rates for the growing season at the Osage Site are 9.3 percent greater, while the growing season rainfall amounts are similar (see Chapter 5). However, the model may overestimate the positive impact of added water on aboveground production.
        The model results showing an increase in warm season perennial grass production with irrigation are verified by data from Owensby et al. (1970). The simulated increases in the cool-season annual grass production, however, are contrary to the results of Owensby et al. (1970), who showed that irrigation decreased the basal area cool-season grasses.
        Model results suggest that irrigation generally has the positive effects of increasing primary production, cattle weight gains, and nutrient uptake by the plants, but nitrogen concentration in the plants (root uptake divided by total net production) was decreased. Adding water stimulates net production more than nitrogen uptake by plants, resulting in decreased nitrogen concentration. Owensby et al. (1970) showed that irrigation had no effect on shoot nitrogen concentration on the 15 July clipping date, but the moisture treatment had a considerably lower nitrogen concentration on the 15 August clipping date.
        More information about the impact of water on the grassland can be gained by comparing the yearly values of the output variables with the annual rainfall. Most of the variables are only slightly correlated with the yearly rainfall, which suggests that the effect

TABLE 10.33 Mean density of grasshoppers on Pawnee stress plots as
determined from sampling populations in spring and summer
1974 on five dates three to four weeks apart (Pfadt and Dodd
1974).

Species	Mean density (mean numbers per 100 sq ft)			
	Control	Fertilized	Irrigated	Fertilized and irrigated
Cordillacris crenulata	11	3	0	0
Trachyrhachys kiowa	6	5	0	0
Trachyrhachys aspera	4	2	0	0
Opeia obscura	3	3	5	4
Melanoplus infantilis	2	2	2	1
Melanoplus gladstoni	0.2	2	3	1
Psoloessa delicatula[a]	19	33	1	0
Eritettix simplex[a]	0	1	10	7
Other species[b]	2	2	3	25

[a] Three samplings done in spring and two in fall.

[b] The number of other species included for the control, fertilized,
irrigated, and fertilized + irrigated treatments are 5, 4, 8, and
6, respectively.

of yearly rainfall is confounded with the impact of the management schemes that were simulated and with the variation of other climatic variables. An increase in yearly rainfall caused an increase in above- and belowground production, shoot uptake of nitrogen and phosphorus, transpired water loss, and standing-crop interception for most of the computer runs.

Based on the model of the Osage Site, the increase in primary production with irrigation was much larger (a 62 to 71 percent increase) than with fertilizer (a 7 to 11 percent increase). This increase suggested that water was the primary limiting factor for primary production at the Osage Site. However, these results are contrary to those of Owensby et al. (1970), who showed that a greater (+16 percent) herbage response to nitrogen (+46 percent) than to water occurred. In the model, cattle weight gains increased more with added fertilizer than with irrigation. These weight gains were probably caused because adding fertilizer increased the nutrient content of the forage, while adding water slightly reduced the nutrient content. Irrigation primarily increased the production of the warm-season perennials, while fertilizer could reduce production of warm-season grasses.

Model results indicated that adding fertilizer was the best strategy for increasing cattle weight gain, and adding water would produce the greatest increase in primary production.

The greatest increases in primary production (+90 percent) and cattle weight gain per head (+19 percent) were achieved in the model by adding both fertilizer and water. Interestingly, the combined effect on primary production of adding both fertilizer and water was roughly approximated by the sum of the independent effects of adding fertilizer and water separately. Most of the increased production resulted from increased warm season perennial grass production. These results compare very well with the results of Owensby et al. (1970) in eastern Kansas, where they showed that the combined effect of irrigation and fertilizer was to increase primary production by 80 percent. The results of Smika et al. (1965) in North Dakota also suggest that the effects of nitrogen and water are additive. Therefore, the combined effect of adding both water and fertilizer to grassland was positive, since primary production, cattle weight gain, and production of warm-season perennial grass were greatly increased.

FIRE

Earliest human records show the familiar use of fire, and, in fact, the most ancient human sites have usually been discovered by finding hearths, broken or crazed stones, baked earth, or accumulations of charcoal. The possession of fire enabled man to move into cold climates, to keep predators at a distance from his camp, and to experiment with foods that otherwise would be unpalatable. Fire and smoke became labor-saving devices for overpowering, trapping, and driving game, which ranged in size from small rodents to cattle and elephants. Fire assisted in the process of collecting such fruits as nuts and acorns and, much later, in preparing land for planting. Fire often escaped control and roamed

unchecked until arrested by rain or some other barrier (Sauer, 1950).

The topography, climate, and vegetation of the Great Plains combined to favor widespreading prairie fires. Before their continuity was interrupted by the plow, the plains had virtually no firebreaks other than a few wide rivers. The climate was characterized by intermittent drought when grass browned, dried, and burned at almost any time but especially during periods of dry, windy weather in autumn and early spring (Jackson, 1965). Lightning was a universal natural cause of vegetation fires (Komarek, 1964). Trail-driving cowboys especially dreaded electrical storms because lightning-started fires frequently caused stampedes.

Prairie fires were referred to frequently throughout the records of early plains travelers and explorers, which indicated that fires were commonplace. In addition to lightning, the nomadic plains Indians used fire against enemies and to control bison and wild mustangs. Essentially no attempt was made to control such fires (Jackson, 1965).

Undoubtedly, the number of fires increased as more people moved onto the plains, some as settlers but others as restless drifters who had little proprietary interest in the land. In addition, the hostility between ranchers using public domain and homesteaders probably resulted in some prairie fires (Jackson, 1965). Fires frequently moved almost as fast as the winds driving them, and records exist of fires traveling 65 km in 4 h (Scott, 1960) and nearly 500 km in about 24 h (Curtis, 1959).

White settlers eventually built roads that served as unintentional firebreaks, and the settlers actively controlled fires to prevent loss of crops, livestock, and buildings. As noted previously, the cessation of these fires led to rapid forest invasion of grasslands along the prairie-forest border (Ahlgren, 1960; Anderson, 1973; Bragg and Hulbert, 1976; Daubenmire, 1968; Kucera et al., 1963).

Prairie fires still occur on the True Prairie and the following account from the 22 May 1977 "Orbit Magazine" in the Daily Oklahoman is of a fire that occurred on 17 February 1976 and passed only a few miles south of the Osage Site:

> It was about 11 A.M. when I heard the first report come over the CB. It came from Fairfax. Mary Hazelbaker was calling for help. She said, "They've got a prairie fire going out on Mule Christensen's place and it's getting away from them. They need some spray rigs out at Lost Man Creek right now."
>
> John Briggs, Osage County commissioner, tried to explain the sequence of events. "Then Mary came back on the air immediately. She said, 'A paper cup or something blew across the road and the fire just jumped Highway 60. Mule and John Sherrill need as much help as they can get. Right now.'"
>
> John continued, "We had a 35 to 45 mile-an-hour south wind that day and we knew if we couldn't stop the fire no telling how far north it could go." He had studied the landscape and decided that if he could get enough help

soon enough the fire could be burned out in the narrow
strip between the Shidler lake and the ravine.

"Later, evidence showed that we could have stopped it
there, too, but no one could get there fast enough. As
the fire hit a pasture the whole 160 acres seemed to
explode at one time. In what seemed like just a few
minutes the fire jumped the lake and the ravine and went
right on."

As the sky was blackened with a fog of smoke so thick
the sun appeared only as a flaming red ball, the first of
the 300 people who would fight the fire the rest of that
smoke-filled day and on into the destructive night, began
to arrive. They came in converted Army surplus
firetrucks, oilfield tank trucks, ranch spray rigs,
pickups, Jeeps and Cadillacs. They were volunteer and
county firefighters, ranchers, cowboys, oilfield workers,
school boys and concerned townspeople. They were armed
with CBs, wet burlap feed sacks and lots of guts.

It was a loosely organized battle. CBs held the
whole thing together. Radios constantly cracked out
instructions, questions, determined locations, conditions.

The fire front covered 6 to 8 miles. Firefighting
units were widely spread out but the action along the
lines of fire was the same. As the high, intense flames
shot over each ridge the men formed lines. Water-soaked
gunny sacks were quickly passed hand to hand. Then the
spray trucks rolled while the firefighters trotted beside
them and tried to keep pace.

Fires lapped at their feet. Flames never slowed down
as they burned to a road and with roaring, fiery leaps
spread into other sections of virgin, tinder-dry grass.
Vehicles carrying precious water were threatened. Tank
trucks, lines of them red-balled a convoy to the critical
firelines.

Now the calls came cracking, "Turn the cattle loose.
If you can't take the time to cut the wires, ram 'em down
with the trucks. But get the cattle out!"

Cowboys in pickup trucks sped to pasture fence rows.
Fences were flattened and bawling cattle herds took off in
mad, frightened disarray. Here and there a cow mother
remembered her baby and slowed down and, bawling, tried to
coax the little one to keep up but the press of the others
forced her on.

Vehicles pressed into immediate service had arrived
at the fire in the condition the SOS had found them in,
many with only marginal gasoline in their tanks. But
again their needs were met. A large gasoline truck owned
by a distributor arrived with gasoline enough to fill
every tank. And once again the firefighters and footmen
raced on ahead of the flames, trying with each mile to
form a barrier against the flaming fury.

Now it was almost midnight and the ones that were on
the fire lines were very tired. They had been raising
those wet gunny sacks over their heads and slamming them

down against the ground with all their strength for more than 12 hours.

And then the fire was over. Burned out.

Acres and acres and acres were blackened. Forty-five square miles lay desolate. White-faced, soot-splattered cattle wandered over the blackened, landscape and here and there carcasses of baby calves lay black and charred. Miles and miles of barbed wire fences lay on the ground. Fence posts, their bottom halves now burned stumps, hung limply from the wires still erect. A few corrals were charred. Some buildings were burned. Two men had been burned seriously enough to require extended hospitalization and a young man had an injury from a pickup mishap.

The Osage County extension agent, Harold Murnan, estimated the damage, conservatively, at over $5 million.

Daubenmire (1968) has written an excellent review of the ecology of fire in grasslands. Early in the history of this country particularly strong feelings abounded about forest fires but not about grassland fires--except that grassland fires were probably not beneficial. Then, by about 1940, the fact that grassland fires were not always detrimental became evident. Although trees were usually killed by fire, grasses had buds and mature seeds that lay just at or below the soil surface, where fire temperatures remained relatively low. Also, during the time fires occurred, most of the exposed plant material was dead, and the grasses redeveloped normal shoots in a short period of time. The changing opinion about the desirability of fires in both grasslands and forests persists today.

Characteristics of a Grassland Fire

The grass fire accelerates quickly to a maximum speed and typically has a rather narrow flame zone. Initially, surface wind velocity, temperature, and relative humidity determine the direction and speed of the fire, but later, as the fire builds, a strong convection column becomes the dominant force. Then, as the burning front expands, the direction and velocity of the wind again become important (Daubenmire, 1968; McArthur, 1966).

In general, large amounts of dry vegetation carry a hotter fire, and wet soil acts as a heat sink but facilitates heat conduction through the soil. Fires travel faster uphill and with the wind, but back-fires, because of increased oxygen, move faster into a strong wind than into a light wind. The temperature of grassland fires may reach 300°C or so in the canopy and at the soil surface. However, the duration of this intense heat is usually very short--a matter of minutes. As a result, perennial grasses usually experience minimum damage from wildfires (Daubenmire, 1968).

After a spring burning, midafternoon soil temperatures during the growing season in a Missouri Tallgrass Prairie were found to be 2.2 to 9.8°C higher than on unburned areas. This difference decreases as plant cover develops (Kucera and Ehrenreich, 1962). In Kansas the maximum annual temperature at a depth of 2.5 cm was elevated 6.7°C and the minimum was raised 1.1°C (Hensel, 1923).

Although Kucera and Ehrenreich reported that minimum soil temperatures were not changed by fire in Iowa, Ehrenreich and Aikman (1963) found that the maximum air temperature at an elevation of 2.5 cm above burned grassland was raised 5.6°C and the minimum temperatures were generally lower at night (Daubenmire, 1968). Peet et al. (1975) also found a 2°C higher temperature on a burned grassland in the University of Wisconsin Arboretum.

Burning, by removing plant cover, exposes soil surface to increased water and wind erosion. There may also be an increase in water loss after burning, especially in the spring, when there is no mulch cover and the soil surface temperature is higher. Later in the growing season, when a large biomass has been stimulated by burning, water loss by transpiration may also reduce soil water. Regardless of the season of burning, regular burning of bluestem-Indiangrass grasslands in Kansas reduced the average water content of the soil to a depth greater than 1 m (Anderson 1964, 1965; Hanks and Anderson 1957; McMurphy and Anderson 1965); the earlier the date of burning in spring and the deeper the soil layer, the greater the reduction in soil water (Table 10.34). Burning also reduced the rate of infiltration (Bieber and Anderson 1961, Hanks and Anderson 1957, McMurphy and Anderson 1965). Anderson (1964) showed that it lowered the field capacity and wilting coefficient on uplands but not on a lowland site, where there was initially more clay in the soil.

Aldous ([1934], as noted by Daubenmire [1968]) reported that the humus content had not declined in the upper 15 cm of the soil, regardless of the season of burning, as a result of six years of annual burning of little bluestem in eastern Kansas. However, in a continuation of the project (Anderson, 1964), annual burning in March reduced the humus content of the upper layer of the soil from 1.77 to 1.47 percent. Because burning volatilizes nitrogen and sulfur and changes other nutrients to water-soluble simple salts, we might expect this ash to act as a fertilizer. Little evidence supports this theory. Only one study (Curtis and Partch, 1950) showed a positive relationship between the ash increment and the number of inflorescences of big bluestem. Old (1969) found no increase in growth on an Illinois prairie where ash was added to unburned or cut and cleared areas. A grass fire in Iowa (Ehrenreich and Aikman, 1963) increased the pH of the upper 18 mm of the soil from 5.8 to 6.1, with the pH increase proportional to the amount of litter. However, within a year, the pH had returned to the former levels, presumably because of the bases that were leached from the ash (Daubenmire, 1968). In spite of the pH change, exchangeable potassium showed no measurable increase.

Considerable discussion has revolved around the relationship between nitrogen and grassland fires (Daubenmire, 1968). Aldous (1934), in a study of the little bluestem of the Flint Hills, found no reduction of nitrogen after six years of annual burning. However, Elwell et al. (1941) calculated that burning ungrazed grassland in eastern Oklahoma volatilized 12.2 kg nitrogen per acre; this calculation assumed that litter was burned to a white ash. In some burning experiments, legumes subsequently increased which either might compensate for the loss of nitrogen or enhance the nitrogen level. Old (1969) showed a higher foliar content of nitrogen and phosphorus on burned than on unburned Illinois prairie,

TABLE 10.34 Average soil water at various depths under burning treatments in the Flint Hills (Anderson, 1965).

| Time of burning | Soil water (cm)[a] | | | | | Average |
	0-30 cm	30-60 cm	60-90 cm	90-120 cm	120-150 cm	
Winter	8.93	8.80	8.45	8.08	7.13	8.27
Early spring	8.93	9.00	8.68	9.13	8.98	8.95
Midspring	9.00	9.05	8.72	8.78	8.58	8.83
Late spring	9.32	9.38	9.00	8.90	8.90	9.10
Control (unburned)	9.30	9.55	9.20	9.38	9.10	9.30
Average	9.10	9.15	8.81	8.85	8.54	8.89

a Tabular values are expressed as cm per 30 cm and are four-year averages for 1960-1963.

which might suggest greater microbial activity. Kucera and Ehrenreich (1962) suggested that burning might reduce the C:N ratio in the soil, making nitrogen more available. Evidence from other parts of the world indicates that burning does not reduce the C:N ratio (Daubenmire, 1968).

Response of Plants

Perennial plants are more susceptible to damage by fire when leaf expansion and growth have occurred to the point where food has been translocated from underground organs. Robocker and Miller (1955), on examining a mixed planting of grass species in Wisconsin, found that, in general, only those species that had already started growth were damaged. Curtis and Partch (1950) at the University of Wisconsin Arboretum showed that both annual weeds and certain prairie plants were able to maintain themselves and spread into a bluegrass sod (Kentucky and Canada bluegrass) when the sod was burned repeatedly. This maintenance happened with either spring or autumn burns, on both annual and biennial burning schedules. Bluegrasses are cool-season grasses that grow early in the spring and late in the fall, so burning at these times selectively harms bluegrasses and gives a competitive advantage to warm-season species that are mostly dormant at these times. Oldfield threeawn (Aristida oligantha) has been effectively controlled by fall burning in the Flint Hills (Owensby and Launchbaugh, 1977).

Anderson et al. (1970) summarized a number of studies on the Flint Hills lasting over seventeen years and concluded that cool-season species were reduced and warm season species were favored by spring burning. The cool-season Kentucky bluegrass might increase under heavy stocking, but it was almost lost from the pastures under all burning treatments. Table 10.35 shows the response of various grasses to three spring burning regimes. The decreasers--big bluestem, little bluestem, Indiangrass, and switchgrass--are part of the native flora but are palatable and selectively grazed; therefore, decreasers lessen in importance with increased grazing pressure. The increasers--sideoats grama, blue and hairy grama, buffalograss, and Kentucky bluegrass--are now part of the climax prairie but are less sought by livestock, less competitive, and less productive than the decreasers. As a result, increasers assume a greater importance under heavy grazing pressure, but under continued heavy grazing they too may decrease. Averages from loamy upland, limestone break, and claypan soils are given in Table 10.35, which indicates marked differences in the percent composition on these three types of sites. For example, big and little bluestem make up 45 to 55 percent of the total biomass on the first two sites but less than 10 on the claypan site. Although these differences are partly masked by the summary data, after ten years of burning the decreasers had increased compared to the unburned treatment, and the late spring burn showed the greatest increase. All the increasers except Kentucky bluegrass, which is really an exotic invader, also increased under the burning regimes.

Hensel (1923) found that early spring burning tended to increase stands of little bluestem, while big bluestem and Indiangrass tended to increase under late spring burning. Hadley

TABLE 10.35   Percent composition of decreaser and increaser grass species
in the Kansas prairie under different burning treatments
(Anderson et al., 1970). The values are ten-year averages
over three sites: loamy upland, limestone breaks, and
claypans.

Species	Treatments			
	Unburned	Early spring	Midspring	Late spring

Decreasers				
Big bluestem (Andropogon gerardi)	16.0	23.0	23.6	28.4
Little bluestem (Schizachyrium scoparius)	13.0	9.4	11.5	13.9
Indiangrass (Sorghastrum nutans)	3.7	4.0	5.8	4.5
Switchgrass (Panicum virgatum)	0.7	0.8	0.6	1.5
Increasers				
Sideoats grama (Bouteloua curtipendula)	8.4	11.5	8.6	9.5
Blue grama, hairy grama (Bouteloua gracilis, B. hirsuta)	3.1	19.3	15.7	12.6
Buffalograss (Buchloe dactyloides)	11.6	8.3	13.1	11.7
Kentucky bluegrass (Poa pratensis)	10.8	0.4	0.5	0.4

and Kieckhefer (1963) also showed that growth of the two bluestems and Indiangrass in an Illinois prairie increased after burning. Kucera (1970) and Kucera and Koelling (1964) found an increase in these species after burning but did not find an increase in legumes. Anderson and Schwegman (1974), working in a burned seral stage in Illinois dominated by Japanese honeysuckle (<u>Lonicera japonica</u>) and trees but containing the three grass species, showed an increase in showy partridgepea (<u>Cassia fasciculata</u>), sensitive senna (<u>C. nictitans</u>), and sidebeak pencilflower (<u>Stylosanthes biflora</u>). McIlvain and Armstrong (1966) found that little bluestem decreased and sand bluestem increased with burning on sandy soils in western Oklahoma.

Hulbert (1969) found that big bluestem showed an increase in yield after removal of litter by either burning or clipping. He indicated that the major short-term effect of fire was caused by the removal of litter rather than by direct heat or by fire-induced nutrient changes. In Nebraska, Weaver and Rowland (1952) reported a total yield reduction of 26 to 53 percent of big bluestem and switchgrass from mulched stands compared with unmulched stands. Hulbert further stated that because soil water was higher in the control plots than on the burned plots, fire resulted in a reduced water supply for seedlings and caused injury or death to young trees and shrubs. In a subsequent study, Bragg and Hulbert (1976) related burning to presettlement amounts of woody vegetation and showed that soil type and topography affected the woody plant invasion subsequent to the cessation of burning.

Graves and McMurphy (1969) in central Oklahoma found western yarrow (<u>Achillea lanulosa</u>), western ragweed (<u>Ambrosia psilostachya</u>), Louisiana sagewort (<u>Artemesia ludoviciana</u>), blackeyed Susan (<u>Rudbeckia hirta</u>), prairie fleabane, and oldfield threeawn to be partly controlled by burning. On the other hand, when the soil was very dry in the Tallgrass Prairie of the Wichita Mountains in southwestern Oklahoma, an August wildfire caused an increase in western ragweed and Canada horseweed (<u>Conyza canadensis</u>) (Penfound, 1968).

Peet et al. (1975) studied the rates of photosynthesis and respiration in big bluestem plants brought to the laboratory from burned and unburned sites at the University of Wisconsin Arboretum. They used gas exchange techniques and found that the temperature dependence of net photosynthesis of plants from the two sites was similar, except that plants from the burned site had a slightly higher maximum carbon dioxide exchange rate: 21 mg CO_2 h^{-1} g^{-1} dry weight as compared to 18 mg CO_2 h^{-1} g^{-1} dry weight for plants from the unburned site. The plants, which were collected in late May, had temperature optima for photosynthesis between 25 and 30°C. Dark respiration responses to temperature for burned and unburned site plants were similar, with a maximum rate of 7 mg CO_2 h^{-1} g^{-1} dry weight at 40°C, the highest temperature of the experiment.

These gas exchange rates were used to predict the net photosynthesis in the field. On 17 May, the predicted net photosynthesis on the unburned and burned sites was 14 and 20 mg CO_2 h^{-1} g^{-1} dry weight, respectively. Year-end biomass production of big bluestem was 531 g m^{-2} for the burned site and 173 g m^{-2} for the unburned.

Several studies have shown that burning increases the amount of flowering. Curtis and Partch (1950) found that litter removal

increased sixfold the number of flowering stems of big bluestem and that burning with addition of ash produced only slightly greater effects than litter removal alone. In 1967, on the Trelease Grassland in Illinois, an early spring burn caused a ten-fold increase in flowering rates over the rates recorded in the unburned control (Old, 1969). Clipping and removal of vegetation also increased flowering, but the increase was only half as great as that caused by burning. The application of litter or mulch decreased the flowering rate, and the response was proportional to the thickness of mulch and the length of time the mulch was in place.

As shown in Table 10.36, Ehrenreich and Aikman (1963) found that early spring burning in the Iowa Hayden Prairie increased flowering of most species. The exceptions, Kentucky bluegrass and sedges, are early blooming, cool-season grasses. Hadley and Kieckhefer (1963), working earlier on the Trelease Grassland in Illinois, found that in 1961 and 1962, the number of flowering stalks of big bluestem and Indiangrass was greatest where fires had occurred during three previous years--1952, 1959, and 1961. Similarly, more flowering stalks were found in the treatment that had been burned two previous years (1952 and 1959) than in the unburned treatment (Table 10.37).

Daubenmire (1968) showed the effect of fire on subsequent numbers of inflorescences for a number of species. Of the twenty-nine species listed, only Indiangrass showed both an increase (Kucera and Ehrenreich, 1962) and a decrease (Dix and Butler, 1954). In many communities, seed production is quite heavy the first year after a fire (Biswell and Lemon, 1943). However, Dix and Butler (1954) found that inflorescence production in sideoats grama in Wisconsin remained constant, while vegetative coverage increased. In prairie dropseed (Sporobolus heterolepis), both inflorescences and vegetative cover increased. Usually, as noted by Daubenmire (1968), the stimulus for seed production is short, lasting two years in Iowa (Ehrenreich and Aikman, 1963). Similar results have been found in Wisconsin by Dix and Butler (1954) and in Illinois by Hadley and Kieckhefer (1963). Seeds harvested from recently burned grassland in Iowa (Ehrenreich and Aikman, 1957) were found to have a higher percentage of germination than seeds on unburned plots. In Wisconsin, big bluestem seedlings were very abundant after fire, and common ragweed (Ambrosia artemisifolia) was also stimulated (Curtis and Partch, 1948; 1950).

A winter burn in Oklahoma increased the areal cover of most tallgrasses (Table 10.38), although big bluestem decreased (Kelting, 1957). Although Aldous (1934) reported lower minimum soil temperatures on burned plots, Kelting did not agree with this finding. Kelting did find higher maximum soil temperatures under the usual conditions that we have been discussing, and he attributed the vegetation response to this temperature differential.

Total Plant Biomass Production

Plants. The impact of rangeland fires on total production has long been of interest, especially in terms of forage production. In Illinois, Hadley and Kieckhefer (1963) found that aboveground primary production increased markedly the first year after a burn,

TABLE 10.36   Frequency of major species' flower stalks on an Iowa prairie after a single burn (Ehrenreich and Aikman 1963; copyright 1965 by The Ecological Society of America).

Species	Frequency	
	Unburned	Burned (1956)
Big bluestem (Andropogon gerardi)	65	100
Little bluestem (Schizachyrium scoparius)	50	75
Indiangrass (Sorghastrum nutans)	55	65
Prairie dropseed (Sporobolus heterolepis)	5	80
Canada wildrye (Elymus canadensis)	15	25
Kentucky bluegrass (Poa pratensis)	100	80
Timothy (Phleum pratense)	25	40
Redtop (Agrostis alba)	25	30
Sedges (Carex spp.)	20	0
Green muhly (Muhlenbergia racemosa)	0	30
Slender wheatgrass (Agropyron trachycaulum)	15	15

and some effect, though decreased, existed two years after the burn (Table 10.39). Aikman (1955) and Ehrenreich (1959) in Iowa and Hulbert (1969) in Kansas obtained similar results. Old (1969) in Illinois and Ehrenreich and Aikman (1963) in Iowa found small increases in production after burning. Koelling and Kucera (1965) and Kucera and Ehrenreich (1962), working on the Tucker Prairie in Missouri, found an increase after burning, though the response was diminished by the second year. Long-term experiments on the Donaldson Pastures in the Kansas Flint Hills have indicated small

TABLE 10.37 Number of big bluestem and Indiangrass treatments (Hadley and Kieckhefer, 1963; copyright 1963 by The Ecological Society of America). Values are number per m². Burning occurred in early spring.

Species	Sample date	Unburned	Burned	
			1952, 1959	1952, 1959, 1961
Big bluestem (Andropogon gerardi)	1961	10.4	15.5	117.7
	1962	27.8	29.5	52.9
Indiangrass (Sorghastrum nutans)	1961	17.3	28.4	132.0
	1962	32.1	38.4	80.8

416

TABLE 10.38   Areal cover on 24 June 1952 of some tallgrasses in Oklahoma after a winter burn (Kelting, 1957; copyright 1957 by The Ecological Society of America).

Species	Area cover (percent)	
	Control	Burn
Little bluestem (Schizachyrium scoparius)	11.9	15.9
Indiangrass (Sorghastrum nutans)	3.6	4.7
Switchgrass (Panicum virgatum)	4.0	6.5
Big bluestem (Andropogon gerardi)	4.3	0.6
Fall witchgrass (Leptoloma cognatum)	3.2	3.3
Scribner panicum (Panicum scribnerianum)	0.2	1.4

decreases (Anderson et al., 1970) or no significant decreases (Owensby and Anderson, 1967) in forage yield after late spring burning on grazed grassland (Table 10.40). Aldous (1934; 1935), Anderson (1965), and Hensel (1923) reported a decrease in forage production on ungrazed grasslands. McMurphy and Anderson (1963) recorded reduced production in sixteen of twenty-six years on ungrazed range and these reductions coincided with years of low precipitation. Perennial grasses increased and perennial forbs decreased (Table 10.41). Eight years of annual burning in central Oklahoma reduced forage production by 53 percent, but Gay and Dwyer (1965) showed no change after a single burning. Results of aboveground biomass production in response to burning in the True Prairie are summarized in Table 10.42.

In general, apparently wherever water is adequate, burning increases production for the subsequent two or three years. However, this postburn increase does not occur when grasslands are burned annually or receive minimal levels of precipitation.

After burning in an Illinois grassland, the average root biomass was 1669 and 1862 g m^{-2} for burned and unburned plots, respectively (Old, 1969). Table 10.43 shows an increase in root biomass after burning, but like aboveground biomass the trend indicates that these levels will revert to the levels of the unburned controls after a sufficient period of nonburning (Hadley and Kieckhefer, 1963).

TABLE 10.39   Primary productivity of Illinois big bluestem and
              Indiangrass community types under three burning
              regimes (Hadley and Kieckhefer, 1963; copyright
              1963 by The Ecological Society of America).

Community type	Years of burn	Year-end biomass (g m^{-2})	
		1961	1962
Big bluestem (Andropogon gerardi)	1952, 1959	364	359
	1952, 1959, 1961	1321	591
	1952, 1959, 1961, 1962	--	1360
	Unburned	302	362
Indiangrass (Sorghastrum nutans)	1952, 1959	502	531
	1952, 1959, 1961	1474	633
	1952, 1959, 1961, 1962	--	1536
	Unburned	489	476

Old (1969) measured the nutrient contents of roots in the
burned and unburned plots (Table 10.44). Nitrogen content from the
burned treatments was significantly lower in May than from the
control plot, perhaps because of root nitrogen depletion by rapidly
growing aerial vegetation. Also, in May, calcium was considerably
higher on the burned areas, perhaps from released calcium from
litter.

Surprisingly, although numerous comments exist in literature
regarding increased palatability of recently burned grasslands
(Allen et al., 1976; Duvall and Whitaker, 1964; Smith and Young,
1959), we have scant information on cattle weight gains (Woolfolk et
al., 1973). Anderson et al. (1970) found that mid- and late spring
burning significantly increased steer gains when compared with
nonburning and that the primary increase occurred early in the
growing season. Late spring burning, preferable to either early or
midspring burning, increased gains in ten of the fourteen summer
seasons. In late summer, unburned pastures permitted a higher gain
than any of the three burned treatments. Studies in the Flint Hills
showed that burning increased crude protein from 0.5 to 2.0 and
slightly increased phosphorus content and protein, crude fiber, and
NFE (nitrogen-free extract) digestibility (Allen et al., 1976; Smith
and Young, 1959; Smith et al., 1960; Vallentine, 1971).

Burning of a degraded Tallgrass Prairie in Oklahoma improved
forage production and botanical composition, but the addition of

TABLE 10.40   Forage grasses and weed forbs yields on grazed grass-
land for three different spring burning times in the
Kansas Flint Hills. Values are an eight-year average
(Anderson et al., 1970).

Time of burning	Loamy uplands		Limestone breaks	
	Forage	Weeds	Forage	Weeds
	$(g\ m^{-2})$		$(g\ m^{-2})$	
Early spring	392	38	237	48
Midspring	363	32	274	30
Late spring	396	18	301	12
Unburned	439	34	387	38

TABLE 10.41   Percentage composition of the Kansas prairie under
three grazing treatments (Anderson et al., 1970).
Values are ten-year averages over three range sites;
loamy uplands, limestone breaks, and claypans.

Ecological category	Unburned	Burned treatments		
		Early spring	Midspring	Late spring
Perennial grasses	74.5	77.2	85.5	86.2
Annual grasses	7.6	2.1	1.6	2.6
Perennial forbs	8.3	10.0	6.0	4.7
Shrubs	0.5	1.2	0.9	0.9

fertilizer did not improve the rate of species change in terms of
forage quality (Graves and McMurphy, 1969). Nitrogen fertilizer
produced an increase of 3.6 kg of forage for each 100 g applied to
the burned plots. This increase is greater than that reported by
Gay and Dwyer (1965). Graves and McMurphy (1969) noted that
phosphorus produced a significant forage yield increase during one
year of the study, but potassium was not effective in changing
forage yield or species composition. They also stated that the
prospect of range fertilization should not be considered if the
range is characterized by low-quality vegetation. The unpalatable

TABLE 10.42   Summary of biomass production in the True Prairie under burned and unburned conditions.

Location	Year-end standing crop (g m^{-2})		References
	Unburned	Burned	
Donaldson Pastures Manhattan, Kansas[a]	473	380	Anderson et al. (1970)
Rock Spring 4-H Ranch Junction City, Kansas	180	340	Hulbert (1969)
Trelease Prairie Urbana, Illinois[b]	395	1397	Hadley and Kieckhefer (1963)
Trelease Prairie Urbana, Illinois	634	756	Old (1969)
Hayden Prairie Howard County, Iowa[c]	364	455	Ehrenreich (1959)
Hayden Prairie Howard County, Iowa[d]	369	447	Ehrenreich and Aikman (1963)

Note: Not all conditions are comparable, so primary references should be consulted for details.

[a] Eight-year average of early, mid- and late spring burns on upland sites.

[b] Average of big bluestem and Indiangrass types on a site that had been burned in the springs of 1952, 1959, and 1961 and sampled in 1961.

[c] Burned previous two springs.

[d] Burned value is average of plots burned in 1954, 1955, and 1954/1955.

TABLE 10.43   Belowground biomass values for Illinois big bluestem
and Indiangrass communities under different burn
conditions (Hadley and Kieckhefer, 1963; copyright
1963 by The Ecological Society of America).

Community type	Years of burn	Average biomass (g m^{-2} to 35 cm)	
		1961	1962
Big bluestem (Andropogon gerardi)	1952, 1959	1100	1123
	1952, 1959, 1961	1212	1262
	1952, 1959, 1962	--	1310
	Unburned	887	900
Indiangrass (Sorghastrum nutans)	1952, 1959	924	965
	1952, 1959, 1961	950	981
	1952, 1959, 1961, 1962	--	1029
	Unburned	782	787

species were capable of responding to fertilization, and no
measurable increase in speed of succession was evident.

Fire has sometimes been used in the prairie-forest border
region to control brush (Darrow and McCully, 1959; Ehrenreich and
Crosby, 1960; Vogl, 1965) and to remove dead wood after trees and
shrubs have been sprayed with an herbicide. Elwell et al. (1970)
examined the effect of burning and the application of 2,4,5-T
(2,4,5-trichlorophenoxyacetic acid) on a post oak (Quercus stellata)
and a blackjack oak (Quercus marilandica) rangeland in central
Oklahoma. They found that a significant increase in the soil water
occurred in plots treated with 2,4,5-T to control brush. Burning
alone or in combination with 2,4,5-T resulted in a slight increase
in grasses, but decreaser species increased markedly in plots
treated only with 2,4,5-T. The combined effect of herbicides plus
fire resulted in an immediate increase in annual weeds, but these
weeds were controlled by perennial grasses in succeeding years.
Plots that were burned several years after brush control by 2,4,5-T
had many tree branches and trunks on the ground; thus, a long, hot
fire resulted. As a result, some perennial grass vegetation was
killed, but weeds did not invade the area. Where 2,4,5-T was used,
yield of herbaceous vegetation increased and provided more fuel for
fire (Elwell et al., 1970). Without herbicides, little fuel was
available for an effective burn.

TABLE 10.44   Nutrient content of roots collected from an Illinois
grassland (Old, 1969; copyright 1969 by The Ecological
Society of America).

Sample date	Nutrient content (percent of oven-dry weight)				
	Nitrogen	Phosphorus	Potassium	Calcium	Magnesium
Control					
May	0.73	0.13	0.13	0.37	0.16
July	0.58	0.14	0.11	0.31	0.12
August	0.49	0.19	0.16	0.30	0.10
September	0.63	0.11	0.16	0.37	0.12
Average	0.61	0.14	0.14	0.34	0.13
Burned treatment					
May	0.65	0.17	0.13	0.53	0.14
July	0.62	0.30	0.12	0.38	0.11
August	0.61	0.17	0.16	0.34	0.12
September	0.48	0.11	0.13	0.33	0.12
Average	0.59	0.18	0.14	0.41	0.12

Burning alone without herbicide control of larger trees
contributed little to woody plant control, although coralberry
(Symphoricarpos orbiculatus) was harmed. In the Wichita Mountains
in southwestern Oklahoma, Penfound (1968) observed that 14 percent
of the post oak and 70 percent of the blackjack oak trees sprouted,
but none of the red cedar (Juniperus virginiana) produced sprouts.

Response of Invertebrates

Knutson and Campbell (1976) reviewed the effects of burning
rangeland on grasshopper populations in the True Prairie in Kansas.
Grasshopper nymphs hatched earlier following burning, and
populations, except of the species that overwinter as nymphs, were
not reduced. Newly hatched nymphs emerge about three weeks earlier
in early-spring burned areas and somewhat later in midspring burned

areas. They feed on early exposed, highly succulent, and probably more highly nutritious new growth, and they cause little noticeable damage since nymphs are small and new growth is abundant. When favorite hosts were damaged by fire, the nymphs apparently fed on other unharmed vegetation.

This headstart in the spring doubtlessly increased grasshopper numbers in successive years by progressively increasing populations and by reducing their vulnerability to predators like quail (Colinus virginianus), which feed heavily on young grasshoppers. In spring, young quail congregate in more protected, unburned areas, where fewer grasshoppers are present. Other predators, parasites, and diseases may also have been adversely affected by burning.

The largest collections of most grasshopper species were made in areas that received an early-spring burn, considerably fewer species were found in areas of midspring burn and even fewer in areas of late-spring burn. In the late-spring burned areas, Phoetaliotes nebrascensis and Orphulella speciosa were among the species collected in largest numbers. Eritettix simplex and Paradalophora haldemanii overwinter as nymphs and only 22 percent of the nymphs and 1.5 percent of the adults of E. simplex and none of P. haldemanii survived burning.

Nagel (1973), using sweep net collections from True Prairie in Kansas, reported that a burned site produced significantly more total numbers and biomass of arthropods. Leafhoppers and tettigoniid grasshoppers accounted for most of the increase in numbers; acridid and tettigoniid grasshoppers produced most of the increase in standing crop biomass. Herbivore biomass on the burned site was significantly greater than on the unburned site, but nonherbivore biomass was not significantly different.

Bertwell and Blocker (1975) used a unit area trapping method in the Kansas True Prairie and found higher Coleoptera biomass in unburned areas than in burned. These results contradicted Nagel's results in the same range area, and the unit area sampling method was probably responsible for the differences. Sweeping samples only the upper vegetation, but the unit area method samples the total population.

Cancelado and Yonke (1970) used Malaise traps and sweep samples to study effects of spring burning on populations of Hemiptera and Homoptera in True Prairie in Missouri. In trap samples, they found significantly greater numbers of leafhoppers and total Hemiptera on the burned area. Sweeping samples collected on the burned area produced nearly as many Miridae (grass bugs) as samples from ten other sites.

Mason (1973) studied leafhopper diversity in a Kansas tallgrass prairie using a unit area trapping method and consistently found fewer leafhoppers in burned treatments. These results contradicted the results of Cancelado and Yonke and Nagel but may be explained by the differences in sampling methods used.

In southwestern Wisconsin, predatory spiders and soil arthropods resistant to dessication were found to increase immediately following a prairie fire (Reichert and Reeder, 1972). However, arthropods, which were unable to escape by getting under the soil surface or other protective objects such as stones, frequently were killed by the fire.

In summary, burning of the True Prairie results in an increase in grasshopper numbers and biomass. Results on such groups as leafhoppers and beetles are reportedly different. Trapping methods showed a decrease in numbers in burned areas, but where sweep net samples were evaluated, burned areas were characterized by higher numbers. Relatively few studies have been made on the response of arthropods (Lussenhop, 1976; Reichert and Reeder, 1972), and no generalities have yet emerged.

Response of Bird Populations

Despite the importance of fire in the True Prairie and its frequent occurrence before the last century, few studies of fire effects on prairie bird populations have gathered anything beyond incidental observations. However, in a study on a True Prairie in Minnesota, Tester and Marshall (1961) used controlled burning as one of several experimental manipulations and documented the responses of several breeding bird species during the year of the burn and the following year. Bobolinks (Dolichonyx oryzivorus) were entirely absent from the burned plots during the year of the burn, as were LeConte's sparrows (Ammospiza leconteii. The following year, bobolink breeding densities on the burned plots compared closely with the densities on nearby control plots. LeConte's sparrows were also present on the burned plots the year after the burn but were absent from the controls. Tester and Marshall attributed these responses to the importance of ground litter to both of these species--the burn-removed litter cover, which by the next year had again begun to accumulate, rendered the habitat more suitable to these species. In addition, sparrows may respond to areas having a substantial growth of new grass, and the prevalence of such vegetation in burned plots in the year following the burn may have made this habitat more suitable than the control plot habitats. A third species, the savannah sparrow (Passerculus sandwichensis), decreased in density on burned plots in the year of burning and further decreased in the following year. Tester and Marshall suggested that these responses might also be tied to litter depth and coverage, because the sparrows apparently prefer areas with more than two years' accumulation of litter.

The results of this study likely apply broadly throughout the True Prairie. Given the response of birds to habitat (especially vegetation) structure, it is little wonder that a stress such as fire that so markedly affects vegetation structure should have rather major effects on breeding bird populations and communities. In many cases, however, the prairie returns rapidly to the former structural configuration following fire, even if some changes take place in plant species composing the vegetation, so the effects of fire on bird populations are probably short term.

Response of Microbes

Wicklow (1975) used laboratory incubation of soil samples from burned prairie stands in Wisconsin and demonstrated the presence of both coprophilous and carbonicolous ascomycetes, which did not occur

on soils in the unburned prairie. Fires may bring about a reduction in microbial competition at the soil surface, thereby permitting the successful development of these postfire fungal colonists. Several of the coprophilous ascomycetes appearing on burned soil were also found on deer and rabbit feces collected from unburned prairie stands. Furthermore, these fires may provide the principal means of initiating growth and development of several noncoprophilous ascomycetes.

ELM Fire Simulations

Fire has been used to control shrub invasion, the accumulation of standing dead and litter, the production of cool-season annual grasses, and to increase cattle weight gains. Computer runs were designed to simulate the effect of varying fire frequency on seasonally grazed pastures (1 May to 30 September) and ungrazed pastures. The seasonally grazed pastures were grazed at moderate intensity (2.4 ha/AU) with steers having an initial weight of 250 kg. The fire frequencies simulated by the model include no fires, spring fires (10 April) once every three years, and annual spring fires. We assumed that approximately 80 percent of the carbon in the standing dead and litter was removed by the fire, that only 10 percent of the nitrogen in the burned plant material was returned to the system (Sharrow and Wright 1977), and that all the phosphorus was returned. The results for the seasonal grazing runs were five-year average values of output variables, while the ungrazed simulations were run for six years. Driving abiotic variables include the Osage Site meteorological variables for 1970 through 1973 (1970 and 1971 were repeated for the fifth and sixth years of the simulation). In the unburned grazing runs we assumed that cattle would only graze 50 percent of the area, with the remaining area being ungrazed (C. E. Owensby, 1977:personal communication; Herbel and Anderson, 1959). On burned pastures the areas were assumed to be grazed evenly (Aldous, 1934; Duvall and Whitaker, 1964; Ewell et al., 1941).

The results (Table 10.45) for the ungrazed runs show that triennial spring fires increase aboveground and belowground production, production of warm-season grasses and forbs, total root biomass, transpiration, and plant uptake of nitrogen and phosphorus but decrease litter biomass, standing-crop interception, and net nitrogen balance in the system. Thus, triennial fires increase primary production and reduce excessive accumulation of litter but decrease production of cool-season annual grasses and produce a net loss of 0.61 g N m^{-2} yr^{-1}. Results suggest that a fire frequency of once every four or five years would result in a neutral nitrogen budget for an ungrazed system. We must note that a considerable amount of uncertainty exists in the estimate of nitrogen input (i.e., dry and wet fall of nitrogen and nitrogen fixation) and nitrogen losses (i.e., NH_4^+ volatilization) from the system. Thus, determination of the exact fire frequency that results in a neutral nitrogen budget is difficult.

Increasing fire frequency from triennial to annual increased above- and belowground production, warm-season grasses and forbs production, total root biomass, bare soil water loss, and plant

TABLE 10.45 Average values of the output variables for the fire simulation.

Fire runs	Primary production (g m⁻² yr⁻¹)		Animal weight gain			Peak live biomass (g m⁻²)			
	Net aboveground	Net belowground	Weight gain per head (kg head⁻¹)	Forage consumption (g m⁻² yr⁻¹)	Average nitrogen content of forage consumed (percent)	Warm-season grass	Warm-season forbs	Cool-season annuals	Total aboveground
			No grazing						
Control	572	257	--	--	--	65	9.1	50	270
1 fire every 3 years	621	306	--	--	--	97	13	25	305
1 fire every year	638	320	--	--	--	130	16	4.6	372
			Moderate steer grazing						
Control	571	211	62.2	90	1.03	61	18	45	266
1 fire every 3 years	629	275	64.4	90	1.01	80	20	24	284
1 fire every year	654	295	63.6	91	0.97	95	31	5.3	320

Note: primary production units: $g\,m^{-2}\,yr^{-1}$; peak live biomass units: $g\,m^{-2}$; weight gain per head: $kg\,head^{-1}$; forage consumption: $g\,m^{-2}\,yr^{-1}$.

TABLE 10.45 continued

Fire runs	Standing crop (g biomass m^{-2})			Abiotic (cm H$_2$O yr^{-1})			Nutrients (g m^{-2} yr^{-1})			Biomass loss by microbial activity (g m^{-2} yr^{-1})
	Standing dead	Litter	Total roots	Transpiration	Bare soil water loss	Standing crop interception	Phosphorus root uptake	Nitrogen root uptake	Net nitrogen balance	
Control	420	433	814	28	12	40	0.52	6.24	1.04	942
1 fire every 3 years	494	292	967	31	12	38	0.60	6.62	-0.61	761
1 fire every year	478	152	1048	32	16	31	0.64	6.51	-2.06	555
Control	360	394	822	28	14	36.0	0.56	6.25	0.62	849
1 fire every 3 years	408	265	924	31	15	34.0	0.62	6.78	-0.84	761
1 fire every year	410	147	991	31	19	28	0.66	6.81	-2.28	576

uptake of phosphorus, but decreased cool-season annual production, litter biomass, root uptake of nitrogen, standing-crop interception, and net nitrogen balance. Thus, increasing fire frequency increased total plant production and warm-season grass production and decreased cool-season annual production. Increased fire frequency, however, also decreased nitrogen uptake by the plants and increased the net loss of nitrogen from 0.61 to 2.06 g m^{-2} yr^{-1}. These results suggest that large nitrogen losses associated with annual spring burning decreased the amount of nitrogen available for plant growth. In fact, nitrogen uptake by the plants during the first three years of the simulation was greatest with the annual burning, while during the last three years of the simulation plant nitrogen uptake was lower in annual than in triennial burning runs. By the end of the sixth year of the simulation, the uptake of nitrogen by the plants was lowest in the annual burning run. These results would also suggest that the long-term effect of nitrogen losses associated with annual burning would result in decreased primary production. In two of the last three years of the six-year simulation, the aboveground production was lower in the annual than in triennial burning runs.

Fire affects microbial activity, nitrogen and phosphorus cycling, and soil water content. Increasing fire frequency causes a marked decrease in total microbial respiration because of the removal of aboveground dead-plant biomass that would normally be broken down by microbial activity in the litter layer. The decrease in litter layer microbial activity is partly compensated for by an increase in the soil microbial activity that results from the increased soil temperatures associated with the removal of standing dead and litter from the system.

As expected, the model results show that fire has a differential effect on the nitrogen and phosphorus cycles. Root uptake of phosphorus is increased as the fire frequency is increased from no fire to annual spring fires (0.52 versus 0.64 g P m^{-2} yr^{-1}), and uptake of nitrogen by the roots is increased with triennial burning then decreased as fire frequency is increased to an annual burn (6.62 versus 6.51 g N m^{-2} yr^{-1}). The nitrogen and phosphorus contents of the live shoot follow a similar pattern, with phosphorus concentration of the shoots increasing from 0.06 percent to 0.08 percent in the unburned and annual burning runs, respectively. Nitrogen concentration in the live shoots is 1.20, 1.22, and 1.19 percent for unburned, triennial burning, and annual burning runs, respectively. Frequent fires apparently have the positive effect of increasing the cycling rate and plant uptake of such nutrients as phosphorus that are returned to the system after burning the mulch layer, while the plant uptake of such nutrients as nitrogen that are volatilized with burning (80 percent loss; Sharrow and Wright, 1977) will be decreased by frequent spring fires.

Model results also showed that increased fire frequency increases average soil water content during the growing season. These results are somewhat surprising, since increased fire frequency increases season-long transpiration water loss. The apparent reason for this anomaly is that the reduction of cool season grasses that results from spring fires causes a decrease in transpiration water loss during the early part of the season, thus increasing the average soil water content for the growing season.

Triennial burning on seasonally grazed pastures increased above- and belowground production, cattle weight gain per head, total root biomass, production of warm-season grasses, and plant uptake of nitrogen and phosphorus, and production of cool-season annuals, litter biomass, and net nitrogen balance are all decreased. Thus, triennial burning has the positive effect of increasing primary production and cattle weight gains and decreasing cool-season annual production, while it also has the negative impact of generating a net loss of 0.84 g N m^{-2} yr^{-1}. The calculated values of the net nitrogen budget for the grazed runs suggest that a fire frequency of once every six to seven years would result in a neutral nitrogen budget. This information compares with a fire frequency of four to five years for the ungrazed pastures and was a result of the volatilization losses of NH_4 from the urine and feces of the cattle.

Increasing the fire frequency from a triennial to an annual rate for the grazed pastures increases aboveground production, production of warm season forbs and grasses, total root biomass, bare soil water loss, and phosphorus root uptake. Cattle weight gains, litter biomass, standing-crop interception, and net nitrogen balance are all decreased. Increasing fire frequency for the grazed runs increases the growing season, and, because it results in decreased transpiration water loss during the spring, fire reduces cool-season grass production. In general, annual burning of grazed pastures seems to negatively affect the system by decreasing cattle weight gains per head and increasing the net nitrogen loss from 0.84 to 2.28 g m^{-2} yr^{-1}. Comparison of the net nitrogen losses for the ungrazed run suggests that recommended time intervals between burning on a grazed pasture should be longer than on an ungrazed pasture and that the negative impacts of the large nitrogen loss associated with annual spring burning should be observable sooner in grazed pastures than in ungrazed pastures.

Short-term, Intermediate, and Long-term Effects

In general, the model results compare very favorably with observed field data. The analysis of model output and field data will be in five subsections: simulated and observed impacts of fire on primary production, cattle weight gains, species composition, standing-crop variables, soil water content, and nutrient uptake by the plants. Short-term (1 year), intermediate (2-5 years), and long-term (>5 years) impacts of fire are also considered.

<u>Primary Production</u>. Model results for primary production show that the short-term impact of a single fire in a grazed or ungrazed pasture is to increase aboveground production by approximately 20 percent the first year after the fire, but production is only 3 percent greater the second year after the fire. Hadley and Kieckhefer (1963) show similar results for a big bluestem prairie in Illinois, with primary production increasing by 1000 and 66 percent the first and second year, respectively. During the fourth year no difference showed between the burned and unburned pastures. The greater response to burning observed in the Illinois prairie was probably because plants experience much less moisture stress there than at the Osage Site. Results for an undisturbed bluestem prairie

in Kansas (Hulbert, 1969) showed that aboveground production is increased by 100 percent during the first year after a spring fire.

The simulated impact of fire over intermediate time periods (Table 10.45) was to increase aboveground production with annual fires, raising production by 12 and 15 percent, respectively, for ungrazed and grazed pastures. Duvall's (1962) results for a slender bluestem (Andropogon tener) range in Louisiana show that annual burning for five years on an ungrazed pasture increased aboveground production by 18 percent, while annual burning for moderately grazed pastures slightly decreased aboveground production. The reason for the discrepancy between the simulated and observed impact of annual fire on the grazed pasture is unknown. The computer simulations were run only for five- and six-year periods and thus do not necessarily represent the long-term impact of different fire frequency on the grassland. However, because aboveground production was lower in the annual burning runs than in the triennial burning run for two years of the last half of the six-year simulation, annual burning will probably cause a long-term decrease in production. These results are verified by results from long-term annual burn experiments in the Kansas Flint Hill region (Owensby and Anderson, 1967; McMurphy and Anderson, 1963), which show that annual burning of seasonally grazed pastures reduces aboveground production. This reduction is also consistent with results for an eight-year burning experiment in Oklahoma (Elwell et al., 1941), where production in an annual burned pasture was reduced by 53 percent.

With the computer model the short-term and intermediate-term impact of burning on cattle weight gains is to increase cattle weight gains per head by 2.2 and 1.4 kg head^{-1} with triennial and annual burning, respectively. These data suggest that triennial burning will maximize cattle weight gains for pastures that are seasonally grazed by steers. Long-term annual spring burning of a bluestem range in Kansas increased cattle weight gain by 9 kg head^{-1} (Anderson et al., 1870; Smith et al., 1963). Thus, the model seems to underestimate the positive effect of spring burning on cattle weight gains. We also must note that the model results showing maximum weight gain per head with triennial spring burning cannot be verified by field data, since this type of experiment has not been performed, which is a topic for future research.

Model results for the effect of fire on species composition shows that increasing the fire frequency promotes the growth of warm-season perennial grasses and decreases the growth of the cool-season annual grasses. Model results are verified by data presented by Anderson et al. (1970) for a bluestem range in Kansas, which showed that annual spring burning increases the percentage of composition and basal cover of the desirable warm-season perennial grasses, while the percentage of composition and basal cover of the less desirable cool-season grasses and annual grasses were greatly decreased. The decrease in cool-season grasses is caused by the burning of emerging cool season shoots with the spring fires. The increase in the warm-season perennial grass is a result of reduced competition for space in the plant canopy and for nutrients in the soil (i.e., nitrogen and phosphorus) during the time the warm-season perennial grasses are just beginning to grow. (Cool-season plants

start growing in March or early April, while the warm-season plants start at the end of April.)

The simulated impact of fire on the standing-crop variables is to decrease the litter biomass by more than 50 percent the first year after the fire. Subsequently, litter accumulated during the second and third years is approximately equal to the unburned litter level at the end of the third year. The total root biomass increased approximately 100 g m^{-2} the first year after the fire and decreased only slightly two years after the burn. This increase in total root biomass with burning is supported by observed data from an Illinois prairie (Hadley and Kieckhefer, 1963), where total root biomass increased by 183 g m^{-2} after two years of annual burning. Litter data from the Illinois prairie (Hadley and Kieckhefer, 1963) and a slender bluestem range in Louisiana (Duvall, 1962) show that it takes two to three years for litter to return to unburned levels after a fire. Thus, model results (see Table 10.45) and observed field data suggest that increasing fire frequency will reduce litter biomass and increase total root biomass.

Model results for the short-term effect of fire on nitrogen and phosphorus cycling show that during the first year after the fire nitrogen and phosphorus root uptakes are increased by 1.2 and 0.12 g m^{-2} yr^{-1}, respectively, while the average seasonal concentration of N in the live shoots is increased from 1.16 to 1.23 percent and phosphorus concentration is increased from 0.096 to 0.109 percent. Field data for a tallgrass prairie in Missouri (Kucera and Ehrenreich 1962) show that the average concentration of phosphorus in the live shoots during the growing season is increased after spring burning, while Campbell et al. (1954) and Neal and Becker (1933) show that nitrogen concentration in the live shoots is increased by spring burning in a prairie in Florida and a bluestem prairie in Louisiana. Model results show that increased uptake of nitrogen and phosphorus is caused by the direct fertilization effect of the two nutrients returned from the burned vegetation and an increased mineralization rate of the nutrients in the soil. The increased mineralization rate resulted directly from increased soil temperature. Data for a tobosa grass (Hilaria mutico) prairie in Texas (Sharrow and Wright, 1977) show that burning or removal of litter causes nitrification and ammonification rates to increased and thus support the results of the model.

Although computer simulations were run for only five or six years, results indicate that the long-term (>5 years) effect of annual burning is to decrease the plant nitrogen uptake, while plant phosphorus uptake continues to be increased by burning. The long-term decrease in plant nitrogen uptake was caused by the large nitrogen losses associated with annual burning ($\sim$2 g N m^{-2} yr^{-1}), while plant phosphorus uptake increased because burning increased the rate at which phosphorus in aboveground plant material was returned to the soil. (In the model, all phosphorus is returned to the soil after a fire.) Unfortunately, field data supporting these results do not exist, because existing studies have evaluated only the short-term effect of burning on nutrient cycling.

Model results indicate that the short- and long-term effects of spring fires are to increase growing season soil water content for both grazed and ungrazed pastures. The increased soil water for both treatments results from decreased transpiration water loss by

cool-season grasses during the spring. (Spring fires reduce cool-season grass production.) Anderson et al. (1970) showed that spring fires increase the soil water content of grazed pastures in eastern Kansas, and their data suggest that the decrease in cool season grass production with fire is a reasonable explanation for the increase in soil water content.

With ungrazed pastures the optimal fire frequency for increasing plant production and minimizing the loss of nitrogen for the system would be a spring fire once every three to four years. Results for grazed pastures suggest that a triennial spring fire maximizes cattle weight gains per head. However, because the nitrogen losses are greater than the losses in an ungrazed pasture, a longer time period might be needed to minimize nitrogen losses from the system. The recommended fire frequency for an ungrazed pasture is consistent with Duvall's (1962) recommendation that ungrazed or lightly grazed slender bluestem pasture in Louisiana should be burned once every three or four years. The results for the grazed pasture are also consistent with Duvall and Whitaker's (1964) results for a slender bluestem pasture, which showed that fire applied at three-year intervals improved forage palatability and nutrient content and allowed cattle to gain weight throughout the growing season (April-November).

PESTICIDES

Response of Plants

The use of pesticides on the True Prairie is somewhat limited, and the targets are usually forbs (Beck and Sosebee, 1975; Elwell and McMurphy, 1973; Faroua, 1975) or insects (Scifres, 1977). However, as discussed previously, herbicides are sometimes used along the eastern border to decrease shrub or brush growth (Dalrymple et al., 1964; Elwell, 1960; McMurphy et al., 1976). Large areas in the Ozark Hills are being converted to improved pasture, often by spraying to remove oaks and hickories. The usual procedure is to spray 0.18 to 0.27 g m^{-2} 2,4,5-T from an airplane or helicopter in the spring. The herbicide is spread as a solution containing about 10 l of diesel oil and 40 l of water per hectare. Approximately 10 days later the land is seeded with 8 g m^{-2} of wheat, 1.8 g m^{-2} of a cool season fescue variety (Festuca spp.), and 0.5 g m^{-2} of a red clover (Trifolium pratense). After a year forage production may reach 280 g m^{-2} and support 1 or 2 AU/ha (Spray, 1974).

Nichols and McMurphy (1969) tested the effect of nitrogen and 2,4-D (2,4-dichlovophenoxyacetic acid) on species composition and forage production in a western wheatgrass and green needlegrass (Stipa viridula) in western South Dakota. Nitrogen was applied at rates of 0.0, 2.7, 5.4, and 10.8 g m^{-2}; 2,4-D at rates of 0.0, 0.04, 0.09, and 0.18 g m^{-2}. These areas were severely overgrazed, and the initial percent frequency of the perennial grasses was only about 8 percent. After two years with 2,4-D, the frequency of the grasses was greater than 40 percent, and the forbs had been reduced proportionately. Spraying to reduce forb competition and

fertilizing to stimulate grass recovery resulted in additional increases in production over the control plots. The 10.8 g m^{-2} rate of nitrogen with 0.18 g m^{-2} 2,4-D increased grass production over the control by 49, 17, and 81 g m^{-2} during the 3-year study. For the same level of nitrogen without 2,4-D, production was not as great, with only 24, 8, and 21 g m^{-2} more than the control for the 3 years. Total perennial grass production for the 3 years increased over the control by 35 g m^{-2} for 2,4-D, 53 g m^{-2} for nitrogen, and 146 g m^{-2} for 2,4-D and nitrogen in combination.

In eastern Nebraska, Klingman and McCarty (1958) reported that 0.09 g m^{-2} 2,4-D applied for 3 years decreased perennial broadleaf weeds by 70 percent and increased forage production by 47 percent. Other studies also have shown that spraying with herbicides results in an increase in grass and a reduction of broadleaf plants (Mitich, 1965).

Response of Invertebrates

In the past high grasshopper populations, particularly in the shortgrass prairie in semiarid areas, have been controlled with insecticides under a cooperative program involving landowners, states, and the federal government. These groups felt that treating incipient outbreaks would reduce extensive use of pesticides should the pest spread into other range areas or cropland without treatment. Once controlled to an average of one or less per square yard (0.84 m^2), grasshoppers usually remain at harmless levels for five to ten years. Since 1951, when modern insecticides came into general use, no widespread outbreaks or severe losses to rangeland have occurred because of grasshoppers. Severe damage has occurred only in small isolated areas. Ecological considerations, however, may restrict such use of insecticides.

Barrett (1968) investigated the effects of a carbamate insecticide, Sevin®, on plant, arthropod, and mammal components within a grain-crop grassland ecosystem in Georgia. The total biomass and numbers of arthropods were reduced more than 95 percent in the treated area and remained well below the control area for five weeks; after seven weeks total biomass but not total numbers returned to the control level (Figure 10.12). Phytophagous insects were more severely affected than predaceous insects and spiders; density of the latter returned to control levels in three weeks.

A highly significant decrease in litter decomposition in the treated area was measured three weeks after spraying. This decrease was presumed to be the result of a reduction in microarthropods and other decomposers.

In summary, although the insecticide remained toxic in the environment for only a few days, long-term side effects on litter decomposition, arthropod density and diversity, and mammal reproduction were demonstrated. No effect could be detected on producer standing crop or net community primary production (Figure 10.13).

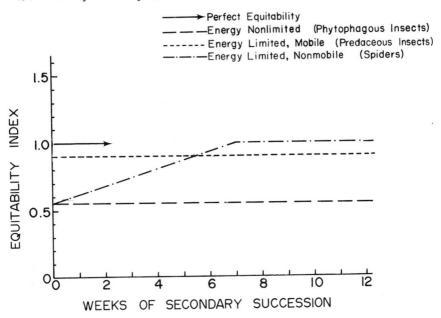

FIGURE 10.12 Arthropod species response to application of an insecticide (Redrawn from Barrett, 1968; copyright 1968 by The Ecological Society of America).

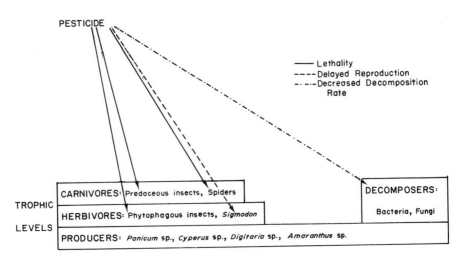

FIGURE 10.13 Response of producers, herbivores, carnivores, and decomposers to the application of an insecticide (Redrawn from Barrett, 1968; copyright 1968 by The Ecological Society of America).

Response of Birds

No studies have directly assessed the influences of pesticides on bird populations in True Prairie habitats. On the basis of avian responses to grazing and burning stress documented earlier in this chapter, one could predict that herbicide applications, which influence the structural composition of vegetation (especially the occurrence of emergent forbs), would produce significant changes in bird densities and species compositions. For insecticides inferences may perhaps be drawn from the consequences of experimental applications of several compounds in the shortgrass prairie. In a study involving applications of various insecticides considered as potential agents of grasshopper control, McEwen et al. (1972) found a range of effects on breeding bird populations. Several substances produced significant decreases in bird populations when applied at moderate to high rates, while low application rates produced no apparent effects. Some effects were immediate and apparent: diazinon (0.0-diethyl 0-[2-isoprophyl-4-menthyl-6-pyrimidinyl] phosphorothioate) applied at a rate of 5.0 to 8.0 oz acre^{-1} (5.67 to 9.07 g ha^{-1}) produced a rapid decrease in total avifauna from 26.6 birds per 0.5 mi (0.8 km) transect to 11.1 birds per transect. Toxaphene (a mixture of chlorinated comphenes [67 to 69 percent combined chlorines]) applied at a rate of 1 lb (2.2 kg) in 0.75 pint fuel oil per acre (0.015 liter ha^{-1}) produced no change in bird populations during the first week following application, but a significant decrease was noted in the second post-spray week. Applications of malathion (0,0-dimethyl S-[1,2,dicarbethoxyethyl] dithiophosphate) produced no discernible effect on bird populations. In a more intensive study on the Pawnee Site of northeastern Colorado, McEwen and Ells (1975) evaluated the effects of application of toxaphene and malathion on several ecosystem components. Populations of the most common bird species did decline by about 30 percent in the toxaphene-treated areas but not in plots treated with malathion.

These latter applications produced significant reductions in insect biomass in all treated areas, which might imply an increase in plant biomass. On the contrary, however, McEwen and Ells found that on control plots the total herbage biomass increased by about 25 percent, but during the same growing season total biomass on the malathion plots increased by only 10 to 11 percent, and on the toxaphene plots there was an increase of only 1.2 to 3.5 percent. Studies by Dyer and Bokhari (1976) suggest that insect herbivory may maintain or even stimulate plant production, and removal of the insects may thus lead to a reduction in total plant production. This observation has many implications. For example, one effect of the insectivorous habits of many of the breeding birds of grasslands might be to reduce insect populations and thus perhaps depress their stimulatory effect on plant production. At this time these relationships remain to be tested, and their applicability to True Prairie ecosystems is unsubstantiated.

Response of Small Mammals

No study of pesticide effects on small mammal communities of the True Prairie is available. Barrett (1968) studied the insecticide effects on a grassland ecosystem in the southeastern United States and discovered profound, long-term effects on litter decomposition and mammalian reproduction. Both of these factors could affect composition and structure of the small mammal community. Cotton rats (Sigmodon hispidus) exhibited delayed reproductivity in the tested area, and density was reduced relative to the control. However, the total small mammal population was not low because a human commensal, the house mouse (Mus musculus), responded positively to the insecticide and reached a higher-than-normal population density on the tested grid.

Because of the high dispersal capacity and reproductive potential of small grassland mammals, it is unlikely that the use of insecticides would result in major extinctions or range reductions within the True Prairie, but the potential for profoundly altering the small mammal community, at least temporarily, could be high when insecticides are added to the environment.

11

Whole System Properties

Relatively few studies, especially in the True Prairie, have sought to examine simultaneously more than one trophic level. Tester and Marshall (1961), however, did examine the vegetation and a number of invertebrates and vertebrates in the Waubun Prairie Research Area, Mahnomen County, Minnesota. The treatments involved grazing as well as spring and fall burning. Vegetation responses were usual: Mowing and burning reduced the litter, which regenerated in two to six years, and grazing reduced big bluestem, Indiangrass, and needlegrass, the latter two of which showed an initial increase after burning.

Bobolinks (Dolichonyx oryzivorus) were not found either in unmowed prairie characterized by deep litter or in burned areas with little or no litter. Breeding pairs occupied habitat with litter depths between these extremes. The savannah sparrow (Passerculus sandwichensis) occurred in all plots, but marked reductions in breeding populations took place in each burned plot. LeConte's sparrows (Ammospizo leconteii) appeared to need a moderate amount of litter combined with a grass canopy rising 30 cm or more above the litter. Although numbers of the masked shrew (Sorex cinereus) seemed to be independent of vegetation characteristics, the meadow vole (Microtus pennsylvanicus) was positively associated with increasing litter, while the prairie deer mouse (Peromyscus maniculatus bairdii) was negatively associated with greater litter depths. Grasshoppers (Orthoptera) were most abundant where light or moderate amounts of litter were found, and large beetle (Coleoptera) populations appeared to be associated with sparse litter.

Results in this book originated from a much larger study which focused on the Osage Site but drew extensively from the literature. By using the whole-system experimental design, we can now identify and examine some ecosystem properties that are documented only in comprehensive studies such as this volume.

SUCCESSIONAL DYNAMICS

The concept of ecological succession was explicitly not investigated at the Osage Site. However, this concept is fundamental to a number of ideas about the structure and function of the True Prairie, thus we need to address the subject here.

Ecological succession involves the transient behavior of ecological systems as they proceed toward a metabolically balanced ecosystem--that is, gross primary production of autotrophs equals utilization and respiration of autotrophs and heterotrophs. When a sequence of transient states arises in response to a perturbation of the mature or climax ecosystem, the process is called secondary succession. Biomass generally accumulates in time and space until a stabilized ecosystem characterized by maintenance of maximum biomass and diversity per unit of energy flow is established (Gutierrez and Fey, 1975). Twenty-four ecosystem attributes have been applied to successional or climax communities (Odum, 1969). These attributes can be divided into six groups: community energetics, community structure, life history, nutrient cycling, selection pressure, and overall homeostasis.

Allogenic (open-loop or outside) environmental factors such as climatic forces, nutrient inputs, erosion, and other geological processes exert a constraining influence on succession. Within the boundary conditions imposed by allogenic factors, the fundamental hypothesis for the past ten years has been that successional dynamics arise from the autogenic (endogenous or closed-loop) feedback structure of the ecosystem (Gutierrez and Fey, 1975). This concept is that the orderly sequence of transient states leading to a climax results from emergent properties at the ecosystem level of organization. An alternate hypothesis is that successional processes can be explained by competitive interactions at the species level (Drury and Nisbet, 1973). Data on hand simply do not permit the resolution of those two points of view. Certainly, the data on allelopathic relations among species in the True Prairie successional sequence suggest powerful species interactions. On the other hand, the community structure characteristic of the same successional sequences clearly conforms to some anticipated responses of a metabolically determined system. The following discussion provides a description of the successional dynamics of the True Prairie ecosystem.

After fields have been retired from cultivation or after an area has been denuded of vegetation, the first invading plants of this secondary succession are not permanent. Usually annual plants appear first and, if subsequent conditions are favorable, are gradually replaced by perennials. Booth (1941b) studied the revegetation pattern in 106 abandoned fields in central Oklahoma and southeastern Kansas and determined that the sequence of the stages of plant succession is weed, annual grass, perennial bunchgrass, and finally fully developed prairie. These patterns may be typical of at least the southern portions of the True Prairie. The weed stage, which lasted about two years, was characterized by an early weed stage--composed of large, widely spaced plants--and a late weed stage--composed of essentially the same species but characterized by a dense growth of stunted plants. The second stage, referred to as the annual stage or the oldfield threeawn stage, received its name

from the most common seed plant at this time, oldfield threeawn (Aristida oligantha). This species indicates poor soil and xeric conditions. It often occurs in pure stands, but actual basal-area plant cover is usually less than 1 percent. This stage, which lasted thirteen years in this study but may persist much longer, produced a limited amount of forage, low in palatability.

As succession proceeds, one of the first species characteristic of the perennial bunchgrass stage is silver bluestem (Andropogon saccharoides). Often, however, the dominant grass of this stage, little bluestem, is not preceded by other perennial grasses except scattered plants or clumps of certain species of lovegrass (Eragrostis spp.), panicum (Panicum spp.), and paspalum (Paspalum spp.). The perennial bunchgrass stage is characterized by a relative paucity of grass species and by a rich array of widely spaced forbs. The length of this stage is indeterminate, but Booth (1941b) reported one thirty-year-old field in this stage.

The fully developed prairie is the final stage and consists of the full complement of species. By this time soil fertility and organic matter have presumably increased and the potential for soil erosion has been markedly reduced. Heavy grazing intensity and burning may delay the time required to attain each successional stage.

The rate of succession can be much more rapid if the soil has not been depleted of nutrients. Rice and Penfound (1954) plowed an acre of prairie and compared it to an unplowed control dominated by little bluestem, big bluestem, Indiangrass, and switchgrass. The first plants to appear in the plowed plots were perennial forbs and grasses. After the first growing season following plowing, western ragweed (Ambrosia psilostachya) was the most important dominant. By the end of the second growing season, all the dominants in the control were prominent in the plowed plots. Aerial cover and dry weight of the aboveground material during the second growing season were significantly greater in the plowed plot. The authors postulated that the increase in yield after plowing was the result of a greater availability of nutrients, particularly nitrogen, that resulted from the decomposition of the large amount of organic matter plowed under during the treatment.

In a study conducted very near the 1954 study, Penfound and Rice (1957) found that a cotton field abandoned ten years earlier and planted to Korean lespedeza (Lespedeza stipulacea) was dominated by oldfield threeawn, Korean lespedeza, and Scribner panicum (Panicum scribnerianum). After the five-year study the dominants were Scribner panicum, western ragweed, heath aster (Aster ericoides), and fall witchgrass (Leptoloma cognatum). When the study began, in 1950, a portion of the abandoned field was replowed, and during the subsequent growing season the dominants were western ragweed, Korean lespedeza, tuber false-dandelion (Pyrrhopappus grandiflorus), and buffalobur nightshade (Solanum rostratum). By the end of the observation period in 1954, the only important plants in the replowed plot were heath aster and fall witchgrass.

In an effort to explain why oldfield threeawn, little bluestem, and switchgrass come in at different stages of succession, Rice et al. (1960) studied seed dispersal of oldfield threeawn and little bluestem. Fruits and even entire plants of oldfield threeawn were observed to carry fairly long distances, but little bluestem fruit

containing viable seeds was dispersed by the wind only slightly over
2 m from the parent plant. The experimental design minimized
estimated dispersal distances, but observation of abandoned fields
showed that little bluestem invasion started around the margin and
moved slowly inward year by year.

In the same study, laboratory experiments were conducted to
establish nutrient requirements for nitrogen, phosphorus, and
potassium. No differences were evident in relative requirements for
potassium, but the order of three species based on increasing
requirements and phosphorus was (1) oldfield threeawn, (2) little
bluestem, and (3) switchgrass, the same order in which the species
invaded abandoned fields.

From 1949-1960, the flora of a southeastern Michigan field
abandoned in 1928 showed little change (Wiegert and Evans, 1964).
The peak standing crop of vegetation remained relatively constant
from year to year, but during the experimental period grasses
decreased in importance from more than 90 percent of standing crop
biomass to less than 50 percent. Using a modified harvest method,
Wiegert and Evans determined that peak aboveground biomass ranged
from 165.0 to 237.9 g m^{-2} during the sample period. The mean
standing crop of roots was 685 and 542 g m^{-2}, respectively, in 1960
and 1961. These data suggest that in the northern True Prairie
total biomass does not change radically, but species composition
consists of different species throughout the sequence.

In a study of the four successional stages described by Booth
(1941b) in the prairie-forest border of Oklahoma, Perino and Risser
(1972) found total aboveground and belowground biomass to be 427,
287, 305, and 787 g m^{-2} for the four stages. The live component of
the aboveground vegetation constituted 51 percent of the total
standing crop in the early stages but amounted to 26 percent in the
last stage, with a concomitant increase in the proportion of litter.
Aerial cover, species richness, and concentration of dominance, as
measured by Simpson's Index (Peet, 1974), tended to increase in this
sere, while species diversity and equitability generally decreased.

A related study by Kapustka and Moleski (1976) utilized the
same fourth stage site but different earlier stage sites. Although
similar concentrations of dominance values for the fourth stage were
found in both studies, Kapustka and Moleski found higher
concentrations of dominance in the earlier stages. The early stages
of the two studies had received different prior treatments of
grazing, fertilization, and cultivation, so these apparent
differences are difficult to interpret. Clearly, though, species
composition and stand structure provide only partial indications of
the rates of succession.

Several investigators have used dominance-diversity curves for
describing community structure (Bazzaz, 1975; Kapustka and Moleski,
1976; Whittaker, 1975). The geometric form of the resulting curve
suggests a community with strong species dominance, while the
lognormal form results from the undisturbed array of normally
distributed species values. Since species with allelopathic
properties may lower total species diversity, scientists have argued
that low species diversity and the geometric form of the early stage
indicate the presence of allelopathic interference. This
interference eventually reduces the importance of the initial
dominants and allows the entry of additional species into the

community, resulting in a lognormal distribution for the later stages (Bazzaz, 1975; Kapustka and Moleski, 1976). Further, Rice (1974) has shown that leachates from oldfield threeawn not only inhibit some early pioneer weed species but also inhibit the growth of several microorganisms associated with symbiotic and nonsymbiotic nitrogen fixations. Since Rice et al. (1960) showed that higher levels of soil nitrogen were required by species of the later seral stages, it appears that the threeawn stage may persist longer than might be expected on the basis of its small stature by allelopathic interference with both microorganisms and higher plants.

Rice (1974) showed that two climax species of the True Prairie, little bluestem and prairie fleabane (<u>Erigeron strigosus</u>), inhibited nitrification but not nitrogen fixation. In other recent studies in Oklahoma, nitrate levels were lowest in soils of the climax vegetation, intermediate in soils of the middle successional stage, and highest in soils of the early successional stage. Ammonium levels were highest in soils of the climax vegetation, intermediate in soils of the middle stage of succession, and lowest in soils of the early stage of succession. The numbers of <u>Nitrobacter</u> and <u>Nitrosomonas</u> were higher in the early and middle stages of succession and lowest in the climax prairie (Rice and Pancholy, 1972), and blue-green algae may be an important nitrogen source in early stages (Kapustka and Rice, 1976). Because these differences in nitrate, ammonium, and microbial levels cannot be explained by pH or textural differences, soils of the climax vegetation may be low in nitrate because of an allelopathic effect by the climax plant species on nitrification.

Successional sequences that follow drought and overgrazing in the True Prairie have been reviewed adequately (Owensby and Anderson, 1965; Sims and Dwyer, 1965; Weaver, 1954; Weaver and Hansen, 1941) and will not be discussed in this volume.

Species Diversity

A currently popular device is to calculate various measures of species diversity, to partition these terms between species richness and equitability, and to relate diversity to niche width, stability, and productivity (McNaughton and Wolf, 1970). The interpretation of these indices is difficult because comparisons are made with terms as ill-defined as <u>diversity</u> and several of the early hypotheses of how diversity relates to other community structural characteristics seem to be of limited generality (McNaughton, 1968; Peet, 1974). Nevertheless, species diversity values were calculated for plants, mammals, and birds using data from the Osage Site and other sites in the True Prairie.

Species diversity was calculated using the Shannon-Wiener function (MacArthur and MacArthur, 1961):

$$H' = - \sum_{i=j}^{S} (P_i \log_{10} P_i) ,$$

where H' is species diversity, S is the number of species, and P_i is the relative density, biomass, frequency, or cover at the ith species. Equitability was calculated (Pielou, 1969):

$$J' = H'/H'_{max} ,$$

where H'_{max} is log S. The actual data used for these calculations are not the most appropriate, because they are from different measures of density or dominance and were collected in different ways from different sites. (Note also that these diversity values were calculated using $\log_{10}$ and thus are not the same as the values presented in Chapter 7, which were calculated using natural logarithms.) However, some cautious but interesting statements regarding the analyses are possible.

The diversity values for plants range from 0.98 to 1.40, with no particular geographical trend evident from these data (Table 11.1). In cases where a comparison is possible, diversity and equitability are higher in grazed treatments than in ungrazed treatments. The number of bird species was only about 15 percent of the number of plant species and both their range of diversity (0.17 to 0.57) and equitability (0.55 to 0.75) were lower than those of the plants. Apparently northern bird communities may be characterized by a higher species diversity than the more southern ones. The number of small mammal species is approximately the same as the number of bird species, and the range of species diversity and equitability is similar (Table 11.2).

If the number of species is plotted against the diversity of each sample, two values of the plant samples are quite separate from the others. From the explanation of the data collection procedures and the collection sites, whether the two points should be included in the further analysis is unclear. If they are not included, the slopes of the regression line through the data points are quite similar among plants, mammals, and birds (regression coefficients of 0.04, 0.09, and 0.07, respectively). If equitability is compared to the number of species, no correlation exists in any of the groups. Species diversity and equitability are closely correlated (r^2) with plants ($r^2 = 0.66$; $r^2 = 0.93$ if two values are omitted) and small mammals ($r^2 = 0.65$) but much less so with birds ($r^2 = 0.27$). All three groups show a similar relationship between diversity and the number of species but for different reasons. The plants and small mammals are strongly related to species equitability, but the bird diversity may be related to dominance of one or a few species.

In terms of the working taxonomic classification categories, considerable disparity exists between the number of species encountered on or near the Osage Site as well as throughout the True Prairie. Our estimates, intended only as rough estimates, of the number of species are tabulated in Table 11.3. However, the differences among groups have fundamental implications for the design of sampling schemes, the sites of areas necessary to sample species composition, and the interpretation of the resulting data.

TABLE 11.1 Plant species diversity (H'), equitability (J'), and number of species (S) for selected sites in the True Prairie.

Location	Treatment	H'	J'	S[a]	Reference
Osage Site, OK	Ungrazed, 1971	1.22	0.80	33	Risser and Kennedy (1972)
	Grazed, 1971	1.31	0.84	36	Risser and Kennedy (1972)
	Ungrazed, 1972	1.09	0.70	36	Risser and Kennedy (1975)
	Grazed, 1972	1.22	0.78	37	Risser and Kennedy (1975)
Noble County, OK	Grazed	1.40	0.87	40	Risser (1976)
Rogers County, OK	Ungrazed	1.04	0.85	17	Ray (1959)
Marshall County, OK	Ungrazed	1.38	0.94	39	Rice (1952)
Union County, SD	Grazed	1.40	0.99	26	Beebe and Hoffman (1968)
Oakville Prairie, ND	Ungrazed	0.98	0.65	31	Wali et al. (1973)

[a] Actual species numbers may be greater since some species that are dormant at the time of sampling are not recorded.

TABLE 11.2   Small mammal species diversity (H'),
             equitability (J'), and number of
             species (S) at the Osage Site, 1970-72.

Date	H'	J'	S
Ungrazed			
May 1970	0.19	0.32	4
August 1970	0.24	0.40	4
May 1971	0.76	0.98	6
August 1971	0.70	0.82	7
October 1971	0.70	0.78	8
May 1972	0.28	0.40	5
August 1972	0.42	0.50	7
October 1972	0.51	0.60	7
Grazed			
May 1972	0.21	0.70	2
August 1972	0.65	0.84	6
October 1972	0.51	0.73	5

TABLE 11.3   Number of species in or near the Osage Site and the
             True Prairie.

	Number of species	
Classification category	True Prairie	Osage Site vicinity
Plant	350	175
Invertebrate	100,000	1,200
Birds	250	12
Mammals	62	14
Bacteria and fungi	90	60

Interseasonal and Intraseasonal Variation
of Ecosystem Components

Not all ecosystem components demonstrate the same variation between years or within years. This variation was measured by examining the peak values for twenty-two components over the three-year sampling period at the Osage Site. Clearly, three years is a very short time over which to measure interseasonal and intraseasonal variations, but the results of Figure 11.1 do show some interesting relationships. The year-to-year peak live plant biomass demonstrated the smallest variation, but the birds also were characterized by low variability. Small mammals were more variable than plants or birds, but less variable than the invertebrates. Within small mammals, carnivores were the most variable. The peak value of total invertebrate biomass had low variability, but some functional groups, such as saprophage and pollen-eating invertebrates, were quite variable.

Intraseasonal variation was measured by calculating the CV between the peak values in spring, summer, and fall, then averaging over the three-year period (Figure 11.2). Of course the category of current-year dead aboveground plant parts is variable, since the value is essentially zero at the beginning of each season and does not reach a high value by the end of spring. The small amount of intraseasonal dynamics of belowground plants and litter is also evident. We might expect that live plants would show more variation, but considerable biomass accumulates by 15 June, the last date of spring in these analyses. Most heterotrophs are more variable than producers and most show greater intraseasonal variation of peak values than interseasonal variation.

Although producers appear less variable than consumers, whether higher trophic-level consumers are more variable than lower trophic-level consumers within each consumer category is unclear. Carnivorous birds are less variable than herbivorous birds, but carnivorous small mammals are more variable than herbivorous and omnivorous small mammals. Sap-feeding invertebrates demonstrate relatively low variability, but pollen-eating invertebrates are quite variable, both within and between years.

Some variability is clearly the result of sampling characteristics, and these analyses may well go beyond what the data really permit. However, not all components are operating under the same constraints and some further analyses should seek to explore the relationship between the variations shown by those groups.

Biomass Pyramids

At a given point in time all the biomass can be placed into defined categories that refer to trophic level or any other of a large series of possible designations. One classical approach has been to divide the biomass or energy equivalents into trophic levels such as producer, primary consumer, secondary consumer. At a gross level this type of classification is satisfactory, but several omissions are inherent in this approach. First, this static view ignores the rate at which energy or materials pass through a compartment--the turnover rate. Second, decomposers are sometimes

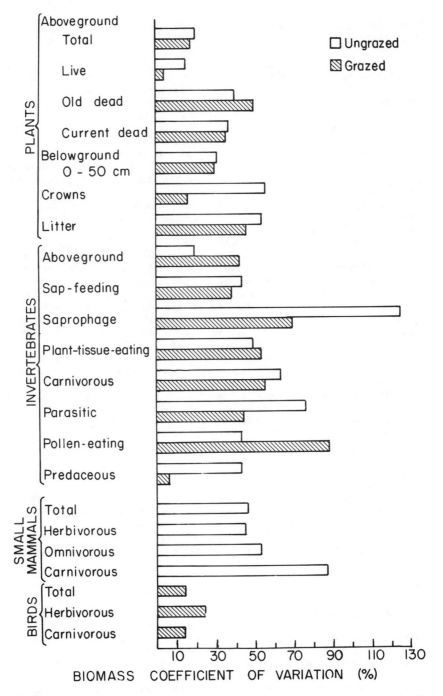

FIGURE 11.1 Interseasonal variation in the ecosystem components of the True Prairie.

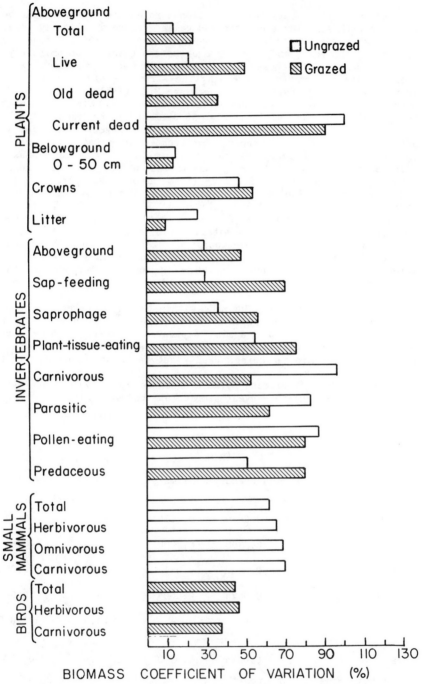

FIGURE 11.2 Intraseasonal variation in the ecosystem components
of the True Prairie.

treated separately, but from a functional role some of them might be combined with some invertebrates and called <u>reducers</u>. Finally, many organisms change their feeding behavior throughout their life cycle or through the year, and their categorization changes concomitantly.

In spite of those limitations, biomass pyramids carry considerable information concerning the structure of ecosystems. A very simple biomass pyramid for the Osage Site (Table 11.4) uses the peak values during the summer (15 July-15 August) for all three years (1970-72). In this analysis, 50 percent of the omnivore small mammals, omnivore invertebrates, and scavenger invertebrates and 44 percent of the soil arthropods are treated as primary consumers. The producer level contains the overwhelming majority of biomass, with approximately 2600 g m^{-2}, while the primary and secondary consumers are about 0.5 g m^2 (less than 0.02 percent) and 0.3 g m^{-2}, respectively. If these same data are compared on the basis of live aboveground and belowground material, the trends are similar but with some differences (Table 11.5). Belowground consumers contribute approximately 0.13 percent of the total belowground biomass and aboveground consumers make up 0.09 percent of the aboveground biomass. Belowground biomass is greater in all three trophic levels--that is, 60, 64, and 85 percent of the producer, primary consumer, and secondary consumer biomass are belowground.

Other comparisons can be made within the trophic-level concept and biomass pyramids (Kozlovsky, 1968; Wiegert and Owen, 1971). In Table 11.6 biomass was again separated between above- and belowground, but the components were also separated into biophage and saprophage. In so doing, dead plant parts and consumers that

TABLE 11.4 Simplest biomass pyramid based on summer values averaged over the three-year samples at the Osage Site.[a]

| Category | Consumers (g m^{-2}) | | Producers (g m^{-2}) | Total |
	Secondary	Primary		
Invertebrates	0.0650	0.1620	--	0.2270
Soil arthropods	0.3590	0.2820	--	0.6410
Small mammals	0.0030	0.0516	--	0.0546
Birds	0.0024	0.0004	--	0.0028
Live and dead plants	--	--	2601.0	2601.0000
Total	0.4294	0.4960	2601.0	2601.9254

[a] Ungrazed treatment.

TABLE 11.5   Biomass pyramid of aboveground and belowground live
             components based on summer values averaged over the
             three-year samples at the Osage Site.

Trophic level	Belowground		Aboveground	
	Component	Biomass (g m^{-2})	Biomass (g m^{-2})	Component
Secondary consumers	Soil arthropods	0.3590	0.0650	Invertebrate
			0.0030	Small mammals
			0.0024	Birds
			0.0704	
Primary consumers	Soil arthropods	0.2820	0.1620	Invertebrate
			0.0516	Small mammals
			0.0004	Birds
			0.2140	
Primary producers	Plant roots	482.0000	323.0000	Plant shoots
Total		482.6410	323.2844	

feed on dead or decaying material were called saprophages. The
complements in each category were termed biophage. Again summer
values were used; these designations would clearly be altered in the
winter. In all comparisons, belowground components were larger than
aboveground components. The most interesting point, however, was
that even in midsummer, when the grassland was at peak live biomass
status, 55 percent of the biomass was in the saprophage components.
However, in both the above- and belowground categories, between 75
and 80 percent of the primary and secondary consumers were
biophagic.

French et al. (1978) used the Osage Site data to construct some
biomass pyramids at different seasons over three years
(Figure 11.3). To simplify the analysis, they assumed that all
scavenger invertebrates were considered saprophagic; primary
consumer mammals, birds, and carnivorous small mammals were
considered biophagic; and raptors, coyotes, deer mice (Peromyscus
maniculatus), and thirteen-lined ground squirrels (Spermophilus
tridecemlineatus) were considered 25, 15, 10, and 25 percent
saprophagic, respectively. In the diagrams the total biomass of
each level was indicated on the logarithmic horizontal scale. On

TABLE 11.6 Biophage and saprophage biomass pyramid of aboveground and belowground live components based on summer values averaged over the three-year samples at the Osage Site.

Trophic level	Belowground (g m^{-2})			Aboveground (g m^{-2})		
	Components	Biophage	Saprophage	Biophage	Saprophage	Components
Secondary consumers	Arthropods	0.2728	0.0862	0.0570	0.0130	Invertebrates Small mammals Birds
Primary consumers	Arthropods	0.2140	0.0680	0.1140 0.0516 0.0004	0.0430	Invertebrates Small mammals Birds
Primary producers	Plant roots	482.0000	922.0000	323.0000	874.0000	Plant shoots and litter
Total		482.4868	922.1542	323.2230	874.0560	

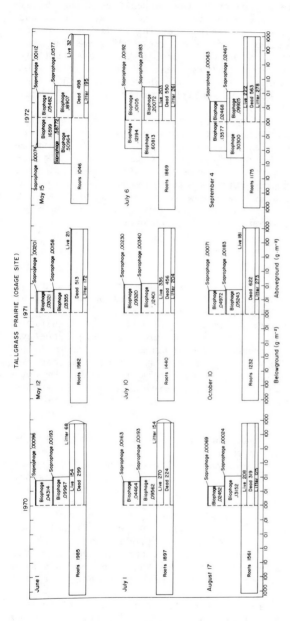

FIGURE 11.3 Biomass pyramids calculated from Osage Site data at three times during each year, 1970-72 (French et al., 1979).

451

the vertical axis, each of the consumer levels of the pyramid was divided on a linear scale according to the percentage of that layer that was biophage or saprophage. The aboveground producer level was subdivided virtually on a linear scale according to live, dead, and litter. The most noteworthy point from these diagrams is that relatively little year-to-year change occurs in the shape of these pyramids, although rainfall for the first year (1970) was well below average and rainfall for the second year was above normal. The same point, that biomass is not particularly sensitive to climatic conditions, was made in Chapter 6.

The True Prairie biomass pyramids show that by far most of the biomass is in the producer component, that most of the biomass is belowground at all trophic levels, and that, while total biomass is greater in the saprophagic part of the pyramid, consumers are greater in the biophagic portion. These conclusions are drawn from summer samples; a greater proportion of biomass would be underground and in the dead categories during the winter. We should note that sampling techniques did not include the top carnivores such as vultures, hawks, coyotes, weasels, skunks, and opossums.

Turnover Rates

Turnover rates can be calculated by comparing the amount of material entering or leaving a compartment to the amount of material in that compartment during the same period of time. Whereas biomass estimates the amount of material in a compartment at a given time, turnover indicates the speed at which a material (e.g., energy, carbon, nutrients) moves through the compartment. Turnover time, the reciprocal of turnover rate, gives the amount of time for the material in the compartment to be completely replaced. In both cases where turnover time deals with individuals and continuous measures such as carbon, a turnover time does not mean that all individuals or all molecules of carbon have been replaced. Rather, the term is an average calculated over a period of time and the total amount of input (or output) is equivalent to the amount in the compartment, under equilibrium conditions.

In Chapter 6 turnover times were calculated on the basis of plant biomass. For example, turnover of the plant shoot compartment, which occurs about once a year, comes from dividing peak live standing crop by annual productivity of plants. In the instances where most plants had a single peak, the two values were about equivalent and turnover time was about one year. In the cases where plants produced two peaks of growth, production was greater than peak standing crop at any one time, so the turnover rate was greater than 1.0. This situation with more than one peak was seen in the grazed treatment, which was characterized by both cool- and warm-season species.

ELM simulation results were used to examine the aboveground plant biomass turnover rates under no grazing and under treatments of moderate seasonal and year-round grazing. Annual aboveground production was compared to average standing crop of both live shoots and standing-dead material (Table 11.7). Like the calculations based on biomass, simulated data show that the ungrazed treatment had a turnover time of about one year. Grazing increased turnover

TABLE 11.7 Aboveground plant biomass turnover rates for
ungrazed, moderate seasonally grazed, and
moderate year-round grazed conditions. Calculations are based on ELM simulations of the
Osage Site.

Year	Turnover rate[a]	Turnover time (years)
Ungrazed		
1970	1.06	0.94
1971	0.95	1.05
1972	0.79	1.27
Average	0.93	1.09
Moderate seasonal grazed		
1970	1.13	0.88
1971	1.21	0.83
1972	1.01	0.91
Average	1.12	0.87
Moderate year-round grazed		
1970	1.35	0.74
1971	1.64	0.61
1972	1.33	0.75
Average	1.44	0.70

[a] The turnover rate is annual aboveground production divided
by average standing crop of live shoots and standing-dead
material.

rates and decreased turnover time. This response to grazing was
greater with year-round grazing than with seasonal grazing. The
latter difference resulted in part from the average standing crop
being higher under seasonal grazing (468 g m^{-2}) than under
year-round grazing (424 g m^{-2}). Also, annual production was greater
under year-round grazing (609 g m^{-2}) than under seasonal grazing
(522 g m^{-2}).

Respiration Budget

Gross production of an ecosystem can be defined as the total amount of carbon (or energy) fixed by the producers. This accumulated carbon or energy may meet one of several fates. For instance, the carbon can be stored, lost from the plants by respiration or death, consumed by herbivores, or utilized by decomposers. Subsequently, carbon or energy in herbivores and decomposers may be stored or lost because of death, respiration, or consumption by higher order consumers.

Using the Osage version of the ELM simulation model based on three years of field data from the Osage Site, we constructed a respiration budget (Table 11.8). The simulated results include three treatments: ungrazed by large herbivores, moderate seasonal grazing, and moderate year-round grazing. Both gross and net production by the whole system are greatest under year-round moderate grazing conditions. The ungrazed treatment is slightly more productive than the grazed. When comparing plant parts, most of the respiration comes from aboveground shoots (70 percent), while roots and crowns contribute 20 and 10 percent, respectively. The amount of biomass consumed by consumers is about twice as much as the biomass respired by consumers.

Table 11.9 compares the amount of energy, in biomass equivalents, released in respiration by plants, animals, and microbial organisms. Under all three treatments, the relative contributions are approximately equal, with the plant releasing about 60 percent, the animals about 1 to 5 percent, and the microorganisms about 35 to 40 percent.

In response to grazing, shoot respiration is increased by 8 and 21 percent under seasonal and year-round grazing, respectively. Plant crown and root crown respiration is decreased about 5 percent under seasonal grazing but increased 25 to 30 percent under year-round grazing. Microbial respiration increases about 3 to 4 percent under seasonal and year-round grazing. Animal respiration increases in order of magnitude between ungrazed and year-round grazing. So the plant respiration contribution is the greatest under all treatments, and animal respiration contributes less than 5 percent even under year-round moderate grazing. If microbial and animal respiration are compared to net production, the system is shown to be not balanced, though nearly so. The percentage net loss, averaged over the three years, is 4.1, 25.7, and 1.4 percent for the ungrazed, seasonally grazed, and year-round grazed, respectively.

Mineral Cycling

Although most biomass is composed of organic materials, a small fraction of the weight consists of inorganic minerals, which are sometimes referred to as ash weight. These minerals are important in ecosystem processes because some are necessary for plant and animal growth. Nutrients held in plant parts, whether dead or alive, are generally not available to other plants but are available to the consumers. Once these plant parts decompose, minerals are

TABLE 11.8 Simulated production, respiration, and consumption for the True Prairie at the Osage Site. Biomass values are an average of three years of simulations, 1970-72.

Treatment	Plant respiration ($g\ m^{-2}\ yr^{-1}$)			Microbial respiration ($g\ m^{-2}\ yr^{-1}$)	Animal ($g\ m^{-2}\ yr^{-1}$)		Production ($g\ m^{-2}\ yr^{-1}$)	
	Shoot	Crown	Root		Respiration	Consumption	Gross	Net
Ungrazed	801	114	242	757	10	14	1890	737
Moderate seasonal grazing	868	108	228	778	54	97	1867	662
Moderate year-round grazing	968	147	299	788	109	209	2299	885

TABLE 11.9   Relative amounts of energy released through respiration under ungrazed, moderate seasonal grazing, and moderate year-round grazing. These calculations are made from simulated data based on a three-year average of Osage data, 1970-72. Values in parentheses are percent of energy in each treatment.

| Treatment | Energy release via respiration (g biomass m^{-2} yr^{-1}) | | | |
	Plants	Animals	Microbial	Total
Ungrazed	1157 (60.0)	10 (0.5)	757 (39.5)	1924
Moderate seasonal	1204 (59.0)	54 (2.7)	778 (38.3)	2036
Moderate year-round	1404 (61.2)	109 (4.7)	788 (34.1)	2311

released where they can be absorbed by other plants, held in the soil, or lost from the system by subsurface or overland flow.

In a study on standing crop and litter of bluestems (Andropogon spp.), Koelling and Kucera (1965) found that considerable leaching or loss from the standing material occurred during the first year. This observation was especially true for potassium, phosphorus, nitrogen, and magnesium, which showed a 96, 72, 68, and 74 percent turnover, respectively, in the vegetation before the end of the first year. The percentage of calcium, iron, and silica dioxide increased by 1.24, 1.66, and 2.63 times during the same period. Nitrogen, calcium, silica dioxide, and iron increased in percent content of litter after two years. Dry matter losses in litter samples were about 50 percent more rapid for foliage than for flower stalks. Total ash after two years of decomposition in litter samples was approximately twice the initial level for both foliage and flower stalks.

The results of applying these mineral contents to the biomass data from the Osage Site are shown on Table 11.10. Even at the time of maximum biomass, ash content is only about 23 g m^{-2}. Table 11.11 shows the distribution of these minerals among aboveground components. Nitrogen is rather evenly distributed among the plant components, although the lowest concentration occurs in live shoots. Very little potassium is in litter and standing dead, but shoots and crowns together contain 72 percent of the potassium. Phosphorus and magnesium are evenly distributed, but the greatest amounts of silica and calcium occur in standing dead and litter. Essentially half the iron is located in standing dead. These data show that, while some patterns of nutrient cycling recur in the True Prairie, the behavior of individual nutrients differs. For example, manipulations that decrease standing dead and litter will have a greater effect on the nutrient cycles of such minerals as calcium and silicon, but alterations of shoots will cause a proportionately greater change in potassium and phosphorus.

TABLE 11.10 Amounts of seven minerals and the ash content of live biomass at the ungrazed plot on the Osage Site at four dates during 1972.

Date	Minerals (g m^{-2})							
	Nitrogen	Potassium	Phosphorus	Calcium	Magnesium	Silica Dioxide	Iron	Ash
15 May	0.78	0.73	0.09	0.13	0.05	1.43	0.01	3.55
6 July	2.03	3.21	0.26	0.81	0.32	11.88	0.07	18.55
9 August	1.85	0.81	0.38	1.22	0.43	16.05	0.07	22.94
15 November	0.02	0.00	0.00	0.01	0.00	0.19	0.00	0.33

457

TABLE 11.11 Mineral content of aboveground producer components at the time of peak live biomass on the ungrazed plot of the Osage Site in 1972. Values in parentheses are the percentage of each mineral in the four components.

| Component | Mineral content (g m^{-2}) | | | | | | | |
	Nitrogen	Potassium	Phosphorus	Calcium	Magnesium	Silica Dioxide	Iron	Ash
Live aboveground	1.85 (18)	0.81 (40)	0.38 (30)	1.22 (17)	0.43 (24)	16.05 (12)	0.07 (8)	22.74 (13)
Standing dead	3.50 (32)	0.37 (18)	0.32 (25)	2.21 (30)	0.51 (29)	48.36 (35)	0.43 (49)	57.63 (34)
Crowns	2.47 (25)	0.66 (32)	0.32 (25)	1.60 (22)	0.46 (26)	27.90 (20)	0.20 (23)	35.74 (21)
Litter	2.71 (25)	0.20 (10)	0.24 (20)	2.34 (31)	0.37 (21)	45.43 (33)	0.17 (20)	55.26 (32)

Effect of Herbivores on the Community Structure

With a few exceptions, the role of predation in regulating vegetational diversity has been largely ignored (French et al., 1976; Harper, 1969; Lee and Inman, 1975). As a result, generalizations are tenuous but are the basis of studies done mostly in England. Harper suggested the following:

1. The introduction of a herbivore that is not regulated by predation or parasitism can take a plant from a dominant to a minority component of the flora, facilitating the invasion by other species.
2. Because all grassland herbivores have palatability preferences, none is truly generalist--that is, none eats every plant species in its feeding path in proportion with frequency of occurrence. Mowed grasslands have a richer flora than undergrazed grasslands, which tend to have monotonous floras. Undergrazing favors less palatable components of the flora and depletes the flora. As the intensity of grazing increases, relatively more unpalatable species are grazed and only totally distasteful species remain. Long, continuous overgrazing leads to the introduction of a new component of the flora in the form of toxic, spiny, or woody species not previously present.
3. An "apostatic" feeder is one that, choosing from among its potential foods, concentrates on the one in greatest supply, and turns elsewhere only when that food becomes rare. This behavior pattern penalizes the food species in greatest supply, favors the minorities, and acts to stabilize mixtures of species.
4. Apparently diversity is encouraged if the introduced herbivore selects the major dominant, and the most species-rich communities are created by continuous grazing with a maintained population of herbivores. The inability of a year-round feeder to adjust its food demand to varying seasonal food supply produces severe overgrazing in seasons of low plant growth and marked undergrazing in periods of rapid plant growth. This cycle leads to a rapid invasion by weedy species.
5. A palatable plant species, heavily consumed by a generalist animal feeder, may be expected to be selected over an evolutionary time scale for decreased palatability. Thus, herbivores prepared to break the palatability barrier would have selective advantage, which would lead to processes of radiative evolution in both the herbivore and the prey. As the palatable plant decreases in palatability and the number of herbivores that can feed on it decreases, the herbivores themselves become increasingly restricted by the plant species on which they can feed. As a result, both the herbivores and the prey would likely experience density-dependent regulation, and plant density is depressed lower than the density expected without predators. This low density permits further species invasions.

Little information of this sort exists for the True Prairie. Numerous studies indicate that heavy cattle grazing deteriorates range condition, promotes undesirable grasses and forbs, and

decreases plant vigor (Hazel, 1967). Five years after an Oklahoma oldfield was abandoned, the Shannon-Wiener diversity index ($\log_{10}$) increased from 1.2 in the first year to 2.4 in the fifth year (Kapustka and Moleski, 1976). On the Osage Site the same index was 1.22 and 1.31 in 1971 and 1.09 and 1.22 in 1972, respectively, in the ungrazed and grazed treatments (Risser and Kennedy, 1972; 1975). So, although these cattle grazing studies are general, none specifically ties the behavior of ungulate herbivores to vegetation structure.

In the ELM simulation runs, net primary production was 55 g m^{-2} greater when small mammals were present than when they were not. This figure represents an increase of 7.1 percent. Approximately 15 g m^{-2} or 2 percent of the net primary production was consumed by small mammals. Increases in net belowground production caused net primary production with small mammal grazing to increase 70 percent. The no-grazing simulations had lower values for warm season grasses, but small mammals decreased cool season peak live biomass by 22%. At the time the cool season annuals were available for food, small mammals were in the period of most rapid growth.

Simulated Responses of the Ecosystem
to Various Management Strategies

Previously, the basic characteristics of the True Prairie were described, various analytical studies were discussed, and a simulation model was presented. This model was built from recognized structural and functional characteristics of the grassland and has been shown to track closely most of the measured attributes of the True Prairie ecosystem. Therefore, we can now use the model as a common basis for comparing ecosystem responses to a number of management strategies. This exercise may produce some insights, but bear in mind that the validation of these results is determined by the soundness of the model structure.

In the simulated cow-calf grazing system, calves were born in February and removed from the range on 1 October, while the cows remained on the range year-round. Light, moderate, and heavy stocking rates were 6.5, 4.0, and 3.2 ha/AU respectively. Ecosystem characteristics of moderately and heavily grazed conditions can be compared to the condition in which there is only small mammal herbivory. In Figure 11.4, the results from a three-year simulation are summarized in comparison to the conditions on the True Prairie when no cattle are grazing. Under both moderate and heavy stocking rates, both the peak live warm-season forb biomass and the water loss from bare soil increase. Grazing the grassland increases the amount of bare soil surface, which not only permits the increase of forbs but also permits more evaporation from the soil. Because nitrogen is lost under grazing conditions, especially by volatilization from feces and urine, the marked depression of the net nitrogen balance is not unexpected. Also, because these grasslands evolved under at least periodic grazing pressures, it is not surprising that many variables demonstrate relatively small changes when grazing is imposed. As compared to moderate grazing, heavy stocking rates accentuate some of the changes in the most

FIGURE 11.4   Responses of various True Prairie ecosystem characteristics to moderate and heavy grazing under a year-long cow-calf grazing system.   These results are based on simulated data from three years at the Osage Site, 1970-72.   The zero point (control) represents the status of each characteristic with only small mammal herbivory.   See text for details.

responsive characteristics, especially characteristics that are negatively affected.

We can compare the ecosystem responses of grasslands grazed with a season-long cow-calf operation with grasslands grazed with a seasonal steer operation. With the latter, yearling steers with an initial weight of 250 kg, were grazed from 1 May through 30 September at a stocking rate of 2.4 ha/AU. When these results are compared to small mammal herbivory as the control (Figure 11.5), again the peak live warm-season forb biomass and the bare soil water loss were the most responsive positive characteristics and the net nitrogen balance was significantly decreased under both treatments. Under seasonal grazing, the cool-season annual biomass increases because of the absence of grazing pressure during early spring. However, the heavier stocking rates during most of the growing season with the seasonal steer operation depresses all estimates of primary production, including net aboveground primary production.

The simulated results of adding nutrients or water to the True Prairie ecosystem are shown in Figure 11.6. The simulated pastures, under a moderate stocking of steers from 1 May to September 30, received either additional nitrogen and phosphorus or water. Twenty g m^{-2} of nitrogen and 3 g m^{-2} of phosphorus were added in the spring (31 March) of the first year of the three-year simulation run. Water was added from May through September if the soil water tension in the 10 to 15 cm soil layer was less than -5 bars. The grassland showed considerable response to both supplemental nutrients and water. When nitrogen and phosphorus were added, root uptake and nitrogen content of the forage increased. Also, because fertilizer was applied early in the spring, the cool season plants demonstrated an increase. As a consequence, cattle weight gains per head were increased.

When the Tallgrass Prairie was irrigated, the primary production characteristics increased, but nutrient uptake was not enhanced proportionately. This resulted in forage of lower nutrient content and a smaller increase in cattle weight gains per head. When nutrients and water were added simultaneously, the effects were largely additive (Figure 11.7). Both nitrogen and phosphorous uptake as well as all primary production values increased.

As described in Chapter 10, the early history of the True Prairie involved repeated burning. Five-year burning simulations of annual and perennial spring fires on a seasonally grazed Tallgrass Prairie are depicted in Figure 11.8. If the grassland is burned every year, the warm-season plants are enhanced, while the cool-season species are decreased. The same responses are evident if the rangeland is burned once every three years, but the magnitudes of the changes are diminished. A most interesting result is the net nitrogen loss under the two burning regimes. While nitrogen is lost under the annual and triennial burning, the loss is much less with only one fire every three years.

These summary model simulations demonstrate that numerous characteristics of the tallgrass prairie respond to various management alternatives. It is now possible to anticipate these changes and, by comparing these results, to make enlightened decisions about the utilization strategies to be employed on the True Prairie.

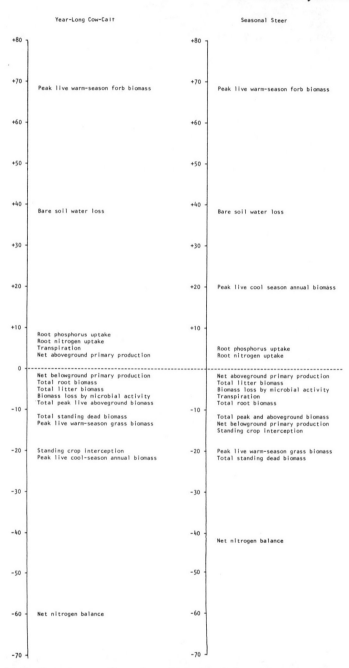

FIGURE 11.5   Responses of various True Prairie ecosystem characteristics to moderate year-long and seasonal grazing regimes.   These results are based on simulated data from three years at the Osage Site, 1970-72. The zero point (control) represents the status of each characteristic with only small mammal herbivory.   See text for details.

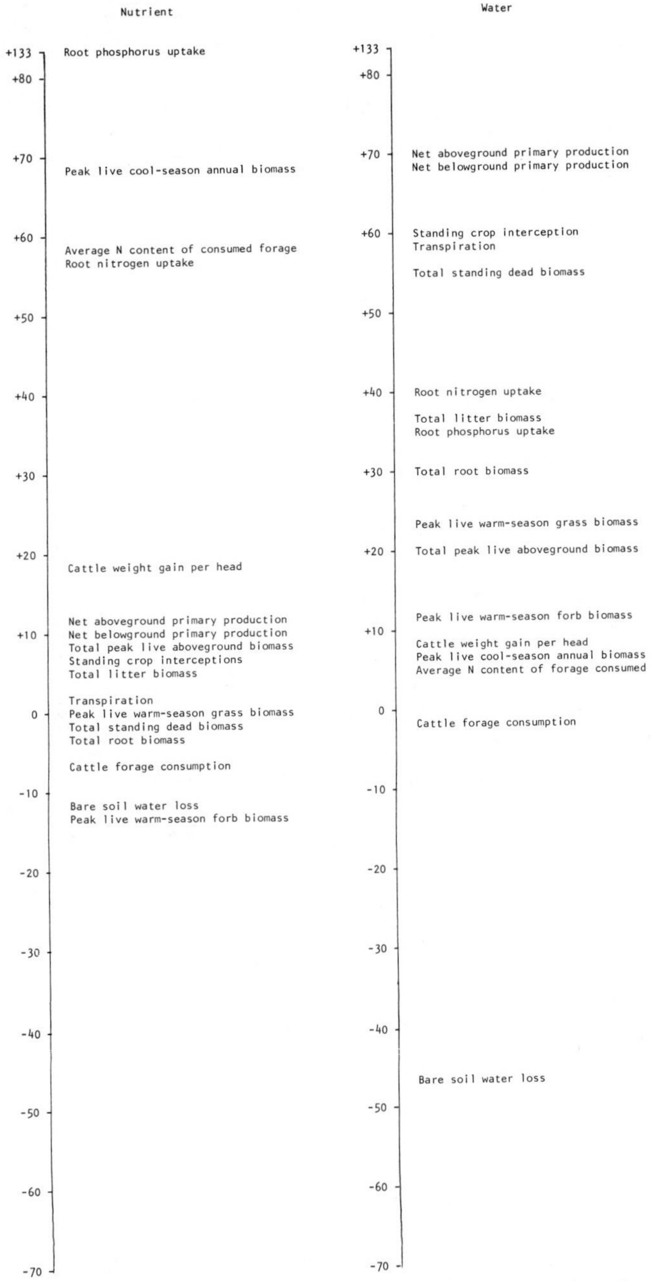

FIGURE 11.6 Responses of various True Prairie ecosystem characteristics to the addition of nutrients and water to seasonally grazed grassland. These results are based on simulated data from three years at the Osage Site, 1970-72. The zero point (control) represents the status of each characteristic on a rangeland with moderate stocking of steers during the growing season. See text for details.

FIGURE 11.7 Responses of various True Prairie ecosystem characteristics to the addition of both nutrients and water to a seasonally grazed grassland. The results are based on simulated data from three years at the Osage Site, 1970–72. The zero point (control) represents the status of each characteristic on a rangeland with moderate stocking of steers during the grazing season. See text for details.

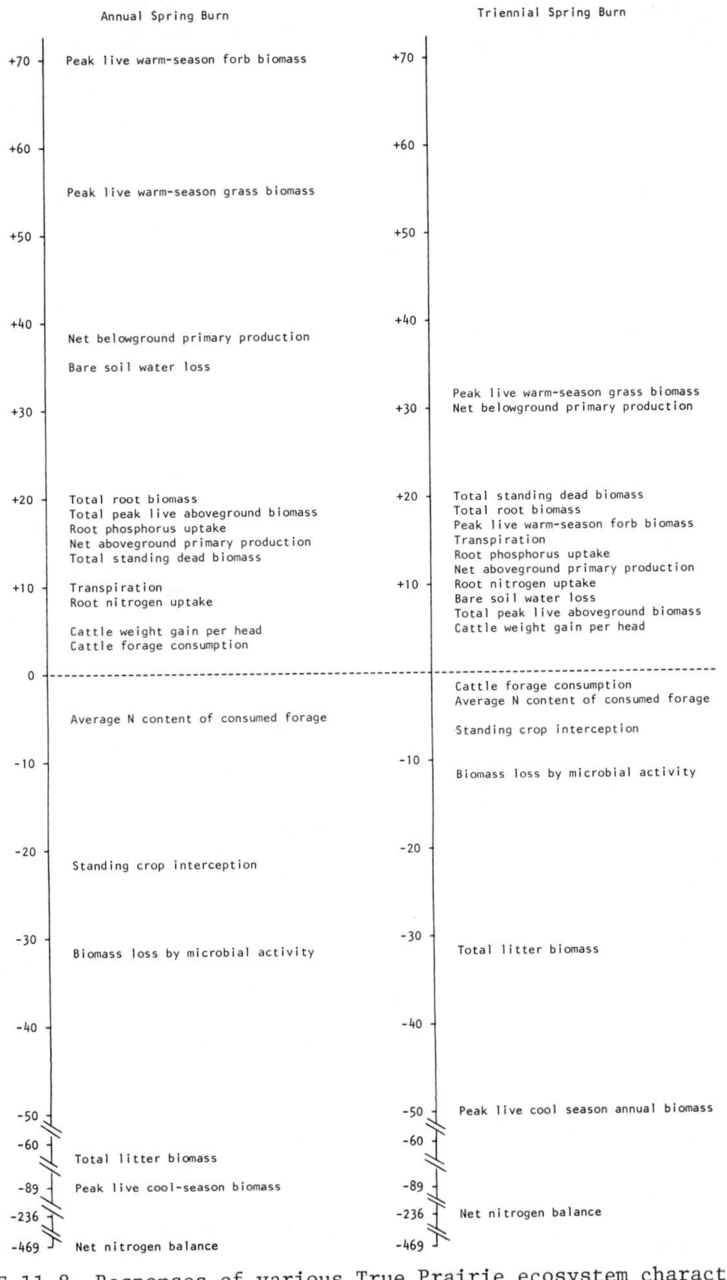

FIGURE 11.8 Responses of various True Prairie ecosystem characteristics to annual and triennial spring burns on a seasonally grazed grassland. These results are based on five years of simulated data for the Osage Site, 1970-72, with 1971 and 1972 repeated for the fourth and fifth year. The zero point (control) represents the status of each characteristic on a rangeland with moderate stocking of steer during the growing season. See text for details.

Summary

GEOLOGY

1. One of the first geological events that contributed to the eventual True Prairie was the Rocky Mountain uplift, which resulted in the deposition of a huge, alluvial, east-sloping plain stretching essentially to the Mississippi River. The mountain range also eventually caused the continental climate characteristic of the grasslands.
2. The northern portions of the True Prairie were glaciated and this process contributed to rounded topography and a heterogeneous geological substrate.
3. Soils of the True Prairie were derived from geological materials of different ages, from Precambrian rocks, which date back more than 600 million years, in the southeast to material in the north deposited subsequent to the latest glaciation (approximately 10,000 years ago).
4. Over geologic time, the True Prairie region supported, in sequence: subtropical forests, warm temperate forests, shrubs and chaparral, grasslands, warm temperate and subtropical forests, grasslands, boreal forests, temperate forests, and grasslands.
5. The origin of the prairie as a recognizable community probably dates from the Oligocene Epoch of the Tertiary Period--about 25 million years ago.

CLIMATE

1. Along a north-south axis through the True Prairie, the growing season climate is substantially homogeneous, with a small continual decrease in maximum and minimum air temperatures that causes average air temperature, number of frost-free days, length of growing season, mixing ratio, and potential evapotranspiration rates to decrease northward. To the north, the number of rain-days greatly increases and the average rainfall amounts decrease.

2.   During the nongrowing season, fairly large north-south changes occur in all climatic variables except wind speed. The maximum north-south difference in climatic variables is generally two to three times larger during the nongrowing season than during the growing season.

3.   During the growing season the wind speed continually increases and the number of frost-free days westward across the True Prairie decreases; during the nongrowing season, cloud cover, minimum air temperature, total precipitation, and rain days increase eastward, and snowfall decreases.

4.   East of the True Prairie, most growing season climatic variables change very little, but during the nongrowing season, windspeed and snowfall decrease significantly and minimum air temperature, total precipitation, and rain days increase.

5.   West of the True Prairie, almost all the growing and nongrowing season climatic variables change rapidly. Some of the most significant changes include increases in wind speed, solar radiation, and potential evapotranspiration and decreases in relative humidity, cloud cover, minimum air temperature, length of growing season, total rainfall, average rainfall, and ratio of precipitation to potential evapotranspiration.

6.   The growing season climate is more variable in the southern portions of the True Prairie, but the nongrowing season climate is more variable in the northern regions.

7.   The maximum diurnal variation in temperature within the True Prairie foliage canopy is at the plane of maximum leaf development. Maximum relative humidity occurs at the lowest levels of the plant canopy.

FLORA

1.   Grasses probably evolved late in the Mesozoic after flowering plants were well diversified, but grasses do not appear frequently in the fossil record until the lower Eocene.

2.   The geographic origin of prairie plants has been from the southeast, southwest, and north. Because environmental conditions favoring migration from these areas would not likely have occurred simultaneously, the origin of the prairies was not a single, unified development of a homogeneous plant formation, but rather a gradual entrance of various elements, each arriving during separate climatic regimes favorable to that particular group.

3.   In general, representative areas of the True Prairie contain about 250 species of higher plants.

4.   Some plant species such as little bluestem occur throughout the entire range of the True Prairie.

5.   Ninety-five percent of the True Prairie species are perennial, many with life spans of twenty years or more.

PRIMARY PRODUCTION

1.   The total amount of energy captured by the grassland is about 20 kcal m^{-2} day^{-1}

2. Maximum rates of carbon fixation for tallgrass species appear to be about 20 to 25 mg CO^2 dm^{-2} h^{-1}.

3. The maximum daily net accumulation rates on the True Prairie range from 4.0 to 6.5 g m^{-2} day^{-1} for total plant biomass, 1.5 to 2.5 g m^{-2} day^{-1} for aboveground biomass, and 1.0 to 3.0 g m^{-2} day^{-1} for roots.

4. Most Tallgrass Prairies produce a peak live standing crop of about 400 g m^{-2}. This figure is probably a 20 percent underestimate of actual peak live biomass, which is probably nearly 480 g m^{-2}.

5. Crown and belowground biomass in the True Prairie range from 100 to 400 and 900 to 1900 g m^{-2}, respectively, and annual root production ranges from 200 to 600 g m^{-2} yr^{-1}.

6. Most True Prairie warm season perennial grasses have rooting depths greater than 1.5 m, and many forbs have even deeper roots.

7. Usually 75 percent or more of the root biomass occurs in the A soil horizon, generally equivalent to the top 25 cm.

8. About 40 percent of total primary production goes into shoots, nearly 40 percent into roots, and the remainder into crowns.

9. The best estimates of total annual primary production range from 850 to 1350 g m^{-2} yr^{-1}.

10. Except at the short time of peak aboveground live biomass, the amount of live producer biomass is greater belowground. When the maximum biomass values achieved during the season are considered, the producer belowground maximum is two to three times greater than the aboveground maximum. If ratios of annual aboveground to belowground productivity are calculated on the rates of production, which includes respiration loss and translocation, values are generally above 1.0.

11. The percent efficiency of energy capture is about 0.75 percent for aboveground net primary, root, and crown productions during the growing season. Light to moderate grazing increases this total net primary production efficiency to 0.87 percent.

12. In the True Prairie the material in the shoot category turns over about once a year, the crown material turns over every 1.25 years, and the roots and litter turn over about every 3 to 4 years.

13. The amount of litter on the soil surface ranges from about 200 to 500 g m^{-2}, although amounts may exceed 700 g m^{-2}

14. Cool-season grasses and most forbs have a higher protein content than warm-season grasses. Crude fiber may be slightly higher in cool-season grasses early in the growing season, but much lower than in warm-season grasses later in the year.

15. Seventy percent or more of the potassium, phosphorus, nitrogen, and magnesium may be leached or lost from the standing-dead plants during the first year.

16. Even at maximum plant biomass, the ash content is only about 23 g m^{-2}. Nitrogen is fairly evenly distributed among the plant components, with the lowest concentration being in shoots. Shoots and crowns contain 72 percent of the potassium, with very little in litter or standing dead. The greatest amounts of silica and calcium are in standing dead and litter, while essentially half the iron is in standing dead.

INVERTEBRATES

1. Insects are an old group and most of the present families evolved prior to the Permian.
2. The grasshopper and leafhopper fauna of the True Prairie is not a distinct unit but has strong ecological and taxonomic affinities with the surrounding biomes.
3. At least some of the invertebrate fauna have apparently evolved in adjacent biomes and spread to the central grasslands. Species numbers and diversity are much greater in seral stages of the deciduous forest than in comparable grass communities of the True Prairie. Because of this proximity to the deciduous and boreal forests, the True Prairie probably has a greater species diversity than other grasslands.
4. Insects and other invertebrates are found in all consumer trophic levels, both above- and belowground. Although the impact of herbivores has commonly been considered destructive, in many instances it may increase primary productivity. Secondary consumers have an impact on population regulation and are primarily beneficial. The increase in the rate of organic matter recycling by saprophages may be the most important activity of invertebrates in the True Prairie.
5. During 3 years of sampling in northeastern Oklahoma, 16 orders and more than 131 families of aboveground insects were determined. More than 3,000 aboveground species are estimated to be at the Osage Site, although belowground species numbers are much lower.
6. Orthoptera, particularly the acridid grasshoppers, are the major aboveground herbivores of the True Prairie ecosystem, while nematodes are the major belowground herbivores.
7. Aboveground invertebrates may reach densities as high as 4,000 m^{-2} and insects as high as 1,850 m^{-2}. Individual groups such as mites, ants, and thrips (1,800, 875, and 700 m^{-2}, respectively) may attain high numbers. Total aboveground invertebrate biomass may reach 270 mg m^{-2}; herbivore biomass reached 150 mg m^{-2}.
8. Belowground numbers of arthropods and nematodes may exceed 135,000 m^{-2} and 6,500,000 m^{-2}, respectively, to a depth of 50 cm. Average belowground biomass for these two groups was consistently estimated to be at least ten times greater than the aboveground invertebrate biomass.
9. Estimates indicate that belowground invertebrates accounted for more than 90, 95, and 93 percent of the herbivorous activity, secondary consumption, and saprophagous activity, respectively.

AVIFAUNA

1. Little is known or can be inferred about bird populations or diversity in North American grasslands during past geological periods. At the present time, however, the prairies are characterized by simple and meager bird faunas.
2. No clearly defined grassland avifauna can be distinguished. While some species clearly exhibit affinities for grassland habitats, only a dozen or so can be considered characteristic

of grasslands, and even these species have distributional patterns that extend beyond the floristically defined borders of the prairies.

3. Assemblages of bird species closely associated biogeographically with the True Prairie are difficult to distinguish. The most abundant breeding birds in True Prairie locations are widespread and also frequent grassland-like agricultural habitats to the east and, to a lesser degree, range into the Mixed-grass Prairie to the west. Distinct north-south patterns in species distributions within the True Prairie are not apparent, although northerly locations do differ in the composition of breeding avifaunas from more southerly locations, and breeding communities to the north are somewhat more diverse.

4. The present bird fauna of North American grasslands bears only faint affinities with its Eurasian steppe counterpart but is closely related to the bird fauna of South American grasslands and savannah.

5. Excluding large game birds and raptors, approximately thirteen breeding species occur in True Prairie habitats, although usually only four or five are present in any single location. Of the breeding species, roughly two-thirds are seasonal migrants that occupy the breeding locations only during the breeding season. Approximately 60 percent of the breeding species are omnivorous during the breeding period, and the remaining 40 percent consume primarily insect prey. No apparent north-south or east-west patterns exist in residency or diet.

6. Densities of breeding birds in the True Prairie average roughly 350 individuals km^{-2}, but may be as high as 600 km^{-2}. Biomass ranges from 0.007 to 0.031 g m^{-2} wet weight (roughly 0.002 to 0.009 g m^{-2} dry weight). These values are generally similar to values recorded in other North American grasslands, although biomass is frequently lower in more arid western grasslands.

7. No patterns of variations in total density or biomass of breeding avifaunas in the True Prairie are apparent. Values from adjacent census plots in the same km^2 may differ more than values from plots 1,000 km apart.

8. True Prairie avifaunas exhibit a well-defined seasonal aspect, with perhaps a 60 to 70 percent change in species composition between summer and winter. In general, wintering avifaunas are less diverse and abundant than breeding avifaunas, although wintering assemblages have received little study.

9. At the Osage Site, meadowlarks were the most common species during spring, then decreased in relative abundance through summer and fall into winter. Small ground-feeding species such as sparrows or horned larks accounted for 10 to 30 percent of the individuals and peaked in relative abundance in summer. During winter, 70 percent of the species present were associated with brushy roadside areas, but none of these species was a native grassland species.

10. Model simulations suggest that the bird populations at the Osage Site process roughly 1 to 3 kcal m^{-2} over the breeding season, which is similar to energy flow estimates from other North American grassland types but approximately 10 percent of

the values recorded for northwestern coniferous forest avifaunas. Daily energy flux at the Osage Site varied from approximately 0.005 to 0.030 kcal m^{-2} day^{-1}. During the breeding season production accounted for 1.5 percent of the total energy flow. Only 25 percent of the breeding season energy flux passes through populations that will later migrate from the area.

11. At the Osage Site, beetles (curculionids, scarabaeids, carabids), grasshoppers and crickets, lepidopterous larvae, and grass seeds are the major diet constituents of breeding birds. Individual variation in dietary composition is marked in meadowlarks but slight in grasshopper sparrows and dickcissels. In general, breeding species feed upon the same pool of prey species and capture similar-sized prey despite considerable differences in body and bill sizes among bird species.

12. Model analyses reveal that breeding bird populations at the Osage Site consume 200 to 535 kg km^{-2} of prey over the breeding season. Chewing phytophagous insects account for the greatest share of the total prey consumed with predaceous invertebrates of secondary importance.

13. Model estimates suggest that energy flow through the bird communities occupying grassland-like agricultural habitats to the east of the True Prairie is roughly equivalent to energy flowing through the bird populations of native True Prairie. However, in hay and fallow fields, because bird densities have increased tremendously since the turn of the century, their energy flow is correspondingly greater there than in native habitats.

SMALL MAMMALS

1. The evolutionary interactions between grass and mammals began late in the Cretaceous and continue today. The radiation of grasses in the early Cenozoic made possible much of the Cenozoic radiation of mammals. Numerous mammalian herbivore lineages shifted from a browsing to a grazing existence as grasslands expanded in response to increasing aridity during the Miocene. Near the close of the Pleistocene about 11,000 years ago, much of the mammalian megafauna of the True Prairie became extinct in North America.

2. The True Prairie includes at least a portion of the distribution of 102 native and 3 introduced mammal species. Patterns of mammalian distribution in central North America do not indicate a unique or even a homogeneous True Prairie mammalian fauna. No species occurs solely within the True Prairie.

3. Small mammal biomass density on a northern True Prairie site in Minnesota averaged only about 14 percent of the density observed on a southern True Prairie site in Oklahoma.

4. The pattern of fluctuation in biomass density at the northern site appears to be seasonal, whereas at the southern site it appeared to be superannual. Winter breeding of voles was detected at the southern site but not at the northern site. Cotton rats sometimes contribute heavily to biomass density at

the southern site, but neither that genus nor an ecological counterpart occurred at the northern site. Variation in small mammal biomass between years is greater at the northern than at the southern locality.

5. Mean number of species captured per trapping effort was higher and less variable at the southern locality. Only eight species were trapped on the northern grid during three years of study, whereas eleven species were captured on the southern grid.

6. At the northern site, mean diversity (H') increased from June to July to August, then decreased from August to September. At the southern site, mean diversity decreased from May to August to October. Variability in diversity on both True Prairie sites was consistently high, with coefficients of variation falling between 43.5 and 48.0. The coefficient of variation for equitability is much greater at the southern site than at the northern site, which reflects that prairie vole and/or cotton rat populations sometimes dominated the small mammal community at the Osage Site. The meadow vole, however, although the most common species in nine of ten samples, never overwhelmingly dominated the small mammal fauna at the northern site.

7. Grassland habitats, including the True Prairie, present essentially a two-dimensional environment to mammals, except for those species small and agile enough to climb on the stems of grasses or species equipped morphologically for burrowing. Creating a third dimension by burrowing is a common habit observed in grassland mammals.

8. Within functional groups of mammals, biomass density tends to be more evenly distributed at a northern True Prairie locality than at the Osage Site.

9. A high mean similarity for life form from grassland site to grassland site indicates that the balance between life forms within the grasslands persists relatively independently of the species that compose a given community. However, small mammal communities have not evolved to be composed of particular arrays of functional groups. Instead, particular arrays of functional groups tend to exist in similar environments because they are able to exploit resources common to those environments. The Osage Site has a distinctive small mammal fauna, while the small mammal fauna of the northern True Prairie site is generally most like the fauna of a montane grassland.

10. Support efficiency of small mammal biomass at the Osage Site is less than the average for grasslands. Low support efficiency resulted at least partly from the two major species there (prairie vole and cotton rat), both of which are nonhibernating herbivores with relatively low assimilation efficiencies.

11. Small mammal production at the Osage Site was relatively very high, indicating that the True Prairie, at least when not being grazed, captures energy in a form palatable for small mammals at a rate capable of supporting exceptionally high secondary productivity, even in the absence of efficient biomass support.

12. Only an average of about 4 percent of the herbage available to the small mammal community at the Osage Site is actually consumed by small mammals. True Prairie mammals appear to

utilize slightly greater-than-average proportions of herbage and seeds and relatively less animal matter than mammals in other grassland types.

13. Movement of energy through small mammals in a True Prairie may be substantial. Small mammals at the Osage Site were estimated to consume as much as 197,168 kcal ha^{-1} yr^{-1}, which indicates that the amount of energy flow through small mammals might be high relative to their biomass but low relative to total standing-crop biomass. Small mammals may have an impact on grassland communities as a result of consumption of standing crop biomass, translocation of soil, enhancement of primary productivity as a result of preventing senescence of vegetation, and perhaps in other ways. Potential effects of grazing by small mammals should be less in the True Prairie than in other North American grasslands because of the greater primary productivity there.

14. Grazing may have a greater effect on small mammal communities in the True Prairie than in other grasslands because the potential for primary productivity is so great in the True Prairie that cessation of grazing allows the environment to change very rapidly in a way that markedly affects small mammal microhabitats. Grazing may alter the relationship between available cover and small mammal populations more drastically than in grasslands where vegetative cover never achieves such a high level.

15. Mean density of small mammal species and its variability increase in the True Prairie when grazing is decreased. Within a single growing season a small mammal fauna in a 1 ha grazing exclosure resembled that of an ungrazed habitat more than the fauna of a grazed habitat in diversity, equitability, species number, species composition, and biomass density.

GRAZING

1. Under the grazed condition at the Osage Site, significantly more energy is translocated belowground than in the ungrazed treatment, the amount of standing dead is much less on the grazed treatment, and (though not shown with the Osage Site data, presumably because of previous management) litter is usually less on a grazed treatment than on an ungrazed treatment.

2. Light to moderate grazing increases turnover rates of the primary producer compartment by about 25 percent.

3. Cattle grazing reduces infiltration and increases compaction, especially on fine-textured soils.

4. The amount of water entering the soil is positively correlated with the amount of standing crop and litter, and both standing crop and litter are negatively associated with grazing intensity.

5. Clipping studies have shown that lower cutting heights and more frequent cutting cause a decrease in both aboveground and belowground production. In warm season grasses, clipping at the time of seedstalk formation causes increased vegetative

reproduction. One or two clippings a year may cause a two- to threefold increase in production.
6. Plant species are differentially adapted for grazing, so some species decrease with grazing and others increase. Grazing also opens the vegetation cover, which permits the entry of some invader species.
7. Total production may be increased with grazing, but overgrazing causes a long-term reduction in primary production.

NUTRIENTS

1. Nitrogen fertilization increases water use, but plants make more efficient use of available water.
2. Both nitrogen and phosphorus can increase yield but only under adequate moisture.
3. Application of nitrogen and phosphorus often selectively enhances the growth of cool-season grasses and forbs.
4. Nitrogen content of the peak standing crop of shoots and roots is about 4.0 and 6.0 to 12.0 g m^{-2}, respectively. Ninety-five percent of the nitrogen in the system is in the soil components, but the total is not likely to exceed 1 to 2 percent by weight of the dry plant material or soil.
5. Total nitrogen released through decomposition is about 6 g m^{-2} yr^{-1}.

WATER

1. Grasslands dominated by cool season grasses have less midsummer soil water than grasslands dominated by warm season grasses.
2. Individual grassland species have specific differences in drought tolerances.
3. Herbage production can be increased with additional water, but in wet years this increase is minimal.
4. Responses to fertilizer are greatest with additional water.
5. Water-use efficiency of the plants decreases with increased water.
6. As soils dry out in the driest part of the year, more water occurs in the lowest soil horizons at the time when both shoot and root growth terminates. Roots stop growing at higher soil water levels, and shoots continue to grow under lower soil water availability.
7. Nutrient uptake by plants is enhanced with increased water availability.
8. All simulated results showed that aboveground and belowground production, live root biomass, and total root biomass increase with irrigation.
9. Invertebrates show a marked species compositional change under irrigated conditions.

FIRE

1. Large amounts of dry vegetation carry a hotter fire, and wet soil acts as a heat sink but facilitates heat conduction through soil. Fires travel faster uphill and with the wind, but because of increased oxygen backfires move faster into a strong wind than into a light wind.
2. The temperature of grassland fires may reach 300°C or so in the canopy and at the soil surface. However, the duration of this intense heat is usually very short--a matter of minutes. Therefore, perennial grasses usually receive minimum damage from wild fires.
3. After spring burning, daytime spring soil temperatures are higher than in unburned areas. Also, especially during the first part of the season, spring burning reduces soil water.
4. Foliar nutrient content may be higher in the year or two following burning.
5. Spring burning reduces the growth of cool-season species.
6. Burning increases flowering in many prairie species.
7. Above- and belowground production may be enhanced by burning, but this effect usually lasts only one to two years after burning.
8. Number and biomass of grasshoppers increase after burning, but the response of other invertebrates is not clear.

PESTICIDES

1. Grass production may be markedly increased by spraying with herbicides that selectively kill forbs.
2. Insecticides may have a short-term effect on the target organisms but have indirect and longer-term effects on processes such as mammal reproductive capacity, microbial decomposition rate, and species diversity.

References

Ahlgren, C. E. 1960. Some effects of fire on reproduction and growth of vegetation in northeastern Minnesota. Ecology 41:431-445.

Ahshapanek, D. 1962. Phenology of a native tallgrass prairie in central Oklahoma. Ecology 43:135-138.

Aikman, J. M. 1955. Burning in the management of a prairie in Iowa. Proc. Iowa Acad. Sci. 62:53-62.

Aldous, A. E. 1930a. Relation of organic food reserves to the growth of some Kansas plants. Agron. J. 22:385-392.

Aldous, A. E. 1930b. Effect of different clipping treatments on the yield and vigor of prairie grass vegetation. Ecology 11:752-759.

Aldous, A. E. 1933. Bluestem pastures. Kansas Agric. Exp. Stn. Bienn. Rep. Dir. 33:184-191.

Aldous, A. E. 1934. Effects of burning on Kansas bluestem pastures. Kans. Agric. Exp. Stn. Bull. 38:3-65.

Aldous, A. E. 1935. Management of Kansas permanent pastures. Kans. Agric. Exp. Stn. Bull. 272:44 p.

Alexander, M. 1961. Introduction to Soil Microbiology. (New York: John Wiley & Sons), 472 p.

Allee, W. C. 1927. Animal aggregations. Q. Rev. Biol. 2:367-398.

Allen, J. A. 1871. The fauna of the prairies. Am. Nat. 5:4-9.

Allen, L. J. 1973. Effects of range burning and nitrogen fertilization on the nutritive value of bluestem grasses. M.S. thesis, Kansas State Univ., Manhattan, 41 p.

Allen, L. J., L. H. Harbers, R. R. Schalles, C. E. Owensby, and E. F. Smith. 1976. Range burning and fertilizing related to nutritive value of bluestem grass. J. Range Manage. 29:306-308.

Allen, O. N. 1957. Experiments in Soil Bacteriology, 3rd ed. (Minneapolis, Minn.: Burgess Pub. Co.), Minneapolis, Minn., 117 p.

Allis, J. A., and A. R. Kuhlman. 1962. Runoff and sediment yields on rangeland watersheds. J. Soil Water Conserv. 17:68-71.

Anderson, K. L. 1940. Deferred grazing of bluestem pastures. Kansas Agric. Exp. Stn. Bull. 291:4-27.

Anderson, K. L. 1960. An effect of fall harvest on subsequent forage yield of true prairie. Agron. J. 52:670-671.

Anderson, K. L. 1964. Burning Flint Hills bluestem ranges. Proc. Ann. Tall Timbers Fire Ecol. Conf. 3:89-103.

Anderson, K. L. 1965. Time of burning as it affects soil moisture in an ordinary upland bluestem prairie in the Flint Hills. J. Range Manage. 18:311-316.

Anderson, K. L., E. F. Smith, and C. E. Owensby. 1970. Burning bluestem range. J. Range Manage. 23:81-91.

Anderson, R. C. 1973. The use of fire as a management tool on the Curtis Prairie. Proc. Ann. Tall Timbers Fire Ecol. Conf. 12:23-35.

Anderson, R. C., and J. Schwegman. 1974. The response of southern Illinois barren vegetation to prescribed burning. Trans. Ill. State Acad. Sci. 63:287-291.

Andrews, R. M., D. C. Coleman, J. E. Ellis, and J. S. Singh. 1974. Energy flow relationships in a shortgrass prairie ecosystem. In: Proc. of the First International Congress of Ecology. (Wageningen, The Netherlands: Centre for Agric. Publ. and Doc.), p. 22-28.

Anway, J. C. 1976. Mammalian consumer submodel. In: G. W. Cole, ed. ELM: Version 2.0. (Fort Collins: Colorado State Univ.), Range Sci. Dep. Sci. Ser. No. 20, p. 137-195.

Ares, J. 1976. Dynamics of the root system of blue grama. J. Range Manage. 29:208-213.

Ares, J., and J. S. Singh. 1974. A model of root biomass dynamics of a shortgrass prairie dominated by blue grama (Bouteloua gracilis). J. Appl. Ecol. 11:727-744.

Armstrong, D. M. 1972. Distribution of mammals in Colorado. Univ. Kans. Mus. Nat. Hist. Monogr. 3:1-415.

Arnold, R. W., and F. F. Riecken. 1964. Grainy gray ped coating in brunizem soils. Proc. Iowa Acad. Sci. 7:350-360.

Asby, W. C., and R. W. Kelting. 1963. A vegetation of the Pine Hills field station in southwestern Illinois. Trans. Ill. State Acad. Sci. 56:188-201.

Atsatt, P. R., and D. J. O'Dowd. 1976. Plant defense guilds. Science 193:24-29.

Auclair, A. N. 1976. Ecological factors in the development of intensive-management ecosystems in the midwestern United States. Ecology 57:431-444.

Babiuk, L. A., and E. A. Paul. 1970. The use of fluorescein isothiocynate in the determination of the bacterial biomass of grassland soil. Can. J. Microbiol. 16:57-62.

Baier, J. D., F. A. Bazzaz, L. C. Bliss, and W. R. Boggess. 1972. Primary production and soil relations in an Illinois sand prairie. Am. Midl. Nat. 88:200-208.

Bailey, C. G., and P. W. Riegent. 1972. Energy dynamics of Encoptolophus sordidus costalis (Scudder) (Orthoptera: Acrididae) in a grassland ecosystem. Can. J. Zool. 51:91-100.

Bailey, V. 1926. A biological survey of North Dakota: I. Physiography and life zones. II. The mammals. N. Am. Fauna 49:1-226.

Barker, W. T. 1969. The flora of the Kansas Flint Hills. Univ. Kans. Sci. Bull. 48:525-584.

Barrett, G. W. 1968. The effects of an acute insecticide stress on a semi-enclosed grassland ecosystem. Ecology 49:1019-1035.

Bauerle, B. 1972. Biological productivity of snakes of the Pawnee Site, 1970-1971. (Fort Collins: Colorado State Univ.), US/IBP Grassland Biome Tech. Rep. No. 207, 72 p.

Baxter, F. P., and F. D. Hole. 1967. Ant (Formicia cinerea) pedoturbation in a prairie soil. Soil Sci. Soc. Am. Proc. 31:425-428.

Bazzaz, F. A. 1968. Succession on abandoned fields in Shawnee Hills, southern Illinois. Ecology 49:924-936.

Bazzaz, F. A. 1975. Plant species diversity in old-field successional ecosystems in southern Illinois. Ecology 56:485-488.

Beaty, E. R., and J. D. Powell. 1976. Response of switchgrass (Panicum virgatum L.) to clipping frequency. J. Range Manage. 29:132-135.

Beck, D. L., and R. E. Sosebee. 1975. Fall application of herbicides for common broomweed control. J. Range Manage. 28:332-333.

Beebe, J. D., and G. R. Hoffman. 1968. Effects of grazing on vegetation and soils in southeastern South Dakota. Am. Midl. Nat. 80:96-110.

Bell, H. M. 1973. Rangeland Management for Livestock Production. (Norman: Univ. Oklahoma Press), 303 p.

Bell, R. H. V. 1971. A grazing ecosystem in the Serengeti. Sci. Am. 225:86-93.

Benninghoff, W. S. 1964. The prairie peninsula as a filter barrier to post-glacial plant migration. Proc. Indiana Acad. Sci. 73:116-124.

Bertwell, R. L. 1972. Coleoptera, especially Curculionidae, of tallgrass prairie. M. S. thesis, Kansas State Univ., Manhattan, 67 p.

Bertwell, R. L., and H. D. Blocker. 1975. Curculionidae from differently managed tallgrass prairie near Manhattan, Kansas. J. Kans. Entomol. Soc. 48(3):319-326.

Bieber, G. L., and K. L. Anderson. 1961. Soil moisture in bluestem grassland following burning. J. Soil Water Conserv. 16:186-187.

Birney, E. C. 1973. Systematics of three species of woodrats (Genus Neotoma) in central North America. Univ. Kans. Mus. Nat. Hist. Misc. Publ. 58:1-173.

Birney, E. C. 1974a. Dynamics of small mammal populations at the Cottonwood and Osage Sites, 1972. (Fort Collins: Colorado State Univ.), US/IBP Grassland Biome Tech. Rep. No. 257, 87 p.

Birney, E. C. 1974b. Twentieth century records of wolverine in Minnesota. Loon 46:78-81.

Birney, E. C., and J. D. Rising. 1968. Notes on distribution and reproduction of some bats from Kansas, with remarks on incidence of rabies. Trans. Kans. Acad. Sci. 70:519-524.

Birney, E. C., W. E. Grant, and D. D. Baird. 1976. Importance of vegetative cover to cycles of Microtus populations. Ecology 57(5):1043-1051.

Biswell, H. H., and P. C. Lemon. 1943. Effect of fire upon seed stalk production of range grasses. J. For. 41:844.

Biswell, H. H., and J. E. Weaver. 1933. Effect of frequent clipping on the development of roots and tops of grasses in prairie sod. Ecology 14:368-390.

Bjorkman, O. 1971. Comparative photosynthetic CO_2 exchange in higher plants. In: M. D. Hatch, C. B. Osmond, and R. O. Slatyer, eds. Photosynthesis and Photorespiration. (New York: John Wiley & Sons), p. 18-32.

Black, C. C. 1971. Ecological implications of dividing plants into groups with distinct photosynthetic capacities. Adv. Ecol. Res. 7:87-114.

Blair, W. F., and T. H. Hubbell. 1938. The biotic districts of Oklahoma. Am. Midl. Nat. 20:425-454.

Blake, A. 1935. Viability of germination of seeds and early life history of prairie plants. Ecol. Monogr. 5:405-460.

Bliss, L. C., and G. W. Cox. 1964. Plant community and soil variation within a northern Indiana prairie. Am. Midl. Nat. 72:115-128.

Blizzard, A. W. 1931. Plant sociology and vegetational change on High Hill, Long Island, New York. Ecology 12:208-231.

Blocker, H. D., and R. Reed. 1971. 1970 insect studies at Osage Comprehensive Site. (Fort Collins: Colorado State Univ.), US/IBP Grassland Biome Tech. Rep. No. 93, 38 p.

Blocker, H. D., and R. Reed. 1976. Leafhopper populations of a tallgrass prairie (Homoptera:Cicadellidae): Collecting procedures and population estimates. J. Kans. Entomol. Soc. 49(2):145-154.

Blocker, H. D., T. L. Harvery, and J. L. Launchbaugh. 1972. Grassland leafhoppers. I. Leafhopper populations of upland seeded pastures at Hays, Kansas. Ann. Entomol. Soc. Am. 65:166-172.

Blocker, H. D., R. Reed, and C. E. Mason. 1971. Leafhopper studies at the Osage Site (Homoptera:Cicadellidae). (Fort Collins: Colorado State Univ.), US/IBP Grassland Biome Tech. Rep. No. 124, 25 p.

Bokhari, U. G., and J. S. Singh. 1974. Effects of temperature and clipping on growth, carbohydrate reserves, and root exudation of western wheatgrass in hydroponic culture. Crop Sci. 14:790-794.

Bokhari, U. G., and J. S. Singh. 1975. Standing state and cycling of nitrogen in soil-vegetation components of prairie ecosystems. Ann. Bot. 39:273-285.

Booth, W. E. 1941a. Algae as pioneers in plant succession and their importance in erosion control. Ecology 22:38-46.

Booth, W. E. 1941b. Revegetation of abandoned fields in Kansas and Oklahoma. Am. J. Bot. 28:415-422.

Borchert, J. R. 1950. Climate of the central North American grasslands. Ann. Assoc. Am. Geogr. 40:1-39.

Botkin, D., and C. R. Malone. 1968. Efficiency of net primary production based on light intercepted during the growing season. Ecology 49:438-444.

Bowles, J. B. 1975. Distribution and biogeography of mammals of Iowa. Spec. Publ. Mus. Texas Tech. Univ. 9:1-184.

Box, T. W., G. M. Van Dyne, and N. E. West. 1969. Range resources of North America. Utah State Univ. Bookstore, Logan, 769 p. (Syllabus)

Bradbury, I. K., and G. Hofstra. 1976. Vegetation death and its importance in primary production measurements. Ecology 57:209-211.

Bragg, T. B., and L. C. Hulbert. 1976. Woody plant invasion of unburned Kansas bluestem prairie. J. Range Manage. 29:19-23.

Branson, F. A., G. F. Gifford, and J. R. Owen. 1972. Rangeland Hydrology Range Sci. Ser. 1, (Denver, Colo.: Soc. Range Manage, 84 p.

Braun, E. L. 1950. Deciduous Forests of Eastern North America. (Philadelphia, Pa.: Blakiston), 596 p.

Bray, J. R. 1958. The distribution of savanna species in relation to light and density. Can. J. Bot. 36:671-681.

Brendel, F. 1887. Flora Peoriana. (Peoria, Ill.: J. W. Franks & Sons Printers), 89 p.

Briggs, L. J., and H. L. Shantz. 1911. The wax seal method for determining the lower limit of available soil moisture. Bot. Gaz. 51:210-219.

Briggs, L. J., and H. L. Shantz. 1916. Hourly transpiration rate in clear days as determined by cyclic environmental factors. J. Agric. Res. 5:583-645.

Brodkorb, P. 1971. Origin and evolution of birds. In: D. S. Farner, J. R. King, and K. C. Parkes, eds. Avian Biology. I. (New York: Academic Press), p. 19-55.

Brody, S. 1945. Bioenergetics and Growth. (New York: Reinhold), 1023 p.

Brownell, P. F., and J. C. Crossland. 1972. The requirement for sodium as a micronutrient by species having the C_4 dicarboxylic photosynthetic pathway. Plant Physiol. 49:794-797.

Bruner, W. E. 1931. The vegetation of Oklahoma. Ecol. Monogr. 1:99-188.

Brunnschweiler, D. H. 1952. The geographic distribution of air masses in North America. Vierteljahrsschr. Naturforsch. Ges. Zür. 97:42-49.

Buchanan, H., W. A. Laycock, and D. H. Price. 1972. Botanical and nutritive content of the summer diet of sheep on a tall forb range in southwestern Montana. J. Anim. Sci. 35:430-432.

Buck, P., and R. W. Kelting. 1962. A survey of the tallgrass prairie in northeastern Oklahoma. Southwest. Nat. 7:163-165.

Buell, N. F., and V. Facey. 1960. Forest-prairie transition west of Itasca Park, Minnesota. Bull. Torrey Bot. Club 87:46-58.

Bulksey, F. S., and J. E. Weaver. 1939. Effect of frequent clipping on the underground food reserves of certain prairie grasses. Ecology 20:246-252.

Buol, S. W., F. D. Hole, and R. J. McCracken. 1973. Soil Genesis and Classification. (Ames: Iowa State Univ. Press), 360 p.

Burris, R. H., and C. C. Black. 1976. CO_2 Metabolism and Plant Productivity. (Baltimore, Md.: Univ. Park Press), 431 p.

Butler, J. E. 1954. Interrelations of autecological characteristics of prairie herbs. Ph.D. thesis, Univ. Wisconsin, Madison.

Byerly, T. C. 1977. Ruminant livestock research and development. Science 195:450-456.

Campbell, J. B., W. H. Arnett, J. D. Lambley, O. K. Jantz, and H. Knutson. 1974. Grasshoppers (Acrididae) of the Flint Hills

native tallgrass prairie in Kansas. Kans. Agric. Exp. Stn. Res. Paper 19, 147 p.

Campbell, R. S., E. A. Epps, C. C. Moreland, J. L. Farr, and F. Bonner. 1954. Nutritive values of native plants on forest range in central Louisiana. La. Agric. Exp. Stn. Bull. 488:18 p.

Canadian Department of Transport. 1947. Climatic Summaries, Vol. 1. (Ottawa, Canada: Meteorol. Div., Canadian Department of Transport.)

Cancelado, R., and T. R. Yonke. 1970. Effect of prairie burning on insect populations. J. Kans. Entomol. Soc. 43:274-281.

Canode, C. L., and A. G. Law. 1975. Seed production of Kentucky bluegrass associated with age of stand. Agron. J. 67:790-794.

Carpenter, J. R. 1940. The grassland biome. Ecol. Monogr. 10:617-684.

Casida, L. E. 1971. Microorganisms in unamended soil as observed by various forms of microscopy and staining. Appl. Microbiol. 21:1040-1045.

Caswell, H. 1976. The validation problem. In: B. C. Patten, ed. Systems Analysis and Simulation in Ecology, Vol. IV. (New York: Academic Press), p. 313-325.

Caswell, H., and F. C. Reed. 1975. Indigestibility of C_4 bundle sheath cells by the grasshoppers, Melanoplus confusus. Ann. Entomol. Soc. Am. 68:686-688.

Caswell, H., F. Reed, S. N. Stephenson, and P. A. Werner. 1973. Photosynthetic pathways and selective herbivory: A hypothesis. Am. Nat. 107:465-479.

Charlevoix, P. F. 1761. Journal of a Voyage to North America, Vol. 2. (London: R. & J. Dodsley), p. 199-200. (Trans. from French.)

Chew, R. M. 1974. Consumers as regulators of ecosystems: An alternative to energetics. Ohio J. Sci. 74:359-370.

Chew, R. M. 1978. The impact of small mammals on ecosystem structure and function. In: D. P. Snyder, ed. Populations of Small Mammals under Natural Conditions, Special Publ. Ser. Vol. 5. (Univ. Pittsburgh, Penn.: Pymatuning Laboratory of Ecology), p. 167-180.

Chiles, R. E. 1968. Bermuda pasture systems for Oklahoma. (Stillwater, Okla. State Univ.: Extension Facts No. 2552).

Choate, J. R. 1970. Systematics on zoogeography of middle American shrews of the genus Cryptotis. Univ. Kans. Publ. Mus. Nat. Hist. 19:195-317.

Choate, J. R., and E. D. Fleharty. 1975. Synopsis of native, recent mammals of Ellis County, Kansas. Occas. Pap. Mus. Texas Tech. Univ. 37:1-80.

Choate, J. R., and H. H. Genoways. 1967. Notes on some mammals from Nebraska. Trans. Kans. Acad. Sci. 69:238-241.

Choate, J. R., and J. E. Krause. 1976. Historical biogeography of the gray fox (Urocyon cinereoargenteus) in Kansas. Trans. Kans. Acad. Sci. 77:231-235.

Choate, J. R., and S. L. Williams. 1978. Biogeographic interpretation of variation within and among populations of the prairie vole, Microtus ochrogaster. Occas. Pap. Mus. Texas Tech. Univ. 49:1-25.

Cholodny, N. G.   1930.   Uber eine neue Methode zur Untersuchung der Bodenmikroflora.   Arch. Mikrobiol. 1:620-652.

Chow, V. T.   1964.   Handbook of Applied Hydrology.   (New York: McGraw-Hill.)

Clark, F. E.   1965.   Agar-plate method for total microbial count. In: C. A. Black, ed. Methods of Soil Analysis: Part 2. Chemical and Microbiological Properties.   (Madison, Wisc.:   Am. Soc. Agron.), p. 1460-1466.

Clark, F. E., and E. A. Paul.   1970.   The microflora of grassland. Adv. Agron. 22:375-435.

Clark, O. R.   1937.   Interception of rainfall by herbaceous vegetation.   Science 86:591-592.

Clark, O. R.   1940.   Interception of rainfall by prairie grasses, weeds, and certain crop plants.   Ecol. Monogr. 10:243-277.

Clements, F. E.   1916.   Plant succession: An analysis of the development of vegetation. Carnegie Inst. Wash. Publ. No. 242, 512 p.

Clements, F. E.   1920.   Plant indicators.   Carnegie Inst. Wash. Publ. 290, 388 p.

Clements, F. E.   1936.   Origin of the Desert Climax and Climate. In: T. H. Goodspell, ed. Essays in geobotany in honor of William Albert Setchell.   (Berkeley: Univ. California Press), p. 87-140.

Clements, F. E., and V. E. Shelford.   1939.   Bioecology.   (New York: John Wiley & Sons), 425 p.

Cockrum, E. L.   1948.   The distribution of the hispid cotton rat in Kansas.   Trans. Kans. Acad. Sci. 51:306-312.

Cockrum, E. L.   1952.   Mammals of Kansas.   Univ. Kans. Publ. Mus. Nat. Hist. 7:1-303.

Cody, M. L.   1966a.   The consistency of intra- and inter-continental grassland bird species counts. Am. Nat. 100:371-376.

Cody, M. L.   1966b.   A general theory of clutch size.   Evolution 20:174-184.

Cody, M. L.   1968.   On the methods of resource division in grassland bird communities.   Am. Nat. 102:107-147.

Cody, M. L.   1974.   Competition and the Structure of Bird Communities.   (Princeton, N.J.:   Princeton Univ. Press), 318 p.

Cole, C. V., G. S. Innis, and J. W. B. Stewart.   1977.   Simulation of phosphorus cycling in semiarid grasslands.   Ecology 58:1-15.

Cole, G. W., ed.   1976.   ELM: Version 2.0.   (Fort Collins: Colorado State Univ.), Range Sci. Dep. Sci. Ser. No. 20, 663 p.

Cole, L. C.   1949.   The measurement of interspecific association. Ecology 30:411-424.

Coleman, D. C.   1973.   Soil carbon balance in a successional grassland.   Oikos 24:195-199.

Coleman, D. C.   1976.   A review of root production processes and their influence on soil biota in terrestrial ecosystems.   In: J. M. Anderson and A. Macfadyen, eds. The Role of Terrestrial and Aquatic Organisms in Decomposition Processes.   (Oxford: Blackwell Scientific Publ.), p. 417-434.

Coleman, D. C., R. Andrews, J. E. Ellis, and J. S. Singh.   1976. Energy flow and partitioning in selected man-managed and natural ecosystems.   Agro-Ecosystems 3:45-54.

Colwell, R. K., and D. J. Futuyma.   1971.   On the measurement of niche breadth and overlap.   Ecology 52:567-576.

Conant, S., and P. G. Risser. 1974. Canopy structure of a tallgrass prairie. J. Range Manage. 27:313-318.

Conard, E. C., and V. E. Youngman. 1965. Soil moisture conditions under pastures of cool-season and warm-season grasses. J. Range Manage. 18:74-78.

Conard, H. S. 1952. The vegetation of Iowa. Univ. Iowa Stud. Nat. Hist. 19:1-66.

Cook, E. D., and C. W. Rector. 1964. Effect of fertilizer and 2,4-D on forage yields of King Ranch bluestem grass, 1960, 1961, and 1963. Texas Agric. Exp. Stn. Prog. Rep. 2307. (Lubbock: Texas Tech. Univ.), 3 p.

Cooper, C. F. 1976. Ecosystem models and environmental policy. Simulation 20:133-138.

Cooper, J. P. 1970. Potential production and energy conversion in temperate and tropical grasses. Herb. Abstr. 40:1-15.

Corbett, E. S., and R. P. Crouse. 1968. Rainfall interception by annual grass and chaparral. U.S. For. Serv. Res. Paper PSW-48. (Berkeley, Calif.: U.S. Forest Service, Pacific Southwest Experiment Sta.), 12 p.

Costello, D. F. 1969. The Prairie World. (New York: T. Y. Cromwell Co.), 242 p.

Coupland, R. T. 1950. Ecology of mixed prairie in Canada. Ecol. Monogr. 20:271-315.

Coupland, R. T. 1961. A reconsideration of grassland classification and the northern Great Plains of America. J. Ecol. 49:135-167.

Coupland, R. T., and T. C. Brayshaw. 1953. The fescue grassland in Saskatchewan. Ecology 34:386-405.

Crider, F. J. 1955. Root-growth stoppage resulting from defoliation of grasses. USDA Tech. Bull. 1102:3-23.

Cummins, K. W., and J. C. Wuycheck. 1971. Caloric equivalents for investigations in ecological energetics. Mitt. Int. Ver. Theor. Angew. Limnol. 18:1-158.

Curtis, J. T. 1956. A prairie continuum in Wisconsin. Ecology 36:558-566.

Curtis, J. T. 1959. Vegetation of Wisconsin. (Madison: Univ. Wisconsin Press), 657 p.

Curtis, J. T., and M. L. Partch. 1948. Effect of fire on the competition between bluegrass and certain prairie plants. Am. Midl. Nat. 39:437-444.

Curtis, J. T., and M. L. Partch. 1950. Some factors affecting flower production in Andropogon gerardi. Ecology 31:488-489.

Dahlman, R. C., and C. L. Kucera. 1965. Root productivity and turnover in native prairie. Ecology 46:84-89.

Dahlman, R. C., and C. L. Kucera. 1967. Carbon-14 cycling in root and soil components of a prairie ecosystem. In: D. J. Nelson and F. C. Evans, eds. Proc. Second Natl. Symp. Radioecology. (Ann Arbor, Mich.), p. 652-660.

Dahlman, R. C., J. S. Olson, and K. Doxtader. 1969. The nitrogen economy of grassland and dune soils. In: Biology and Ecology of Nitrogen. (Washington, D.C.: National Academy of Sciences), p. 54-82.

Dale, M. B. 1970. Systems analysis and ecology. Ecology 5:1-16.

Dalgarn, M. C., and R. E. Wilson. 1975. Net productivity and ecological efficiency of Andropogon scoparius growing in an Ohio relic prairie. Ohio J. Sci. 75:194-197.

Dalrymple, R. L., and D. D. Dwyer. 1967. Root and shoot growth of the range grasses. J. Range Manage. 20:141-145.

Dalrymple, R. L., D. D. Dwyer, and P. W. Santelmann. 1964. Vegetational response following winged elm and oak control in Oklahoma. J. Range Manage. 17:249-253.

Daniel, H. A. 1935. A study of the magnesium content of grasses and legumes and the relation between this element and the total Ca, P, and Na in the plants. J. Am. Soc. Agron. 27:922-927.

Daniel, H. A., and H. J. Harper. 1934. The relation between the mineral composition of mature grass and available plant food in the soil. J. Am. Soc. Agron. 26:986-992.

Daniel, H. A., and H. J. Harper. 1935. The relation between effective rainfall and total calcium and phosphorous in alfalfa and prairie hay. J. Am. Soc. Agron. 27:644-651.

Darrow, R. A., and W. G. McCully. 1959. Brush control and range improvement in the post oak-blackjack area of Texas. Tex. Agric. Exp. Stn. Bull. 942. (Lubbock: Texas Tech. Univ.), p. 1-15.

Daubenmire, R. F. 1942. An ecological study of the vegetation of southeastern Washington and adjacent Idaho. Ecol. Monogr. 12:53-79.

Daubenmire, R. F. 1968. The ecology of fire in grasslands. Adv. Ecol. Res. 5:209-266.

Daubenmire, R. F. 1970. Steppe vegetation of Washington. Wash. Agric. Exp. Stn. Tech. Bull. 62:131 p. (Pullman: Washington State Univ.)

Davis, A. M. 1977. The prairie-deciduous ecotone in the upper middle west. Ann. Assoc. Am. Geogr. 67:204-213.

Davis, D. E., and F. B. Golley. 1963. Principles in Mammalogy. (New York: Reinhold Publ. Co.), 348 p.

Dee, R. F., and T. W. Box. 1967. Influence of commercial fertilizer on the seasonal protein content of range grasses. J. Range Manage. 20:96-99.

Dhillon, B. S., and N. H. E. Gibson. 1962. A study of the Acarina and Collembola of agricultural soils. 1. Numbers and distribution in undisturbed grassland. Pedobiologia 1:189-209.

Dice, D. R. 1943. The Biotic Provinces of North America. (Ann Arbor: Univ. Michigan Press), 78 p.

Dickinson, C. E., and J. L. Dodd. 1976. Phenological pattern in the shortgrass prairie. Am. Midl. Nat. 96:367-378.

Dickinson, C. H., and G. J. F. Pugh. 1974. Biology of Plant Litter Decomposition, Vols. 1 and 2. (London: Academic Press), 321 p.

Dickinson, J. C. 1964. Chemical composition of natural grass as influenced by nitrogen fertilization. M.S. thesis, Oklahoma State Univ., Stillwater.

Dilz, K., and J. W. Woldendorp. 1960. Distribution and nitrogen balance of ^{15}N labeled nitrate applied on grass sods. In: Eighth International Grassland Congress. (Reading, England), p. 150-153.

Dix, R. L. 1964. A history of biotic and climatic changes within the North American grassland. In: D. J. Crisp, ed. Grazing in

Terrestrial and Marine Environments. (Oxford: Blackwell Scientific Publ.), p. 71-90.

Dix, R. L., and J. E. Butler. 1954. The effects of fire on a dry, thin soil prairie in Wisconsin. J. Range Manage. 7:265-268.

Dorf, E. 1960. Climatic changes of the past and present. Am. Sci. 48:341-364.

Downton, W. J. S. 1971. Adaptive and evolutionary aspects of C_4 photosynthesis. In: M. D. Hatch, C. B. Osmond, and R. O. Slatyer, eds. Photosynthesis and Photorespiration. (New York: John Wiley & Sons), p. 3-17.

Dragonn, F. J. 1969. Effects of cultivation and grass on surface runoff. Water Resour. Res. 5:1078-1083.

Dragonn, F. J., and A. R. Kuhlman. 1968. Effect of pasture management practices on runoff. J. Soil Water Conserv. 23:55-57.

Drury, W. H., and C. T. Nisbet. 1973. Succession. J. Arnold Arbor. Harv. Univ. 54:311-368.

Duvall, V. L. 1962. Burning and grazing increase herbage on slender bluestem range. J. Range Manage. 15:14-16.

Duvall, V. L., and L. B. Whitaker. 1964. Rotation burning: A forage management system for longleaf pine-bluestem ranges. J. Range Manage. 17:322-326.

Dwyer, D. D. 1958. An annotated plant list for Adam's Ranch, Osage County, Oklahoma. M.S. thesis, Fort Hays Kansas State Coll., Hays.

Dwyer, D. D. 1961. Activities and grazing preferences of cows with calves in northern Osage County, Oklahoma. Okla. Agric. Exp. Stn. Bull. B-588:61 p. Oklahoma State Univ., Stillwater.

Dwyer, D. D., W. C. Elder, and G. Singh. 1963. Effects of height and frequency of clipping of pure stands of range grasses in northcentral Oklahoma. Okla. Agric. Exp. Stn. Bull. B-614:10 p. Oklahoma State Univ., Stillwater.

Dyer, M. I., and U. G. Bokhari. 1976. Plant-animal interactions: Studies of the effects of grasshopper grazing on blue grama grass. Ecology 57:762-772.

Dyksterhuis, E. J. 1946. The vegetation of the Fort Worth Prairie. Ecol. Monogr. 16:3-29.

Dyksterhuis, E. J. 1948. The vegetation of the Western Cross Timbers. Ecol. Monogr. 18:327-376.

Dyksterhuis, E. J. 1949. Condition and management of range land based on quantitative ecology. J. Range Manage. 2:104-115.

Eden, A. 1940. Coprophagy in the rabbit. Nature 145:36-37.

Edwards, E. E. 1948. The settlement of grasslands. In: A. Stefferud, ed. Grass: Yearbook of Agriculture, 1948. (Washington, D.C.: U.S. Department of Agriculture), p. 16-25.

Edwards, N. T., and P. Sollins. 1973. Continuous measurement of carbon dioxide evolution from partitioned forest floor components. Ecology 52:406-412.

Egler, F. E. 1951. A commentary on American plant ecology based on the textbooks of 1947. Ecology 32:673-695.

Ehrenreich, J. H. 1959. Effect of burning and clipping on growth of native prairie in Iowa. J. Range Manage. 12:133-137.

Ehrenreich, J. H., and J. M. Aikman. 1957. Effect of burning on seedstalk production of native prairie grasses. Proc. Iowa Acad. Sci. 64:205-212.

Ehrenreich, J. H., and J. M. Aikman. 1963. An ecological study of certain management practices on native plants in Iowa. Ecol. Monogr. 33:113-130.

Ehrenreich, J. H., and J. S. Crosby. 1960. Forage production on sprayed and burned areas in the Missouri Ozarks. J. Range Manage. 13:68-70.

Ellis, J. E., J. A. Wiens, C. F. Rodell, and J. C. Anway. 1976. A conceptual model of diet selection as an ecosystem process. J. Theor. Biol. 60:93-108.

El-Sharkawy, M. A., and J. D. Hesketh. 1965. Photosynthesis among species in relation to characteristics of leaf anatomy and CO_2 diffusion resistances. Crop Sci. 5:517-521.

Elwell, H. M. 1960. Land improvement through brush control. Soil Conserv. 26:56-59.

Elwell, H. M., and W. E. McMurphy. 1973. Weed control with phenoxy herbicides on native grasslands. Okla. Agric. Exp. Stn. Bull. B-706:14 p. Oklahoma State Univ., Stillwater.

Elwell, H. M., H. A. Daniel, and F. A. Fenton. 1941. The effect of burning pasture and native woodland vegetation. Okla. Agric. Exp. Stn. Bull. B-247:22 p. Oklahoma State Univ., Stillwater.

Elwell, H. M., W. E. McMurphy, and P. W. Santelmann. 1970. Burning and 2,4,5-T on post and blackjack oak rangeland in Oklahoma. Okla. Agric. Exp. Stn. Bull. B-675:11 p. Oklahoma State Univ., Stillwater.

Emlen, J. T. 1971. Population densities of birds derived from transect counts. Auk 88:323-342.

Emlen, J. T., and J. A. Wiens. 1965. The Dickcissel invasion of 1964 in southern Wisconsin. Passenger Pigeon 27:51-59.

England, C. M., and E. L. Rice. 1957. A comparison of the soil fungi of an tallgrass prairie and of an abandoned field in central Oklahoma. Bot. Gaz. 118:186-190.

Erdman, L. W. 1959. Legume inoculation: What it is--what it does. U.S. Dep. Agric., Farmers' Bull. 2003. (Washington, D.C.: USDA).

Evans, J. 1970. About nutria and their control. Resour. Publ. Bur. Sports Fish. Wildl. Denver 86:7-65.

Evers, R. A. 1955. Hill prairies of Illinois. Ill. Nat. Hist. Surv. Bull. 26:367-446.

Farnell, D. R., M. C. Futrell, V. H. Watson, W. E. Poe, and R. E. Coats. 1975. Field studies on etiology and control of fescue toxicosis. J. Environ. Qual. 4:120-122.

Faroua, H. 1975. Interseeding and paraquat effects on central and eastern Oklahoma rangeland vegetation. M. S. thesis, Oklahoma State Univ., Stillwater.

Fenneman, N. M. 1938. Physiography of Eastern United States. (New York: McGraw-Hill), 714 p.

Fleharty, E. D., and J. R. Choate. 1973. Bioenergetic strategies of the cotton rat, Sigmodon hispidus. J. Mammal. 54:680-692.

Fleharty, E. D., J. R. Choate, and M. A. Mares. 1972. Fluctuations in population density of the hispid cotton rat: Factors influencing a "crash." Bull. South. Calif. Acad. Sci. 71:132-138.

Forrester, J. W. 1968. Principles of Systems. (Cambridge, Mass.: Wright-Allen Press), 374 p.

French, N. R., ed. 1971. Preliminary analysis of structure and function in grasslands. Range Sci. Dep. Sci. Ser. No. 10. (Fort Collins: Colorado State Univ.), 387 p.

French, N. R., R. K. Steinhorst, and D. M. Swift. 1979. Grassland trophic biomass pyramids. In: N. R. French, ed. Perspectives in Grassland Ecology. (New York: Springer-Verlag), 204 p.

French, N. R., W. E. Grant, W. Grodzinski, and D. M. Swift. 1976. Small mammal energetics in grassland ecosystems. Ecol. Monogr. 46:201-220.

Fretwell, S. D. 1972. Populations in a seasonal environment. Monogr. Popul. Biol. No. 5. (Princeton, N.J.: Princeton Univ. Press), 217 p.

Fretwell, S. D. 1973. The regulation of bird populations on Konza Prairie: The effects of events off of the prairie. In: L. C. Hulbert, ed. Third Midwest Prairie Conference Proceedings. (Kansas State Univ., Manhattan), p. 71-76.

Fryrear, D. W., and W. G. McCully. 1972. Development of grass root systems as influenced by soil competition. J. Range Manage. 25:254-256.

Gadgil, M., and O. T. Solbrig. 1972. The concept of r and K selection: Evidence from wild flowers and some theoretical considerations. Am. Nat. 106:461-471.

Gallup, W. D., and H. M. Briggs. 1948. The apparent digestibility of prairie hay of variable protein content, with some observations of fecal nitrogen excretion by steers in relation to their dry matter intake. J. Anim. Sci. 7:110-116.

Gardner, C., J. A. Jewell, H. Dunn, and I. Y. Mahmand. 1957. Effects of mowing a native tallgrass prairie in central Oklahoma. Proc. Okla. Acad. Sci. 38:30-31.

Gardner, R. G. 1958. A vegetational analysis of the Phillips Agricultural Demonstration Project Ranch, Foraker, Oklahoma. M. S. thesis, Texas A & M College, College Station.

Garnert, W. B. 1936. Native grass behavior as affected by periodic clipping. J. Am. Soc. Agron. 28:447-456.

Gay, C. W., and D. D. Dwyer. 1965. Effect of one year's nitrogen fertilization on native vegetation under clipping and burning. J. Range Manage. 18:273-277.

Geiger, R. 1965. The Climate Near the Ground. (Cambridge, Mass.: Harvard Univ. Press), 611 p. (Trans. from German.)

Geist, V. 1974. On the relationship of ecology and behaviour in the evolution of ungulates: Theoretical considerations. In: V. Geist and F. Walther, eds. The Behaviour of Ungulates and Its Relation to Management. (Morges, Switzerland), Int. Union Conserv. Nature Nat. Resour. Publ. No. 24 (New Series), p. 235-246.

Genoways, H. H., and D. A. Schlitter. 1967. Northward dispersal of the hispid cotton rat in Nebraska and Missouri. Trans. Kans. Acad. Sci. 69:356-357.

Gessaman, J. A. 1973. Ecological energetics of homeotherms: A view compatible with ecological modeling. (Logan: Utah State Univ. Press), Monogr. Ser., Vol. 20, 155 p.

Getz, L. L. 1971. Microclimate, vegetative cover, and local distribution of the meadow vole. Trans. Ill. Acad. Sci. 63: 9-21.

Gibson, J. S., and J. W. Batten. 1970. Soils. (University: Univ. Alabama Press), 296 p.

Gleason, H. A. 1901. The flora of the prairies. B.S. thesis, Univ. Illinois, Urbana.

Gleason, H. A. 1917. A prairie near Ann Arbor, Michigan. Rhodora 19:163-165.

Gleason, H. A. 1923. Botanical observations in northern Michigan. Bull. Torrey Bot. Club 24:276-283.

Golley, F. B. 1960. Energy dynamics of an old field community. Ecol. Monogr. 30:187-206.

Gould, F. W. 1967. The grass genus Andropogon in the United States. Brittonia 19:70-76.

Gould, F. W. 1968. Grass Systematics. (New York: McGraw-Hill), 382 p.

Graber, R. R., and J. W. Graber. 1963. A comparative study of bird populations in Illinois, 1906-1909 and 1956-1958. Ill. Nat. Hist. Surv. Bull. 28:383-528.

Grant, W. E. 1974. The functional role of small mammals in grassland ecosystems. Ph.D. Dissertation, Colorado State Univ., Fort Collins. 179 p.

Grant, W. E., and E. C. Birney. 1979. Small mammal community structure in North American grasslands. J. Mammal. 60:23-36.

Grant, W. E., N. R. French, and D. M. Swift. 1977. Response of a small mammal community to water and nitrogen treatments in a shortgrass prairie ecosystem. J. Mammal. 58(4):637-652.

Graves, J. E., and W. E. McMurphy. 1969. Burning and fertilization for range improvement in central Oklahoma. J. Range Manage. 22:165-169.

Gray, F., and M. H. Roozitalab. 1976. Benchmark and key soils of Oklahoma. Okla. Agric. Exp. Stn. Bull. MP-97:36 p. Oklahoma State Univ., Stillwater.

Gray, T. R. G., and S. T. Williams. 1971. Soil Micro-organisms. (New York: Hafner), 240 p.

Grayson, D. K. 1977. Pleistocene avifaunas and the overkill hypothesis. Science 195:691-693.

Greene, G. L. 1970. Seasonal occurrence of Chrysomelidae in a native prairie near Manhattan, Kansas. J. Kans. Entomol. Soc. 43:95-101.

Guilday, J. E., and P. W. Parmalee. 1972. Quaternary periglacial records of voles of the genus Phanacomys merriam (Cricetidae:Rodentia). Quat. Res. 2:170-175.

Gunderson, H. L. 1955. Nutria, Myocaster coypus, in Minnesota. J. Mammal. 36:465.

Gutierrez, L. T., and W. R. Fey. 1975. Feedback dynamics analysis of secondary successional transients in ecosystems. Proc. Natl. Acad. Sci. 72:2733-2737.

Hadley, E. B. 1970. Net productivity and burning responses of native eastern North Dakota prairie communities. Am. Midl. Nat. 84:121-135.

Hadley, E. B., and B. J. Kieckhefer. 1963. Productivity of two prairie grasses in relation to fire frequency. Ecology 44:389-395.

Hagmeier, E. M., and C. D. Stults. 1964. A numerical analysis of the distributional patterns of North American mammals. Syst. Zool. 13:125-155.

Haines, H. 1971. Characteristics of a cotton rat (Sigmodon hispidus) population cycle. Tex. J. Sci. 23:3-27.

Hall, E. R., and K. R. Kelson. 1959. The Mammals of North America, Vols. I and II. (New York: Ronald Press), 1083 p.

Halloran, A. F., and C. A. Shrader. 1960. Longhorn cattle management on the Wichita Mountain Wildlife Refuge. J. Wildl. Manage. 24:191-196.

Hanks, R. J., and K. L. Anderson. 1957. Pasture burning and moisture conservation. J. Soil Water Conserv. 12:228-229.

Hanson, H. C. 1955. Characteristics of the Stipa comata-Bouteloua gracilis-Bouteloua curtipendula association of northern Colorado. Ecology 36:269-280.

Hanson, H. C., and W. Whitman. 1938. Characteristics of major grassland types in western North Dakota. Ecol. Monogr. 8:57-114.

Harder, W. 1949. Zur Morphologie und Physiologie des Blinddarmes der Nagetiere. Verh. Dtsch. Zool. Mainz 2:95-109.

Harlan, J. R. 1956. Theory and Dynamics of Grassland Agriculture. (New York: D. Van Nostrand), 281 p.

Harlan, J. R. 1960a. Grasslands of Oklahoma. (Agronomy Dep., Stillwater: Oklahoma State Univ.), (Memo. rep.).

Harlan, J. R. 1960b. Production characteristics of Oklahoma forages, native range. Okla. Agric. Exp. Stn. Bull. B-547:34 p. Oklahoma State Univ., Stillwater.

Harner, R. F., and K. J. Harper. 1973. Mineral composition of grassland species of the eastern Great Basin in relation to stand productivity. Can. J. Bot. 51:2037-2046.

Harper, H. J. 1957. Effects on fertilization and climatic conditions on prairie hay. Okla. Agric. Exp. Stn. Bull. 492:23 p. Oklahoma State Univ., Stillwater.

Harper, H. J., H. A. Daniel, and H. F. Murphy. 1933. The total nitrogen, phosphorous and calcium content of common weeds and native grasses of Oklahoma. Proc. Okla. Acad. Sci. 14:36-44.

Harper, J. L. 1969. The role of predation in vegetational diversity. Brookhaven Symp. Biol. 22:48-62.

Harper, J. L., P. H. Lovell, and K. G. Moore. 1970. The shapes and sizes of seeds. Annu. Rev. Ecol. Syst. 1:327-356.

Harper, R. M. 1943. Forests of Alabama. Monogr. 10, Geol. Surv. Alabama, 230 p.

Harris, J. O. 1971. Microbiological studies at the Osage Site, 1970. US/IBP Grassland Biome Tech. Rep. No. 102. (Fort Collins: Colorado State Univ.), 39 p.

Harris, P. 1974. A possible explanation of plant yield increasing following insect damage. Agro-Ecosystems 1:219-225.

Harrison, C. M. 1939. Greenhouse studies of the effect of clipping of various heights on the production of roots, reserve carbohydrates and top growth. Plant Physiol. 14:505-516.

Hartley, W. 1950. The global distribution of tribes of Gramineae in relation to historical and environmental factors. Aust. J. Agric. Res. 1:355-373.

Hartley, W. 1964. The distribution of the grasses. In: C. Barnard, ed. Grasses and Grasslands. (London: MacMillan and Co.), p. 29-46.

Hawks, R. J. 1965. Estimating infiltration from soil moisture properties. J. Soil Water Conserv. 20:49-51.

Haynes, J. L. 1940. Ground rainfall under vegetation canopy of crops. J. Am. Soc. Agron. 32:176-184.

Hays, H. A., and P. H. Ireland. 1967. A big free-tailed bat (Tadarida macrotis) taken in southeast Kansas. Southwest. Nat. 12:196.

Hazard, E. B. 1963. Records of the opossum in northern Minnesota. J. Mammal. 44:118.

Hazel, D. B. 1967. Effect of grazing intensity on plant composition, vigor, and production. J. Range Manage. 20:249-253.

Heady, H. F. 1975. Rangeland Management. (New York: McGraw-Hill), 460 p.

Hensel, R. L. 1923. Recent studies on the effect of burning on grassland vegetation. Ecology 4:183-188.

Henzell, E. F., and P. J. Ross. 1973. The nitrogen cycle of pasture ecosystems. In: G. W. Butler and R. W. Bailey, eds. Chemistry and Biochemistry of Herbage. (New York: Academic Press), p. 227-246.

Herbel, C. H., and K. L. Anderson. 1959. Response of true prairie vegetation on major Flint Hills range sites to grazing treatment. Ecol. Monogr. 29:171-186.

Herferd, L. R. 1951. The effect of different intensities and frequencies of clipping on forage yield of Andropogon scoparius Michx. and Paspalum plicatulum Michx. M.S. thesis, Texas A&M Univ., College Station.

Herman, R. P., and C. L. Kucera. 1975. Vegetation management and microbial function in a tallgrass prairie. Iowa St. J. Res. 50:255-260.

Hess, S. L. 1959. Introduction to Theoretical Meteorology. (New York: Henry Holt & Co.),

Hewitt, G. B., and W. H. Burleson. 1976. An inventory of arthropods from three rangeland sites in central Montana. J. Range Manage. 29:232-237.

Hibbard, C. H. 1970. Pleistocene mammalian local faunas from the Great Plains and central lowland provinces of the United States. In: W. Dort, Jr. and J. K. Jones, Jr., eds., Pleistocene and Recent Environments of the Central Great Plains. (Lawrence: Univ. Press Kansas), p. 395-433.

Hibbard, E. A. 1956. Range and spread of the gray and fox squirrels in North Dakota. J. Mammal. 37:525-531.

Hibbard, E. A. 1970. Additional Minnesota opossum records. Loon 42:77-78.

Hill, R. T. 1901. Geography and geology of the Black and Grand Prairies, Texas. U.S. Geolog. Surv. 21st Annu. Rep., Part 7, 666 p.

Hitchcock, A. S. 1971. Manual of the Grasses of the United States, 2nd ed., Vols. I and II. (New York: Dover Publ. Inc.), 1051 p.

Hitchcock, A. S., and A. Chase. 1950. Manual of the grasses of the United States. U.S. Dep. Agric. Spec. Publ. No. 200. Washington, D.C., 1051 p.

Hodgson, H. J. 1976. Forage crops. Sci. Am. 234:61-75.

Hoffmann, R. S., and J. K. Jones, Jr. 1970. Influence of late-glacial and post-glacial events on the distribution of recent mammals on the northern Great Plains. In: W. Dort, Jr. and J. K. Jones, Jr., eds. Pleistocene and Recent Environments of the Central Great Plains. (Lawrence: Univ. Press Kansas), p. 355-394.

Hoffmann, R. S., J. K. Jones Jr., and H. H. Genoways. 1971. Small mammal survey on the Bison, Bridger, Cottonwood, Dickinson, and Osage Sites. (Fort Collins: Colorado State Univ.), US/IBP Grassland Biome Tech. Rep. No. 109, 69 p.

Hole, F. D., and G. A. Nielson. 1970. Some processes of soil genesis under prairie. In: P. Schramm, ed. Proc. Symp. on Prairie and Prairie Restoration. (Galesburg, Ill.: Knox College Biol. Field Stn. Spec. Publ. No. 3), p. 28-34.

Holm-Hansen, O. 1973. The use of ATP determinations in ecological studies. In: T. Rosswall, ed. Modern Methods in the Study of Microbial Ecology. (Stockholm: Swedish Nat. Sci. Res. Council) Bull. Ecol. Res. Comm. (Stockholm) No. 17, p. 215-222.

Holmes, R. T., and F. W. Sturges. 1975. Bird community dynamics and energetics in a northern hardwoods ecosystem. J. Anim. Ecol. 44:175-200.

Holzman, B. 1937. Sources of moisture for precipitation in the United States. U.S. Dep. Agric. Tech. Bull. No. 589. Washington, D.C., 41 p.

Horn, H. S. 1966. Measurement of "overlap" in comparative ecological studies. Am. Nat. 100:419-424.

Horwitz, B. 1941. Dynamic Meteorology. (New York: McGraw-Hill), 365 p.

Hsiao, T. C., and E. Aceredo. 1974. Plant responses to water deficits, water-use efficiency, and drought resistance. Agric. Meteorol. 14:59-84.

Huffine, W. W., and W. C. Elder. 1960. Effect of fertilizer on native grass pastures in Oklahoma. J. Range Manage. 13:34-36.

Hulbert, L. C. 1969. Fire and litter effects in undisturbed bluestem prairie in Kansas. Ecology 50:874-877.

Humfeld, H. 1930. A method for measuring carbon dioxide evolution from the soil. Soil Sci. 30:1-11.

Hungate, R. E. 1975. The rumen microbial ecosystem. Annu. Rev. Ecol. Syst. 6:39-66.

Hunt, H. W. 1977. A simulation model for decomposition in grasslands. Ecology 58(3):469-484.

Hurley, R. J., and E. C. Franks. 1976. Changes in the breeding ranges of two grassland birds. Auk 93:108-115.

Hutchison, B. A., and D. R. Mott. 1976. Forest meteorology research within the Oak Ridge Site, Eastern Deciduous Forest Biome. US/IBP Eastern Deciduous Forest Biome Contribution No. 76/4.

Hyde, R. M., and C. E. Owensby. 1973. Fertilizing native grassland. Kans. Agric. Exp. Stn. Publ. No. L-296. Kansas State Univ., Manhattan, 8 p.

Hyder, D. N. 1972. Defoliation in relation to vegetative growth. In: V. B. Youngner and C. M. McKell, eds. The Biology and Utilization of Grasses. (New York: Academic Press), p. 304-317.

Ingeburg, F., U. Hadden, and W. Harder. 1951. Zur Erhahrangrsphysologie der Nagetiere: Uber die Bedeutung der Coecotrophie und der Zusammensetzung der Coecotrophie. Pflugers Arch. Eur. J. Physiol. 253:173-180.

Innis, G. S., ed. 1978. Grassland Simulation Model. Ecol. Studies, Vol. 26. (New York: Springer-Verlag), 298 p.

Jackson, A. S. 1965. Wildfires in the Great Plains grasslands. Proc. Annu. Tall Timbers Fire Ecol. Conf. 4:241-259.

Jameson, D. A., and D. L. Huss. 1959. The effect of clipping leaves and stems on number of tillers, herbage weights, root weights, and food reserves of little bluestem. J. Range Manage. 12:122-126.

Jansson, S. L. 1958. Tracer studies on nitrogen transformations in soil with special attention to mineralization-immobilization relationships. K. Lantbrukshogsk. Ann. 24:101-361.

Jansson, S. L. 1960. On the establishment and use of tagged microbial tissue in soil organic matter research. Proc. 7th Int. Congr. Soil Sci. Trans. 2:635-642.

Jarman, P. J. 1974. The social organization of antelope in relation to their ecology. Behaviour 58:215-267.

Jenkinson, D. S. 1965. Studies on the decomposition of plant material in soil. I. Losses of carbon from ^{14}C-labelled rye grass incubated with soil in the field. J. Soil Sci. 16:104-111.

Jenkinson, D. S. 1971. Studies on the decomposition of ^{14}C-labelled organic matter in soil. Soil Sci. 111:64-70.

Jenness, R. 1974. The composition of milk. In: B. L. Larson and V. R. Smith, eds., Lactation: Nutrition and Biochemistry of Milk/ Maintenance, Vol. 3. (New York: Academic Press), p. 3-107.

Jenny, H. 1941. Factors of soil formation. McGraw-Hill Book Co., Inc., New York, 281 p.

Jenny, H., S. P. Gessel, and F. T. Bingham. 1949. Comparative study of decomposition rates of organic matter in temperate and tropical regions. Soil Sci. 68:419-432.

Jones, A. S., and E. G. Patton. 1966. Forest "prairies" and soils in the Black Belt of Sumpter County, Alabama, in 1832. Ecology 47:75-80.

Jones, C. H. 1944. Studies in Ohio floristics. III. Vegetation of Ohio prairies. Bull. Torrey Bot. Club 71:536-548.

Jones, J. K., Jr. 1964. Distribution and taxonomy of mammals of Nebraska. Univ. Kans. Publ. Mus. Nat. Hist. 16:1-356.

Jones, J. K., Jr., E. D. Fleharty, and P. B. Dunnigan. 1967. The distributional status of bats in Kansas. Univ. Kans. Mus. Nat. Hist. Misc. Publ. 46:1-33.

Jones, R. E. 1962. The quantitative phenology of two plant communities in Osage County, Oklahoma. Proc. Okla. Acad. Sci. 42:31-38.

Jordan, C. F. 1971. A world pattern in plant energetics. Am. Sci. 59:425-433.

Kale, H. W., II. 1965. Ecology and bioenergetics of the Long-billed Marsh Wren in Georgia salt marshes. (Cambridge, Mass.: Harvard Univ.), Publ. Nuttall Ornithol. Club No. 5. 142 p.

Kanai, R., and C. C. Black. 1972. Biochemical basis for net CO_2 assimilation in C_4-plants. In: C. C. Black, ed. Net Carbon Dioxide Assimilation in Higher Plants: Proc. Symp. South Sect. Am. Soc. Plant Physiol. (Raleigh, N. C.: Cotton Inc.), p. 75-85.

Kapustka, L. A., and F. L. Moleski. 1976. Changes in community structure in Oklahoma old field succession. Bot. Gaz. 137:7-10.

Kapustka, L. A., and E. L. Rice. 1976. Acetylene reduction (N_2-fixation) in soil and old field succession in central Oklahoma. Soil Biol. Biochem. 8:497-503.

Kearney, T. H., and H. L. Shantz. 1911. The water economy of dryland crops. U.S. Dep. Agric. Yearb. Agric. 1911:351-361.

Kelly, J. M., G. M. Van Dyne, and W. F. Harris. 1974. Comparison of three methods of assessing grassland productivity and biomass dynamics. Am. Midl. Nat. 92:357-369.

Kelting, R. W. 1954. Effects of moderate grazing on the composition and plant production of a native tallgrass prairie in central Oklahoma. Ecology 35:200-207.

Kelting, R. W. 1957. Winter burning in central Oklahoma grassland. Ecology 38:520-522.

Kendeigh, S. C. 1941. Birds of a prairie community. Condor 43:165-174.

Kendeigh, S. C. 1970. Energy requirements for existence in relation to size of bird. Condor 72:60-65.

Kendeigh, S. C. 1974. Ecology with Special Reference to Animals and Man. (Englewood Cliffs, N.J.: Prentice-Hall), 474 p.

Kendeigh, S. C., V. R. Dol'nik, and V. M. Gavrilov. 1977. Avian energetics. In: J. Pinowski and S. C. Kendeigh, eds. Granivorous Birds in Ecosystems. (Cambridge: Cambridge Univ. Press), p. 127-204.

Kennedy, R. K. 1972. Preliminary network evaluation on methods of primary producer biomass estimation. In: P. G. Risser, ed. Preliminary Producer Data Synthesis, 1970 Comprehensive Network Sites. (Fort Collins: Colorado State Univ.), US/IBP Grassland Biome Tech. Rep. No. 161, p. 30-46.

Kennedy, W. 1841. The Rise, Progress, and Prospects of the Republic of Texas. London. (Reprinted by Molyneaux Craftsman, Inc., Fort Worth, Texas).

Kibbe, A. L. 1952. A Botanical Study and Survey of a Typical Midwestern County, Hancock County, Illinois, Covering a Period of 119 Years, from 1883 to 1952. (Carthage, Ill.: Private printed).

Kilburn, P. D. 1959. The forest-prairie ecotone in northeastern Illinois. Am. Midl. Nat. 62:206-217.

Kilburn, P. D., and C. D. Ford. 1963. Frequency distribution of hill prairie plants. Trans. Ill. State Acad. Sci. 56:94-97.

Kilburn, P. D., and D. K. Warren. 1963. Vegetation-soil relationships in hill prairies. Trans. Ill. State Acad. Sci. 56:142-145.

Kilgore, D. L., Jr. 1969. An ecological study of the swift fox (Vulpes velox) in the Oklahoma panhandle. Am. Midl. Nat. 81:512-534.

Kilgore, D. L., Jr. 1970. The effects of northward dispersal on growth rate of young, size of young at birth, and litter size in Sigmodon hispidus. Am. Midl. Nat. 84:510-520.

Kilgore, G. L., and F. W. Boren. 1972. Effects of fertilizer applications on yield, protein content and stand composition of a tallgrass meadow. Agron. Abstr. 1972:46.

Kincer, J. B. 1923. The climate of the Great Plains as a factor in their utilization. Ann. Assoc. Am. Geogr. 13:67-80.

Kirita, H., and K. Hozumi. 1966. Re-examination of the absorption method of measuring soil respiration under field conditions. Physiol. Ecol. 14:23-31.

Klingman, D. L., and M. K. McCarty. 1958. Interrelations of methods of weed control and pasture management at Lincoln, Nebraska, 1949-1955. (Washington, D.C.), U.S. Dep. Agric. Tech. Bull. 1180:49 p.

Klipple, G. E., and D. F. Costello. 1960. Vegetation and cattle responses to different intensities of grazing on short-grass ranges on the Central Great Plains. (Washington, D.C.: U.S. Dep. Agric.), 82 p.

Knutson, H., and J. B. Campbell. 1976. Relationships of grasshoppers (Acrididae) to burning, grazing, and range sites of native tallgrass prairie in Kansas. Proc. Annu. Tall Timbers Conf. Ecol. Anim. Control Habitat Manage. 6:107-120.

Koelling, M. R., and C. L. Kucera. 1965. Dry matter losses and mineral leaching in bluestem standing crop and litter. Ecology 46:529-532.

Komarek, E. V. 1964. The natural history of lightning. Proc. Annu. Tall Timbers Fire Ecol. Conf. 3:139-183.

Koopman, K. F. 1968. Artiodactyls. In: S. Anderson and J. K. Jones, Jr., eds. Recent Mammals of the World: A Synopsis of Families. (New York: Ronald Press), p. 385-406.

Köppen, W., and R. Geiger, eds. 1930. Handbuch der Klematologie, Vol. I. (Berlin: Gebrüder Borntraeger).

Kozlovsky, D. G. 1968. A critical evaluation of the trophic level concept. Ecology 49:48-60.

Kramer, J. P. 1967. A taxonomic study of the brachypterous North American leafhoppers of the genus Lonatura (Homoptera:Cicadellidae: Deltocephalinae). Trans. Am. Entomol. Soc. 93:433-462.

Krebs, C. J., and J. H. Myers. 1974. Population cycles in small mammals. Adv. Ecol. Res. 8:267-399.

Krebs, C. J., B. L. Keller, and R. H. Tamarin. 1967. Microtus population biology: Demographic changes in fluctuating populations of M. ochrogaster and M. pennsylvanicus in southern Indiana. Ecology 50:587-607.

Kucera, C. L. 1956. Grazing effects on the composition of virgin prairie in north central Missouri. Ecology 37:389-391.

Kucera, C. L. 1970. Ecological effect of fire on tallgrass prairie. In: P. Schramm, ed. Proc. Symp. on Prairies and Prairie Restoration. (Galesburg, Ill.: Knox College Biol. Field Stn. Spec. Publ. No. 3, p. 12.

Kucera, C. L., and J. H. Ehrenreich. 1962. Some effects of annual burning on central Missouri prairie. Ecology 43:334-336.

Kucera, C. L., and D. R. Kirkham. 1971. Soil respiration studies in tallgrass prairie in Missouri. Ecology 52:912-915.

Kucera, C. L., and M. R. Koelling. 1964. The influence of fire on the composition of central Missouri prairie. Am. Midl. Nat. 72:142-147.

Kucera, C. L., R. C. Dahlman, And M. R. Koelling. 1967. Total net productivity and turnover on an energy basis for tallgrass prairie. Ecology 48:536-541.

Kucera, C. L., J. H. Ehrenreich, and C. Brown. 1963. Some effects of fire on tree species in Missouri prairie. Iowa State College J. Sci. 38:179-185.

Küchler, A. W. 1964. Potential natural vegetation of the conterminous United States. (New York: Am. Geogr. Soc.), Am. Geogr. Soc. Spec. Publ. 36 (Manual), 116 p.

Küchler, A. W. 1971. A biogeographical boundary: the Tatschl line. Trans. Kans. Acad. Sci. 73:298-391.

Kumar, R., R. J. Lavigne, J. E. Lloyd, and R. E. Pfadt. 1976. Insects of the Central Plains Experiment Range, Pawnee National Grassland. Agric. Exp. Stn. Sci. Monogr. 32. Univ. Wyoming, Laramie, 74 p.

Kurten, B. 1972. The Age of Mammals. (New York: Columbia Univ. Press), 250 p.

Küster, E. 1967. The actinomycetes. In: N. A. Burges and F. Raw, eds. Soil Biology. (New York: Academic Press), p. 111-127.

Laetsch, W. M. 1974. The C_4 syndrome: A structural analysis. Ann. Rev. Plant Physiol. 25:27-52.

Lampe, R. P. 1976. Aspects of the predatory strategy of the North American badger, Taxidea taxus. Unpubl. Ph.D. thesis, Univ. Minnesota, Minneapolis, 103 p.

Lanyon, W. E. 1956. Ecological aspects of the sympatric distribution of meadowlarks in the north-central states. Ecology 37:98-108.

Larsen, E. C. 1947. Photo-periodic responses of geographical strains of Andropogon scoparius. Bot. Gaz. 109:132-149.

Lathrop, E. W. 1958. The flora and ecology of the Chatauqua Hills of Kansas. Univ. Kans. Sci. Bull. 39:97-209.

Latter, P. M., J. B. Cragg, and O. W. Heal. 1967. Comparative studies on the microbiology of four moorland soils in the northern Pennines. J. Ecol. 55:445-482.

Lee, J. J., and D. L. Inman. 1975. The ecological role of consumers: An aggregated systems view. Ecology 56:1455-1458.

Lengkeek, V. H., and R. M. Pengra. 1973. Carbon dioxide evolution and cellulose, root and litter decomposition in soils at the Cottonwood Site, 1972. (Fort Collins: Colorado State Univ.), US/IBP Grassland Biome Tech. Rep. No. 233, 20 p.

Lieth, H. 1968. The determination of plant dry matter production with special emphasis on the underground part. In: F. E. Eckardt, ed. Functioning of Terrestrial Ecosystems at the Primary Production Level. (Paris: UNESCO), p. 179-186.

Livingston, B. E., and F. Shreve. 1921. The distribution of vegetation in the United States, as related to climatic conditions. Carnegie Inst. Wash. Publ. No. 284, 590 p.

Long, C. A.   1972.   Taxonomic revision of the American badger, Taxidea taxus.   J. Mammal. 53:725-759.

Lowery, G. H., Jr.   1974.   The Mammals of Louisiana and Its Adjacent Waters.   (Baton Rouge: Louisiana State Univ. Press), 578 p.

Lussenhop, J.   1976.   Soil arthropod response to prairie burning. Ecology 57:88-98.

Lyles, L.   1975.   Possible effects of wind erosion on soil productivity.   J. Soil Water Conserv. 30:279-283.

MacArthur, R. H.   1971.   Patterns of terrestrial bird communities. In: D. S. Farmer and J. R. King, eds. Avian Biology, Vol. I. (New York: Academic Press), p. 189-221.

MacArthur, R. H., and J. W. MacArthur.   1961.   On bird species diversity.   Ecology 42:594-598.

Macfadyen, A.   1963.   The contribution of the microfauna to total soil metabolism.   In: J. Doeksen and J. van der Drift, eds. Soil Organisms.   (Amsterdam: North Holland Publ.), p. 3-16.

MacNamara, C.   1924.   The food of Collembola.   Can. Entomol. 56:99-105.

Majerus, M. E.   1975.   Response of root and shoot growth of three grass species to decrease in soil water potential.   J. Range Manage. 28:473-476.

Maksimov, N. A.   1929.   The Plant in Relation to Water: A Study of the Physiological Basis of Drought Resistance.   (London: G. Allen & Unwin, Ltd.), 451 p.   (Trans. from Russian with notes by R. H. Vapp.)

Martin, A. C., H. S. Zim, and A. L. Nelson.   1961.   American Wildlife and Plants: A Guide to Wildlife Food Habits.   (New York: Dover Publ.), 500 p.

Martin, P. S.   1958.   Pleistocene ecology and biogeography of North America.   In: C. L. Hubbs, ed. Zoogeography.   (Washington, D.C.: Am. Assoc. Adv. Sci.), p. 375-420.

Martin, P. S.   1973.   The discovery of America.   Science 179:969-974.

Martin, P. S.   1975.   Vanishings and future of the prairie.   Geosci. Man 10:39-49.

Martin, S. G.   1974.   Adaptations for polygynous breeding in the Bobolink, Dolichonyx oryzivorus.   Am. Zool. 14:109-119.

Mason, C. E.   1973.   Leafhopper and plant diversity in a Kansas bluestem prairie.   Ph.D. thesis, Kansas State Univ., Manhattan.

Maschmeyer, J. R., and J. A. Quinn.   1976.   Copper tolerance in New Jersey populations of Agrostis stolonifera and Paronychia fastigiata.   Bull. Torrey Bot. Club 103:244-251.

May, L. H.   1960.   The utilization of carbohydrate reserves in pasture plants after defoliation.   Herb. Abstr. 30:239-245.

May, S. W., and P. G. Risser.   1973.   Microbial decomposition and carbon dioxide evolution at the Osage Site, 1972.   (Fort Collins: Colorado State Univ.), US/IBP Grassland Biome Tech. Rep. No. 222, 22 p.

May, S. W.   1974.   Microbial decomposition of cellulose and native plant litter in a True Prairie.   Ph.D. dissertation, Univ. Oklahoma, Norman.

Maynard, L. A., and J. K. Loosli.   1956.   Animal Nutrition.   (New York: McGraw-Hill), 484 p.

Mayr, E.   1946.   History of the North American bird fauna.   Wilson Bull. 58:3-41.

McArthur, A. G. 1966. Weather and grassland fire behavior. Aust. For. Res. Inst. Leafl. 100:23.

McEwen, L. C., and J. O. Ells. 1975. Field ecology investigations of the effects of selected pesticides on wildlife populations. (Fort Collins: Colorado State Univ.), US/IBP Grassland Biome Tech. Rep. No. 289, 55 p.

McEwen, L. C., C. E. Knittle, and M. L. Richmond. 1972. Wildlife effects from grasshopper insecticides sprayed on short-grass range. J. Range Manage. 25(3):188-194.

McIlvain, E. H., and C. G. Armstrong. 1966. A summary of fire and forage research on shinnery oak rangelands. Proc. Annu. Tall Timbers Fire Ecol. Conf. 5:127-129.

McIlvain, E. H., and M. C. Shoop. 1962. Calf weight can guide proper range stocking. West. Livestock J. 40:83-85.

McIlvain, E. H., and M. C. Shoop. 1969. Grazing systems in the southern Great Plains. In: 22nd Annu. Meeting, Am. Soc. Range Manage., Denver, Colo., p. 21-22.

McIntosh, R. P. 1974. Plant ecology, 1947-1972. Ann. Mo. Bot. Gard. 61:132-165.

McKeena, M. G. 1969. The origin and early differentiation of therian mammals. Ann. N. Y. Acad. Sci. 167:217-240.

McKendrick, J. D., C. E. Owensby, and R. M. Hyde. 1975. Big bluestem and Indiangrass vegetative reproduction and annual reserve carbohydrate and nitrogen cycles. Agro-Ecosystems 2:75-93.

McMillan, C. 1959. The role of ecotypic variation in the distribution of the central grassland of North America. Ecol. Monogr. 29:285-305.

McMillan, C. 1965. Grassland community fractions from central North American under simulated climates. Am. J. Bot. 52:109-116.

McMurphy, W. E. 1970. Fertilization and deferment of a native hay meadow in north central Oklahoma. Okla. Agric. Exp. Stn. Bull. B-678:10 p. Oklahoma State Univ., Stillwater.

McMurphy, W. E. 1976. Management of introduced grasslands of Oklahoma. In: J. R. Estes and R. J. Tyrl, eds. The Grasses and Grasslands of Oklahoma. Ann. Okla. Acad. Sci. Publ. No. 6. (Ardmore, Okla.: Samuel Roberts Noble Foundation), p. 75-89.

McMurphy, W. E., and K. L. Anderson. 1963. Burning bluestem range: Forage yields. Trans. Kans. Acad. Sci. 66:49-51.

McMurphy, W. E., and K. L. Anderson. 1965. Burning Flint Hills range. J. Range Manage. 18:265-269.

McMurphy, W. E., and B. B. Tucker. 1975. Midland Bermudagrass pasture research. Okla. Agric. Exp. Stn. Prog. Rep. 715:14-20. Oklahoma State Univ., Stillwater.

McMurphy, W. E., J. F. Stritzke, B. B. Webb, and L. M. Rommann. 1976. From brush range to tall fescue pasture with aerial treatments. Rangeman's J. 3:119-121.

McNaughton, S. J. 1968. Structure and function in California grasslands. Ecology 49:962-972.

McNaughton, S. J. 1976. Serengeti migratory wildebeest: Facilitation of energy flow by grazing. Science 191:92-94.

McNaughton, S. J., and L. L. Wolf. 1970. Dominance and the niche in ecological systems. Science 167:131-139.

Mengel, R. M.   1970.   The North American Central Plains as an isolating agent in bird specialization. In: W. Dort, Jr. and J. K. Jones, Jr., eds. Pleistocene and Recent Environments of the Central Great Plains. Lawrence: Univ. Press Kansas), p. 279-340.

Menhenick, E. F.   1967.   Structure, stability, and energy flow in plants and arthropods in a Sericea lespedeza stand. Ecol. Monogr. 37:255-272.

Mitchell, J. E., and R. E. Pfadt.   1974.   A role of grasshoppers in a shortgrass prairie ecosystem. Environ. Entomol. 3:358-360.

Mitich, L. W.   1965.   Pasture renovation with 2,4-D in North Dakota. Down to Earth 20:26-28.

Moir, R. J.   1968.   Ruminant digestion and evolution. Handb. Physiol. 5:2673-2694.

Moir, W. H.   1969.   Energy fixation and the role of primary producers in energy flux of grassland ecosystems. In: R. L. Dix and R. G. Beidleman, eds. The Grassland Ecosystem: A Preliminary Synthesis. Range Sci. Dep. Sci. Ser. No. 2. (Fort Collins: Colorado State Univ.), p. 125-147.

Moreau, R. E.   1954.   The main vicissitudes of the European avifauna since the Pliocene. Ibis 96:411-431.

Morrissey, T.   1956.   The flora of the pine hill prairie relict. Proc. Iowa Acad. Sci. 63:201-213.

Moser, L. E., and K. L. Anderson.   1965.   Nitrogen and phosphorous fertilization of bluestem range. Trans. Kans. Acad. Sci. 67(4): 613-616.

Moss, E. H.   1952.   Grassland of the Peace River region, western Canada. Can. J. Bot. 30:98-124.

Moss, E. H., and J. A. Campbell.   1947.   The fescue grassland of Alberta. Can. J. Res. Sect. C, Bot. Sci. 25:209-227.

Mott, G. O.   1974.   Nutrient recycling in pastures. In: D. A. Mays, ed. Forage Fertilization. (Madison, Wisc.: Am. Soc. Agric., Crop Sci. Soc. Am., and Soil Sci. Soc. Am.), p. 323-339.

Moyer, L. R.   1910.   The prairie flora of southwestern Minnesota. Bull. Minn. Acad. Sci. 4:357-378.

Mueller, I. M.   1941.   An experimental study of rhizomes of certain prairie plants. Ecol. Monogr. 11:165-188.

Murphy, H. F.   1933.   Recovery of phosphorous from prairie grasses growing on central Oklahoma soils treated with super phosphate. J. Agric. Res. 47:911-917.

Murphy, H. F., and H. A. Daniel.   1936.   The composition of some Great Plains grasses and the influence of rainfall on plant composition. Proc. Okla. Acad. Sci. 17:37-40.

Munn, R. E.   1966.   Descriptive micrometeorology. (New York: Academic Press), 245 p.

Murray, C. L.   1974.   A vegetation analysis of a pimpled prairie in northeastern Oklahoma. M.S. thesis, Univ. Tulsa, Oklahoma.

Nagel, H. G.   1973.   Effect of spring prairie burning on herbivorous and non-herbivorous arthropod populations. J. Kans. Entomol. Soc. 46:485-496.

National Academy of Science.   1975.   An evaluation of the International Biological Program. Committee to Evaluate the IBP, Natl. Acad. Sci., Washington, D.C. (memo report).

Neal, W. M., and R. B. Becker. 1933. The composition of feed stuffs in relationship to nutritional anemia in cattle. J. Agric. Res. 47(4):249-255.

Neiland, B. M., and J. T. Curtis. 1956. Differential responses to clipping of six prairie plants in Wisconsin. Ecology 37:355-365.

Nero, R. W. 1974. Cougars in Manitoba. Blue Jay 32:55-56.

Newbould, P. J. 1968. Methods of estimating root production. In: F. E. Eckardt, ed. Functioning of Terrestrial Ecosystems at the Primary Production Level. (Paris: UNESCO), p. 187-190.

Nichols, J. T., and W. E. McMurphy. 1969. Range recovery and production as influenced by nitrogen and 2,4-D treatments. J. Range Manage. 22:116-119.

Nielson, G. A., and F. D. Hole. 1963. A study of the natural processes of incorporation of organic matter into soil in the University of Wisconsin Arboretum. Trans. Wis. Acad. Sci. Arts Lett. 52:213-227.

Nixon, E. S., and C. McMillan. 1964. The role of soil in distribution of four grass species in Texas. Am. Midl. Nat. 71:114-140.

Odum, E. P. 1969. The strategy of ecosystem development. Science 164:242-270.

Odum, E. P. 1971. Fundamentals of Ecology. (Philadelphia: W. B. Saunders), 574 p.

Old, S. M. 1969. Microclimate, fire, and plant production in an Illinois prairie. Ecol. Monogr. 39:355-384.

Olmsted, C. E. 1941. Growth and development in range grasses. I. Early development of Bouteloua curtipendula in relation to water supply. Bot. Gaz. 102:499-519.

Olmsted, C. E. 1945. Growth and development in range grasses. V. Photoperiodic responses of clonal divisions of three latitudinal strains of side-oats grama. Bot. Gaz. 106:382-401.

Oosting, H. J. 1942. An ecological analysis of the plant communities of Piedmont, North Carolina. Am. Midl. Nat. 28:1-126.

Oosting, H. J. 1956. The Study of Plant Communities. (San Francisco: W. H. Freeman & Co.), 440 p.

Orpurt, P. A., and J. T. Curtis. 1957a. Microfungi in relation to the prairie continuum in Wisconsin. Ecology 37:355-365.

Orpurt, P. A., and J. T. Curtis. 1957b. Soil microfungi in relation to the prairie continuum in Wisconsin. Ecology 38:628-637.

Osborn, H. 1939. Meadow and Pasture Insects. (Columbus, Ohio: The Educators' Press), 288 p.

Ovington, J. D., D. Heitkamp, and D. B. Lawrence. 1963. Plant biomass and productivity of prairie, savanna, oakwood, and maize field ecosystem in central Minnesota. Ecology 44:52-63.

Owensby, C. E., and K. L. Anderson. 1965. Reseeding "go-back" land in the Flint Hills of Kansas. J. Range Manage. 18:224-225.

Owensby, C. E., and K. L. Anderson. 1967. Yield responses to time of burning in the Kansas Flint Hills. J. Range Manage. 20(1):1216.

Owensby, C. E., and J. L. Launchbaugh. 1977. Controlling prairie threeawn (Aristida oligantha Michx.) in central and eastern Kansas with fall burning. J. Range Manage. 30:337-339.

Owensby, C. E., and E. F. Smith. 1977. Carbohydrate and nitrogen reserve cycles for continuous season-long and intensive early stocked Flint Hills bluestem range. J. Range Manage. 30: 258-261.

Owensby, C. E., R. M. Hyde, and K. L. Anderson. 1970. Effects of clipping and supplemental nitrogen and water on loamy upland bluestem range. J. Range Manage. 23:341-346.

Paris, O. H. 1969. The function of soil fauna in grassland ecosystem. In: R. L. Dix and R. G. Beidleman, eds. The Grassland Ecosystem: A Preliminary Synthesis. Range Sci. Dep. Sci. Ser. No. 2. (Fort Collins: Colorado State Univ.), p. 331-360.

Parton, W. J. 1976. Abiotic submodel. In: G. W. Cole, ed. ELM: Version 2.0. Range Sci. Dep. Sci. Ser. No. 20. (Fort Collins: Colorado State Univ.), p. 12-61.

Parton, W. J., and P. G. Risser. 1976. Osage Site version of the ELM grassland model. In: Proceedings of the 1976 Summer Computer Simulations Conference. (La Jolla, Calif.: Simulation Councils, Inc.), p. 536-543.

Parton, W. J., and P. G. Risser. 1979. Simulated impact of management practices upon the tallgrass prairie. In: N. R. French, ed. Perspectives in Grassland Ecology. (New York: Springer-Verlag), 204 p.

Parton, W. J., and J. S. Singh. 1976. Simulation of plant biomass on a shortgrass and a tallgrass prairie with emphasis on belowground processes. (Fort Collins: Colorado State Univ.), US/IBP Grassland Biome Tech. Rep. No. 300, 76 p.

Parton, W. J., J. S. Singh, and D. C. Coleman. 1978. A model of production and turnover of roots in shortgrass prairie. J. Appl. Ecol. 44:515-542.

Pavlychenko, T. 1942. Root systems of certain forage crops in relation to the management of agricultural soils. (Ottawa: National Res. Council, Canada), Publ. No. 1088.

Peden, D. G., G. M. Van Dyne, R. W. Rice, and R. M. Hansen. 1974. The trophic ecology of Bison bison L. on shortgrass plains. J. Appl. Ecol. 11:489-498.

Peet, M. R., R. C. Anderson, and M. S. Adams. 1975. Effect of fire on big bluestem production. Am. Midl. Nat. 94:15-26.

Peet, R. K. 1974. The measurement of species diversity. Ann. Rev. Ecol. Syst. 5:285-307.

Penfound, W. T. 1964. The relation of grazing to plant succession in the tallgrass prairie. J. Range Manage. 5:256-260.

Penfound, W. T. 1968. Influence of a wildfire in the Wichita Mountains Wildlife Refuge, Oklahoma. Ecology 49:1003-1005.

Penfound, W. T., and E. L. Rice. 1957. Effects of fencing and plowing on plant succession in a revegetating field. J. Range Manage. 10:21-22.

Penman, H. L. 1948. Natural evaporation from open water, bare soil, and grass. R. Soc. (Lond.), Proc. A 193:120-145.

Perino, J. V., and P. G. Risser. 1972. Some aspects of structure and function in Oklahoma old-field succession. Bull. Torrey Bot. Club 99:233-239.

Peterson, M. L., and R. M. Hagan. 1953. Production and quality of irrigated pasture mixtures as influenced by clipping frequency. Agron. J. 45:283-287.

Petryszyn, Y., and E. D. Fleharty. 1972. Mass and energy of detritus clipped from grassland vegetation by the cotton rat (Sigmodon hispidus). J. Mammal. 53:168-175.

Pfadt, R. E., and J. L. Dodd. 1974. Invasion and segregation of grasshopper species among four stress treatments of the Pawnee shortgrass plains. Presented at the Entomological Society of America Annual Meeting, Minneapolis, Minnesota, December 2-5.

Phillips, J. F. 1931. The biotic community. J. Ecol. 19:1-24.

Phillips, P. D. 1939. Rodent distribution in overgrazed and normal grasslands. M. S. thesis, Univ. Oklahoma, Norman.

Pianka, E. R., and W. S. Parker. 1975. Age-specific reproductive tactics. Am. Nat. 109:453-464.

Pielou, E. C. 1969. An Introduction to Mathematical Ecology. (New York: Wiley-Interscience), 286 p.

Pieper, R. D., M. Ellstrom, E. Staffeldt, and R. Raitt. 1972. Primary producers, invertebrates, birds, and decomposers on the Jornada Site, 1971. (Fort Collins: Colorado State Univ.), US/IBP Grassland Biome Tech. Rep. No. 200, 73 p.

Pomeroy, L. R. 1970. The strategy of mineral cycling. Ann. Rev. Ecol. Syst. 1:171-191.

Pool, R. J. 1914. A study of the vegetation of the sandhills of Nebraska. Minn. Bot. Stud. 4:189-312.

Porter, C. 1965. An analysis of variation between upland and lowland switchgrass, Panicum virgatum, in central Oklahoma. Ecology 47:980-991.

Pound, R., and F. E. Clements. 1900. The Phytogeography of Nebraska. (Lincoln: Botanical Survey of Nebraska, Univ. Nebraska), 442 p.

Price, D. W. 1973. Abundance and vertical distribution of microarthropods in the surface layers of a California pine forest soil. Hilgardia 42:121-147.

Price, P. W. 1975. Insect Ecology. (New York: John Wiley & Sons), 514 p.

Quinn, J. A. 1969. Variability among high plains populations of Panicum virgatum. Bull. Torrey Bot. Club 96:20-41.

Quinn, J. A., and D. F. Hervey. 1970. Trampling losses and travel by cattle on sandhills range. J. Range Manage. 23:50-55.

Ramakrishnan, P. S., and R. Kumar. 1976. Adaptive responses of an alkaline soil population of Cynodon dactylon (L.) Pers. to NPK nutrition. J. Ecol. 64:187-194.

Rankin, H. T., and D. E. Davis. 1971. Woody vegetation in the Black Belt prairie of Montgomery County, Alabama, 1845-46. Ecology 52:716-719.

Rao, M. R., L. H. Harbers, and E. F. Smith. 1973. Seasonal change in nutritive value of bluestem pastures. J. Range Manage. 26:419-422.

Rasmussen, J. L. 1971. Abiotic factors in grassland ecosystem analysis and function. In: N. R. French, ed. Preliminary Analysis of Structure and Function in Grasslands. Range Sci. Dep. Sci. Ser. No. 10. (Fort Collins: Colorado State Univ.), p. 11-34.

Rathcke, B. J. 1976. Competition and coexistence within a guild of herbivorous insects. Ecology 57:76-87.

Rauzi, F. 1963. Water intake and plant composition as affected by differential grazing on rangeland. J. Soil Water Conserv. 18:114-116.

Rauzi, F., and C. L. Hanson. 1966. Water intake and runoff as affected by intensity of grazing. J. Range Manage. 19:351-356.

Rauzi, F., and F. Smith. 1973. Infiltration rates: Three soils with three grazing levels in northeastern Colorado. J. Range Manage. 26:126-129.

Raven, P. H., and D. I. Axelrod. 1974. Angiosperm biogeography and past continental movements. Ann. Mo. Bot. Gard. 61:539-673.

Ray, R. J. 1959. A phytosociological analysis of the tallgrass prairie in northeastern Oklahoma. Ecology 40:56-61.

Reardon, P. O., and D. L. Huss. 1965. Effects of fertilization on little bluestem community. J. Range Manage. 18;238-241.

Reardon, P. O., C. L. Leinweber, and L. B. Merrill. 1972. The effect of bovine saliva on grasses. J. Anim. Sci. 34:897-898.

Redmann, R. E. 1971. Carbon dioxide exchange by native Great Plains grasses. Can. J. Bot. 49:1341-1345.

Reed, R. C. 1972. Insects and other major arthropods of tallgrass prairie. M. S. thesis, Kansas State Univ., Manhattan, 133 p.

Rehm, G. W., W. J. Moline, and E. J. Schwartz. 1972. Response of a seeded mixture of warm season grasses to fertilization. J. Range Manage. 25:452-456.

Reichert, S. E., and W. G. Reeder. 1972. Effects of fire on spider distribution in southwestern Wisconsin prairies. In: J. H. Zimmerman, ed. Proceedings of the Second Midwest Prairie Conference. (Madison: Univ. Wisconsin), p. 73-90.

Reiners, W. A. 1968. Carbon dioxide evolution from the floor of three Minnesota forests. Ecology 49:471-483.

Reuss, J. O., and G. S. Innis. 1977. A grassland nitrogen flow simulation model. Ecology 58:379-388.

Rhoades, E. D., L. F. Locke, H. M. Taylor, and E. H. McIlvain. 1964. Water intake on a sandy range as affected by 20 years of differential cattle stocking rates. J. Range Manage. 17:185-190.

Rice, E. L. 1950. Growth and floral development of five species of range grasses in central Oklahoma. Bot. Gaz. 3:361-377.

Rice, E. L. 1952. Phytosociological analysis of a tall-grass prairie in Marshall County, Oklahoma. Ecology 33:112-116.

Rice, E. L. 1964. Inhibition of nitrogen-fixing and nitrifying bacteria by seed plants. I. Ecology 45:824-837.

Rice, E. L. 1974. Allelopathy. (New York: Acadamic Press), 353 p.

Rice, E. L., and S. K. Pancholy. 1972. Inhibition of nitrification by climax ecosystems. Am. J. Bot. 59:1033-1040.

Rice, E. L., and W. T. Penfound. 1954. Plant succession and yield of living plant material in a plowed prairie in central Oklahoma. Ecology 35:176-180.

Rice, E. L., W. T. Penfound, and L. M. Rohrbaugh. 1960. Seed dispersal and mineral nutrition in succession in abandoned fields in central Oklahoma. Ecology 41:224-228.

Ricklefs, R. E. 1973. Ecology. (Newton, Mass.: Chiron Press), 861 p.

Risser, P. G. 1970. Comprehensive Network Site Description, Osage. (Fort Collins: Colorado State Univ.), US/IBP Grassland Biome Tech. Rep. No. 44, 5 p.

Risser, P. G. 1971. Osage Site, 1970 report, primary production. (Fort Collins: Colorado State Univ.), US/IBP Grassland Biome Tech. Rep. No. 80, 41 p.

Risser, P. G., ed. 1972a. A preliminary compartment model of a tallgrass prairie, Osage Site, 1970. (Fort Collins: Colorado State Univ.), US/IBP Grassland Biome Tech. Rep. No. 159, 21 p.

Risser, P. G., ed. 1972b. Preliminary producer data synthesis, 1970 Comprehensive Sites. (Fort Collins: Colorado State Univ.), US/IBP Grassland Biome Tech. Rep. No. 161, 148 p.

Risser, P. G. 1975. Environmental Report, Black Fox Station. (Tulsa, Okla.: Public Service Company).

Risser, P. G. 1976. Biological Field Studies. (Oklahoma City: Sooner Generating Station, Oklahoma Gas and Electric Co.).

Risser, P. G., and F. L. Johnson. 1973. Carbon dioxide exchange characteristics of some prairie grass seedlings. Southwest. Nat. 18:85-91.

Risser, P. G., and R. K. Kennedy. 1972. Herbal dynamics of a tallgrass prairie, Osage, 1971. (Fort Collins: Colorado State Univ.), US/IBP Grassland Biome Tech. Rep. No. 173, 75 p.

Risser, P. G., and R. K. Kennedy. 1975. Herbage dynamics of an Oklahoma tallgrass prairie, Osage, 1972. (Fort Collins: Colorado State Univ.), US/IBP Grassland Biome Tech. Rep. No. 273, 116 p.

Robbins, C. S., and W. T. Van Velzen. 1967. The breeding bird survey, 1966. (Washington, D.C.: Fish Wildl. Serv., U.S. Dep. Interior), Bureau Sport, Fish. Wildl., Spec. Sci. Rep. Wildl. No. 102, 43 p.

Robertson, J. H. 1939. A quantitative study of the true prairie vegetation after three years of extreme drought. Ecol. Monogr. 9:431-492.

Robinson, J. B., and G. M. MacDonald. 1964. The aerobic bacterial flora of a New Zealand tussock-grassland soil. N. Z. J. Agric. Res. 7:146-157.

Robocker, W. C., and B. J. Miller. 1955. Effects of clipping, burning, and competition on establishment and survival of some native grasses in Wisconsin. J. Range Manage. 8:117-121.

Robocker, W. C., J. T. Curtis, and H. L. Ahlgren. 1953. Some factors affecting emergence and establishment of native grass seedlings in Wisconsin. Ecology 34:194-199.

Rodell, C. F. 1977. A grasshopper model for a grassland ecosystem. Ecology 58:227-245.

Rogler, G. A., and R. J. Lorenz. 1957. Nitrogen fertilization of northern Great Plains rangelands. J. Range Manage. 10:156-160.

Rogler, G. A., and R. J. Lorenz. 1970. Fertilization for Range Improvement: Range Research and Range Problems. (Madison, Wisc.: Crop Sci. Soc. Am.), p. 81-86.

Roos, F. H., and J. A. Quinn. 1977. Phenology and reproductive allocation in Andropogon scoparius (Gramineae) populations in communities of different successional stages. Am. J. Bot. 64:535-540.

Root, R. B. 1967. The niche exploitation pattern of the Blue-gray gnatcatcher. Ecol. Monogr. 37:317-350.

Rosenzweig, M. L. 1968. Net primary productivity of terrestrial communities: Prediction from climatological data. Am. Nat. 102:67-74.

Ross, D. T. 1967. The AED approach to generalized computer-aided design. In: Proc. ACM (Assoc. Comput. Machinery) Computer National Meeting, 22nd National Conf., p. 367-385.

Ross, H. H. 1970. The ecological history of the Great Plains: Evidence from grassland insects. In: W. Dort, Jr. and J. K. Jones, Jr., eds. Pleistocene and Recent Environments of the Central Great Plains. (Lawrence: Univ. Press Kansas), p. 225-240.

Rossi, J., and S. Riccardo. 1927. Primi saggi di un metodo diretto per l'esame batteriologico de suolo. Nuovi Ann. dell Agric. 7:457-470.

Rotenberry, J. T., and J. A. Wiens. 1976. A method for estimating species dispersion from transect data. Am. Midl. Nat. 95:64-78.

Ruggles, D. 1835. Geological and miscellaneous notice of the region around Ft. Winnebago Michigan territory. Am. J. Sci. 30:1-80.

Ruthven, A. G. 1908. The faunal affinities of the prairie region of central North America. Am. Nat. 42:388-393.

Samish, Y., and D. Coller. 1968. Estimation of photorespiration of green plants on their mesophyll resistance to CO_2 uptake. Ann. Bot. 32:687-697.

Sampson, H. S. 1921. An ecological survey of the prairie vegetation of Illinois. Bull. Ill. Nat. Hist. Surv. 13:523-577.

Samtsevich, S. A. 1965. Active excretions of plant roots and their significance. Sov. Plant Physiol. 12:837-846.

Sargeant, A. B., and D. W. Warner. 1972. Movements and denning habits of a badger. J. Mammal. 53:207-210.

Sauer, C. O. 1950. Grassland climax, fire, and man. J. Range Manage. 3:16-21.

Sauer, R. H. 1978. A simulation model for grassland primary producer phenology and biomass dynamics. In: G. S. Innis, ed. Grassland Simulation Model. Ecol. Studies, Vol. 26. (New York: Springer-Verlag), p. 55-87.

Savage, D. A., and V. G. Heller. 1947. Nutritional qualities of range forage plants in relation to grazing with beef cattle on the Southern Plains Experimental Range. U.S. Dep. Agric. Tech. Bull. 943, 61 p.

Schaffner, J. H. 1926. Observations on the grasslands of the central United States. Ohio State Univ. Stud., Contrib. Bot. 178:1-56.

Schuchert, C., and C. O. Dunbar. 1941. Textbook of Geology: Part 2. Historical Geology. (New York: John Wiley & Sons), p. 443-444.

Schultz, C. B., and L. D. Martin. 1970. Quaternary mammalian sequence in the central Great Plains. In: W. Dort, Jr. and J. K. Jones, eds. Pleistocene and Recent Environments of the Central Great Plains. (Lawrence: Univ. Press Kansas), p. 341-353.

Schulz, E. D. 1967. Soil respiration of tropical vegetation types. Ecology 48:652-653.

Schumacher, C. M. 1969. White grubs in bluestem hills. Kans. Stockman 1969(May):12-13.

Schuster, M. F. 1967. Response of forage grasses to rhodesgrass scale. J. Range Manage. 20:307-309.

Scifres, C. J. 1977. Herbicides and the range ecosystem: Residues, research, and the role of the rangeman. J. Range Manage. 30:86-91.

Scott, R. F. 1960. The red buffalo. Frontier Times 34:4.

Scott, J. A., N. R. French, and J. W. Leetham. 1979. Patterns of consumption in grasslands. In: N. R. French, ed. Perspectives in Grassland Ecology. (New York: Springer-Verlag), 204 p.

Sealy, S. G. 1971. The irregular occurrences of the Dickcissel in Alberta, Manitoba and Saskatchewan. Blue Jay 29:12-16.

Selander, R. K. 1966. Sexual dimophism and differential niche utilization in birds. Condor 68:113-151.

Sellers, W. D. 1965. Physical Climatology. (Chicago: Univ. Chicago Press), 272 p.

Senter, W. R. 1975. Establishment of warm and cool season grass pastures on wooded sites in east central Oklahoma after aerial spraying and burning. M. S. thesis, Oklahoma State Univ., Stillwater.

Sesták, Z., J. Catský, and P. G. Jarvis, eds. 1971. Plant Photosynthetic Production: Manual of Methods. (The Hague, Netherlands: Dr. W. Junk N.V. Publishers), 818 p.

Seton, E. T. 1929. Lives of Game Animals, Part 1. Hoofed Animals, Vol. III. (New York: Doubleday & Co.), 412 p.

Shackleford, M. W. 1929. Animal communities of an Illinois prairie. Ecology 10:126-154.

Shantz, H. L. 1954. The place of grasslands in the earth's cover of vegetation. Ecology 35:143-145.

Shantz, H. L., and R. Zon. 1924. The physical basis of agriculture: Natural vegetation. In: C. E. Baker, supervisor. Atlas of American Agriculture. (Washington, D.C.: USDA Bur. Agric. Econ.), p. 1-29.

Sharrow, S. H., and H. A. Wright. 1977. Effect of fire, ash, and litter on soil nitrate temperature, moisture, and Tobosagrass production in the Rolling Plains. J. Range Manage. 30:266-270.

Shelford, V. E. 1926. Naturalist's Guide to the Americas. (Baltimore, Md.: The Williams & Wilkins Co.), 761 p.

Shelford, V. E. 1931. Some concepts of bioecology. Ecology 12:455-467.

Shelford, V. E., and W. P. Flint. 1943. Populations of the chinch bug in the upper Mississippi Valley from 1923 to 1940. Ecology 24:435-455.

Shelford, V. E., and G. S. Winterringer. 1959. The disappearance of an area of prairie in the Cook County, Illinois forest preserve district. Am. Midl. Nat. 61:89-95.

Shields, J. A., E. A. Paul, W. E. Lowe, and D. Parkinson. 1973. Turnover of microbial tissue in soil under field conditions. Soil Biol. Biochem. 5:753-764.

Shimek, B. 1911. The prairies. Bull. Lab. Nat. Hist. Univ. Iowa 6:169-240.

Shimek, B. 1925. Papers on the prairie. Univ. Iowa Stud. Nat. Hist. 11(5):1-36.

Shimek, B. 1931. The relationship between the migrant and native flora of the prairie region. Univ. Iowa Stud. Nat. Hist. 14:10-16.

Shoop, M. C., and E. H. McIlvain. 1971. Why some cattlemen overgrazed--and some don't. J. Range Manage. 24:252-257.

Short, L. L., Jr. 1968. Sympatry of red-breasted meadowlarks in Argentina, and the taxonomy of meadowlarks (Ares: Leistes, Pezites, and Sturnella). Am. Mus. Novit. 2349:1-30.

Simpson, G. G. 1947. Holarctic mammalian faunas and continental relationships during the Cenozoic. Bull. Geol. Soc. Am. 53:613-688.

Simpson, G. G. 1964. Species density of North American Recent mammals. Syst. Zool. 13:57-74.

Sims, P. L., and D. D. Dwyer. 1965. Pattern of retrogression of native vegetation in northcentral Oklahoma. J. Range Manage. 18:20-24.

Sims, P. L., J. S. Singh, and W. K. Lauenroth. 1978. The structure and function of ten western North American grasslands. I. Abiotic and vegetational characteristics. J. Ecol. 66:251-285.

Singh, J. S., and D. C. Coleman. 1973. A technique for evaluating functional root biomass in grassland ecosystems. Can. J. Bot. 51:1867-1870.

Singh, J. S., and D. C. Coleman. 1974. Distribution of photoassimilated ^{14}carbon in the root system of a shortgrass prairie. J. Ecol. 62:359-365.

Singh, J. S., and D. C. Coleman. 1977. Evaluation of functional root biomass and translocation of photoassimilated ^{14}C in a shortgrass prairie ecosystem. In: J. K. Marshall, ed. The Belowground Ecosystem: A Synthesis of Plant-associated Processes. Range Sci. Dep. Sci. Ser. No. 26. (Fort Collins: Colorado State Univ.), p. 123-131.

Singh, J. S., W. K. Lauenroth, and R. K. Steinhorst. 1975. Review and assessment of various techniques for estimating net aerial primary production in grasslands from harvest data. Bot. Rev. 41:181-232.

Sloan, R. E. 1972. The evolution of horses. Hist. Geol. Invest. 8:1-12.

Smeins, F. E., and D. E. Olsen. 1970. Species composition and production of a native northwestern Minnesota tall grass prairie. Am. Midl. Nat. 84:398-410.

Smika, D. E., J. H. Haas, and J. F. Power. 1965. Effects of moisture and nitrogen fertilizer on growth and water use by native grass. Agron. J. 57:483-486.

Smith, A. 1975. Sward productivity within micro-pattern and height and frequency of defoliation. J. Br. Grassl. Soc. 30:279-288.

Smith, C. C. 1940. The effect of overgrazing and erosion upon the biota of a mixed-grass prairie of Oklahoma. Ecology 21:381-397.

Smith, E. F. 1953. Bluestem pasture in summer and winter for making beef. J. Range Manage. 6:347-349.

Smith, E. F. and V. A. Young. 1959. The effect of burning on the chemical composition of little bluestem. J. Range Manage. 12:139-140.

Smith, E. F., V. A. Young, K. L. Anderson, W. S. Ruliffson, and S. N. Rogers. 1960. The digestibility of forage on burned and

nonburned bluestem pasture as determined with grazing animals. J. Anim. Sci. 19:388-392.

Smith, R. C. 1954. An analysis of 100 years of grasshopper populations in Kansas (1854 to 1954). Trans. Kans. Acad. Sci. 57:397-433.

Smith, R. E. 1958. Natural history of the prairie dog in Kansas. Univ. Kans. Mus. Nat. Hist. Misc. Publ. 16:1-36.

Smith, S. J., and L. B. Young. 1975. Distribution of nitrogen forms in virgin and cultivated soils. Soil Sci. 120:354-360.

Smolik, J. D. 1974. Nematode studies at the Cottonwood Site. (Fort Collins: Colorado State Univ.), US/IBP Grassland Biome Tech. Rep. No. 251, 80 p.

Smolik, J. D., and L. E. Rogers. 1976. Effects of cattle grazing and wildfire on soil-dwelling nematodes of the shrub-steppe ecosystem. J. Range Manage. 29:304-306.

Soil Conservation Service. 1954. A Guide to Plant Names in Texas-Oklahoma-Louisiana-Arkansas. (Washington, D.C.: Soil Conservation Service, USDA), 91 p. (Rev. by C. A. Rechentnin.)

Soil Survey Staff. 1975. Soil Classification. (Washington, D.C.: Soil Conservation Services, U.S. Dep. Agric.).

Sokal, R. R., and P. H. A. Sneath. 1963. Principles of Numerical Taxonomy. (San Francisco: W. H. Freeman).

Sorenson, L. H. 1967. Duration of amino acid metabolites formed in soils during decomposition of carbohydrates. Soil Sci. 104:234-241.

Sparrow, E. B., and K. B. Doxtader. 1973. Adenosine triphosphate (ATP) in grassland soil: Its relationship to microbial biomass and activity. (Fort Collins: Colorado State Univ.), US/IBP Grassland Biome Tech. Rep. No. 224, 161 p.

Spedding, C. R. W. 1971. Grassland Ecology. (Oxford: Clarendon Press), 221 p.

Sperry, T. M. 1935. Root systems in Illinois prairie. Ecology 16:178-202.

Sprague, M. A. 1959. Microclimate as an index of site adaptation and growth potential. Am. Assoc. Adv. Sci. Publ. 53:49-57.

Spray, N. 1974. Brush spraying converts Ozark hills into profitable cow-calf country. Practicing Nutritionist 8:1-4.

Stains, H. J., and R. H. Baker. 1958. Furbearers in Kansas: A guide to trapping. Univ. Kans. Mus. Nat. Hist. Misc. Bull. 18:1-100.

Stansel, R. H., E. B. Reynolds, and J. H. Jones. 1939. Pasture improvement in the Gulf Coast prairie of Texas. Texas Agric. Exp. Stn. Bull. No. 570. Texas Tech. College, Lubbock.

Stebbins, G. L. 1972. The evolution of the grass family. In: V. B. Youngner and C. M. McKell, eds. The Biology and Utilization of Grasses. (New York: Academic Press), p. 1-17.

Steiger, T. L. 1930. Structure of prairie vegetation. Ecology 11:170-217.

Stepanich, R. M. 1975. Major soil arthropods of an Oklahoma tallgrass prairie. (Fort Collins: Colorado State Univ.) US/IBP Grassland Biome Tech. Rep. No. 275, 70 p.

Stoddard, L. A. 1941. The Palouse grassland association of northern Utah. Ecology 22:158-163.

Stoddard, L. A., and A. D. Smith. 1955. Range Management. (New York: McGraw-Hill), 433 p.

Swanson, C. P. 1957. Cytology and Cytogenetics. (Englewood Cliffs, N.J.: Prentice-Hall), 596 p.

Swift, D. M., and N. R. French, coordinators. 1972. Basic field data collection procedures for the Grassland Biome 1972 season. (Fort Collins: Colorado State Univ.), US/IBP Grassland Biome Tech. Rep. No. 145, 86 p.

Swinbank, W. C. 1963. Long-wave radiation from clear skies. Q. J. R. Meteorol. Soc. 89:339-348.

Syvertsen, J. P., G. L. Nickell, R. W. Spellenberg, and G. L. Cunningham. 1976. Carbon reduction pathways and standing crop in three Chihuahuan desert plant communities. Southwest. Nat. 20:311-320.

Taber, F. W. 1939. Extension of the range of the armadillo. J. Mammal. 20:489-493.

Taylor, H. M. 1960. Moisture relationships of some rangeland soils of the southern Great Plains. J. Range Manage. 13:77-84.

Terman, M. R. 1974. Behavioral interactions between Microtus and Sigmodon: A model for competitive exclusion. J. Mammal. 55:705-719.

Tester, J. R., and W. H. Marshall. 1961. A study of certain plant and animal interrelations on a native prairie in northwestern Minnesota. Univ. Minn. Mus. Nat. Hist. Occas. Paper No. 8, 51 p.

Thayer, D. W. 1972. Microbiological studies at Pantex Site, 1971. (Fort Collins: Colorado State Univ.), US/IBP Grassland Biome Tech. Rep. No. 184, 29 p.

Thomas, J. A., and E. C. Birney. 1979. Parental care and mating system of the prairie vole, Microtus ochrogaster. Behav. Ecol. Sociobiol. 5:1-16.

Thompson, I. 1939. Geographic affinities of the flora of Ohio. Am. Midl. Nat. 21:730-751.

Thomson, J. W. 1940. Relic prairie areas in central Wisconsin. Ecol. Monogr. 10:685-717.

Thornthwaite, C. W. 1933. The climates of the earth. Geogr. Rev. 23:433-440.

Thornthwaite, C. W. 1948. An approach to a rationale classification of climate. Geogr. Rev. 38:55-94.

Tjepkema, J. D., and R. H. Burris. 1976. Nitrogenase activity associated with some Wisconsin prairie grasses. Plant Soil 45(1):81-94.

Todd, R. L., K. Cromack, Jr., and R. M. Knutson. 1973. Scanning electron microscopy in the study of terrestrial microbial ecology. In: T. Rosswall, ed. Modern Methods in the Study of Microbial Ecology. Bull. Ecol. Res. Comm. (Stockholm) No. 17. (Stockholm: Swedish Nat. Sci. Res. Council), p. 109-118.

Tomanek, G. W. 1948. Pasture types in western Kansas in relation to the intensity of utilization in past years. Trans. Kans. Acad. Sci. 51:171-196.

Tomanek, G. W., E. P. Martin, and F. W. Albertson. 1958. Grazing preference comparisons of six native grasses in mixed prairie. J. Range Manage. 11:191-194.

Torrence, J. D. 1975. Response of Acridid grasshoppers to differential nitrogen treatments on tallgrass prairie. M. S. thesis, Kansas State Univ., Manhattan, 86 p.

Transeau, E. N.   1905.   Forest centers of eastern America.   Am. Nat.
    39:875-889.
Transeau, E. N.   1935.   The prairie peninsula.   Ecology 16:423-437.
Trewartha, G. T.   1954.   An Introduction to Climate.   (New York:
    McGraw-Hill), 402 p.
Trewartha, G. T.   1961.   The Earth's Problem Climates.   (Madison:
    Univ. Wisconsin Press), 334 p.
Tribe, H. T.   1961.   Microbiology of cellulose decomposition in
    soil.   Biol. Sci. 92:61-77.
Trolldenier, G.   1973.   The use of fluorescence microscopy for
    counting soil microorganisms.   In: T. Rosswall, ed.   Modern
    Methods in the Study of Microbial Ecology.   Bull. Ecol. Res.
    Comm. (Stockholm) No. 17.   (Stockholm: Swedish Nat. Sci. Res.
    Council), p. 53-59.
Troughton, A.   1957.   The underground organs of herbage grasses.
    (Berkshire, England: Hurley), Commonw. Bur. Pastures Field
    Crops Bull. No. 44.
Turnbull, A. L., and C. F. Nicholls.   1966.   A "quick trap" for area
    sampling of arthropods in grassland communities.   J. Econ.
    Entomol. 59:1100-1104.
Udvardy, M. D. F.   1963.   Bird faunas of North America.   In: Proc.
    of the XIII Int. Ornithological Congr.   (Ithaca, New York), p.
    1147-1167.
U.S.   Department   of   Commerce.   1950-1970.   Climatological
    Data--National Summary, Vols. 1-21.   (Washington, D.C.:   U.S.
    Government Printing Office).
U.S. Soil Conservation Staff.   1970.   Soil Taxonomy of the National
    Cooperative   Soil   Survey.   (Washington,   D.C.:   USDA   Soil
    Conservation Serv.).
Vallentine, J. F.   1971.   Range Developments and Improvements.
    (Provo, Utah: Brigham Young Univ. Press), 516 p.
Van Amburg, G. L., and J. D. Dodd.   1970.   Soil characteristics and
    grass rooting habits of two soil types in east-central Texas.
    Southwest. Nat. 14:337-338.
Van Dyne, G. M., and Z. Abramsky.   1975.   Agricultural systems
    models and modelling: An overview.   In: G. E. Dalton, ed.
    Study of Agricultural Systems.   (Barking, Essex, England:
    Applied Sci. Publ., Ltd.), p. 23-106.
Van Dyne, G. M., and H. F. Heady.   1965.   Botanical composition of
    sheep and cattle diets on a mature annual range.   Hilgardia
    36:465-492.
Van Hook, R. I.   1971.   Energy and nutrient dynamics of spider and
    orthopteran populations in a grassland ecosystem.   Ecol.
    Monogr. 41:1-26.
Van Valen, L., and R. E. Sloan.   1966.   The extinction of the
    multituberculates.   Syst. Zool. 15:261-278.
Van Velzen, W. T., ed.   1972.   Thirty-sixth breeding-bird census.
    Am. Birds 26:937-1006.
Van Velzen, W. T., ed.   1973.   Thirty-seventh breeding-bird census.
    Am. Birds 27:955-1019.
Van Velzen, W. T., ed.   1974.   Thirty-eighth breeding bird census.
    Am. Birds 28:987-1054.
Vaughan, T. A.   1972.   Mammalogy.   (Philadelphia: W. B. Saunders &
    Co.), 463 p.

Vestal, A. G. 1914. A black-soil prairie station in northeastern Illinois. Bull. Torrey Bot. Club 41:351-364.

Vickery, P. J. 1972. Grazing and net primary production of a temperate grassland. J. Appl. Ecol. 9:307-314.

Vogl, R. J. 1965. Effects of spring burning on yields of brush prairie savanna. J. Range Manage. 18:202-205.

Voight, J. W., and J. E. Weaver. 1951. Range condition classes of native midwestern pasture: An ecological analysis. Ecol. Monogr. 21:39-60.

Waksman, S. A., and F. C. Gerretsen. 1931. Influence of temperature and moisture upon the nature and extent of decomposition of plant residues by microorganisms. Ecology 12:33-60.

Wali, M. K., G. W. Dewald, and S. M. Jalal. 1973. Ecological aspects of some bluestem communities in the Red River Valley. Bull. Torrey Bot. Club 100:339-348.

Walker, E. P. 1964. Mammals of the World, Vol. 1. (Baltimore, Md.: The Johns Hopkins Press), p. 150.

Waller, S. S., C. M. Britton, and J. D. Dodd. 1975. Soil fertility and production parameters of Andropogon scoparius tillers. J. Range Manage. 28:476-479.

Walter, H. 1973. Vegetation of the Earth in Relation to Climate and the Eco-physiological Conditions. (New York: Springer-Verlag), 237 p.

Warcup, J. H. 1957. Studies on the occurrence and activity of fungi in a wheat field. Trans. Br. Mycol. Soc. 40:237-259.

Ward, R. T. 1956. Vegetational change in a southern Wisconsin township. Proc. Iowa Acad. Sci. 63:321-326.

Watts, D. 1971. Principles of Biogeography. (New York: McGraw-Hill), 402 p.

Watts, J. G. 1963. Insects associated with black grama grass, Bouteloua eripoda. Ann. Entomol. Soc. Am. 56:374-379.

Weaver, J. E. 1919. The ecological relations of roots. Carnegie Inst. Wash. Publ. No. 6, 128 p.

Weaver, J. E. 1930. Underground plant development in its relation to grazing. Ecology 11:543-557.

Weaver, J. E. 1950a. Effects of different intensities of grazing on depth and quantity of roots and grasses. J. Range Manage. 2:100-113.

Weaver, J. E. 1950b. Stabilization of midwestern grassland. Ecol. Monogr. 20:251-270.

Weaver, J. E. 1954. North American Prairie. (Lincoln, Nebr.: Johnsen Publ. Co.), 348 p.

Weaver, J. E. 1958. Native grassland of southwestern Iowa. Ecology 39:733-750.

Weaver, J. E. 1960. Flood plain vegetation of the central Missouri Valley and contacts of woodland with prairie. Ecol. Monogr. 30:37-64.

Weaver, J. E. 1961. The living network in prairie soils. Bot. Gaz. 123:16-28.

Weaver, J. E. 1968. Prairie Plants and Their Environment. (Lincoln: Univ. Nebraska Press), 276 p.

Weaver, J. E., and F. W. Albertson. 1936. Effects of the great drought on the prairies of Iowa, Nebraska, and Kansas. Ecology 17:567-639.

Weaver, J. E., and F. W. Albertson. 1956. Grasslands of the Great Plains. (Lincoln, Nebr.: Johnsen Publ. Co.), 395 p.

Weaver, J. E., and W. E. Bruner. 1954. Nature and place of transition from true prairie to mixed prairie. Ecology 35:117-126.

Weaver, J. E., and F. E. Clements. 1938. Plant Ecology. (New York: McGraw-Hill), 601 p.

Weaver, J. E., and R. W. Darland. 1948. Changes in vegetation and production of forage resulting from grazing lowland pasture. Ecology 29:1-29.

Weaver, J. E., and R. W. Darland. 1949. Soil-root relationships of certain native grasses in various soil types. Ecol. Monogr. 19:304-338.

Weaver, J. E., and T. J. Fitzpatrick. 1934. The prairie. Ecol. Monogr. 4:109-295.

Weaver, J. E., and W. W. Hansen. 1941. Native midwestern pastures: Their origin, composition, and degeneration. Univ. Nebr. Conserv. Surv. Div. Bull. 22, 93 p.

Weaver, J. E., and P. J. Hemmil. 1931. The environment of the prairie. Contrib. Bot. Surv. Nebr. 6:1-50.

Weaver, J. E., and V. H. Hougen. 1939. Effect of frequent clipping on plant production in prairie and pasture. Am. Midl. Nat. 21:396-414.

Weaver, J. E., and N. W. Rowland. 1952. Effects of excessive natural mulch on development, yield, and structure of native grassland. Bot. Gaz. 114:1-19.

Weaver, J. E., and G. W. Tomanek. 1951. Ecological studies in a midwestern range; the vegetation and effect of cattle on its composition and distribution. Univ. Nebr. Conserv. Surv. Div. Bull. 31, 82 p.

Weaver, J. E., and E. Zink. 1946. Length of life of roots of ten species of perennial range and pasture plants. Physiol. Plant. 21:201-217.

Weinmann, H. 1952. Carbohydrate reserves in grasses. In: Proc. 6th International Grassland Congress, Vol. 1. (State College: Pennsylvania State Univ.), p. 655-660.

Welch, W. H. 1929. Forest and prairie, Benton County, Indiana. Proc. Indiana Acad. Sci. 39:67-72.

Wellman, P. I. 1939. The Trampling Herd. (New York: Carrick & Evans), 433 p.

Whitaker, J. O., Jr. 1972. Zapus hudsonius. Mamm. Species 11:1-7.

Whitaker, J. O., Jr. 1974. Cryptotis parva. Mamm. Species 43:1-8.

White, E. M. 1961. A possible relationship of little bluestem distribution to soils. J. Range Manage. 14:243-247.

White, T. C. R. 1976. Weather, food and plagues of locusts. Oecologia 22:119-134.

Whitehead, D. C. 1970. The role of nitrogen in grassland productivity. (Berkshire, Hurley, England: Commonw. Bur. Pastures Field Crops Bull. 48.

Whitford, P. B. 1958. A study of prairie remnants in southeastern Wisconsin. Ecology 39:727-753.

Whitman, W. C. 1969. Microclimate and its importance in grassland ecosystems. In: R. L. Dix and R. G. Beidleman, eds. The Grassland Ecosystem: A Preliminary Synthesis. Range Sci. Dep.

Sci. Ser. No. 2. (Fort Collins: Colorado State Univ.), p. 40-64.

Whittaker, R. H. 1975. Communities and Ecosystems. (New York: Macmillan Co.), 385 p.

Whyte, R. O. 1960. Crop Production and Environment. (London: Faber & Faber), 392 p.

Wicklow, D. T. 1975. Fire as an environmental cue initiating ascocarp development in a tallgrass prairie. Mycologia 67:852-862.

Wiegert, R. G. 1964. Population energetics of meadow spittlebugs (Philaenus spumarius L.) as affected by migration and habitat. Ecol. Monogr. 34:217-241.

Wiegert, R. G. 1975. Simulation models of ecosystems. Ann. Rev. Ecol. Syst. 6:311-338.

Wiegert, R. G., and F. C. Evans. 1964. Primary production and disappearance of dead vegetation on an old field in southeastern Michigan. Ecology 45:49-63.

Wiegert, R. G., and F. C. Evans. 1967. Investigations of secondary productivity in grasslands. In: K. Petrusewicz, ed. Secondary Productivity of Terrestrial Ecosystems (Principles and Methods), Vol. II. (Warszawa: Państwowe Wydawnictwo Naukowe), p. 499-518.

Wiegert, R. G., and D. F. Owen. 1971. Trophic structure, available resources, and population density in terrestrial vs. aquatic ecosystems. J. Theor. Biol. 30:69-81.

Wiegert, R. G., D. C. Coleman, and E. P. Odum. 1970. Energetics of the litter-soil subsystem. In: J. Phillipson, ed. Methods of Study of Soil Ecology: Proceedings of the Paris Symposium. (Paris: IBP-UNESCO), p. 93-98.

Wiens, J. A. 1969. An approach to the study of ecological relationships among grassland birds. Ornithol. Monogr. 8:1-93.

Wiens, J. A. 1971. Avian ecology and distribution in the Comprehensive Network, 1970. (Fort Collins: Colorado State Univ.), US/IBP Grassland Biome Tech. Rep. No. 77, 49 p.

Wiens, J. A. 1972a. Predictability of patterns and variability of precipitation in grasslands. (Fort Collins: Colorado State Univ.), US/IBP Grassland Biome Tech. Rep. No. 168, 23 p.

Wiens, J. A. 1973. Pattern and process in grassland bird communities. Ecol. Monogr. 43(2):237-270.

Wiens, J. A. 1974a. Habitat heterogeneity and avian community structure in North American grasslands. Am. Midl. Nat. 91:195-213.

Wiens, J. A. 1974b. Climatic instability and the "ecological saturation" of bird communities in North American grasslands. Condor 76:385-400.

Wiens, J. A. 1975. Avian communities, energetics, and functions in coniferous forest habitats. In: D. R. Smith, coordinator. Proceedings of the Symposium on Management of Forest and Range Habitats for Nongame Birds. (Washington, D.C.: USDA Forest Serv. Gen. Tech. Rep. WO-1), p. 226-265.

Wiens, J. A. 1976. Population responses to patchy environments. Ann. Rev. Ecol. Syst. 7:81-120.

Wiens, J. A. 1977a. On competition and variable environments. Am. Sci. 65:590-597.

Wiens, J. A. 1977b. Model estimation of energy flow in North American grassland bird communities. Oecologia 31:135-151.

Wiens, J. A., and M. I. Dyer. 1975. Rangeland avifaunas: Their composition, energetics, and role in the ecosystem. In: D. R. Smith, coordinator. Proceedings of the Symposium on Management of Forest and Range Habitats for Nongame Birds. (Washington, D.C.: USDA Forest Serv. Gen. Tech. Rep. WO-1), p. 146-182.

Wiens, J. A., and M. I. Dyer. 1977. Assessing the potential impact of granivorous birds in ecosystems. In: J. Pinowski and S. C. Kendeigh, eds. Granivorous Birds in Ecosystems. (Cambridge: Cambridge Univ. Press), p. 205-266.

Wiens, J. A., and J. T. Emlen. 1966. Post-invasion status of the Dickcissel in southern Wisconsin. Passenger Pigeon 27:63-69.

Wiens, J. A., and G. S. Innis. 1973. Estimation of energy flow in bird communities. II. A simulation model of activity budgets and population bioenergetics. In: Proceedings of the 1973 Summer Computer Simulation Conference, Vol. II. (La Jolla, Calif.: Simulation Councils, Inc.), p. 739-752.

Wiens, J. A., and G. S. Innis. 1974. Estimation of energy flow in bird communities: A population bioenergetics model. Ecology 55:730-746.

Wiens, J. A., and R. A. Nussbaum. 1975. Model estimation of energy flow in northwestern coniferous forest bird communities. Ecology 56:547-561.

Wiens, J. A., J. T. Rotenberry, and J. F. Ward. 1974. Bird populations at ALE, Pantex, Osage, and Cottonwood, 1972. (Fort Collins: Colorado State Univ.), US/IBP Grassland Biome Tech. Rep. No. 267. 107 p.

Wight, J. R. 1976. Range fertilization in the northern Great Plains. J. Range Manage. 29:180-185.

Wildung, R. E., T. R. Garland, and R. L. Buschbom. 1975. The interdependent effects of soil temperature and water content on soil respiration rate and plant root decomposition in arid grassland soils. Soil Biol. Biochem. 7:373-378.

Wilson, J. R. 1975. Comparative response to nitrogen deficiency of a tropical and temperate grass in the interrelationship between photosynthesis, growth, and the accumulation of non-structural carbohydrate. Neth. J. Agric. Sci. 23:104-112.

Wilson, J. W., III. 1974. Analytical zoogeography of North American mammals. Evolution 28:124-140.

Wistendahl, W. A. 1975. Buffalo Beats, a relic prairie within a southeastern Ohio forest. Bull. Torrey Bot. Club 102:178-186.

Witkamp, M. 1966. Rates of CO_2 evolution from the forest floor. Ecology 47:492-494.

Witkamp, M. 1971. Soils as components of ecosystems. Ann. Rev. Ecol. Syst. 2:85-110.

Witkamp, M., and M. L. Frank. 1969. Evolution of carbon dioxide from litter, humus, and subsoil of a pine stand. Pedobiologia 9:358-365.

Witkamp, M., and J. S. Olson. 1963. Breakdown of confined and nonconfined oak letter. Oikos 14:138-147.

Wood, T. G. 1967. Acari and Collembola of moorland soils from Yorkshire, England. II. Vertical distribution in four grassland soils. Oikos 18:137-140.

Woodwell, G. M., and R. H. Whittaker. 1968. Primary production in terrestrial ecosystems. Am. Zool. 8:19-30.

Woolfolk, J. S., C. E. Owensby, R. R. Schalles, L. H. Harbers, L. J. Allen, and E. F. Smith. 1973. Response of yearling steers to burning, fertilization, and intensive early season stocking of bluestem pastures. Kans. Agric. Exp. Stn. Bull. 568:10-12.

Wright, H. E. 1974. Landscape development, forest fires, and wilderness management. Science 186:487-495.

Wright, J. C., and E. A. Wright. 1948. Grassland types of south central Montana. Ecology 29:449-460.

Wuenscher, M. L., and G. C. Gesloff. 1971. Growth of _Andropogon scoparius_ (little bluestem) in phosphorous deficient soils. New Phytol. 70:1035-1042.

Youngner, V. B. 1972. Physiology of defoliation and regrowth. In: V. B. Youngner and C. M. McKell, eds. The Biology and Utilization of grasses. (New York: Academic Press), p. 292-303.

Youngner, V. B., F. Nudge, and R. Ackerson. 1976. Growth of Kentucky bluegrass leaves and tillers with and without defoliation. Crop Sci. 16:110-113.

Zelitch, I. 1971. Photosynthesis, Photorespiration, and Plant Productivity. (New York: Academic Press), 347 p.

Zimmerman, J. L. 1965. Breeding bird census, grassland. Audubon Field Notes 19:614.

Zimmerman, J. L. 1966. Breeding bird census, grassland. Audubon Field Notes 20:665.

Zimmerman, J. L. 1967. Breeding bird census, grassland. Audubon Field Notes 21:667.

Author Index

Subject Index

Taxonomic Index